DATE DUE

CONCEPTS OF
INORGANIC PHOTOCHEMISTRY

CONCEPTS OF INORGANIC PHOTOCHEMISTRY

Edited by

Arthur W. Adamson

Department of Chemistry
University of Southern California
Los Angeles, California

Paul D. Fleischauer

Chemistry and Physics Laboratory
The Aerospace Corporation
Los Angeles, California

A Wiley-Interscience Publication

JOHN WILEY & SONS

NEW YORK / LONDON / SYDNEY / TORONTO

Library of Congress Cataloging in Publication Data

Adamson, Arthur W
  Concepts of inorganic photochemistry.

  "A Wiley-Interscience publication."
  Includes bibliographical references.
  1. Photochemistry.   2. Chemistry, Inorganic.
I. Fleischauer, Paul D., joint author.   II. Title.

QD708.2.A3   541'.35   74-30198
ISBN 0-471-00795-1

Printed in the United States of America

10 9 8 7 6 5 4 3 2 1

CONTRIBUTORS

Arthur W. Adamson
Department of Chemistry
University of Southern California
Los Angeles, California

John F. Endicott
Department of Chemistry
Wayne State University
Detroit, Michigan

Paul D. Fleischauer
Chemistry and Physics Laboratory
The Aerospace Corporation
El Segundo, California

Peter C. Ford
Department of Chemistry
University of California
Santa Barbara, California

Leslie S. Forster
Department of Chemistry
University of Arizona
Tucson, Arizona

Malcolm Fox
School of Chemistry
City of Leicester Polytechnic
Leicester, LE1 9BH
United Kingdom

Ray E. Hintze
Department of Chemistry
University of California
Santa Barbara, California

Richard L. Lintvedt
Department of Chemistry
Wayne State University
Detroit, Michigan

John D. Petersen
Department of Chemistry
University of California
Santa Barbara, California

Gerald B. Porter
Department of Chemistry
University of British Columbia
Vancouver, British Columbia

Arnd Vogler
Fachbereich Chemie
Universität Regensburg
84 Regensberg, Germany

Edoardo Zinato
Instituto di Chimica Generale ed
  Inorganica
Universitá di Perugia
06100 Perugia, Italy

v

FOREWORD

The term *coordination chemistry* has come to include much of inorganic chemistry so that now, in general usage, it can be said to embrace the chemistry of inorganic particles in compact environments. It is not restricted to orthodox coordination complexes but is concerned also with systems in which the atoms or ions are in direct contact with solvent molecules—the interactions can range from the very selective to the very labile and nonspecific—as well as to the particles in either nonmolecular or molecular solids.

The present volume encompasses photoeffects of coordination chemistry, if the latter term is given its generalized meaning. Although a wide range of systems is dealt with, a selection has been made from the vast body of knowledge comprising the whole of the photochemistry of inorganic systems. In a book of convenient size, selection of subject matter is not only desirable but also necessary. Choices have been made in the interests of both coherence and timeliness, and those aspects of inorganic photochemistry that have experienced the most rapid growth in the last two decades are treated.

A stage in the development of a subject in chemistry which can properly be designated as its beginning is usually difficult to identify. Often, when the origin is traced to what might be called a beginning, the disparity in sophistication between the earliest intimations and the subject in maturity can be so great that the later manifestations are not illuminated by the earliest history. But, though it may not be possible or useful—apart from historical interest—to trace the beginning of a subject, there is a later critical phase that can usually be identified and that is worth examining. It is the stage in which the growth of knowledge becomes autocatalytic, so that discoveries as they are reported stimulate participation by others. Autocatalytic growth takes place when the descriptive foundations have been laid, when the theoretical ideas are sufficiently advanced to invest the experimental results with general significance, and when the experimental methods are competent to provide rigorous tests of theory. When the critical phase is reached, the opportunities, fairly commonly, are recognized simultaneously and independently in different laboratories.

Of the range of subjects dealt with, the photochemistry of coordination complexes, which constitutes a major part of this book, conforms most closely to the scenario outlined. The stability of a new compound in the dark and in light (laboratory light or, for the purposes of a more severe test, sunlight) has always been a matter of practical importance. In addition to this practical background interest, sporadic attempts at the systematic study of photochemical effects in inorganic complexes were made over a period of several decades. However, until recently, neither the experimental approaches nor the theoretical machinery were well enough developed for the studies to be other than qualitatively descriptive. This appraisal is in no sense a disparagement of the competence, inventiveness, or industry of early workers, but it does recognize the limitations in the development of both the "photo" and "chemistry" aspects of photochemistry several decades ago. Among those who contributed to the early systematic studies of photoeffects are O. Baudisch, who began studies of iron cyanide complexes about 1921, and R. Schwarz and K. Tede, who began work on a variety of metal complexes, including those of cobalt(III), about 1923.

The period of autocatalytic growth set in during the late 1950s with three laboratories—those of H. L. Schläfer, R. A. Plane and J. P. Hunt, and A. W. Adamson—who, independently and within a time span of two or three years, reported their first results on the photochemistry of transition-metal complexes. By the late 1950s, some major principles governing the substitution and redox reactions of complex ions had been enunciated. In addition, considerable progress had been made in the descriptive aspects of the reactions, and, thus, basic information on the thermally activated reactions was available. Furthermore, by virtue of the contributions of Hartmann, Orgel, Jørgensen, Ballhausen, and others, the implications of ligand-field theory had begun to be appreciated by most of those interested in transition-metal complexes. Ligand-field theory provided a means of understanding absorption spectra and the nature of the electronically excited state, and constituted the theoretical framework that guided the choice of system to study. Finally, thanks to the efforts of physical chemists who had continued to work in the field of photochemistry and also to the burgeoning of effort in the photochemistry of organic compounds, convenient quantitative experimental methods were at hand, including methods for characterizing short-lived intermediates. In short, the time was ripe for the developments that ensued, and several people, including the senior editor of this volume, were early aware of the opportunities.

The chapters in this book, although up-to-date and authoritative, will probably not be final statements on the subjects they cover. The field is still in a stage of active growth. In fact, an important phase—that of using photochemistry for the preparation of unusual inorganic compounds—has only recently begun. By appealing to the example set by the work in organic photochemistry, we can be certain that the photo-preparative aspect has enormous scope in organometallic chemistry, and there is little doubt that important opportunities are inherent also in the purely inorganic systems.

The forces motivating research in the subject of this book are powerful. Chief among them are the appreciation of the fact that, for the purposes of understanding reactivity in a fundamental way, excited states are no less significant than ground states and the recognition of the advantages of using light to control the energy content of the excited state. Apart from the intrinsic motivation, a component arises also from the opportunities foreseen for applications of the knowledge in such areas as protection of materials against the action of light, photo-imaging, laser technology, and the conversion of light to electrical energy. I am confident that the present volume, containing as it does contributions by leaders in the field under the supervision of editors who are active in the field and who between them view the field in wide perspective, will do much to stimulate research both in the basic aspects of the subject and in the applications of the basic knowledge.

HENRY TAUBE

Stanford, California
August 1974

PREFACE

New fields in science tend to pass through certain characteristic stages of development, and the subject of inorganic photochemistry has been no exception. As Professor Taube notes in his Foreword, there was an induction period during which important but isolated studies were made. Next came a "takeoff" stage which extended through the 1960s and early 1970s. During this second period the photochemistry of coordination compounds, especially of Cr(III) and Co(III), was consolidated to the degree that some rules of behavior became apparent and some preliminary theoretical and mechanistic rationalizations became possible. Sensitization was reported late in the period and then increasingly studied. It became appropriate to summarize a rapidly growing field and several important review articles appeared, as well as the excellent monograph by V. Balzani and V. Carrasiti.

We are now into a third characteristic stage. On the one hand, one sees second- and third-generation types of investigations in which experiments of high sophistication are carried out, often with very high capability equipment. These studies are designed to reach even further toward the understanding of the intimate details of excited-state processes. On the other hand, wide new areas of exploratory investigation have appeared and inorganic photochemistry is noticeably forming subfields or specialities, each with a growing priesthood.

The post World War II development of the general field of inorganic chemistry passed through similar stages, reaching the third one perhaps 15 years ago. A need developed for a book that would organize critically the idea as well as the data content of the various fields of coordination chemistry. *Modern Coordination Chemistry*, edited by J. Lewis and R. Wilkens, did this very well indeed. It is our hope that *Concepts of Inorganic Photochemistry* will prove to be similarly timely.

The selection of topics is designed to allow a comprehensive account of the developments in coordination photochemistry through emphasis on the application of summarizing principles to the wide variety of fascinating contemporary observations. The chapter authors have emphasized informative, interpretive discussion of their material. There has been much intercommunication with the aim of achieving good articulation between chapters, consistency in terminology, and a minimum of duplication. The result, we believe, is that many desirable features both of a review monograph and of a textbook are present.

It is intended that this book will be useful as resource material for courses in photochemistry and in inorganic chemistry. We hope especially that investigators in the field, both established and newly entering, will find the book stimulating in the concepts presented and useful for the critical summaries of data that are included.

The initial chapters discuss the physical properties of the excited states of coordination compounds. Chapters 3 and 4 deal with the charge-transfer and substitution photochemistries of the first row transition-metal ions. Chapters 5 through 9 treat the photochemistry of the heavier elements, carbonyl and related complexes, certain families of chelates, "simple" ions in solution, and complexes in the solid state. A final chapter deals briefly with some concepts regarding photochromism and chemiluminescence.

We are, of course, vastly indebted to the chapter contributors for making the book possible, but wish here especially to acknowledge the exceptionally fine degree of cooperation and of willingness to expend much effort. We and the chapter authors are very grateful to many individuals who gladly helped with comments and with preprints of unpublished material. Many thanks are due to Molly Ewart and to Pamela Hoff for their assistance with the preparation of the manuscript and the coordination of the whole operation.

ARTHUR W. ADAMSON

PAUL D. FLEISCHAUER

Los Angeles, California

CONTENTS

xiii

CONCEPTS OF
INORGANIC PHOTOCHEMISTRY

1

PHOTOPHYSICAL PROCESSES–
ENERGY LEVELS AND SPECTRA

Leslie S. Forster

Department of Chemistry
University of Arizona
Tucson, Arizona

I. INTRODUCTION

Absorption of a photon by a species, A, leads to the formation of a short-lived electronically excited molecule, A*. The electronic energy can be dissipated chemically or physically according to the generalized reaction scheme:

$$A + h\nu \rightarrow A^* \tag{1-1}$$

$$A^* \rightarrow \text{primary products} \rightarrow \text{final products} \tag{1-2}$$

$$A^* \rightarrow A + h\nu' \tag{1-3}$$

$$A^* \rightarrow A + \text{heat} \tag{1-4}$$

From the photophysical point of view, there are two questions of interest to photochemists:

1. What is the state initially populated in step (1-1)?
2. By what pathway(s) does the excitation energy lead to a chemical reaction [step (1-2)] or to the ground state of the unchanged species [steps (1-3) and (1-4)]?

Not only is the photochemical quantum yield often dependent on the exciting wavelength, but, in some cases, the entire course of the photochemical reaction is a function of excitation wavelength. An example of this phenomenon is provided by the photochemistry of Co(III) complexes. The absorption spectrum and associated energy level scheme of $Co(NH_3)_6^{3+}$ are shown in Fig. 1-1. Excitation in the visible and near-ultraviolet regions leads to aquation with a small quantum yield [mode

1

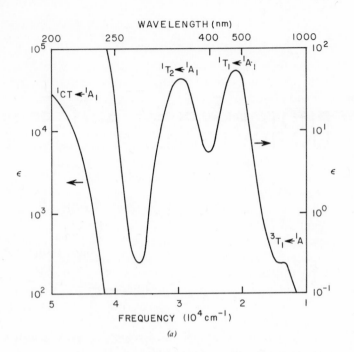

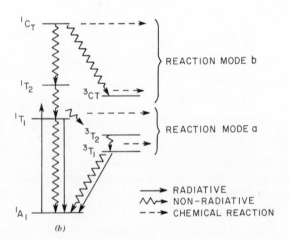

Fig. 1-1. (*a*) Absorption spectrum of Co(NH₃)₆³⁺ [after A. W. Adamson, *Pure Appl. Chem.* **24**, 451 (1970)]. (*b*) Energy levels of Co(NH₃)₆³⁺ and possible photophysical and photochemical steps.

(a)], while at 2537 Å reduction to Co(II) occurs with much higher efficiency [mode (b)]. Figure 1-1*b* is highly schematic, but it serves to indicate the manifold of radiative and nonradiative processes designated in steps (1-3) and (1-4), respectively. This so-called *Jablonski diagram* is analogous to the one commonly encountered in organic systems, where the spin-allowed transitions are usually singlet–singlet ($S_1 \leftrightarrow S_0$) while the spin-forbidden transitions are singlet–triplet ($T_1 \leftrightarrow S_0$). It is not surprising that a detailed specification of reaction patterns is hard to obtain, even in the case of such a thoroughly studied system as Co(III).

A full understanding of the photochemical primary process requires an interpretation of the absorption spectrum of the molecule in question and the assessment of the rates for the several photophysical processes, radiative and nonradiative. In this chapter, the first of these areas is treated synoptically with emphasis on those features of importance to the photochemistry of metal complexes. The connection between the various rates and photochemistry is explored in Chapter 2.

II. SPECTRA OF METAL COMPLEXES

Light absorption can induce changes in molecular electronic, vibrational, and rotational energies. The total energy, E_k, corresponds to a solution of the Schrödinger equation

$$H\Psi_k = E_k\Psi_k \qquad (1\text{-}5)$$

To a good approximation, the electronic (E_i^e) and vibrational ($E_{i\alpha}^v$) energies and wave functions can be separated

$$E_{i\alpha} = E_i^e + E_{i\alpha}^v \qquad (1\text{-}6)$$

$$\Psi_{i\alpha} = \Theta_i(q, Q)\phi_\alpha^i(Q) \qquad (1\text{-}7)$$

where q and Q represent the totality of electronic and nuclear coordinates, respectively. As long as this, the *Born–Oppenheimer approximation*, is valid it is possible to describe the molecular energy by a potential energy diagram (Fig. 1-2) in which the vibrational energies are superimposed upon the electronic "curves." The absorption spectrum is a summation of the individual vibronic transitions, $\Psi_{j\beta} \leftarrow \Psi_{i\alpha}$. Transition energies are then governed by the Bohr frequency condition

$$h\nu = E_{j\beta} - E_{i\alpha} \qquad (1\text{-}8)$$

In like manner, the vibronic transitions, $\Psi_{j\beta} \rightarrow \Psi_{i\alpha}$, contribute to the emission spectrum. Since rotational levels are, in polyatomic molecules, too closely spaced to be resolved into individual lines, they are ignored

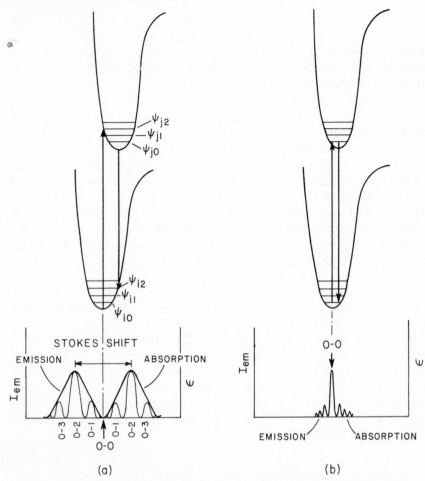

Fig. 1-2. Potential curves and corresponding idealized spectra for (*a*) different equilibrium geometries in the ground and excited states, (*b*) same equilibrium geometries.

here. In many cases, especially in solution at ambient temperature, the individual vibronic components are also unresolved. A typical spectrum (Fig. 1-1) consists of broad bands, each of which is associated with one or more electronic transitions. It is desirable to label (assign) these transitions in a meaningful way in order to use the spectra of any class of molecules, for example, Cr(III) or Co(III) complexes, to systematize the photochemical results. These assignments are based on a number of approximations or models which may or may not be closely related to reality.

III. ENERGY LEVELS OF FREE IONS[1-3]

The energy levels of ions in the gas phase are collected in the extensive tabulation of Moore.[4] For a free ion it is convenient to express the Hamiltonian operator as

$$H = H_0 + H_R + H_{SO} \qquad (1\text{-}9)$$

where H_0 represents the energy of a single electron in the field of the nucleus, H_R corresponds to the electrostatic repulsion between electrons, and H_{SO} includes the effect of spin-orbit coupling. In one-electron systems (for example, H, He^+, Li^{2+}), $H_R = 0$, and the Schrödinger equation can be solved exactly. The Ψ_k are the well-known atomic orbitals, $1s$, $2s$, $2p$, $3d$, and so forth, with s, p, d, f designating the magnitude of the orbital angular momentum quantum number, $l = 0, 1, 2, 3$, respectively. There are $2l + 1$ degenerate orbitals for each l value. Thus, s orbitals are nondegenerate, while the degeneracies of p, d, and f orbitals are 3, 5, and 7, respectively.

When a second electron is added to an ion, $H_R \neq 0$, and exact solution of the Schrödinger equation is no longer possible. Nevertheless, it is convenient to describe the electronic structure of the ion in terms of the hypothetical atomic orbitals, that is, by an electron configuration. The ground-state configuration of He is $1s^2$, while that of Cr(III) is $1s^2 2s^2 2p^6 3s^2 3p^6 3d^3$. The one-to-one correspondence between orbital energies and the "real" energy differences that are observed in spectra now disappears since a configuration may give rise to more than one term (Fig. 1-3). The term symbols are analogous to the orbital symbols with S,

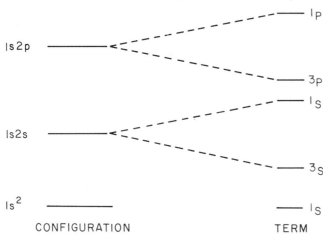

Fig. 1-3. Configurations and terms for He.

P, D, and so forth, designating the orbital angular momentum for the atom, $L = 0, 1, 2, \ldots$. The spin angular momentum for the atom, S (not to be confused with the term symbol for $L = 0$), is then represented by the multiplicity, $2S + 1$, the superscript in the term symbol.

The terms arising from any configuration can be determined by a straightforward, but often tedious, procedure in which the influence of the Pauli principle must be included.[3] Only electrons outside filled shells need be considered for this purpose. For example, the V(III) configuration can be specified as d^2. Electrons within a set of degenerate orbitals are called equivalent. The terms for the transition metal ions are compiled in Table 1-1.

Table 1-1.   Terms for Equivalent d Electrons[a]

Configuration	Terms[b]
d^1, d^9	$^2\underline{D}$
d^2, d^8	$^1S, {}^1D, {}^1G, {}^3P, {}^3\underline{F}$
d^3, d^7	$^2P, {}^2D(2), {}^2F, {}^2G, {}^2H, {}^4P, {}^4\underline{F}$
d^4, d^6	$^1S(2), {}^1D(2), {}^1F, {}^1G(2), {}^1I,$ $^3P(2), {}^3D, {}^3F(2), {}^3G, {}^3H, {}^5\underline{D}$
d^5	$^2S, {}^2P, {}^2D(3), {}^2F(2), {}^2G(2), {}^2H,$ $^2I, {}^4P, {}^4D, {}^4F, {}^4G, {}^6\underline{S}$

[a] The ground-state terms are underlined.
[b] The numbers in parentheses indicate the number of terms with the same term symbol.

The total degeneracy (spin and orbital) of a term is $(2L + 1)(2S + 1)$. For example, there are $(2(3) + 1)(2(3/2) + 1) = 28$ states of equal energy in the ^4F term, when $H_{so} = 0$. Whenever a term is split by a perturbation, the total degeneracy is conserved, that is, the sum of the degeneracies of the split components must equal the degeneracy of the original (unsplit) term. Furthermore, the sum of the orbital degeneracies of all the components arising from a given term equals $2L + 1$.

For elements in the first transition series ($Z \leq 30$), it is usually possible to ignore spin-orbit coupling in making assignments. In this approximation, the *Russell–Saunders coupling* scheme, L and S are "nearly" good quantum numbers. When Z increases, that is, for the second and third transition series, the rare earths and the actinides, the Russell–Saunders coupling approximation becomes progressively less valid.

IV. IONIC ENERGY LEVELS IN CUBIC SYMMETRY—LIGAND-FIELD THEORY[1,5,6]

When an ion is transferred from the gas phase to a condensed phase, liquid or solid, interaction with the environment alters the energy levels and consequently the transition energies. The environmental perturbation, H_{LF}, which must be added to Eq. (1-9), can be separated into two parts, a symmetric component, whose effect is to change the separation between the free ion terms, and the ligand-field perturbation. Although this latter interaction affects the term intervals, it also serves to remove all or part of the degeneracy of each term, and it is this splitting that is of interest in describing the spectra of metal complexes. The number of components into which any term is split is determined by the symmetry of the ligand field, and the methods of group theory are employed for the analysis.

Although the interaction between the electrons on the transition-metal ion and those on the ligands constituting the immediate environment should properly be treated quantum mechanically, that is, as chemical bonds, the classical, electrostatic model provides a convenient basis for this discussion. This picture can be replaced by a more comprehensive model, for example, molecular orbital (MO) theory, but the simpler description is quite adequate for many spectral interpretations, and covalency effects can be introduced as required.

The electrostatic model is often called crystal-field theory, while the MO extension is termed ligand-field theory. However, here ligand-field theory is used to designate all symmetry-dependent perturbations.

Two cubic symmetries will be treated explicitly, octahedral (O_h) for MX_6 and tetrahedral (T_d) for MX_4 complexes. Either of these symmetries leads to a partial removal of the fivefold d orbital degeneracy, and the order of the resulting t_2 and e orbital sets can be obtained from the electrostatic model. Although alternative descriptions are possible, it is usual to designate the t_2 orbitals as d_{xy}, d_{xz}, d_{yz}, and the e orbitals as $d_{x^2-y^2}$, d_{z^2} (Fig. 1-4).

The parameter, Dq (or Δ), is a measure of the ligand-field strength, and the e_g–t_{2g} separation in O_h is defined as $10\,Dq$. The ligand-field transition

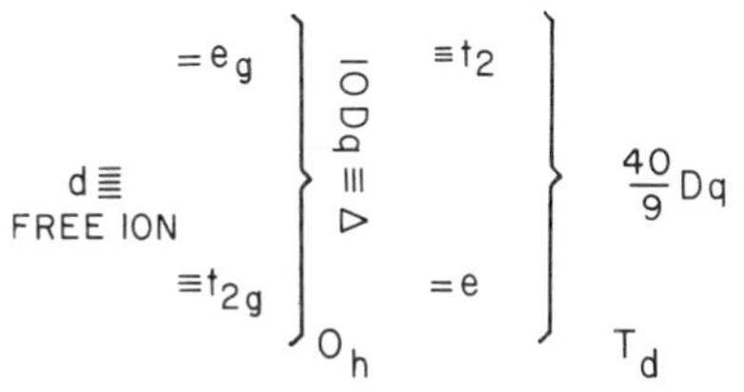

Fig. 1-4. d orbital splitting patterns in O_h and T_d fields.

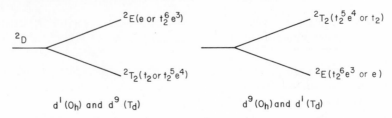

Fig. 1-5. Splitting patterns for d^1 and d^9 complexes in O_h and T_d fields.

in a d^1 complex, namely, $Ti(H_2O)_6^{3+}$, is then assigned as $^2E \leftarrow {}^2T_2$ ($t_2 \rightarrow e$) (Fig. 1-5). The g subscript, appropriate to centrosymmetric complexes, will only be used when required for intensity considerations.

The configurations of octahedral d^2 and d^3 ions are t_2^2 and t_2^3, respectively. There are, however, two possibilities for a d^4 ion. In the free ion, the four d electrons occupy different d orbitals with parallel spins ($S = 2$), in accord with Hund's rule of maximum multiplicity. When two electrons are placed in the same orbital, the spins must be paired and energy is required for the pairing. For small values of Dq, the spin-pairing energy exceeds $10 Dq$, the configuration is $t_2^3 e$, and the free-ion spin ($S = 2$) is maintained. As the ligand-field strength is increased, a point is reached where $10 Dq$ exceeds the spin-pairing energy, and the configuration becomes $t_2^4(S = 1)$. These configurations are designated high and low spin, respectively. This same duality exists for d^5, d^6, and d^7 ions, but only one configuration is possible for octahedral d^8 and d^9 ions. Magnetic measurements can be employed to determine which description, spin-free or spin-paired, is appropriate. Although the decision is usually clear, occasionally an equilibrium between the two types of species is encountered.

In determining the term separations produced by ligand fields, two different, but completely equivalent, approaches have been employed for many-electron systems, the weak- and strong-field formalisms.

A. Weak Fields

In this limit, the effect of the ligand field is assumed to be small compared to the repulsion between the d electrons. Since the splitting of the terms arising from a free-ion d^n configuration is due to electron repulsion, the weak-field condition means that the ligand-field splitting of the term is small compared to the free-ion separation between the terms. Consequently, the analysis is based on free-ion wave functions, whose energies are modified by the symmetric perturbation (nephelauxetic effect[7]). The number of terms and their degeneracies are obtained from group theory, but the actual energies must be computed from the weak-field matrices.[5,6] Since the perturbation does not involve spin, only the reduction in spatial

Table 1-2. Term Splittings in Cubic Fields—Weak Field Approximation

Free Ion Term	Orbital Degeneracy	Cubic Terms[a]
S	1	A_1
P	3	T_1
D	5	$T_2 + E$
F	7	$A_2 + T_2 + T_1$
G	9	$A_1 + E + T_2 + T_1$

[a] The degeneracies of A, E, T terms are 1, 2, and 3, respectively.

symmetry need be considered. The splitting of free-ion terms in octahedral and tetrahedral fields is shown in Table 1-2.

B. Strong Fields

In the strong-field limit, the ligand-field perturbation is much greater than the interelectronic repulsion energy. In this case, it is not appropriate to speak of free-ion terms, but rather to utilize the d orbitals appropriate to the environmental symmetry, for example, t_2 and e for cubic symmetry. Terms corresponding to the configuration are then derived according to group theory (Table 1-3).

Table 1-3. Term Systems for Strong Field Limit

Configuration	Terms
t_2^6	$^1A_1$
t_2^5, t_2	$^2T_2$
t_2^4, t_2^2	$^3T_1, {}^1T_2, {}^1E, {}^1A_1$
t_2^3	$^4A_2, {}^2E, {}^2T_2, {}^2T_1$
e^4	$^1A_1$
e^3, e	^2E
e^2	$^3A_2, {}^1E, {}^1A_1$
$t_2e, t_2^5e^3, t_2e^3, t_2^5e^3$	$^3T_1, {}^3T_2, {}^1T_1, {}^1T_2$
$t_2^2e, t_2^4e, t_2^2e^3, t_2^4e^3$	$^4T_1, {}^4T_2, {}^2A_1, {}^2A_2, {}^2E(2),^a {}^2T_1(2), {}^2T_2(2)$
$t_2^3e, t_2^3e^3$	$^5E, {}^3A_1, {}^3A_2, {}^3E(2), {}^3T_1(2), {}^3T_2(2),$ $^1A_1, {}^1A_2, {}^1E, {}^1T_1(2), {}^1T_2(2)$
$t_2e^2, t_2^5e^2$	$^4T_1, {}^2T_1(2), {}^2T_2(2)$
$t_2^2e^2, t_2^4e^2$	$^5T_2, {}^3A_2, {}^3E, {}^3T_1(3), {}^3T_2(2), {}^1A_1(2),$ $^1A_2, {}^1E(3), {}^1T_1, {}^1T_2(3)$
$t_2^3e^2$	$^6A_1, {}^4A_1, {}^4A_2, {}^4E(2), {}^4T_1, {}^4T_2, {}^2A_1(2),$ $^2A_2, {}^2E(3), {}^2T_1(4), {}^2T_2(4)$

[a] When a terms of a given symmetry occurs more than once, the appropriate number is indicated.

The interelectronic repulsion is then incorporated as a perturbation that serves to remove some of the degeneracy within each configuration. In the free ion, the electron repulsion is expressed in terms of three Condon–Shortley parameters, F_0, F_2, and F_4, or, alternatively, by the set of Racah parameters, A, B, C,[5,8] but B and C are sufficient when only energy differences are required. The nephelauxetic effect of the symmetric perturbation is included by reduction of B and C from the free-ion values.[7] Unfortunately, the two parameters are not reduced by the same fraction, and the variation of C/B may introduce interpretive difficulties.[9]

C. The Real Situation—Intermediate Fields

Neither the weak- nor the strong-field limits are encountered in actual practice. In other words, an intermediate case obtains. Either limit can be used as a starting point for calculations that must converge to the same result. In the weak-field approach, terms of the same free-ion symmetry, for example, the two ^4P terms in Cr(III), are mixed by the ligand field. Similarly, in the strong-field approach, terms of the same symmetry but derived from different configurations, namely, $^4T_1(t_2^2e)$ and $^4T_1(t_2e^2)$, are mixed by the electrostatic repulsion. Each term in the weak field limit must be correlated, that is, joined continuously, with a term of the same symmetry in the strong-field limit. All of the relevant information is contained in the *Tanabe–Sugano diagrams*[10] (Fig. 1-6), in which term energies are plotted as a function of Dq. In d^1 and d^9 ions the distinction between strong and weak fields is not involved, and the only difference between the two configurations is an inversion of the term order.

In using the Tanabe–Sugano diagrams for d^4, d^5, d^6, and d^7, the apparent discontinuity in slope is an artifact of the plotting technique. In these systems, this discontinuity occurs when the ground-state changes from low spin to high spin. It must be kept in mind that the diagrams in Fig. 1-6 were constructed for particular values of $C/B (\sim 4)$. This quantity may vary markedly for a given ion.[9] Also, the ratio will be quite different in octahedral and tetrahedral complexes. Consequently, the Tanabe–Sugano diagrams are only to be used for qualitative purposes.

Experience has shown that the common ligands can be arranged as a function of increasing ligand-field strength. This ordering, the spectrochemical series, is

$$CN^- > NO_2^- \sim phenanthrene > 2{,}2'\text{-bipyridine}$$

$$> SO_3^{2-} \sim ethylenediamine > NH_3 > pyridine$$

$$> NCS^- > H_2O \sim oxalate^{2-} > ONO^- \sim OH^- > F^-$$

$$> SCN^- > Cl^- > Br^- > I^-$$

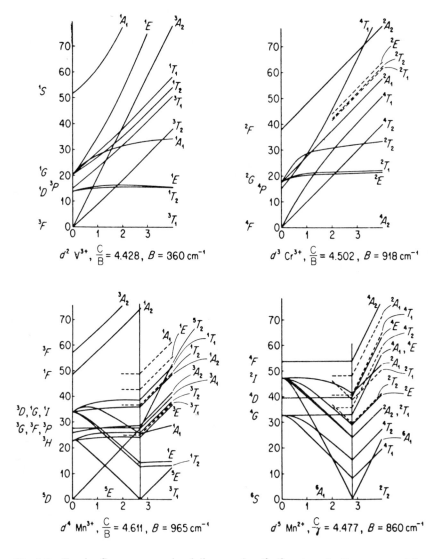

Fig. 1-6. Tanabe–Sugano energy level diagrams for d^2–d^8 systems in O_h symmetry (after Ref. 10).

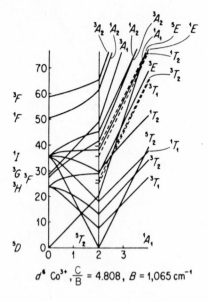

d^6 Co^{3+}, $\frac{C}{B}$ = 4.808, B = 1,065 cm^{-1}

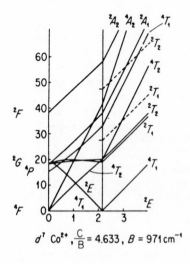

d^7 Co^{2+}, $\frac{C}{B}$ = 4.633, B = 971 cm^{-1}

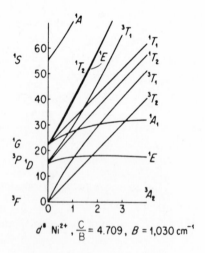

d^8 Ni^{2+}, $\frac{C}{B}$ = 4.709, B = 1,030 cm^{-1}

Fig. 1-6 (*Continued*)

Thus, $Fe(CN)_6^{3-}$ is a strong-field (low-spin) complex, while $Fe(H_2O)_6^{3+}$ is a weak-field (high-spin) complex.

V. INTENSITIES OF ELECTRONIC TRANSITIONS

In solution, the light transmitted (I) by a sample of pathlength l cm and molar concentration, c, at a particular wavelength, λ, can be represented by the equation

$$I = I_0 10^{-\varepsilon cl} \tag{1-10}$$

where I_0 is the incident intensity and ε the molar extinction coefficient. The fraction of light absorbed, $(I_0 - I)I_0^{-1} = 1 - 10^{\varepsilon cl} \approx \varepsilon cl$. This last approximation is valid when only a small fraction of the light is absorbed ($<10\%$). In crystalline systems, a different expression is sometimes employed

$$I = I_0 e^{-\sigma Nl} \tag{1-11}$$

where N denotes molecules cm^{-3}. The molecular absorption cross section, σ, in cm^2, is related to ε by the equation

$$\sigma = 3.82 \times 10^{-21} \varepsilon \tag{1-12}$$

The oscillator strength for a band with appreciable absorption only between $\bar{\nu}_1$ and $\bar{\nu}_2$ is

$$f = 4.32 \times 10^{-9} \int_{\bar{\nu}_1}^{\bar{\nu}_2} \varepsilon(\bar{\nu}) \, d\bar{\nu} \approx 4.6 \times 10^{-9} \varepsilon_{max} \Delta\bar{\nu}_{1/2} \tag{1-13}$$

where $\bar{\nu}$ is the energy in cm^{-1} and $\Delta\bar{\nu}_{1/2}$ the half-width of the absorption band. In strong transitions, $f \approx 1$ and $\varepsilon_{max} = 10^5 - 10^6$ $M^{-1} cm^{-1}$. Sometimes the "strength" of a transition is cataloged in terms of ε_{max}. If only broad bands are involved ($\Delta\bar{\nu}_{1/2} \sim 2$–$4{,}000$ cm^{-1}) such a criterion is useful. However, in the spectra of transition-metal complexes, bands of greatly different widths are encountered, and the oscillator strength is more meaningful.

The intensity of a transition between any two states, Ψ_1 and Ψ_2, is proportional to the square of the transition moment integral, $\int \Psi_2 \mu_e \Psi_1 \, dt$, where μ_e is the operator corresponding to the interaction between the electric vector of the exciting light and the electrons in the molecule. The integration is performed over all coordinates, electronic, nuclear, and spin. The analogous equation for a transition induced by the magnetic component of the incident light is $\int \Psi_2 \mu_m \Psi_1 \, dt$. If either of these integrals is zero, the transition is said to be electric or magnetic-dipole forbidden, respectively. Otherwise, it is electric and magnetic-dipole allowed (we ignore the very weak electric-quadrupole transitions here).

The conditions under which the transition moment integral is zero are given by the selection rules, which can be divided into two classes, symmetry and spin.

A. Spin-Forbidden Transitions

The spin-selection rule, $\Delta S = 0$, prohibits transitions between terms of different S. For example, in Co(III), $^3T_1 \nleftrightarrow {}^1A_1$ ($\nleftrightarrow$ means not allowed) is spin forbidden. The $\Delta S = 0$ selection rule is rigorous only if spin-orbit coupling is completely absent, a situation never encountered. Consequently, transitions that violate the selection rule are often observed, albeit with smaller intensity than spin-allowed transitions. As Z increases, the selection rule becomes less restrictive. Spin-forbidden transitions are more intense in the second and third transition series. The well-known 2537 Å line in the Hg emission spectrum is due to the $^3P_1 \rightarrow {}^1S_0$. This line is, however, less intense than the 1850 Å line ($^1P_1 \rightarrow {}^1S_0$). Spin-forbidden transitions in the spectra of the first-row transition elements typically have $\varepsilon_{max} \leq 1 \text{ M}^{-1} \text{ cm}^{-1}$, but in the second and third transition series,[11] this value may approach $100 \text{ M}^{-1} \text{ cm}^{-1}$.

B. Symmetry-Forbidden Transitions

For a transition between individual vibronic components of the ground state ($\Psi_{i\alpha}$) and excited state ($\Psi_{j\beta}$), the transition moment integral is

$$M_{i\alpha}^{j\beta} = \int \Psi_{j\beta} \mu_e \Psi_{i\alpha} \, d\tau \tag{1-14}$$

where the volume element, $d\tau$, includes both electronic and nuclear coordinates. This can be written as

$$M_{i\alpha}^{j\beta} = \int \Theta_j(q, Q)\phi_\beta^j \mu_e \Theta_i(q, Q)\phi_\alpha^i(Q) \, d\tau_e \, d\tau_n = \int \phi_\beta^j M_i^j \phi_\alpha^i \, d\tau_n \tag{1-15}$$

Since the integration can be carried out separately over the electronic coordinates ($d\tau_e$) and nuclear coordinates ($d\tau_n$), the electronic contribution can be separated, and

$$M_i^j = \int \Theta_j \mu_e \Theta_i \, d\tau_e \tag{1-16}$$

$$M_{i\alpha}^{j\beta} = M_i^j \int \Phi_\beta^j \phi_\alpha^i \, d\tau_n \tag{1-17}$$

If M_i^j vanishes for symmetry reasons, the transition is said to be symmetry forbidden. The symmetry-selection rules are obtained according to group theory.[5]

A particular example of symmetry-selection rules is the Laporte rule, which is operative in all systems with a center of symmetry, that is, whenever g or u subscripts can be included in the term symbol. The Laporte rule prohibits electric dipole transitions between terms of like symmetry, $g \leftrightarrow g$, $u \leftrightarrow u$, but magnetic-dipole transitions are allowed between these terms. All transitions between terms derived from the same free-ion configuration, for example, d^n, are *Laporte forbidden* in centrosymmetric complexes.

Just as spin-forbidden transitions are observed, so do symmetry-forbidden transitions occur, often with intensities comparable to those of symmetry-allowed transitions. To understand this apparent anomaly, it is necessary to consider the way in which intensities arise. When a transition is symmetry allowed, the intensity of all transitions originating in the ground-state vibrational level (α), is obtained from $\Sigma_\beta (M_{i\alpha}^{i\beta})^2$, where the summation extends over all vibrations of the upper state. The Franck–Condon factors, $\int \phi_\beta^i \phi_\alpha^i \, d\tau_n$, then determine the relative intensities of the vibronic transitions. These factors embody, quantitatively, the Franck–Condon principle, that is, transitions are vertical (nuclei do not move during an electronic transition). If the equilibrium nuclear positions are the same in both the ground and excited states, transitions in which the vibrational quantum numbers are unchanged. $0 \leftarrow 0$, $1 \leftarrow 1$, and so forth, are the strongest (Fig. 1-2b). More commonly the potential minima are displaced, and the $0 \leftarrow 0$ transition is not the most intense. In Fig. 1-2a, the *Franck–Condon factor* is greatest for $2 \leftarrow 0$.

The above analysis applies to an allowed mechanism. When $\int \Theta_i \mu_e \Theta_i \, d\tau_e = 0$, the intensity is derived by "stealing" from an allowed transition, that is,

$$M_i^j = (M_i^j)_0 + \sum_k \left(\frac{\partial M_i^j}{\partial Q_k}\right) Q_k \tag{1-18}$$

where the Q_ks are normal coordinates. In symmetry-allowed transitions, $(M_i^j)_0 \neq 0$, but the magnitude need not be large. The second term in this expression describes the vibronically induced intensity, that is, the forbidden mechanism. Normally, the intensities derived by the forbidden mechanism are less than those obtained by the allowed mechanism. However, in the spectra of metal complexes, this is not necessarily true. Consider a centrosymmetric octahedral complex. In this case, the forbidden mechanism applies $[(M_i^j)_0 = 0]$. When the center of symmetry is destroyed, say by chelation ($O_h \rightarrow D_3$), the transition may become group-theoretically allowed, yet the intensification effect of the distortion on

$(M_1^j)_0$ may be quite small, and the actual intensity obtained by the allowed mechanism insignificant. Thus, the oscillator strengths of the $^4T_2 \leftarrow {}^4A_2$ Cr(III) transition are not correlated with departures from O_h symmetry.[12] For example, they are 1.1×10^{-3} for both $Cr(CN)_6^{3-}$ and $Cr(C_2O_4)_3^{3-}$. Values of ε_{max} typically range from 10 to 100 $M^{-1} cm^{-1}$ for $d-d$ transitions.

In one respect, allowed and forbidden transitions differ, that is, the 0–0 band (no-phonon line) is absent in symmetry-forbidden transitions (actually a weak magnetic dipole 0–0 band does appear). Unless the vibrational structure is well resolved, this criterion will not be of value in assessing the symmetry of the complex. Also, the total intensity of an allowed transition is temperature independent, while the forbidden mechanism leads to an intensity enhancement with temperature.[13]

C. Radiative Relaxation Rates

If the absorption $(j \leftarrow i)$ and emission $(j \rightarrow i)$ transitions are identical, the radiative rate, k_r, can be computed from the oscillator strength of the absorption

$$k_r = 2.88 \times 10^{-9} \bar{\nu}_{ji}^2 n^2 \int \varepsilon(\bar{\nu}) \, d\bar{\nu} \frac{g_l}{g_u} \tag{1-19}$$

where n is the refractive index of the medium and g_l/g_u is the ratio of the degeneracies in the lower and upper states, respectively. Although this equation is rigorous for line-like resonance fluorescence, it is not valid when broad absorption and emission are involved. In this case, the equation

$$k_r = 2.88 \times 10^{-9} n^2 \langle \nu_f^{-3} \rangle^{-1} \int \frac{\varepsilon(\bar{\nu}) \, d\bar{\nu}}{\bar{\nu}} \frac{g_1}{g_n} \tag{1-20}$$

has been suggested.[14] When the absorption and emission spectra are "mirror images," Eq. (1-20) yields good results, but large excited-state distortions (Section X) lead to deviation from mirroring, and Eq. (1-20) may be a very poor approximation.[15] Even if the excited-state distortion is not large, the transition probabilities for vibronic transitions in absorption and emission may differ, because the vibronically induced mixing with the perturbing states is different in the ground and excited states (the mixing coefficient decreases with energy separation).

Equation (1-19) can be applied to the 0–0 transition and k_r^{0-0} computed. If the 0–0 transition were well resolved, then the following expression could be used to calculate k_r

$$k_r = \frac{k_r^{0-0}}{\eta} \tag{1-21}$$

where η is the fraction of quanta emitted in the 0–0 line. Unfortunately, the 0–0 transition is often buried in the broad absorption band, and this approach is then inapplicable, but it has been used for ruby[16] and $Cr^{3+} : NaMgAl(C_2O_4)_3 \cdot 9H_2O$.[17]

VI. CLASSIFICATION AND ASSIGNMENT OF ELECTRONIC TRANSITIONS

It has been useful to classify the spectra of organic molecules according to the "orbital jump" involved. Thus, there are $\pi-\pi^*$, $n-\pi^*$, $\sigma-\sigma^*$, and so forth, transitions. In like fashion, it is convenient to describe transition-metal-complex spectra in terms of the orbital change.

As long as only $d-d$ transitions are involved, it is possible to assign the spectra of metal complexes without explicitly introducing molecular orbitals. The effect of covalent bonding can be formally included within the ligand-field model by adjustment of the parameters, Dq, B, and C. However, it is sometimes necessary, and often useful, to treat the ligands more explicitly, that is, employ molecular orbitals[1,5,6]

$$\Psi = a\phi_M + b\phi_L \qquad (1\text{-}22)$$

where ϕ_M and ϕ_L refer to metal ion and ligand orbitals, respectively.

Both σ- and π-bonding can be encompassed within this formalism by utilizing combinations of metal and ligand orbitals of appropriate symmetry (Fig. 1-7). If only σ-bonding is important, the t_2 orbital set is

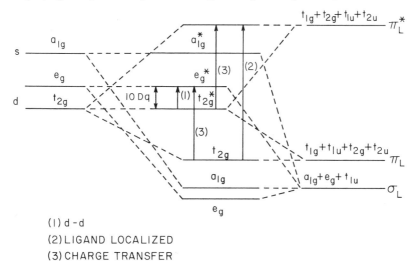

(1) d–d
(2) LIGAND LOCALIZED
(3) CHARGE TRANSFER

Fig. 1-7. Molecular orbital diagram for O_h symmetry.

nonbonding, and the interval between the antibonding e^* and the t_2 orbitals is 10 Dq. When π-bonding is possible, the t_2 orbitals are involved, and now 10 Dq is the $t_2^*-e^*$ separation.

If the interaction between ligand and metal ion orbitals is not too large, that is, if the metal ion electrons are mainly localized on the metal and the ligand electrons are localized on the ligands, it is useful to classify electronic transitions as (1) metal localized, (2) ligand localized, and (3) charge transfer.

Metal-localized transitions include $d-d$ transitions and $f-f$ transitions in rare-earth ions. Ligand-localized transitions encompass the perturbed $\sigma-\sigma^*$, $\pi-\pi^*$, and $n-\pi^*$ transitions of the free ligands. Intramolecular charge-transfer transitions are of two types, metal to ligand (CTTL) and ligand to metal (CTTM). An intermolecular charge transfer to solvent (CTTS) transition has also been recognized.[18] Although this classification has considerable utility, it is based on the weak-interaction assumption and would be valueless if in Eq. (1-22) $a \simeq b$.[19] However, it is often possible to separate transitions into these three classes.

If the CTTL transition involves the promotion of an electron from an orbital largely localized on the metal, that is, $a \gg b$, to an orbital in which $a \ll b$, an electron is essentially "transferred" from the metal to the ligand, and the metal is oxidized. Conversely, a transition from a ligand-localized orbital ($a \ll b$) to a metal-localized orbital ($a \gg b$) results in reduction of the metal (CTTM).

Several criteria are employed in the assignment of the bands in an absorption spectrum. The strategy can be illustrated for $Co(NH_3)_6^{3+}$ (Fig. 1-1). This species is d^6, and magnetic measurements indicate that it is a low-spin complex. There are two low-energy spin-allowed $d-d$ transitions arising from $t_2^6 \rightarrow t_2^5e$ in the spectrum of the spin-paired d^6 ion, $^1T_1 \leftarrow {}^1A_1$ and $^1T_2 \leftarrow {}^1A_1$. It is reasonable to assign the two bands between 350 and 500 nm with ε_{max} between 10 and 100 $M^{-1} cm^{-1}$ as these metal-localized $d-d$ transitions. There are two spin-forbidden $d-d$ transitions (also from $t_2^6 \rightarrow t_2^5e$) that should lie at lower energy than $^1T_1 \leftarrow {}^1A_1$; these are $^3T_1 \leftarrow {}^1A_1$ and $^3T_2 \leftarrow {}^1A_1$. The weak long-wavelength shoulder at 740 nm is assigned to the former with the latter hidden in the tail of the more intense $^1T_1 \leftarrow {}^1A_1$ band.

The very intense absorption below 250 nm is assigned as CTTM since no intraligand NH_3 absorption is expected in this region. This CT band moves to a longer wavelength as the ligand-ionization potential is decreased (the σ_L, and/or π_L orbital energies are raised) and the CT transition is described as ligand to metal. A similar situation is found in $Cr(NH_3)_6^{3+}$ (Fig. 1-8). This assignment is consistent with the observation that a redox reaction is induced by CT absorption in Co(III) complexes. An extensive

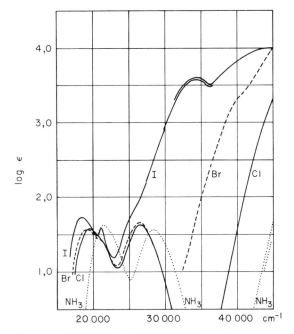

Fig. 1-8. Spectra of $Cr(NH_3)_6^{3+}$ and $Cr(NH_3)_5X^{2+}$ complexes (after Ref. 24).

tabulation of CTTM spectra is presented in Chapter 3 (Table 3-1). The high intensity of the CT band is ascribed to the $g \leftrightarrow u$ character of the transition.

Although spectral assignments are often made in this straightforward manner, one must be prepared for more challenging cases. This is especially true in complexes of second and third transition series metals,[19] but complexities are also present in the first transition series. For example, in the photoaquation of $Cr(NH_3)_5Cl^{2+}$, two reaction modes are probably involved. In one pathway, Cl^- is replaced, while the solvent is exchanged for NH_3 in the other mode.[20] The relative importance of the two pathways is wavelength dependent. This result has been interpreted as indicating a mixing of ligand-field and CTTM states.[21]

The spectrum of $[Ru(bipy)_3]Cl_2$ (Fig. 1-9) exemplifies the complications that can arise in the second and third transition series.[22] As the atomic number of the metal ion increases, charge-transfer transitions move to lower energies. The d–d and CT transitions then fall in the same region of the spectrum, and the d–d transitions may be obscured by more intense CT transitions. If, in addition, the ligand-localized transitions also fall in the same spectral region, the assignments do indeed present a challenge. The spin-allowed ligand-localized transitions (π–π^*) were identified by

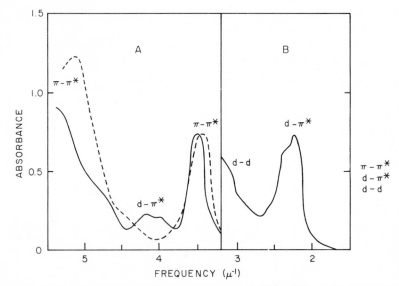

Fig. 1-9. (A) ———, [Ru(bipy)₃]Cl₂, 1×10^{-4} M; – – –, 2,2′-bipyridine, 2.5×10^{-5} M in 50% H_2SO_4; (B) ———, [Ru(bipy)₃]Cl₂, 5×10^{-4} M; 1 cm path length in all spectra. After ref. 22.

comparing the spectra of the complex and the protonated ligand (broken line). The spin-allowed CT transitions were assigned on the basis of previous work,[23] while the *d–d* transition assignments are tentative. The identity of the low intensity tail below 20,000 cm⁻¹ has been the subject of considerable uncertainty. It was initially ascribed to a spin-forbidden *d–d* transition and later assigned as a spin-allowed weak CT transition. On the basis of emission lifetimes it has now been labeled as a spin-forbidden CT transition. The history of this last assignment clearly demonstrates the difficulties that can be encountered.

VII. LOWER SYMMETRY LIGAND FIELDS

Few, if any, octahedral complexes conform exactly to O_h symmetry in all electronic states. Departures from O_h (or T_d in the case of tetrahedral complexes) symmetry may arise from several sources:

1. Intramolecular
   a. Nonidentical ligands, for example, MX_4Y_2.
   b. Coordinating atoms do not conform to the symmetry group, for example, bidentate ligands reduce the symmetry to D_3 (or lower).
   c. Coordinating atoms conform to the symmetry group, but other ligand atoms do not, for example, $M(NH_3)_6$.
   d. Jahn–Teller effect.

2. Lattice effects
   a. Lattice forces distort geometry.
   b. Charge-compensating ions that are not part of the complex but change the effective ligand field.

Terms that are degenerate in a given symmetry can be split into two or more components when the symmetry is reduced. Two common distortions from O_h symmetry are tetragonal (D_{4h} and C_{4v}) and trigonal (D_3). Examples of tetragonal complexes are: *trans*-[Cr(NH$_3$)$_4$Cl$_2$]$^+$ (D_{4h}) and Cr(NH$_3$)$_5$Cl^{2+} (C_{4v}). A typical D_3 complex is Cr(en)$_3^{3+}$. The effect of the reduction of symmetry on O_h terms is summarized in Table 1-4.

Table 1-4. Splitting of Terms Upon Reduction of Symmetry

O_h	D_{4h}, C_{4v} [a]	D_3, C_{3v}
A_{1g}	A_{1g}	A_1
A_{2g}	B_{1g}	A_2
E_g	$A_{1g} + B_{1g}$	E
T_{1g}	$A_{2g} + E_g$	$A_2 + E$
T_{2g}	$B_{2g} + E_g$	$A_1 + E$

[a] Drop g subscript in C_{4v}.

The effect of departure from O_h symmetry due to ligands such as NH$_3$ is generally small, and it is appropriate to treat the complexes as O_h species.

Symmetry reduction is manifested in absorption spectra by band splittings. When the bands are broad, the separation between the components must be greater than 1000 cm^{-1} to be detectable, but a smaller splitting may lead to an observable broadening of the bands. For example, there are two bands in the spectrum of Cr(NH$_3$)$_5$X^{2+} ($^4B_2 \leftarrow {}^4A_2$ and $^4E \leftarrow {}^4A_2$) corresponding to the Cr(NH$_3$)$_6^{3+}$ $^4T_2 \leftarrow {}^4A_2$ transition (Fig. 1-8).[24] When X = Cl$^-$, the second band appears as a shoulder, but when X = Br$^-$ or I$^-$, the two components are well resolved.

Although the magnitude of a distortion-induced splitting can often be estimated from the solution absorption spectrum (for example, in Fig. 1-8 the $^4T_2$ state in Cr(NH$_3$)$_5$Br^{2+} is split by 2700 cm^{-1}), the relative ordering of the split levels is best determined by recording the polarized spectra of the species suitably oriented in a crystalline matrix. An example of such an assignment is shown for Cr(ox)$_3^{3-}$ in Fig. 1-10. The solid curves were obtained with light polarized along the threefold axis, while the dotted curves refer to light polarized perpendicular to that axis.[25] In a recent theoretical treatment of tetragonal complex photochemistry, the order of the levels is necessary for predictive purposes.[26]

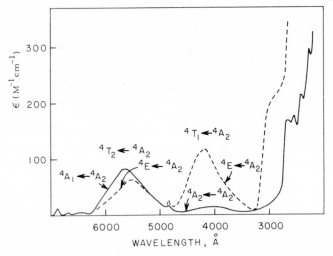

Fig. 1-10. The polarized crystal spectrum of $Cr(ox)_3^{3-}$ in $NaMgAl(ox)_3 \cdot 9H_2O$ at 25°C (after Ref. 25).

The magnitude of the splitting produced by a tetragonal field depends on the "difference" between the ligands on and off the fourfold axis. This difference is conveniently expressed in terms of the relative position in the spectrochemical series. Thus, the $Cr(H_2O)_6^{3+}$ $^4T_2 \leftarrow {}^4A_2$ transition is hardly split in $Cr(H_2O)_5F^{2+}$ because F^- and H_2O are not widely separated in the spectrochemical series. On the other hand, the splitting of $^4T_2 \leftarrow {}^4A_2$ into two components is readily apparent in $Cr(NH_3)_5Br^{2+}$. NH_3 has a relatively large Dq value, while Dq is rather small for Br^-.

The extreme case of tetragonality is obtained when no ligands are present on the fourfold axis and the complex is four-coordinated (square-planar), for example, $Ni(CN)_4^{2-}$. Such an extreme perturbation makes the O_h description a poor starting point for the analysis, and $Ni(CN)_4^{2-}$ is diamagnetic, whereas all O_h d^8 systems must be paramagnetic. The spectral assignments in square-planar complexes have proven to be quite difficult. For example, the assignment of the $PtCl_4^{2-}$ spectrum is still controversial in spite of much effort.[27]

In principle, nonlinear molecules distort in such a manner as to remove the electronic degeneracy of all degenerate states. This distortion is a consequence of the *Jahn–Teller theorem*.[28] In practice, this splitting is significant on a low-resolution spectral scale only for orbital degeneracy. For O_h symmetry, E, T_1, and T_2 states are split, but the magnitudes of the Jahn–Teller splittings may vary considerably. In condensed media, the spectral consequences of a *static Jahn–Teller* distortion are the same as that of a low-symmetry ligand field, that is, a band is split into two or more components.

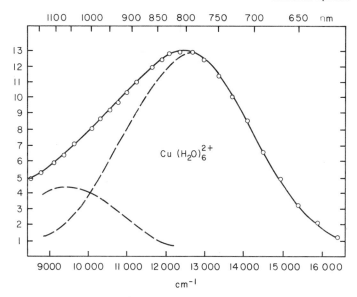

Fig. 1-11. Spectrum of $Cu(H_2O)_6^{2+}$ (after Ref. 29). The dotted curves indicate a possible resolution of the $^2T_{2g} \leftarrow {}^2E_g$ transition into components.

The results of the Jahn–Teller effect are readily observed in the spectrum of $Cu(H_2O)_6^{2+}$ complexes[29] (Fig. 1-11). The ground state of this d^9 system is ^2E. The spectral band corresponding to $^2T_2 \leftarrow {}^2E$ is broad and asymmetric, indicating the presence of more than one transition under the band envelope. These bands arise from transitions between the $^2A_1$ and $^2B_1$ components of ^2E and the $^2B_2$ and ^2E components of $^2T_2$.

It should be kept in mind that both ground and excited states are subject to Jahn–Teller distortions, so that band splitting can occur in transitions originating in a nondegenerate gound state, for example, A_2. Generally speaking, E states are split to a much greater extent than T_1 or T_2 states.

Surprisingly, the Jahn–Teller effect can lead to a diminution of splittings that arise from other sources, namely, trigonal fields and spin-orbit coupling.[30] This, the *Ham effect*, leads to a splitting in the V^{3+} $^3T_2 \leftarrow {}^3T_1$ band of only $4\,cm^{-1}$ rather than an expected trigonal splitting of some $400\,cm^{-1}$.

VIII.   EMISSION SPECTRA[31]

Emission spectra, when observable, are useful in two respects: (1) as an aid to locating the $v = 0$ level of the lowest excited states; and (2) to provide information about the distortion of the lowest excited states relative to the ground state, that is, by means of the Stokes shift.

Emission spectra are less complex than absorption spectra since, with few exceptions, substantial emission occurs only from the lowest level of a given multiplicity. Consequently, the luminescence spectrum of an organic molecule can contain two contributions, a spin-allowed fluorescence band ($S_1 \rightarrow S_0$) and a spin-forbidden phosphorescence band ($T_1 \rightarrow S_0$). By analogy, in Cr(III) $^4T_2 \rightarrow {}^4A_2$ is fluorescence and $^2E \rightarrow {}^4A_2$ phosphorescence.

The relation between the absorption and emission spectra for transitions between two electronic states is illustrated in Fig. 1-2 for two cases. (a) Ground- and excited-state equilibrium geometries differ, and (b) ground- and excited-state geometries are identical. The excited-state distortion can preserve the molecular symmetry or lead to a different symmetry.

The difference in energy between the broad maxima in Fig. 1-2a is the *Stokes shift*. When the symmetry is preserved, it is possible to predict qualitatively the extent of the Stokes shift from the Tanabe–Sugano diagrams. If the d electron distribution is essentially the same in both the ground and excited states, that is, both states are derived from the same configuration, the bond strengths and the equilibrium internuclear distances are nearly unchanged (Fig. 1-2b). At the same time, the energies of the two terms would depend on Dq in a parallel fashion. This is exemplified by the $^2E \rightarrow {}^4A_2$ spectra of d^3 systems. $^2E_2$ and $^4A_2$ are "parallel" (both arise from t_2^3), and the spectra, if symmetry allowed (e.g., D_3 complexes), exhibit a prominent 0–0 band (Fig. 1-2b). Conversely, the $^4T_2 \leftrightarrow {}^4A_2$ spectra exhibit Stokes shifts of 3–5×10^3 cm^{-1}, as expected from the absence of parallelism between $^4T_2(t^2e)$ and $^4A_2(t_2^3)$. It should be noted that not all spin-forbidden transitions are sharp. Thus, in Mn^{2+} (d^5), $^4A_1(t_2^3e^2) \leftarrow {}^6A_1(t_2^3e^2)$ is sharp but $^4T_1(t_2^4e) \leftarrow {}^6A_1(t_2^3e^2)$ is broad. If the excited-state distortion is marked, a large Stokes shift will result. When a transition is symmetry forbidden, the 0–0 transition is weak, and the Stokes shift in low-resolution spectra appears to be substantial even for a transition like $^2E \rightarrow {}^4A_2$, for example, Cr(CN)$_6^{3-}$.[32]

In order to assess the energetic requirements for various photochemical pathways, it is of considerable importance to determine the energy separation between the excited states. If the 0–0 transitions of any two states can be located, this problem is solved, since the difference in the 0–0 energies gives the desired information directly. When the absorption and emission spectra for a given transition are sharp, the 0–0 energy can generally be evaluated without difficulty. However, the problem is more difficult when the spectra are broad. The emission spectrum intensities should be corrected for the wavelength dependence of the detector sensitivity and multiplied by $1/\bar{\nu}^2$.[15] Both the absorption and emission spectra should be recorded at the same temperature and plotted on a $\bar{\nu}$ scale. Ideally, the two spectra would intersect at the 0–0 band, a situation most likely to obtain when both spectra are recorded at low temperatures. However, the 0–0 energy obtained from low-temperature spectra in this manner does not apply to the room temperature photochemistry that takes place in solvent-relaxed excited states (see Section X). Several alternative strategies have been offered for estimating the 0–0 energies from emission and absorption spectra,[21] but evaluation of this quantity is still subject to some uncertainty.

IX. EXCITATION SPECTRA

The fraction of incident light absorbed at any wavelength is

$$(I_0 - I)I_0^{-1} \approx \varepsilon c l = \alpha \qquad (1\text{-}23)$$

if this fraction is less than 0.10. When a sample luminesces, the intensity of light emission, I_r, is proportional to the amount of light absorbed at the excitation wavelength (λ)

$$I_r = k\Phi\alpha I_0 \qquad (1\text{-}24)$$

where Φ is the luminescence quantum yield (Chapter 2) and k is a constant for a given instrumental geometry. The determination of I_0 as a function of λ over the range of interest is a requisite for this procedure. If Φ is also constant, the excitation and absorption spectra are identical. Although this technique is of limited use in the spectroscopy of transition-metal complexes because a luminescent species is required, there are circumstances when it can be useful. These include the location of transitions with small α (due to limited solubility and/or small ε), the recording of spectra at low temperatures where the detection of weak transitions is sometimes difficult, and the spectra of opaque crystals. Spin-forbidden transitions can often be detected in this way. Another useful application of excitation spectroscopy involves a comparison of the excitation and absorption spectra of the same system under identical conditions. If the spectra are not congruent this indicates a wavelength dependence for Φ, an observation with important consequences in the elucidation of the pathway for electronic energy degradation (Chapter 2). One source for a reduction in Φ would be a photochemical process with a significant quantum yield that originates in a higher state and is signalled by a relative decrease in the excitation spectral intensity as the photochemically active region is entered. It should be emphasized that accurate correction of the excitation spectrum for the wavelength variation of I_0 is essential.

X. EXCITED STATE DISTORTIONS AND STOKES SHIFTS

The validity of the potential curves in Fig. 1-2 is dependent only on the applicability of the Born–Oppenheimer separability of the wave functions. For diatomic molecules in the gas phase, the coordinate is the internuclear distance, r. If the molecule has N atoms, the potential "curve" is actually a $3N - 5$ dimensional hypersurface. In the two-dimensional diagram, the coordinate can be one of the normal coordinates, or for pictorial simplicity, merely be some unspecified generalized coordinate.

When a molecule is embedded in a solid or liquid matrix, potential curves no longer, strictly speaking, represent intramolecular energies, but rather the total energy of the molecule plus the environment. That the interaction between solute and solvent can be energetically significant is evident from solvent-induced shifts in absorption and emission spectra.[33]

The interaction between an absorption center and its environment has, in solids, been described by a configuration-coordinate diagram.[34] This is identical in appearance to a potential-energy diagram, but the distance coordinate is a configuration coordinate that need not be a real coordinate.[35] The configuration-coordinate model has been widely used to describe spectral characteristics such as Stokes shifts, and to interpret the luminescent behavior of crystalline solids.

The configuration-coordinate diagrams can also be employed in liquid systems, but complicating factors arise in this context. Although vibrational relaxation times are comparable in liquid and solid media, the rotational relaxation of the molecular environment is relatively slow in liquid solutions. The increase in rotational relaxation time, coupled with a reduction in translational mobility, leads to a marked change in the configuration-coordinate diagram when the viscosity is increased. This effect is particularly noticeable as the glass point is passed and is evidenced by a decrease in the Stokes shift when the environment becomes rigid.[36] This result can be interpreted in terms of the potential curves shown in Fig. 1-12. In both fluid and rigid solutions absorption is to a configuration corresponding to the ground-state orientation of the solvent molecules (solid curves). If the viscosity is small, reorientation of the solvent molecules precedes emission, which originates in the relaxed configuration (dashed curves). However if the viscosity is high, no solvent reorientation occurs, and emission takes place from the ground-state configuration (upper solid curve). Consequently, the emission spectrum is blue-shifted in a glassy solution, while the absorption spectrum is relatively unaffected. When the dipole moment changes magnitude and especially direction upon excitation, the displacement between the non-relaxed and relaxed curves is significant. Indole is an example where this

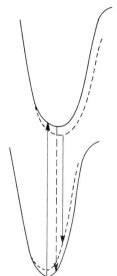

Fig. 1-12. Potential curves in rigid and fluid media; (——) ground-state solute-solvent orientation; (– – –) relaxed excited-state solute-solvent orientation.

effect is so important that a specific complex (exciplex) is found in the excited state.[37] The effect should be rather small in nonpolar molecules, but in C_{4v} complexes this phenomenon may be more important. This relaxation shift must be recognized when emission spectra recorded in rigid media at low temperatures are used to interpret photochemical behavior in fluid solvents.

It must continually be borne in mind that a two-dimensional configuration coordinate diagram can be very misleading. The need to include more than one coordinate in treating distortions has been detailed by Sturge.[28]

We confine our attention here to broad transitions where the vibrational structure is poorly resolved. Spin-allowed transitions of many organic molecules fall into this category. The Stokes shifts in these molecules span a large range, for example, naphthalene ($3700 \, \text{cm}^{-1}$), quinoline ($8000 \, \text{cm}^{-1}$).[38]

The data on fluorescence spectra of metal complexes are not extensive and consequently not much information about the Stokes shifts of such systems is available. Fluorescence ($^4T_2 \rightarrow {}^4A_2$) has been reported for Cr^{3+} in several crystalline lattices, but in only one case $Cr(urea)_6^{3+}$, has solution fluorescence been observed.[39] Some representative Stokes shifts are listed in Table 1-5.

In terms of the configuration coordinate model,

$$\text{Stokes shift} = \tfrac{1}{2}(k_e + k_g)X_0^2 \qquad (1\text{-}25)$$

where k_e and k_g are the force constants of the vibration in the excited and ground states, respectively, and X_0 is the displacement of the minima of

Table 1-5.   Representative Stokes Shifts for Some Transition Metal Complexes

System	Stokes Shift (cm^{-1})	Ref.
$^4T_2 \leftrightarrow {}^4A_2$		
$Cr^{3+}: Al_2O_3$ (ruby)	3,600	40
$Cr^{3+}: Be_3Al_2Si_6O_{18}$ (emerald)	3,300	40
$Cr^{3+}: LiNbO_3$	5,050	41
$Cr^{3+}: LiTaO_3$	4,630	41
$Cr(urea)_6^{3+}$ (sol'n)	3,700	39
CrF_6^{3-} in $(NH_4)_3[CrF_6]$	2,240	42
$CrF_5(H_2O)^{2-}$ in $K_2[CrF_5(H_2O)]$	3,200	42
$Cr(H_2O)_6^{3+}$ in $[Cr(H_2O)_6]F_3$	4,170	42
$^3T_1 \leftrightarrow {}^1A_1$		
$[Rh(NH_3)_5N_3]^{2+}$	11,300	43
$[Rh(NH_3)_5Br]^{2+}$	9,500	43
$[Rh(NH_3)_5I]^{2+}$	9,000	43
$[Ir(en)_2Cl_2]^+$	9,400	44
$Ir(en)_3^{3+}$	17,200	44
$Rh(en)_3^{3+}$	11,600	44
$trans$-$[Rh(py)_4Cl_2]^+$	6,100	44
$[Rh(diphenylphosphinoethane)_2Cl_2]^+$	5,400	45
$[Ir(diphenylphosphinoethane)_2Cl_2]^+$	7,800	45

the potential curves along the configuration coordinate.[34] If we accept the notion that X_0 is likely to be largest for the radial vibrations (the force constants for these are likely to be most affected by $t_2 \rightarrow e$ excitation), the data in Table 1-5 can be used for crude inferences. The stretching vibrations for a MX_6 species are a_{1g}, e_g, and t_{2u}. Although metal-ligand vibrational frequencies for a given ligand vary considerably with the metal,[46] the a_{1g} frequencies, which might be used as a measure of k_g, are larger in MF_6 than in MO_6 complexes. Yet, the Stokes shifts show the opposite trend. Within the framework of this model, X_0 is much larger in $Cr(H_2O)_6^{3+}$ than in CrF_6^{3-}. Of course, this approach is likely to be oversimplified, but it is important to recognize that radial vibrational frequencies are much less in metal complexes than in aromatic organic molecules. The comparable Stokes shifts in the two molecular classes would indicate greater excited-state distortions in metal complexes. This point is important in the theoretical interpretation of nonradiative transitions (Chapter 2).

The Stokes shifts in the d^6 complexes of the second and third transition series appear to be larger than those encountered in Cr^{3+} (Table 1-5). A note of caution is warranted here, however. The reported shifts for Rh^{3+} and Ir^{3+} complexes refer to the spin-forbidden $^3T_1 \leftrightarrow {}^1A_1$ transition, which is broad in low-spin d^6 systems because $t_2 \rightarrow e$ excitation is involved. It is possible that lower-energy, weak spin-forbidden transitions were unobserved in absorption and that the Stokes shifts are in reality smaller. The very large shift in $Ir(en)_3^{3+}$ (17,200 cm^{-1}) is unprecedented in molecules where exciplexes or strong solvent–solute interactions are absent. Several overlapping weak transitions are in fact observed in the 80°K absorption spectrum of $[Rh(diphenylphosphinoethane)Cl_2]^+$.[45] It is especially difficult to estimate the Stokes shift accurately when the absorption in question is weak and partially overlapped by a strong transition.

The greater the difference between the geometries in the ground and excited state, the more invalid Eq. (1-20) becomes for the calculation of natural lifetime,[15] $\tau_0 = k_r^{-1}$. Considerable discussion has centered around the possibility that lifetimes are much longer than expected because of large excited-state distortions.[47] It is difficult to discuss this problem in terms of a single configuration coordinate that roughly represents a symmetric breathing vibration. Chemical reaction along this coordinate would correspond to

$$ML_6^* \rightarrow M + 6L \qquad (1\text{-}26)$$

that is, all the metal-ligand bonds are broken simultaneously (Fig. 1-13), a

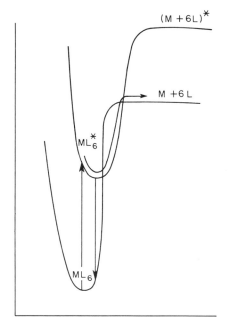

CONFIGURATION COORDINATE
("SYMMETRIC STRETCH")

Fig. 1-13. The pathway for dissociation along the symmetric configuration coordinate $(ML_6^* \rightarrow (M + 6L)^*)$.

process requiring much energy. If a dissociative mechanism is involved

$$ML_6^* \rightarrow ML_5 + L \qquad (1\text{-}27)$$

is the more likely process. The "reaction" coordinate is very different from the "configuration" coordinate (Fig. 1-14).

In this connection the discussion of Hammond is instructive.[48] He distinguishes two reaction modes by which an excited molecule can react. For a dissociative process these are

$$
\begin{aligned}
1. \quad & ML_6^* \rightarrow (ML_5 + L)^* \\
2. \quad & ML_6^* \rightarrow ML_5 + L
\end{aligned}
\qquad (1\text{-}28)
$$

The first mechanism requires an activation energy to surmount the barrier along the reaction coordinate whereas in the second, the dissociation proceeds in a manner analogous to intramolecular radiationless processes (Chapter 2).

Even in the absence of large excited-state distortions, the actual radiative lifetime may be very different from that calculated from Eq. (1-20).

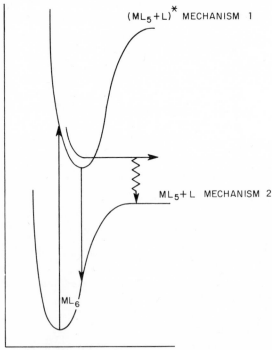

Fig. 1-14. The pathways for $ML_6^* \rightarrow ML_5 + L$ photodissociation.

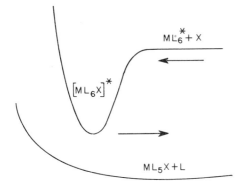

Fig. 1-15. The pathway for an associative mechanism.

As a final possibility, consider a potential diagram for an associative reaction (Fig. 1-15). This is visualized as a two-step process

$$ML_6^* + X \rightarrow ML_6X^*$$

$$ML_6X^* \rightarrow ML_5X + L$$

The equilibrium geometries of ML_6X^* and ML_6X are very different (ML_6X is in a repulsive ground state), and emission from ML_6X^*, if it occurs at all, would appear to have a very large Stokes shift. Of course, the lifetime of ML_6^* would have to be sufficiently long for $ML_6^* + X \rightarrow ML_6X^*$ to take place.

XI. MISCELLANEOUS ENVIRONMENTAL EFFECTS

As we have already observed, there are dangers inherent in the use of spectral data obtained under one set of conditions (e.g., in a low-temperature solid) for interpreting processes taking place under entirely different conditions (e.g., room-temperature photochemistry). Some environmental influences have been described above. In this section, additional perturbations that can be induced by environment are detailed.

A. Solvent Effects

Ligand-localized bands exhibit the same type of solvent dependence as the spectra of the free ligands. For example, $\pi-\pi^*$ transitions are shifted slightly toward the red with increasing solvent polarity, while $n-\pi^*$ transitions are blue shifted.

The d–d transition energies are little affected (<12%) by changes in solvent composition, assuming of course, that ligand exchange with the solvent does not occur. Intensities, however, may change by as much as

30%.[49] In very viscous or rigid solvents, where the solvent relaxation time is not short compared to the excited-state lifetime, the emission spectrum may be sensitive to the excitation wavelength.[50] This effect is caused by the different solvent-solute "species" that are present at the instant of absorption.

B. Lattice Effects—Solid State Spectra

In assessing the influence of a lattice on the spectra of transition-metal ions, we restrict ourselves to the spectra of well-defined complexes, for example, $Cr(H_2O)_6^{3+}$, of sufficient stability to exist in liquid solution as well as in crystals. Many of the solid-state spectral studies are thereby excluded, for example, $Cr^{3+}:Al_2O_3$.

There are a number of advantatges in recording the spectrum of a complex embedded in a crystalline matrix. The regularity of the environment plus the ease with which the temperature can be reduced by employing liquid N_2 (77°K) or He (4°K) often leads to a much sharper spectrum in which individual vibronic transitions can be identified. In this way 0–0 bands can be accurately located and weak spin-forbidden transitions detected, even when masked by the more intense spin-allowed bands in solution spectra. This provides more precise information about the energy difference between the several excited states than is normally possible from solution spectra. In addition, spectral assignments from polarized spectra require single crystals.

Environmentally induced spectral features unique to the crystalline environment include

1. Lattice vibrations (phonons)—it is usually possible to avoid errors from this source by a careful assignment of the intramolecular vibronic structure.

2. Lattice-induced distortions—these may be due to crystal-packing requirements or arise from electrostatic fields produced by the counter ions. The ^2E state in $Cr(CN)_6^{3-}$ is split some 56 cm^{-1} when this species is present in $K_3[Co(CN)_6]$, but the amount of this splitting varies with the crystalline host.[51] One would expect no ^2E splitting when the O_h complex is in a fluid solution.

Another example of this type of lattice effect is $Cr(urea)_6^{3+}$, where the $^4T_2 \leftarrow {}^4A_2$ absorption maximum is nearly invariant to host, but the $^4T_2 \rightarrow {}^4A_2$ emission maximum changes by some 1600 cm^{-1} from one lattice to another.[52]

3. Intermolecular interactions—in pure crystals or doped crystals many complexes are in close proximity. This gives rise to strong intermolecular coupling that results in spectral changes.

4. Transitions due to lattice defects—few crystals are perfect, and spectral transitions associated with defects can appear in the same region of the spectrum as the spectrum of interest.

XII. RARE-EARTH ION SPECTRA

The absorption and emission spectra of rare-earth ions have been described by Dieke.[53] Rare-earth ion spectra differ significantly from transition-metal ion spectra in several respects. In the first place, the spin-orbit coupling is not negligible compared to electron repulsion and terms such as the Eu^{3+} ^7F are split into many components spanning a range of many thousand cm^{-1}. In addition, the spin-selection rule breaks down badly in these systems. Also, the $4f$ electrons are shielded, and very sharp spectral lines are common in f–f transitions. This shielding also reduces the ligand-field splittings that seldom exceed 250 cm^{-1} compared to 20,000 cm^{-1} in transition-metal ions. Finally, the generalization that only the lowest energy state of a given multiplicity emits, is often violated.

REFERENCES

1. A. B. P. Lever, *Inorganic Electronic Spectroscopy*, Elsevier, Amsterdam, 1968.
2. G. Herzberg, *Atomic Spectra and Atomic Structure*, Dover, New York, 1944.
3. J. C. Slater, *Quantum Theory of Atomic Structure*, Vol. I, McGraw-Hill, New York, 1960.
4. C. Moore, *Atomic Energy Levels*, Vols. I–III, National Bureau of Standards Circular 467, 1949.
5. B. N. Figgis, *Introduction to Ligand Fields*, Wiley-Interscience, New York, 1966.
6. C. J. Ballhausen, *Introductions to Ligand Field Theory*, McGraw-Hill, New York, 1962.
7. C. K. Jørgensen, *Progr. Inorg. Chem.* **4**, 73 (1962).
8. J. S. Griffith, *The Theory of Transition-Metal Ions*, Cambridge U.P., London, 1961.
9. D. L. Wood, J. Ferguson, K. Knox, and J. F. Dillon, *J. Chem. Phys.* **39**, 890 (1963).
10. Y. Tanabe and S. Sugano, *J. Phys. Soc. Jap.* **9**, 753 (1954).
11. G. S. Arnold, W. L. Klotz, W. Halper and M. K. DeArmond, *Chem. Phys. Letters* **19**, 546 (1973).
12. L. S. Forster, *Transition Metal Chemistry* **5**, 1 (1969).
13. T. M. Dunn, in *Modern Coordination Chemistry* (J. Lewis and R. G. Wilkins, Eds.), Chap. 4, Wiley-Interscience, New York, 1960.

14. S. J. Strickler and R. A. Berg, *J. Chem. Phys.* **37**, 814 (1962).
15. J. B. Birks and D. J. Dyson, *Proc. Roy. Soc.* A **275**, 135 (1963).
16. D. F. Nelson and M. D. Sturge, *Phys. Rev.* A **137**, 1117 (1965).
17. F. Castelli and L. S. Forster, *J. Luminescence*, **8**, 252 (1974).
18. H. H. Schmidtke, *Z. Phys. Chem. (Frankfort)* **38**, 170 (1963).
19. H. H. Schmidtke, in *Physical Methods in Advanced Inorganic Chemistry* (H. A. O. Hill and P. Day, Eds.), p. 147, Wiley-Interscience, London, 1968.
20. H. F. Wastgestian and H. L. Schläfer, *Z. Phys. Chem. (Frankfort)* **62**, 127 (1968).
21. P. D. Fleischauer, A. W. Adamson, and G. Sartori, in *Inorganic Reaction Mechanisms Part II* (J. O. Edwards, Eds.), p. 1, Wiley, New York, 1972.
22. F. E. Lytle and D. M. Hercules, *J. Amer. Chem. Soc.* **91**, 253 (1969).
23. C. K. Jørgensen, *Acta Chem. Scand.* **11**, 166 (1957).
24. M. Linhard and M. Weigel, *Z. Anorg. Chem.* **266**, 49 (1951).
25. O. S. Mortensen, *J. Chem. Phys.* **47**, 4215 (1967).
26. J. I. Zink, *J. Amer. Chem. Soc.* **94**, 8039 (1972).
27. D. S. Martin, Jr., *Inorg. Chim. Acta Rev.*, **1971**, 107.
28. M. D. Sturge, *S. S. Physics* **20**, 91 (1967).
29. J. Bjerrum, C. J. Ballhausen, and C. K. Jørgensen, *Acta Chem. Scand.* **8**, 1275 (1954).
30. D. S. McClure and P. J. Stephens, in *Coordination Chemistry* (A. E. Martel, Ed.), Vol. 1, Chap. 2, Van Nostrand Reinhold, New York, 1971.
31. P. D. Fleischauer and P. Fleischauer, *Chem. Rev.* **70**, 199 (1970).
32. R. Condrate and L. S. Forster, *J. Chem. Phys.* **48**, 1514 (1968).
33. R. S. Becker, *Theory and Interpretation of Fluorescence and Phosphorescence*, Wiley-Interscience, New York, 1969.
34. C. C. Klick and R. Schulman, *S. S. Phys.* **5**, 108 (1957).
35. M. Lax, *J. Chem. Phys.* **20**, 1752 (1952).
36. D. M. Hercules and L. B. Rogers, *J. Phys. Chem.* **64**, 397 (1960).
37. M. S. Walker, T. W. Bednar, and R. Lumry, *J. Chem. Phys.* **47**, 1020 (1966).
38. I. Berlman, *Handbook of Fluorescence Spectra of Aromatic Molecules*, 2nd Edition, Academic, New York, 1971.
39. G. B. Porter and H. L. Schläfer, *Z. Physik. Chem. (Frankfort)* **37**, 109 (1963).
40. P. Kisliuk and C. A. Moore, *Phys. Rev.* **160**, 307 (1967).
41. A. M. Glass, *J. Chem. Phys.* **50**, 1501 (1969).
42. H. L. Schläfer, H. Gausmann, and H. U. Zander, *Inorg. Chem.* **6**, 1528 (1967).
43. T. R. Thomas and G. A. Crosby, *J. Mol. Spectrosc.* **38**, 118 (1971).
44. M. K. DeArmond and J. E. Hillis, *J. Chem. Phys.* **54**, 2247 (1971).
45. G. S. Arnold, W. L. Klotz, W. Halper, and M. K. DeArmond, *Chem. Phys. Letters* **19**, 546 (1973).
46. J. R. Ferraro, *Low-Frequency Vibrations of Inorganic and Coordination Compounds*, Plenum, New York, 1971.
47. A. Adamson, *J. Phys. Chem.* **71**, 798 (1967).
48. G. S. Hammond, *Advan. Photochem.* **7**, 373 (1969).
49. J. Bjerrum, A. W. Adamson, and O. Bostrup, *Acta. Chem. Scand.* **10**, 329 (1956).

50. F. Castelli and L. S. Forster, *J. Amer. Chem. Soc.*, **95,** 7223 (1973).

51. H. L. Schläfer, H. Wagener, F. Wasgestian, G. Herzog, and H. Ludi, *Ber. Bunsunges, Physik. Chem.* **75,** 879 (1971).

52. J. L. Laver and P. W. Smith, *Aust. J. Chem.* **24,** 1807 (1971).

53. G. H. Dieke, *Spectra and Energy Levels of Rare Earth Ions in Crystals,* Wiley-Interscience, New York, 1968.

2

KINETICS OF
PHOTOPHYSICAL PROCESSES

Gerald B. Porter

Department of Chemistry
University of British Columbia
Vancouver, British Columbia

I. INTRODUCTION

In any photochemical reaction, a chemical change occurs while the reactant species is in an excited state, usually but not necessarily an electronically excited state. The reactive species may be in the same state as was initially populated by absorption of radiation, or it may be in some other state reached by rapid intermolecular or intramolecular conversions. To understand the detailed mechanism by which a photochemical reaction occurs, we must establish, as fully as possible, the identity of the excited state in which the primary photochemical change occurs and the energetic and kinetic relationships among any particular excited state, other excited states, and the ground state.

There are four major sources of information about the excited states of a molecule: (1) normal absorption spectra, (2) photochemistry, (3) absorption and emission spectra of excited species, and (4) intermolecular quenching or sensitization. Certainly the bulk of the information available on excited states, especially of transition-metal coordination compounds, comes to us from studies of the normal ground-state absorption spectra. As has been established in Chapter 1, such spectra, backed up by ligand-field theory (again, see Chapter 1 for discussion and references), have provided extensive information about energy levels, nature of excited states, and even some kinetic parameters. Explorations of photochemistry alone do not provide explicit information concerning the detailed nature of excited states, except by inference, such as, for example, tentative association of a charge-transfer state with a photo-chemical redox reaction. But, photochemical data taken together with other supplementary information about excited states constitute an important area of mechanistic investigations.

That supplementary information can be provided by the remaining two techniques (3) and (4) mentioned above. Of these, studies of emission spectra have provided the bulk of the available data about excited-state lifetimes and corroboration of their energies. Excited-state absorption spectra, which have been used extensively in organic photochemistry, have received much less attention than they deserve for transition-metal complexes. The last of these techniques, (4) the study of intermolecular interactions between excited-state species and quenching molecules, is a powerful method for identification of reactive species. Energy-transfer processes such as these have now become an established part of any photochemical investigation.

In this chapter, these latter two techniques, spectra of excited-state species and energy transfer, are discussed with particular reference to the establishment of the nature of reactive species and to the overall kinetic scheme that applies to excited-state mechanisms of coordination compounds of transition-metal ions.

II.   LUMINESCENCE SPECTRA

Much of our understanding of luminescence phenomena is based on studies of organic molecules, mostly aromatic hydrocarbons, by chemists, and of inorganic phosphors by physicists. Each group has developed its own language, scarcely intelligible to the other.[1] This division has, to some extent, been carried over into research on the spectra of transition-metal complex ions.

Chemists have tended to study molecular coordination compounds, often in solution in a rigid glass, with the intent to learn about molecular energy levels and, perhaps, more importantly, about the kinetic processes associated with the excited species involved in the luminescence. Physicists, on the other hand, have generally studied crystalline systems, including semiconductors and ions doped into host ionic lattices; their concern has been principally the investigation of the energy and identity of excited states interpreted in terms of band theory and crystal-field theory. In supplement of photochemical investigations, studies of molecular complexes are most useful. However, data on ionic crystals reveal details beyond those usually obtainable with molecular complexes about the excited states. Therefore, we must look at both kinds of systems in order to have a clear picture of the excited state patterns and behaviors of transition-metal ion complexes.

It is interesting to note that one of the earliest detailed descriptions of a luminescent compound in the literature is that of a transition-metal coordination compound: $K_2[Pt(CN)_4]$.[2] Although the work was done before 1936, it included such data as absolute quantum yields and an estimate of the excited-state lifetime (0.25 nsec), as well as measurements of temperature and wavelength effects.

A.   Luminescence of Organic Molecules

Despite the early work on $K_2[Pt(CN)_4]$, the principles of molecular luminescence have been established largely through studies of aromatic molecules.[3,4] Some of these principles may be applicable to the luminescence of inorganic molecules and ions, but the species themselves and the nature of the excitation are quite different. Each principle must be examined carefully for possible modification to inorganic systems.

An important point to keep in mind, however, is that, according to Einstein's theory,[5] *any molecule that absorbs light must be able subsequently also to emit light spontaneously.* Furthermore, this statement applies equally well to each excited state reached by absorption of radiation. That is, luminescence occurs from each level of each such excited state that can be reached by absorption of light. Einstein showed the quantitative relationship between the absorption coefficients and the transition probability for the emission, both spontaneous and stimulated. Whether or not such emission can be *observed*, however, depends on the relative values of the radiative transition probability and the rate constants for other processes that degrade the excitation energy.

In 1950, Kasha[6] stated an empirical rule, based on studies of aromatic molecule luminescence, concerning which excited state of a molecule emits:

> The emitting level of a given multiplicity is the lowest excited level of that multiplicity.

With organic molecules that have singlet ground states, fluorescence originates from the lowest excited singlet state, and phosphorescence from the lowest triplet state. Among the many hundreds of organic molecules examined, only azulene and its derivatives provide exceptions to this rule.[7] They fluoresce from the second excited singlet state. Kasha's rule suggests that "internal conversion" processes, that is, nonradiative transitions among the excited states of a given spin multiplicity, normally occur so rapidly that the energy above that of the lowest excited state is degraded to thermal energy before emission can occur. It also follows that internal conversion of the lowest excited singlet state to the ground state occurs at a rate only competitive with the radiative rate. Kasha estimated a rate constant for internal conversion among upper electronic states of about 10^{13} sec^{-1}. More recent experimental data[8] confirm the order of magnitude of that estimate. The anomalous case of azulene has been explained by the large energy gap between the first and second excited singlet states and the relatively small gap between the first singlet and ground states.

That all levels of a molecule reached by absorption must emit is amply demonstrated by recent work[8,9] using picosecond (10^{-12} sec) pulses and sensitive detection. Emission quantum yields as small as 10^{-6} have been studied by this means. It is in this sense that *Kasha's rule* must be modified to: The *principal* emitting level is the lowest level of one multiplicity.

Another important principle derived from the study of organic molecules is that due to Vavilov[10]: In condensed media, emission from any given excited electronic state occurs from the lowest vibrational levels of that state. This comes about because molecular "collisions" in condensed media occur faster than emission by many orders of magnitude. Thus there is usually sufficient time before emission occurs for a Boltzmann equilibrium of the vibrational energy to be established with the surrounding medium. This concept applies to only "normal" emission. When luminescence of short-lived states is examined in solution with picosecond resolution or in the gas phase at low pressure, *Vavilov's rule* may not then apply. Furthermore, of course, the rate constants of nonradiative processes such as *internal conversion* and photochemical change are not limited to the maximum of 10^9 sec^{-1} of radiative processes. Hence Vavilov's rule does not necessarily apply to processes other than emission. However, recent measurements indicate that vibrational equilibration is complete within 6×10^{-12} sec.[11] To compete effectively with such relaxation, a process must have a rate constant of approximately $>10^{10}$ sec^{-1}.

A nonradiative transition between two states of different multiplicity is called *intersystem crossing*. Kasha[6] suggests a rate constant of the order of 10^7 sec^{-1}; smaller than that for internal conversion because it is dependent on spin-orbit coupling. The prohibition factor of approximately 10^6 applies to organic molecules composed only of light atoms. Both intersystem crossing and absorption and emission between states of different multiplicity are enhanced when heavy atoms are incorporated into the molecule,[12] supporting the idea of spin-orbit perturbation in such cases. Organic molecules with $\pi-\pi^*$ or $n-\pi^*$ excited states often exhibit phosphorescence from the lowest triplet state, especially at low temperatures where nonradiative processes degrading that triplet state are less important.

In general, then, an aromatic molecule or an organic molecule with nonbonding or "n"-type electrons may exhibit fluorescence and/or phosphorescence from the vibrational levels, which are in equilibrium with the surrounding medium, of the lowest excited singlet state and the lowest triplet state, respectively. Studies in the gas phase at low pressures[13] show that both internal conversion and intersystem crossing occur intramolecularly, that is, electronic energy is converted to vibrational energy within the molecule. In condensed media the isoenergetic processes are followed by rapid vibrational equilibration.

Considerable effort has been made during the last decade to formulate theoretical models of those nonradiative processes that lead to relaxation of electronic states.[14] Most of this work has relied on data obtained in organic systems. How the same models can be extended to inorganic systems cannot really be stated yet, mainly because of the lack of firm information on internal conversion and intersystem crossing. We return to this subject in a later section.

B.   Luminescence of Inorganic Compounds

Of all the possible transition metal coordination compounds, only those with d^2, d^3, d^5, d^6, and d^8 configurations have been reported to luminesce (aside from those cases in which emission is localized on a ligand, such as with the oxinates[15]). As Table 2-1 shows, however, if we include studies of crystals containing transition-metal ions as substitutional or interstitial impurities in other lattices, for example, Os^{4+} in $Cs_2[HfCl_6]$[16] or Co^{2+} in MgO,[17] the list can be extended to include d^4 and d^7 ions as well. So far, the luminescence only of d^1 and d^9 ions has escaped observation.

Some emission spectra, because of their sharp line character, are relatively easy to identify. This is the case for phosphorescence from d^3 ions, for example. In other systems, it may be far more difficult to make an unequivocal assignment of the luminescence and absorption bands. Those of $Ru(dipy)_3^{2+}$ have had at least three different labels.[18] It is, in general, necessary to take into account the spectral position, width, and shape of the emission, as well as its quantum yield, decay lifetime, and temperature dependence, together with the absorption spectrum, theoretical considerations, and perhaps a bit of luck!

The kinds of emission that have been characterized so far are: phosphorescence localized on metal orbitals with a sharp line spectrum, both phosphorescence and fluorescence localized on metal orbitals with

Table 2-1. Examples of Transition Metal Ions with Configuration d^n, Known to Luminesce

	As a Discrete Complex Ion	As an Ion in a Lattice
d^1	—	—
d^2	$[V(urea)_6]^{3+}$	V^{3+} in ZnS
d^3	$[Cr(NH_3)_6]^{3+}$	Cr^{3+} in Al_2O_3
d^4 (low spin)	—	Os^{4+} in Cs_2HfCl_6
d^4 (high spin)	—	—
d^5 (low spin)	—	—
d^5 (high spin)	$[MnBr_2(tpp)_2]$	Mn^{2+} in ZnS
d^6 (low spin)	$[Pt(NCS)_6]^{2-}$	—
d^6 (high spin)	—	—
d^7	—	Co^{2+} in Cs_2ZnCl_4
d^8	$[Pt(CN)_4]^{2-}$	Ni^{2+} in $KMgF_3$
d^9	—	—

broad structureless spectra, and charge-transfer phosphorescence that is characterized by a structured but not sharp spectrum. In the following sections each of these types of luminescence is examined in detail.

1. Sharp-Line Phosphorescence. Easily the best known example of sharp line emission is that of ruby, Cr^{3+} in Al_2O_3. The so-called R lines of ruby have been extensively investigated[19] in emission and absorption, and with a ruby laser in stimulated emission. Emission and absorption are mirror images reflected around the R lines near 14,400 cm^{-1} (Fig. 2-1), except that absorption involves transitions to other states as well as the phosphorescent state. All of the d^3 ions, when examined in an appropriate ionic

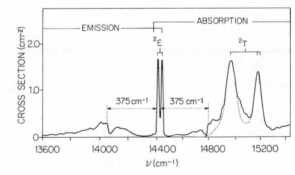

Fig. 2-1. Absorption and phosphorescence spectra of ruby at 77°K. The "R" lines near 14,000 cm^{-1} are the split origin of the $^2E \leftrightarrow {}^4A_2$ transition.[19]

matrix such as beryl, MgO, Al_2O_3, or GASH (guanidine aluminum sulfate hydrate), exhibit a similar sharp emission when irradiated with visible or ultraviolet light,[20] except that with a small ligand field, fluorescence may occur instead, as is described in the next section.

So characteristic is this emission that most coordination compounds of d^3 ions also can be made to phosphoresce under appropriate conditions, especially at low temperature in a rigid matrix.[21] In some cases, notably for $Cr(CN)_6^{3-}$ in DMF solvent,[22] emission is observed at room temperature. Sharp emission has also been found from d^2, d^4, d^5, d^7, and d^8 ions[16,17,23,24] often, however, in ionic matrices rather than in coordination compounds as such.

The sharp-line spectrum arises from the transition: $^2E_g \rightarrow {}^4A_{2g}$ for d^3 ions in O_h symmetry (octahedral complexes). Since this transition connects two states that, in the strong-field limit (see Chapter 1), have the same orbital configuration: t_{2g}^3, the equilibrium configurations are very similar. According to the Franck–Condon principle the spectral distribution is such that most of the intensity is in lines corresponding to simultaneous excitation of only one or two vibrational quanta with the change in electronic state.

At low temperatures where the Boltzmann population of vibrational levels other than the lowest is small, the emission spectrum consists of a single sharp line (under high resolution it may be split) that is the origin, zero-zero line, or "no-phonon" line (in crystals), and a number of other sharp lines lying on the long-wavelength side of the origin. The separations of these latter lines from the origin match vibration frequencies or combinations thereof of the ground electronic state. The $^2E_g \leftarrow {}^4A_{2g}$ transition in absorption has the same origin (however, see Chapter 1 on medium effects), but with the vibronic components lying on the blue side of the origin. The separations of these lines from the origin represent vibrations and combinations of vibrations of the ^2E_g state.

Idealized spectra are given in Fig. 1-2 of Chapter 1 with a state diagram showing how those spectra arise. To the extent that the strong field approximation is valid, the vibrational frequencies of the molecule are the same in the ^2E_g as in the $^4A_{2g}$ state. A good mirror image between absorption and emission would then be observed (with the spectrum linear in cm^{-1}). Such is often, but not always the case. The mirror image observed for MnF_6^{2-} in $Cs_2[GeF_6]$ crystal matrix[25] shown in Fig. 2-2 was not found for $Cr(CN)_6^{3-}$ in $K_3[Co(CN)_6]$ crystal matrix.[26] In the fluoride compound, the pattern of vibronic structure is relatively simple because there are only a few vibrational modes, while the cyanide compound has many more normal modes, and its spectrum is correspondingly complex (Fig. 2-3). It is also found that an apparent mirror-image relationship

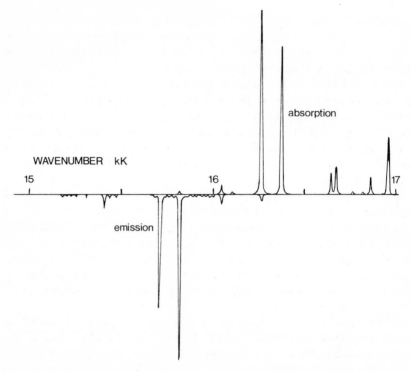

Fig. 2-2. Absorption and phosphorescence spectra of MnF_6^{2-} in $Cs_2[GeF_6]$ at 80°K.[61]

between absorption and emission observed at low resolution no longer holds under higher resolution.

For d^3 complexes, the separation of the ^2E and $^4A_2$ states is given approximately by[27]

$$E(^2E) - E(^4A_2) = 9B + 3C - \frac{90B^2}{10Dq} \tag{2-1}$$

where B and C are the Racah parameters and $10Dq$, the ligand-field strength. The dependence on ligand-field strength is relatively small as is evident from the data in Table 2-2, but depends instead on the values of B and C according to the nephelauxetic effect.

In a sharp phosphorescence spectrum the intensity distribution among the various lines is determined by the Franck–Condon principle, by the symmetry of the vibration(s) contributing to the line, and by the mechanism by which this nominally forbidden transition becomes allowed. Often the origin itself is magnetic-dipole allowed, while the rest of the spectrum is permitted through vibration-electronic mixing of an electric-dipole interaction (details in Chapter 1).

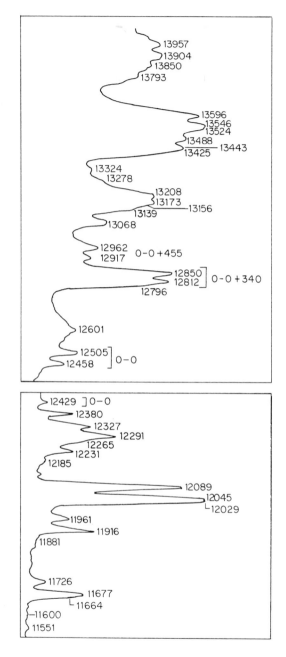

Fig. 2-3. Absorption and phosphorescence of $K_3[Cr(CN)_6]$ at 80°K. Absorption measured in the pure crystal, emission in a dilute crystal of $K_3[Co(CN)_6]$.[26]

45

Table 2-2. Frequencies (in kiloKaysers) of the Observed d–d Transitions in Cr(III) Complexes with the Best Values of the Racah Parameters[27]

	$10Dq = \bar{\nu}(^4T_2 \leftarrow {}^4A_2)$	$\bar{\nu}(^2E \leftarrow {}^4A_2)$	B	C
CrBr₃	13.43	14.00	0.440	3.23
CrCl₃	13.62	14.33	0.575	3.17
$[CrF_6]^{3-}$	15.06	15.67(?)	0.910(?)	2.91
$[Cr(exan)_3]$	16.00	12.73	0.436	2.94
$[Cr(urea)_6]^{3+}$	16.15	14.19	0.697	2.87
$[Cr(ox)_3]^{3-}$	17.50	14.44	0.664	2.92
$[Cr(H_2O)_6]^{3+}$	18.00	14.90	0.695	2.93
Ruby	18.15	14.43	0.822	2.62
$[Cr(CH_3)_6]^{3-}$	20.80	14.02	0.577	3.02
$[Cr(NH_3)_6]^{3+}$	21.45	15.12	0.711	3.12
$[Cr(CN)_6]^{3-}$	26.70	12.46	0.589	2.39

One of the important features of the sharp-line phosphorescence of transition-metal complex ions is that, being forbidden by both spin and symmetry, the transition probability is small and the lifetime of the phosphorescing state is long. Although the actual lifetime is necessarily shorter than the radiative lifetime (because of dissipative processes) if phosphorescence can be observed at all, the observed lifetime will also be relatively long. For example, $[Cr(en)_3]^{3+}$ has a radiative lifetime of 6 msec [28]; at 77°K the phosphorescence decays with a lifetime of 0.12 msec. and is easily observed.[28] Phosphorescence can still just be seen near room temperature, where the lifetime has decreased to about 0.001 msec.[29] It was this relatively long lifetime of the ^2E_g state that prompted Schläfer to suggest it as the precursor of photoaquation in Cr(III) complexes (see Chapter 4).[30]

2. Fluorescence. Probably only in d^3 systems has fluorescence been observed so far, and then only in special circumstances. The transition is spin allowed but symmetry forbidden according to the Laporte rule.

In a d^3 ion, fluorescence concerns the transition from lowest quartet state to the ground state: $^4T_{2g} \rightarrow {}^4A_{2g}$. These two states have the strong-field configurations, $t_{2g}^2 e_g$ and t_{2g}^3, respectively. That is, one electron is transferred from an antibonding e_g orbital to a nonbonding or only weakly bonding t_{2g} orbital. On a schematic potential-coordinate diagram (Fig. 1-2), the effect of the antibonding orbital occupation is to increase the equilibrium internuclear separation in the upper state relative to that of the ground state. The origin of the transition does not appear in the spectrum because the Franck–Condon overlap is small. Absorption reaches high-vibrational levels of the $^4T_{2g}$ state at wavelengths near the maximum of the absorption curve. Correspondingly, fluorescence populates high-vibrational levels of the ground state. A Stokes shift results, in that the emission occurs at lower energy than the absorption.

An example is illustrated in Fig. 2-4. The absorption maximum of $CrCl_3$ is at 13,900 cm^{-1}, and the fluorescence maximum is at 11,300 cm^{-1}, hence the Stokes shift is 2600 cm^{-1}. Both bands seem to be devoid of vibronic structure, even in the region of about 12,000 to 13,000 cm^{-1}, where the origin might be expected to lie. Had the spectrum been taken at low temperature to avoid any hot bands and under high resolution, some structure might have been observed in this region. This is the case, at least in absorption, with ruby[31] and especially with MnF_6^{2-},[25] shown in Fig. 2-5. So far, however, no structure has been seen in the fluorescence spectrum of a d^3 ion.

In a few cases, notably CrF_6^{3-}, $CrCl_6^{3-}$ and some of the aquo fluoro Cr(III) complexes, fluorescence is the only emission observed.[33] Again, there are only a few examples of d^3 that exhibit both fluorescence and phosphorescence. The total emission and absorption spectra[32] for one of these, $Cr(urea)_6^{3+}$, given in Fig. 2-6, make a classic (but nearly unique) case. The large Stokes shift of the fluorescence, approximately 3700 cm^{-1}, contrasts strongly with the essentially zero shift in the low resolution of Fig. 2-6 of the phosphorescence, with respect to their corresponding absorption bands. A peculiarity of the emission spectrum is that phosphorescence lies to the blue of the fluorescence maximum, a phenomenon unknown in organic molecules.

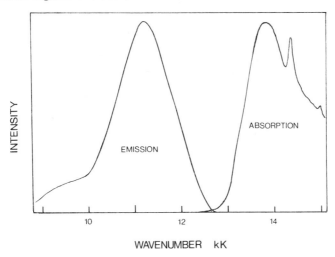

Fig. 2-4. Absorption and fluorescence spectra of CrCl$_3$, which has Cr(III) in nearly octahedral sites, at 80°K.[27]

OPTICAL DENSITY⎯⎯⎯⟶

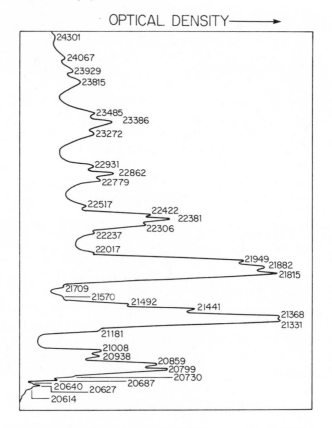

Fig. 2-5. Absorption $(^4T_2 \leftarrow {}^4A_2)$ of MnF_6^{2-} in $Cs_2[GeF_6]$ at 80°K.[25]

Schläfer et al.[33] make the point that the type of emission that occurs for d^3 complexes depends on the relative positions of the absorption maxima for $^4T_2 \leftarrow {}^4A_2$ and $^2E \leftarrow {}^4A_2$. If there is only a small separation, the origin of $^4T_2$ lies below that of ^2E, and only fluorescence occurs. The more common case is with $10Dq \geq 17{,}000 \text{ cm}^{-1}$ so that the ^2E lies lowest in energy and only phosphorescence can be detected. [A report[34] that fluorescence could be seen from $Cr(CN)_6^{3-}$ $(10Dq = 26{,}700 \text{ cm}^{-1})$ is apparently in error.[35]] Intermediate cases show both fluorescence and phosphorescence. However, in those examples, the lifetime of fluorescence is the same as that of phosphorescence.[36] Furthermore, Laver and Smith[37] have shown that, for $Cr(urea)_6^{3+}$, the fluorescence disappears at low temperatures. These data show that the origin of $^4T_2 \leftarrow {}^4A_2$ lies at higher energy than that of $^2E \leftarrow {}^4A_2$. The observed fluorescence occurs subsequent to back-intersystem crossing from the ^2E state to $^4T_2$, at least for those complexes showing both fluorescence and phosphorescence. However, part of the fluorescence (in an amount dependent on temperature and the specific counter ion) is prompt fluorescence,[38,39] with short lifetime (ca. 50 μsec), and part is delayed and has the same lifetime as the phosphorescence.

The question of the actual lifetime of the $^4T_2$ state is an important one. Wasgestian[22] and others[40,41] have demonstrated the role of this state rather than the relatively long-lived ^2E state in photoaquation reactions. (See Chapter 4 for details concerning this question.) From the integrated absorption band, the oscillator strength for the $^4T_2 \leftarrow {}^4A_2$ transition for various Cr(III) complexes is found[20] to lie in the range of 0.7×10^{-3} to 2×10^{-3}. This may be compared with values of the order of 10^{-7} to 10^{-6} for $^2E \leftarrow {}^4A_2$. Such oscillator strengths correspond to radiative lifetimes of the order of 5×10^{-6} sec. Although few data are available on experimental lifetimes for complexes that fluoresce only, Zander's[36] measurements are consistent with the calculations: 1.5×10^{-6} sec for $(NH_4)_3[CrF_6]$.

A number of measurements have been attempted with ruby to establish the $^4T_2$ lifetime. The results point to a value less than 5×10^{-12} sec.[42] Even fewer measurements have been made for complex ions of Cr(III). These

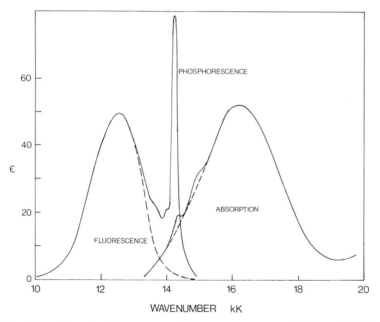

Fig. 2-6. Absorption, phosphorescence, and fluorescence spectra of $Cr(urea)_6^{3+}$. Emission at 80°K, absorption at 300°K in water/methanol/ethylene glycol (3:2:2). Dotted lines represent approximate resolution of the spectra into the several transitions.[32]

aspects are explored in a later section on the kinetic schemes relating to excited states.

3. Broad Phosphorescence. Unlike the phosphorescence of d^3 systems, in particular, phosphorescence of d^6 octahedral complexes originates from an excited state with strong-field electron configuration different from that of the ground state. The broad phosphorescence observed in some d^6 compounds has been assigned as the transition: $^3T_1 \rightarrow {}^1A_1$.[43] These states have the strong-field configurations: $t_{2g}^5 e_g^1$ and t_{2g}^6, respectively. Except for the spin quantum-number change, the phosphorescence transition in these d^6 complexes represents the same orbital transition as the fluorescence transition in d^3 complexes. Both have spectra that are broad and structureless and lie considerably to the long-wavelength side of the main absorption bands. However, the Stokes shifts between absorption and emission are even less well known for d^6 phosphorescence than for d^3 fluorescence. Fluorescence can be observed only in a few d^3 complexes with $\Delta\bar{\nu} \approx 2000$ to 5000 cm^{-1}. In d^6 complexes, the absorption spectrum is so weak that it is difficult to observe, and there is often a large apparent Stokes shift from the lowest energy observed band, often as much as $16,000 \text{ cm}^{-1}$.[44] The band seen in absorption in such cases is almost certainly not the inverse of the phosphorescence transition. The difficulty in observing $^3T_1 \leftarrow {}^1A_1$ comes about not so much because it is a weak transition, but because it will be broad and structureless, and lies on the long-wavelength tail of considerably strong bands. In Fig. 2-7 are shown several examples of what is called "d–d phosphorescence," together with the relevant parts of the absorption spectra.

The assignment of a particular absorption band as $^3T_1 \leftarrow {}^1A_1$, especially when there are apparent Stokes shifts up to $16,000 \text{ cm}^{-1}$, must be regarded as suspect. A simple calculation using data for cis-$[Rh(Cl)_2(phen)_2]^+$ will suffice to demonstrate the point. The phosphorescence lifetime is observed to be 4.2×10^{-5} sec as measured in a rigid glass at 77°K.[44,45] Demas and Crosby[46] find the quantum yield to be 0.042; thus the minimum value of the radiative lifetime

$$\tau_0 = \frac{\tau}{\phi} \qquad (2\text{-}2)$$

is 1.0×10^{-3} sec, which should hold if $\phi_{isc} = 1$ (see later discussion, p. 57). The oscillator strength will be given approximately by

$$f = \frac{g_u}{g_l \tau_0 (\bar{\nu})^2} = 4.5 \times 10^{-5} \qquad (2\text{-}3)$$

in which $\bar{\nu}$ is the emission maximum (here $14,100 \text{ cm}^{-1}$) and the ratio of

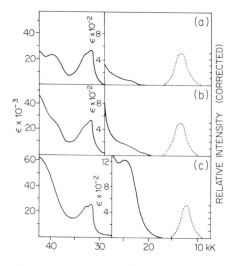

Fig. 2-7. Absorption (300°K) and broad phosphorescence (77°K) of (a) cis-[Rh(Cl)₂(bipy)₂]Cl·2H₂O, (b) cis-[Rh(Br)₂(bipy)₂]NO₃, and (c) cis-[Rh(I)₂(bipy)₂]I.[45]

the degeneracies, g_u/g_l, is 9. The oscillator strength is related to the integrated extinction coefficient, which we here approximate as $\varepsilon_{max}\Delta\bar{\nu}$:

$$f = 4.6 \times 10^{-9}\varepsilon_{max}\Delta\bar{\nu} \tag{2-4}$$

Using the bandwidth of the emission spectrum, 2800 cm^{-1}, and the value calculated above, we find $\varepsilon_{max} \simeq 3.5$ M^{-1} cm^{-1}.

It can be argued that the approximation in Eq. (2-3) has validity only when applied to nearly fully allowed transitions and that there is a more elegant formulation.[47] However, it has been used with some success for weak transitions[20] and is useful at least for order-of-magnitude estimates, such as that given here. The absorption spectrum of [Rh(Cl)₂(phen)₂]$^+$ is shown in Fig. 2-8. We look then for an absorption band approximately 3000 cm^{-1} wide with maximum extinction coefficient of about 3.5 M^{-1} cm^{-1}. It is apparent that we would not be able to observe such a band unless either it had a very much larger ε_{max} than we have calculated, perhaps larger by as much as a factor of over ten, or it was well separated

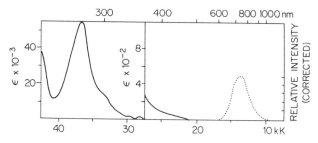

Fig. 2-8. Absorption (300°K) and phosphorescence (77°K) of cis-[Rh(Cl)₂(phen)₂]$^+$.[45]

from the tail of the other absorption bands. There have been suggestions in the literature[18] that the calculation of τ_0 by equations similar to (2-3) and (2-4) might easily be in error by a factor of almost 100. Nevertheless, in those cases that the calculation is suspected, assignments of the bands in the absorption spectra are by no means unequivocal.

4. Structured Phosphorescence. A second type of phosphorescence is also observed from some d^6 compounds; however, the two types are mutually exclusive. In no case do both kinds of emission occur in the same substance.

This phosphorescence exhibits vibrational structure in its spectrum, but is not as sharp a spectrum as d^3 phosphorescence and, therefore, seems less complex. Each band in the spectrum is at least 25 cm^{-1} wide, and generally at least three prominent vibrational bands stand out.[48] Some examples are given in Fig. 2-9 with the corresponding absorption spectra.

Phosphorescence of this type has so far been observed only with d^6 ions of the second or third transition series and *only* with the ligands; 1,10-phenantholine (phen):

and 2,2'-bipyridine (bipy):

or their derivatives, occupying four of the coordination positions. The vibrational spacings of about 1450 cm^{-1} (bipy) and 1300 (phen) found in the luminescence spectra are also observed in the infrared spectra of other complexes containing these ligands and have been assigned as ring vibrations.[49] However, there are d^6 complexes with these ligands that show the broad phosphorescence, without vibrational structure.[45]

Crosby and his colleagues have carried out analyses of d^6 phosphorescence spectra and conclude that the structured phosphorescence arises from charge-transfer transitions between ligand and metal. Charge-transfer bands can readily be seen in absorption for almost all transition-metal complexes in the ultraviolet portion of the spectrum. With bipy and phen as ligands these charge-transfer absorption bands often move into the visible region (as, for example, the well-known $Fe(phen)_3^{2+}$, formed in ferrioxalate actinometry). Charge-transfer bands are generally very intense, with ε_{max} of the order of 10^4 M^{-1} cm^{-1}, representing nearly fully

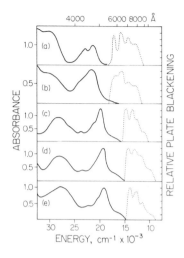

Fig. 2-9. Absorption (82°K) and structured phosphorescence (77°K) of (*a*) [Ru(py)$_2$(bipy)$_2$]I$_2$, (*b*) [Ru(CN)$_2$(bipy)$_2$], (*c*) [Ru(en)(bipy)$_2$]I$_2$, (*d*) Ru(Cl)$_2$(bipy)$_2$], and (*e*) [Ru(ox)(bipy)$_2$].[48]

allowed transitions. Unlike the d–d bands, they need not be Laporte forbidden and are assigned in this instance as $^1T_{1u} \leftarrow {}^1A_{1g}$ transitions. The $^1T_{1u}$ state for d^6 complexes has the electron configuration, $t_{2g}^5 t_{1u}^1$, in which the t_{1u} orbital is an antibonding orbital largely localized on the ligands. That is, unlike the e_g antibonding orbital, t_{1u} is not antibonding in the sense of the metal-ligand bond. The transition, $^1T_{1u} \leftrightarrow {}^1A_{1g}$, has an oscillator strength near unity and thus a radiative lifetime for the $^1T_{1u}$ state of approximately 10^{-9} sec. This is far too short to represent the structured emission, for which the radiative lifetimes are of the order of 10 to 50 μsec.[18] From Eq. (2-3) the oscillator strength expected for the transition responsible for phosphorescence is $f = 2 \times 10^{-4}$, using $\bar{\nu} = 16{,}000$ cm^{-1} and $\tau_0 = 2 \times 10^{-5}$ sec. Then the maximum extinction coefficient of the absorption band inverse to phosphorescence is expected to be of the order of $\varepsilon_{\max} = 100$ M^{-1} cm^{-1}, using Eq. (2-4) for the estimate with a bandwidth of 500 cm^{-1}. Weak bands of this intensity are occasionally located in the tail of the strong *CT* band,[50] as shown in Fig. 2-10. From their position relative to the first main *CT* band and relative to the phosphorescence, this weak absorption is assigned as $^3T_{1u} \leftarrow {}^1A_{1g}$ and is also proposed as the inverse of the phosphorescence.

5. Ligand Phosphorescence. In some instances, the intense absorption in the visible is largely due to intraligand transitions. Such is the case with bipy- or phen-type ligands when, for example, with Rh(III), the CT transition is shifted to high energy. These bands, assigned as π–π* transitions, are only slightly shifted to lower energy (ca. 1000 cm^{-1}) relative to those of the free ligand itself.[44] The luminescence observed

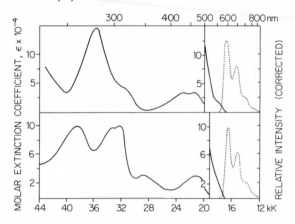

Fig. 2-10. Absorption (300°K) and structured phosphorescence (77°K) spectra of (*a*) [Ru(diphenylphen)₃]Cl₂ and (*b*) [Ru(diphenylbipy₃]Cl₂.[50]

from a complex such as $Rh(phen)_3^{3+}$ is also remarkably similar to the phosphorescence of the free ligand. Therefore, the emission of the complex ion is also assigned as a triplet-singlet $\pi-\pi^*$ intraligand transition. The lifetime of the phosphorescence is considerably shorter than that of the free ligand (48 msec and 1.5 sec for $Rh(phen)_3^{3+}$ and phenanthroline, respectively), an effect readily accounted for on the basis of enhanced spin-orbit coupling in the case of the heavy-atom complex ion. The corresponding singlet-triplet transitions would be too weak to be seen in absorption.

C.   Which Kind of Luminescence is Observed?

The pattern of emission observed from any manifold of excited states of a molecule depends on what are essentially kinetic parameters as well as on the technique used for study of the emission.

Kasha's rule has been modified by Demas and Crosby[46] to read:

> In the absence of photochemistry from upper excited states, emission from a transition metal complex with an unfilled *d* shell will occur from the lowest electronic state in the molecule or from those states which can achieve a significant Boltzmann population relative to the lowest excited state.

A further modification would be along the lines mentioned in a previous section, that the *principal* emission is from the lowest excited state. The rule given certainly does cover the existing data for d^3 and d^6 complexes.

Even then, however, there are some apparent exceptions known to this rule, just as there are exceptions to Kasha's rule for organic molecules.

One of these exceptions concerns the d^8 complexes of Ni^{2+}. Rosiello[51] has reported emission of $Ni(en)_3^{2+}$ as a broad band centered at about 23,000 cm^{-1}, while there are absorption bands at 11,000, 18,000, and 32,000 cm^{-1}. Ni^{2+} in an MgO lattice luminesces at 8000 and 8180 cm^{-1} in the infrared[17] and at 21,300 cm^{-1} in the visible,[52] with absorption bands at 8600, 14,000, 21,500, and 24,500 cm^{-1}, as well as additional bands at higher energies.[53] The emission at 21,300 cm^{-1} has a lifetime of 12 μ sec and thus has been assigned[52] as the phosphorescence, $^1T_1 \rightarrow {}^3A_2$. Ralph and Townsend[17] excited the infrared luminescence by electron bombardment and assigned the emission as fluorescence: $^3T_2 \rightarrow {}^3A_2$. Both these emission spectra and their assignments have been confirmed more recently by Manson,[54] using optical as well as X-ray and electron-beam excitation. It seems strange that the $^1T_1$ state lying above three other excited states of the same multiplicity, $^1T_2$, $^1A_1$, and ^1E, as well as two others of different multiplicity, $^3T_1$ and $^3T_2$, would undergo intersystem crossing with a rate constant of only 10^5 sec^{-1} (the reciprocal of the measured lifetime). This state certainly cannot have a significant Boltzmann population relative to the lowest excited state, because there is a gap of some 13,000 cm^{-1}. However, as in the other case discussed below, all of the spectra display very sharp vibronic lines, often characteristic of ions dissolved in a crystalline matrix. This, together with the relatively large gaps among the various electronic states in d^8 ions, may contribute to the slow nonradiative processes.

In the d^4 system of Os^{4+} in $Cs_2[HfCl_6]$, extremely sharp spectra have been observed in absorption[55] from the ultraviolet through the visible region. Spin-orbit splitting of the $^3T_1$ ground state into four components is quite large, approximately 5000 cm^{-1}. Emission is seen to each of these four ground-state levels from several excited electronic states.[16] The energy-level scheme proposed is shown in Fig. 2-11 with the observed transitions among the various levels. Emission bands at 11,000, 7500, and 5400 cm^{-1} represent transitions from ^1E and

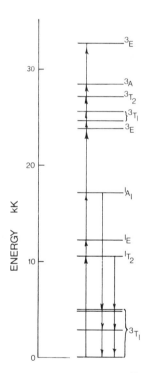

Fig. 2-11. Energy levels of Os^{4+} in $Cs_2[HfCl_6]$ at 2°K, showing the observed absorption and phosphorescence transitions.[16,55]

$^1T_2$ to the various ground-state components, while those bands at 17,000, 14,000, and 12,000 cm^{-1} are attributed to emission from the $^1A_1$ state. Here again, then, emission is observed from an excited state lying above other states of the same multiplicity by too large a gap for any equilibrium population to be important. Although lifetimes were not measured, the radiative lifetime of the $^1A_1$ state must be of the order of 50 μ sec, as judged by the linewidths and intensities reported,[55] together with Eqs. (2-2) and (2-3); therefore, it seems unlikely that the nonradiative transitions from $^1A_1$ to other states could have rate constants in excess of 10^6 sec^{-1}. These problems are discussed further in a later section (see p. 66).

Are we in a position to answer the question posed as the title of this section? Only to the extent that, so far, the only exception to Crosby's rule given above is that of Ni(en)$_3^{2+}$, as long as we restrict our consideration to complex ions of well-defined identity as opposed to ions imbedded in an essentially ionic matrix. For the latter, the vibronic structure is often particularly simple, and intersystem crossing is slow.

III. KINETIC SCHEMES FOR PRIMARY PROCESSES

In the analysis of the various interstate processes occurring subsequent to absorption of light, a display of the distribution in energy of the excited electronic states relative to the ground state is especially instructive. This takes the form of the Jablonski diagram of Fig. 2-12, familiar to organic photochemists with singlet and triplet manifolds of states. The multiplicities appropriate to the present discussion may not be the same, and the

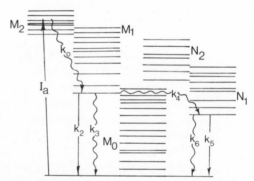

Fig. 2-12. Jablonski diagram (schematic) with two manifolds of states: M and N, having different spin multiplicities. Nonradiative transitions are represented by wavy lines, radiative transitions by straight lines.

detailed patterns of states will certainly differ from those of organic molecules.

A. Simplified Mechanism

Each electronic state is, in principle, convertible to each of the other states by radiative and by nonradiative processes. To a first approximation, we restrict our consideration to the especially simple case that absorption populates a particular electronic state and that emission, assumed to be phosphorescence, occurs only from the single lowest excited state. This basic mechanism is amplified as appropriate in later discussion (see p. 59). In the Jablonski diagram of Fig. 2-12, M_0, M_1, M_2, and so forth, are electronic states of increasing energy of one multiplicity, M, and N_1, N_2, and so forth, are electronic states of multiplicity, $N = M \pm 1$. Radiative transitions are represented by straight lines and nonradiative processes by wavy lines. The following mechanism accompanies this diagram (* denotes vibrational excitation):

$$M_0 + h\nu \longrightarrow M_2^* \qquad\qquad I_a \qquad\qquad (2\text{-}5)$$

$$M_2^* \xrightarrow{\text{solvent}} M_1 \qquad\qquad k_0[M_2^*] \qquad\qquad (2\text{-}6)$$

$$M_1 \longrightarrow M_0 \qquad\qquad k_3[M_1] \qquad\qquad (2\text{-}7)$$

$$M_1 \longrightarrow N_1 \qquad\qquad k_4[M_1] \qquad\qquad (2\text{-}8)$$

$$N_1 \longrightarrow M_0 + h\nu_{\text{phos}} \qquad\qquad k_5[N_1] \qquad\qquad (2\text{-}9)$$

$$N_1 \longrightarrow M_0 \qquad\qquad k_6[N_1] \qquad\qquad (2\text{-}10)$$

The numbering of rate constants matches current usage for a more detailed mechanism considered later. With a steady-state treatment of this mechanism, the following quantum yields can be calculated:

$$\phi_{\text{isc}} = \frac{k_4}{k_3 + k_4} \qquad\qquad (2\text{-}11)$$

$$\phi_{\text{phos}} = \frac{k_4}{(k_3 + k_4)} \frac{k_5}{(k_5 + k_6)} \simeq \phi_{\text{isc}} \frac{k_5}{(k_5 + k_6)} \qquad\qquad (2\text{-}12)$$

and, for pulse excitation, the time dependence of the phosphorescence intensity is

$$I(t) = \left\{ \frac{k_4 k_5 [M_1]_0}{k_3 + k_4 - k_5 - k_6} \right\} \{ e^{-(k_5+k_6)t} - e^{-(k_3+k_4)t} \} \qquad\qquad (2\text{-}13)$$

The second term of Eq. 2-13 represents the "grow-in" of the phosphorescence, the first term gives its decay, and the quantity, $[M_1]_0$, represents the integrated absorbed intensity of the pulse. The decay of phosphorescence occurs with a lifetime,

$$\tau = \frac{1}{k_5 + k_6} \tag{2-14}$$

It follows that

$$\frac{\phi_{phos}}{\tau} = \phi_{isc} k_5 \tag{2-15}$$

which cannot be further separated without independent information about either the intersystem crossing quantum yield or the radiative rate constant [see the approximation in Eq. (2-3) for $\tau_0 = 1/k_5$]. In this scheme, the internal conversion processes that convert M_2^* into M_1 are assumed to be fast, with no alternative competing paths. Under these circumstances, the quantum yield of phosphorescence will be independent of which state of M_1, M_2, and so forth, is reached by absorption. If, however, absorption occurs directly to the state, N_1, the net effect is the same as if ϕ_{isc} were unity, and

$$\phi_{phos} = \frac{k_5}{k_5 + k_6} \qquad \text{with } \phi_{isc} = 1 \tag{2-16}$$

or

$$\frac{\phi_{phos}}{\tau} = k_5 \qquad \text{excitation to } N_1 \tag{2-17}$$

A comparison of the excitation spectrum, quantum yield versus absorbed wavelength, with the corresponding absorption spectrum gives direct information about ϕ_{isc} and, thus, about k_5 itself.

B. Wavelength Dependence

Should ϕ_{phos} not be independent of the wavelength absorbed in the region of $M_1 \leftarrow M_0$, $M_2 \leftarrow M_0$, and so forth, there must be processes that compete with the process, $M_2^* \rightarrow M_1$. This competition must occur *before* vibrational equilibration has led to a state such as M_1 common to the degradation path at all wavelengths. Since vibrational equilibration in a condensed medium is unlikely to require more than about 10^{-12} sec, the rate constant (first order) must exceed approximately 10^{10} sec^{-1} for any such competing reaction. One possibility, of course, is photochemical reaction. Alternately, there may be yet another manifold of states, readily accessible to M_2^*, M_3^*, and so forth, but not to M_1, and that internal conversion to this manifold is the competing process.

To include these features in the mechanism, one additional reaction is needed; either

$$M_2^* \rightarrow \text{Products} \tag{2-18}$$

or

$$M_2^* \rightarrow X \rightarrow M_0 \qquad k_7^*[M_2^*] \tag{2-19}$$

in which k_1^* is assumed to be a function of the wavelength absorbed, that is, of the vibronic level reached in absorption. The phosphorescence quantum yield, Eq. (2-12), now has the form

$$\phi_{phos} = \frac{k_0}{k_0 + k_1^*(\lambda)} \, \phi_{isc} \, \frac{k_5}{k_5 + k_6} \tag{2-20}$$

The phosphorescence lifetime remains as in Eq. (2-14), so that the ratio, ϕ_{phos}/τ, is also dependent on wavelength.

C. Back Intersystem Crossing

A further complication is added if molecules in the state, N_1, can also undergo intersystem crossing *back* to the state M_1

$$N_1 \rightarrow M_1 \qquad k_{-4}[N_1] \tag{2-21}$$

The steady-state phosphorescence quantum yield then becomes

$$\phi_{phos} = \frac{k_4 k_5}{(k_3 + k_4)(k_5 + k_6) + k_3 k_{-4}} \tag{2-22}$$

and is smaller than that given by Eq. (2-12) by an amount that depends on the relative importance of k_{-4} compared with $k_5 + k_6$. The phosphorescence lifetime becomes

$$\frac{1}{\tau} = \frac{1}{2}\{(k_M + k_N) - [(k_M - k_N)^2 + 4k_4 k_{-4}]^{1/2}\} \tag{2-23}$$

where

$$k_M = k_3 + k_4 \tag{2-24}$$

$$k_N = k_{-4} + k_5 + k_6 \tag{2-25}$$

In the case that $4k_4 k_{-4} \ll (k_M - k_N)^2$, Eq. (2-23) reduces to the form,[56]

$$\frac{1}{\tau} \simeq k_N - \frac{k_4 k_{-4}}{k_M - k_N} \simeq k_5 + k_6 + (1 - \phi_{isc})k_{-4} \tag{2-26}$$

and ϕ/τ is still given by Eq. (2-15), with ϕ_{isc} as defined in Eq. (2-11) and $k_N \ll k_M$.

It is well known that the phosphorescence quantum yield and lifetimes decrease with increasing temperature. An example[29] is given in Fig. 2-13 for $Cr(en)_3^{3+}$. The effect in this and other similar cases[57] is a direct result of the temperature dependence of k_{-4}, which has associated with it an activation energy equal to the separation in energy of the states, M_1 and N_1. It is on the basis of this assumption that estimates have been made for the position of the origin of state, M_1. Specifically, the technique has been

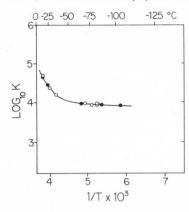

Fig. 2-13. Lifetime of the phosphorescence decay of Cr(en)$_3^{3+}$ in water/methanol/ethylene glycol (1:2:1) as a function of temperature.[29]

used to locate the origin of the $^4T_{2g}$ state[58] of d^3 complexes, a piece of information vital to interpretation of both photochemical and energy transfer data.

D. Thermal Equilibration

An alternative and different treatment for the case of back intersystem crossing, $N_1 \rightarrow M_1$, is the assumption that the two states are in thermal equilibrium with one another. The kinetic requirement is essentially: $k_4 \gg k_3$. Equation (2-23) then reduces to the approximate form:

$$\frac{1}{\tau} = \frac{k_5 + k_6 + (k_3 k_{-4}/k_4)}{1 + (k_{-4}/k_4)} = \frac{k_5 + k_6 + k_3 K}{1 + K} \tag{2-27}$$

in which K is the equilibrium constant for the states M_1 and N_1:

$$K = \frac{k_{-4}}{k_4} = \frac{[N_1]_{ss}}{[M_1]_{ss}} \tag{2-28}$$

This equilibrium constant is valid as long as the species in question, which here are excited-state molecules, come to a Boltzmann equilibrium with the medium in all but electronic energy in a time short compared with their lifetimes. The quantum yield in Eq. (2-22) simplifies also to:

$$\phi_{phos} = \frac{k_5}{k_5 + k_6 + k_3 K} \tag{2-29}$$

and the ratio:

$$\frac{\phi_{phos}}{\tau} = \frac{k_5}{1 + K} \tag{2-30}$$

Thus, $1/(1 + K)$ becomes the equivalent of an intersystem-crossing quantum yield [compare Eq. (2-15)], but is now strongly temperature dependent, because K is given by

$$K = \frac{g_M}{g_N} e^{-\Delta\varepsilon/RT} \tag{2-31}$$

where g_M and g_N are the degeneracies of M_1 and N_1, respectively, and $\Delta\varepsilon$ is the difference in energy of M_1 and N_1.

Crosby[59] has generalized Eqs. (2-27) and (2-29) for the case of three excited states, a, b, and c, which luminesce and are in equilibrium with each other, characterized by the energy differences, $\Delta\varepsilon_{ab}$ and $\Delta\varepsilon_{ac}$, the radiative rate constants, k_{ar}, k_{br}, and k_{cr}, and the total radiative plus nonradiative rate constants, k_a, k_b, and k_c, excluding that for intersystem crossing. The lifetime is given by

$$\frac{1}{\tau} = \frac{k_a + k_b K_{ab} + k_c K_{ac}}{k_a + k_b K_{ab} + k_c K_{ac}} \tag{2-32}$$

where the equilibrium constants are given by Eq. (2-31). These expressions were fitted[59] to the data for the τ and ϕ variation with temperature as shown in Fig. 2-14 with the parameters in Table 2-3, for the compound, $Ru(bipy)_3^{2+}$. This treatment satisfactorily rationalizes the fact that, although the lifetime increases with decreasing temperature, the quantum yield for total emission decreases. Nevertheless, a spectral resolution of the emission spectra could not be made because of overlap of the different types. It should be noted that the three different emissions have the same lifetime, that given in Eq. (2-32).

This same interpretation can be made of those Cr(III) complexes such as $Cr(urea)_6^{3+}$, which both phosphoresce and fluoresce. The fluorescence:

$$M_1 \rightarrow M_0 + h\nu_{fluor} \qquad k_2[M_1] \tag{2-33}$$

is "delayed" fluorescence, having a lifetime:

$$\frac{1}{\tau} = \frac{(k_5 + k_6) + (k_2 + k_3)K}{1 + K} \tag{2-34}$$

Table 2-3. Energy Gaps, Emission Quantum Yields, and Radiative and Nonradiative Lifetimes (μsec) of Emitting States of $Ru(bipy)_3^{2+}$ in PMM,[59] Relative to Lowest Level at Approximately 18,000 cm^{-1}

State at ν_0+	ϕ	τ_r	τ_{nr}
a. 0 cm^{-1}	0.17	1091	183
b. 10 cm^{-1}	0.23	82	19
c. 61 cm^{-1}	0.40	1.7	0.7

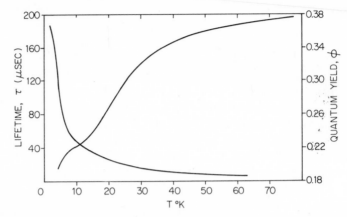

Fig. 2-14. Luminescence-decay lifetime and total-emission quantum yield from Ru(bipy)$_3^{2+}$ in polymethylmethacrylate matrix.[59]

in common with phosphorescence. If only a steady state, rather than equilibrium, is assumed, the phosphorescence lifetime is as given by Eq. (2-26):

$$\frac{1}{\tau} = k_5 + k_6 + (1 - \phi_{\text{isc}})k_{-4} \tag{2-35}$$

and since both k_{-4} and K have the same temperature dependence, it is obviously difficult to choose between the two mechanisms. Dingle[38] has measured a short-lived emission with Cr(urea)$_6^{3+}$, which has been corroborated by Forster.[60] However, there may still be delayed fluorescence having a longer lifetime.

The quantum yields corresponding to Eq. (2-34) are

$$\phi_{\text{phos}} = \frac{k_5}{(k_5 + k_6) + (k_2 + k_3)K} \tag{2-36}$$

$$\phi_{\text{fluor}} = \frac{k_2 K}{(k_5 + k_6) + (k_2 + k_3)K} \tag{2-37}$$

In this way, the energy separation of ^2E and $^4T_2$ and, therefore, the origin of $^4T_2$ can be established even if $^4T_2 \leftarrow {}^4A_2$ absorption is structureless.

The reciprocal lifetime is $k_5 + k_6$ at low temperatures where $K \approx 0$ and decreases to $k_2 + k_3$ at high temperatures, except that at extremely high temperatures, other processes degrading M_1 and N_1 may intrude. The data for Mn(IV) Cs$_2$[MnF$_6$] provide a good example of this pattern.[61] In Figs. 2-5 and 2-15 are shown the emission spectra at low and high temperatures. At low temperatures only the characteristic sharp line spectrum of $^2E \rightarrow {}^4A_2$ of the phosphorescence is seen. At high temperatures that spectrum is broadened in the individual lines width and in the addition of

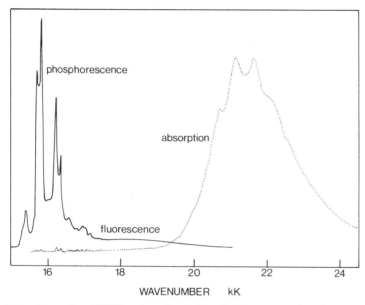

Fig. 2-15. Absorption (300°K) and total emission (590°K) spectra of $Mn^{4+}:Cs_2[GeF_6]$.[61]

hot bands from higher vibrational levels of ^2E. Above about 400°K, a new broad emission band occurs, centered at 18,000 cm^{-1}, which is the delayed fluorescence. The quantum yields in Fig. 2-16 show the onset of fluorescence and the eventual decrease of both fluorescence and phosphorescence. A simple function of K is obtained by taking the ratio of ϕ_{fluor} and ϕ_{phos}:

$$\frac{\phi_{fluor}}{\phi_{phos}} = \frac{k_2}{k_5} K \qquad (2\text{-}38)$$

which should hold even if new degradative steps come in at high temperatures. Since K has the form of Eq. (2-31), we have:

$$\ln \frac{\phi_{fluor}}{\phi_{phos}} = \ln \left[\frac{k_2}{k_5} \frac{g(^4T_2)}{g(^2E)} \right] - \frac{\Delta\varepsilon}{RT} \qquad (2\text{-}39)$$

and the plot of the data in Fig. 2-17 yields $\Delta\varepsilon = 3900\ cm^{-1}$ and $k_2/k_5 = 6 \times 10^3$.

Based on integrated absorption coefficients, k_2 and k_5 are $7.2 \times 10^4\ sec^{-1}$ and 58 sec^{-1}, respectively, and the ratio, 1.2×10^3, is smaller by a factor of five than that obtained from the temperature dependence. In addition, for MnF_6^{2-}, because the $^4T_2 \leftarrow\ ^4A_2$ absorption shows a rich vibronic structure, the origin can be located. It is found to be 4600 cm^{-1} above that of the $^2E \leftarrow\ ^4A_2$ transition, again in moderate agreement with $\Delta\varepsilon$ obtained in Fig. 2-17.

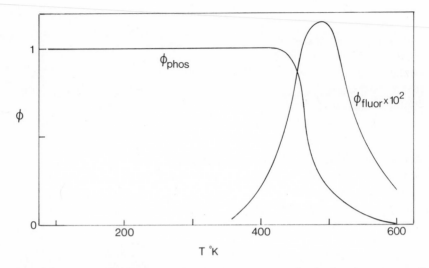

Fig. 2-16. Quantum yields of delayed fluorescence and of phosphorescence of $Mn^{4+}:Cs_2[GeF_6]$ as a function of temperature.[61]

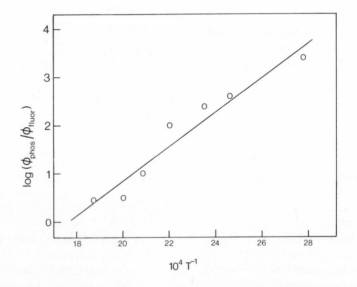

Fig. 2-17. Temperature dependence of the ratio of phosphorescence to fluorescence quantum yields for $Mn^{4+}:Cs_2[MnF_6]$.

It is apparent, however, that this treatment gives no information at all about the actual rate of intersystem crossing. The lifetime, Eq. (2-34), contains neither k_4 nor k_{-4}, since they are assumed here to be simply much faster than $k_2 + k_3$ and $k_5 + k_6$, respectively. And, although Eq. (2-35) does contain k_{-4}, it is as a product with $(1 - \phi_{isc})$.

E. Radiative Relaxation

According to Strickler and Berg,[47] the radiative rate constant for a given emission process can be evaluated from the absorption spectrum representing the same transition. Their theory, however, applies to broad bands with good mirror relationship between emission and absorption and for a transition which is strongly allowed. Under these conditions

$$k = \frac{1}{\tau_0} = 2.88 \times 10^{-9} n^2 \langle \bar{\nu}^{-3} \rangle^{-1} \frac{g_1}{g_u} \int \varepsilon \, d \ln \bar{\nu} \qquad (2\text{-}40)$$

in which n is the refractive index of the medium, g_1 and g_u are the degeneracies of the lower and upper states, $\langle \bar{\nu}^{-3} \rangle$ is a mean of the cube of the reciprocal of the emission frequency, and the integration is carried out over the absorption band reciprocal to emission. Their formulation specifies $\varepsilon_{max} \simeq 8000 \, M^{-1} \, cm^{-1}$ and a relatively small Stokes shift. Obviously, these are quite stringent conditions that virtually deny the use of Eq. (2-40) for transition-metal complexes. The treatment has more often been used to decide if certain emission and absorption bands are indeed reciprocal. In those circumstances, the approximations mentioned earlier in Eqs. (2-3) and (2-4) are useful. Thus

$$k = \frac{1}{\tau} \simeq 5 \times 10^{-9} \bar{\nu}^{-2} \varepsilon_{max} \Delta \bar{\nu} \qquad (2\text{-}41)$$

gives an order-of-magnitude estimate for the radiative rate constant.

When the absorption band corresponding to phosphorescence cannot be measured directly, as is often the case for d^6 broad phosphorescence, the oscillator strength can still be estimated. The transition is assumed to steal intensity from nearby spin-allowed transitions by mixing the phosphorescing state with other states by spin-orbit coupling.

The oscillator strength, for example, of the $^2E \leftarrow {}^4A_{2g}$ transition in a d^3 ion, is proportional to that of the $^4T_{2g} \leftarrow {}^4A_{2g}$ transition according to second-order perturbation theory[20]

$$\frac{f(^2E - {}^4A_2)}{f(^4T_2 - {}^4A_2)} = \frac{4}{9} \frac{\nu(^2E - {}^4A_2)}{\nu(^4T_2 - {}^4A_2)} \frac{\zeta^2}{\nu(^4T_2 - {}^2E)} \qquad (2\text{-}42)$$

where ζ is the one electron spin-orbit coupling parameter. Reasonable agreement has been obtained for k_5 obtained in this way for a number of d^3 complex ions.

The ways of extracting radiative rate constants from experimental data have been explored in the previous section. Provided that phosphorescence yields are not dependent on the wavelength of absorbed light, and the value of ϕ_{isc} is known independently, the radiative rate constant is simply given by

$$k_5 = \frac{\phi_{phos}}{\tau \phi_{isc}} \tag{2-43}$$

In the general case, instead of ϕ_{isc}, the quantum yield for formation of the emitting state is required.

F. Nonradiative Relaxation

A distinction must be made between intramolecular relaxation and intermolecular relaxation. The latter can be quenching by a second species, with or without electronic energy transfer.

Intramolecular relaxation processes can be characterized as:

1. Vibrational relaxation of an excited species within one electronic state by collisional interaction with the surrounding medium.

2. Internal conversion from one electronic state to another state of the same multiplicity.

3. Intersystem crossing between states of different multiplicity.

Process (1) is, of course, not intramolecular, but in condensed media, the excited species is in immediate contact with the solvent or crystalline host molecules. The "isolated" molecule case is seldom applicable to transition-metal complexes. Vibrational relaxation is generally assumed to be rapid compared with other intramolecular and intermolecular relaxation processes in condensed media. However, this need not necessarily be the case, as there is evidence, for example, that while vibrational relaxation occurs within approximately 10^{-12} sec, intersystem crossing might take place with rate constants greater than 10^{10} sec^{-1}.

Processes (2) and (3) take place between isoenergetic vibronic levels of the two states, followed by vibrational relaxation removing excess vibrational energy. Such interstate processes need not represent relaxation, although that is the usual situation envisaged; the case for back intersystem crossing involving an energy barrier of as much as $10\,kT$ is well established.

1. Theoretical Treatment of Nonradiative Rates. Much of the theoretical work has been applied to aromatic molecules, to which the bulk of the experimental data apply. These treatments have been critically explored by Jortner et al.[62] and Henry and Kasha.[14] Most of the approaches are based on the idea that the accessible vibronic levels of the initial state are coupled to a near continuum of isoenergetic vibronic levels of the final state.

In the "statistical" limit,[63] the density of levels in the final state is so high that crossing is irreversible and, hence, intramolecular. The vibrational relaxation by the medium is rapid. In the "resonance" limit, generally applying to small molecules, the density of levels is low, crossing occurs only at a small number of accidentally degenerate levels of the two states, and relaxation cannot occur without interaction with the surrounding medium. In the gas phase at low pressures with moderately large molecules, the situation seems to be intermediate to these extremes.

We use here the unified analysis of Englman and Jortner[64] to describe the theory concerning rate constants of nonradiative processes. The rate constant of crossing from state i to state j is the sum over all levels of the individual transition probabilities, k_{ij}, each determined by matrix elements for the coupling of the states, which include both electronic factors, C, and Franck–Condon factors, S_{ij}:

$$k_{ij} = \langle \psi_j | V_{ij} | \psi_i \rangle; \qquad V_{ij} = C S_{ij} \qquad (2\text{-}44)$$

If the two states involved have the same multiplicity, for example, if the rate constant is for internal conversion, vibronic coupling largely determines C; otherwise, for intersystem crossing, spin-orbit coupling must also be included in its evaluation. Only a few "promoting" vibrational modes contribute to C, whereas nearly all molecular vibrations are included in evaluating S_{ij}. Two quantities: ΔE, the difference in the zero-point energies of i and j, and ΔQ^0, the displacement of one potential minimum relative to the other along an appropriate coordinate(s), are useful in describing the dependence of k_{ij} on molecular structure. If only one vibration is effective in coupling i to j, for example, a symmetric breathing vibration,[65] ΔQ^0 is readily interpreted by means of a two-dimensional potential diagram.

The pertinent conclusions are:

1. The number of equienergetic states, ψ_j, that can couple with ψ_i increases rapidly with increasing ΔE, the energy gap.

2. The magnitude of S_{ij} increases with increasing ΔQ^0 and decreases with increasing ΔE.

3. C varies with ΔE for all states important in the coupling of ψ_i and ψ_j; hence, no general statement can be made.

The values of k_{ij} depend on ΔE and ΔQ^0 in a complicated fashion, and trends in k_{ij} are difficult to predict on an a priori basis.

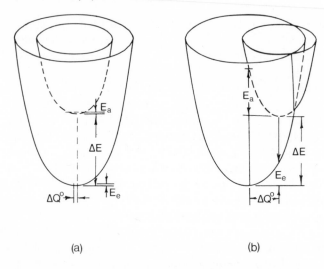

(a) (b)

Fig. 2-18. Schematic potential energy surfaces for (a) strong coupling and (b) weak coupling. The Stokes shift, E_s, is represented by the sum: $E_e + E_a$.[64]

Two limiting cases are envisaged by Englman and Jortner: weak coupling, when ΔQ^0 is small, and strong coupling, when ΔQ^0 is large. These are illustrated in Fig. 2-18 schematically as the crossing of two potential surfaces. When there is a large displacement of the two surfaces, the crossing point will not be far from the minimum of the upper potential surface.

The coupling strength can be related to the Stokes shift between absorption and emission, E_s. In the strong-coupling case, E_s must be large:

$$E_s \gg 2\langle \bar{\nu} \rangle \tanh \langle \bar{\nu} \rangle / 2kT \tag{2-45}$$

where $\langle \bar{\nu} \rangle$ is the mean vibrational energy in the molecule (which might be expected to be ca. 500 cm^{-1} or less). In the weak-coupling limit, E_s is small:

$$E_s \simeq 2\langle \bar{\nu} \rangle \tag{2-46}$$

It is expected that the Stokes shift would certainly have to exceed $10,000 \text{ cm}^{-1}$ for the strong coupling to apply. Therefore it is conceded to apply principally to relaxation via photochemical rearrangement, rather than interstate crossing. The weak-coupling case, to which the theories of Robinson and Frosch[66] and of Ross et al.[67] apply, is presumed to be the situation for the small configurational changes involved in internal conversion and intersystem crossing, where $E_s \simeq 4000 \text{ cm}^{-1}$ or less.

At low temperatures, the transition probability for the weak-coupling case becomes

$$k_{ij} = \frac{C^2 (2\pi)^{1/2}}{h (\bar{\nu}_M \, \Delta E)^{1/2}} \exp \left(\frac{-\gamma \, \Delta E}{\bar{\nu}_M} \right) \qquad (2\text{-}47)$$

where $\bar{\nu}_M$ is the energy of the highest frequency vibrational modes, and the parameter

$$\gamma = \log \frac{\Delta E}{\bar{\nu}_M \, \Delta_M^2} - 1 \simeq 1 \qquad (2\text{-}48)$$

where Δ_M is the dimensionless coordinate corresponding to ΔQ_M^0. Equation (2-47) exemplifies the exponential energy-gap dependence so often cited in $T_1 \to S_0$ aromatic molecule relaxation processes.[68] Equation (2-47) also predicts the absence of any major temperature dependence for the crossing-rate constant. The so-called deuterium effect[69] on $T_1 \to S_0$ in aromatic molecules is rationalized since $\bar{\nu}_M$ corresponds to C—H stretching modes.

At higher temperatures, the expression for k_{ij} becomes more complex, but still only a weak temperature dependence is involved. In this treatment, the electronic coupling matrix elements, C, are assumed to be temperature independent. If this assumption is invalid, a temperature dependence could be introduced into the weak-coupling limit.

The results for the strong-coupling case include a temperature dependence at other than very low temperatures, in terms of the energy difference between the crossing point and the minimum of the state, i.

The $^4T_2 \rightsquigarrow {}^2E$ transition has been treated quantitatively by Barnett and Englman[65] with a somewhat different formalism, but the conclusions are consistent with those described above.

2. Experimental Aspects of Nonradiative Rate Constants. In the experimental sense, there is not a great deal of information available about intersystem crossing and internal conversion in transition-metal complexes. That intersystem crossing from the lowest excited state to the ground state (involving a spin change) is slow at low temperatures is evident in the fact that phosphorescence does occur. Since a strong deuterium effect increases the lifetime and quantum yield of phosphorescence in both d^3 systems[28] and in d^6 systems,[70] intersystem crossing must be competitive with the radiative process. Yet there is a strong effect of the medium in some systems, and, more importantly, a strong temperature dependence of both lifetime and quantum yield.

The case of $Cr(CN)_6^{3-}$ is pertinent here. In a host crystal of $K_3[Co(CN)_6]$, the phosphorescence has a lifetime of 0.12 sec at $T < 50°K$, decreasing to approximately 0.05 sec at $300°K$.[58] Since the quantum yield is approximately constant over this range, this lifetime probably represents the reciprocal of k_5, the radiative rate constant, and the small temperature dependence is that expected for a forbidden but vibronically allowed transition.[31] When $K_3[Cr(CN)_6]$ is dissolved in a solvent composed of degassed methanol, water, and ethylene glycol (2:1:1), the limiting low-temperature lifetime is 4×10^{-3} sec, but at room temperature it drops to less than 10^{-6} sec.[29] However, dissolved in degassed DMF,[22] its phosphorescence has a lifetime at room temperature longer than the low-temperature limit in the aqueous solvent. In each case, in the region of temperature that the phosphorescence lifetime does become a strong function of temperature: above 400°K, 160°K, and 250°K, for the crystal, for the methanol/water/glycol solvent and for DMF solvent, respectively, the apparent activation energy is the same: approximately 9 ± 1 kcal mole^{-1}. There are clearly environmental effects of major importance that are not predictable by any of the current theories of nonradiative processes.

As has been mentioned earlier, attempts to measure prompt fluorescence lifetimes in d^3 systems and so to estimate the sum, $k_2 + k_3 + k_4$, have generally ended in "less than" results. Thus, for $Cr(acac)_3$ and $Cr(CN)_6^{3-}$, the phosphorescence grows in with a lifetime less than 10^{-9} sec.[71] In the case of ruby, there have been several estimates, such as $10^{-12} < \tau < 5 \times 10^{-8}$ sec by Kisliuk and Moore,[72] based on information from a variety of sources; $\tau < 2 \times 10^{-9}$ sec by direct observation of phosphorescence grow-in, and $\tau < 45 \times 10^{-12}$ sec, inferred by absence of fluorescence, both by Everett[42]; and $\tau < 4 \times 10^{-9}$ sec by Anson and Smith.[73]

When the $^4T_2$ state of Cr(III) lies lower in energy than the ^2E state, fluorescence is the favored emission process. Thus, lifetime measurements at low temperature should give the value of $k_2 + k_3$ (presumably, in such cases, k_4 would be appreciably smaller than usual because of an energy barrier). Such measurements have been reported by Glass[74] for Cr(III) in $LiNbO_3$ and $LiTaO_3$. At 4°K phosphorescence is observed, but very weakly, less than 10^{-3} of the very broad (2500 cm^{-1}) fluorescence, and the lines are relatively broad (50 cm^{-1}) compared with the R lines of ruby, for example. The measured fluorescence lifetime at 4°K in both matrices is 10 μsec, which Glass[74] attributes completely to the radiative lifetime, because the quantum yield of fluorescence is approximately 1.

It is apparent that, in the case of transition-metal ions in fluoride-ion lattices, internal conversion and intersystem crossing among upper electronic states may be relatively slow, as deduced from the fact that other than the lowest-energy excited states are observed to luminesce. This could simply be a "super" deuterium effect, in that the promoting modes associated with the nonradiative processes are low frequency, $\bar{\nu} \simeq 600$ cm^{-1} compared with 2500 and 3300 cm^{-1} for C—D and C—H vibrations, respectively.[75]

Most of the available information about nonradiative processes in transition-metal complexes comes from lifetime measurements. The straightforward application of Eq. (2-15), with $\phi_{isc} = 1$, gives the values of k_5 (and k_2) listed in Table 2-4; k_6 follows from the reciprocal phosphorescence lifetime:

$$k_6 = \frac{1}{\tau} - k_5 \qquad (2\text{-}49)$$

Table 2-4. Phosphorescence of Transition-Metal Complex Ions, Including Spectral Position, $\bar{\nu}$ (in kiloKaysers), Nonradiative, k_{nr}, and Radiative, k_r, Rate Constants in 10^3 sec^{-1}, and Quantum Yields of Emission, ϕ.

Complex[a]	Transition	$\bar{\nu}$	ϕ	k_{nr}	k_r[b]	Ref.
$[Cr(NH_3)_6]^{3+}$	$^2E \rightarrow {}^4A_2$	15.06	0.0033	18.0	0.059	28
$[Cr(NH_3)_5(H_2O)]^{3+}$	$^2E \rightarrow {}^4A_2$	15.00	0.0030	14.0	0.042	28
$[Cr(D_2O)_6]^{3+}$	$^2E \rightarrow {}^4A_2$	14.75	0.0010	7.7	0.0077	28
$[Cr(NCS)_6]^{3-}$	$^2E \rightarrow {}^4A_2$	12.89	0.23	0.016	0.194	28
$trans$-$[Cr(NH_3)_2(NCS)_4]^-$	$^2E \rightarrow {}^4A_2$	13.33	0.011	3.2	0.035	28
$[Cr(en)_3]^{3+}$	$^2E \rightarrow {}^4A_2$	15.06	0.0090	9.9	0.090	28
$[Cr(CN)_6]^{3-}$	$^2E \rightarrow {}^4A_2$	12.35	0.0042	0.299	0.001	28
$[Cr(acac)_3]$	$^2E \rightarrow {}^4A_2$	12.80	0.021	0.24	0.050	28
$Cr^{3+}:Al_2O_3$[c]	$^2E \rightarrow {}^4A_2$	14.43	>0.7	0.026	0.213	19
$Mn^{4+}:Cs_2GeF_6$[c]	$^2E \rightarrow {}^4A_2$	16.04	~1	—	0.060	61
cis-$[RhCl_2(bipy)_2]^+$	$^3T_1 \rightarrow {}^1A_1$	13.7	0.037	30.3	1.16	45, 46
cis-$[RhCl_2(phen)_2]^+$	$^3T_1 \rightarrow {}^1A_1$	13.7	0.042	28.3	1.73	45, 46
cis-$[RhBr_2(bipy)_2]^+$	$^3T_1 \rightarrow {}^1A_1$	13.7	0.18	45.8	9.8	45, 46
cis-$[RhBr_2(phen)_2]^+$	$^3T_1 \rightarrow {}^1A_1$	13.3	0.19	50.4	12.1	45, 46
$[Rh(NH_3)_5Cl]^{2+}$	$^3T_1 \rightarrow {}^1A_1$	14.97	0.0013	82.0	0.107	70
$[Rh(ND_3)_5Cl]^{2+}$	$^3T_1 \rightarrow {}^1A_1$	14.97	0.093	1.4	0.15	70
$[Rh(NH_3)_5Br]^{2+}$	$^3T_1 \rightarrow {}^1A_1$	14.87	0.0091	95.0	0.87	70
$[Rh(ND_3)_5Br]^{2+}$	$^3T_1 \rightarrow {}^1A_1$	14.87	0.34	1.7	0.89	70
$[Rh(NH_3)_5I]^{2+}$	$^3T_1 \rightarrow {}^1A_1$	14.34	0.031	127.0	4.06	70
$[Rh(ND_3)_5I]^{2+}$	$^3T_1 \rightarrow {}^1A_1$	14.34	0.40	6.1	4.1	70
$[Ru(bipy)_3]^{2+}$	$^3CT \rightarrow {}^1A_1$	17.12	0.38	120.0	79.0	18
$[Ru(phen)_3]^{2+}$	$^3CT \rightarrow {}^1A_1$	17.62	0.58	42.6	59.5	18
$[Ru(tripy)_2]^{2+}$	$^3CT \rightarrow {}^1A_1$	16.62	0.48	49.0	44.8	18
cis-$[Ru(CN)_2(bipy)_2]$	$^3CT \rightarrow {}^1A_1$	17.12	0.27	184.0	68.0	18
$[Ru(en)(bipy)_2]^{2+}$	$^3CT \rightarrow {}^1A_1$	14.78	0.022	1020.0	23.2	18
$[Ru(ox)(bipy)_2]$	$^3CT \rightarrow {}^1A_1$	14.19	0.012	1620.0	20.3	18
$[Os(bipy)_3]^{2+}$	$^3CT \rightarrow {}^1A_1$	14.09	0.035	1080.0	39.1	18
$[Os(phen)_3]^{2+}$	$^3CT \rightarrow {}^1A_1$	14.48	0.13	360.0	51.8	18
$[Os(tripy)_2]^{2+}$	$^3CT \rightarrow {}^1A_1$	14.28	0.12	226.0	32.0	18
cis-$[IrCl_2(phen)_2]^+$	$^3CT \rightarrow {}^1A_1$	19.8	0.50	73.1	71.4	50
$[IrCl_2(bipy)_2]^+$	$^3CT \rightarrow {}^1A_1$	19.9	0.52	81.5	86.9	50

[a] Measured in rigid glass at or near 77°K, except where noted.
[b] Calculated from ϕ/τ.
[c] Crystalline matrix.

with the assumption that $k_{-4} = 0$. This should be valid only at low temperature. The strong temperature dependence observed for the phosphorescence lifetime and quantum yield has been attributed mainly to that of k_{-4}.[57,76]

IV. ENERGY TRANSFER

Selective population and depopulation of specific excited states can often be accomplished by energy transfer with suitable donor or acceptor molecules. Quenching and sensitization have played an important part in recent developments in the photochemistry of inorganic compounds.

The radiative intermolecular process, absorption by one molecule of the emission of another, is a trivial case not really considered as energy transfer. Its contribution can be made negligible by appropriate choice of conditions.

There are otherwise two principal mechanisms for energy transfer: long-range resonance transfer and collisional energy transfer. The first is generally recognized by the fact that it occurs even in a rigid medium at dilutions such that donor and acceptor may be an average of some 50 to 100 Å apart. Collisional energy transfer is the most important for transition metal complexes.

A. Long-Range Transfer

Although this mechanism has seldom been invoked in inorganic systems, its occurrence is a possibility that should be considered, at least when the medium is a rigid glass. The resonance transfer takes place by a coulombic interaction, and is described by a distance parameter, R_0, which may be, in turn, related to concentration. In the dipole approximation, R_0 is given by[77]

$$R_0^6 = \frac{8.79 \times 10^{-25} K^2 \Phi_D \Omega}{n^4} \tag{2-50}$$

where R_0 is the distance in centimeters at which the energy-transfer rate is the same as the sum of all the intramolecular rates of degradation of the donor, K is an orientation factor, Φ_D the donor luminescence yield, n the refractive index of the medium, and Ω:

$$\Omega = \int_0^\infty \varepsilon_A(\bar{\nu}) f_D(\bar{\nu}) \frac{d\bar{\nu}}{\bar{\nu}^4} \tag{2-51}$$

is an integral expressing the overlap of absorption spectrum of the acceptor, in terms of its extinction coefficient, $\varepsilon_A(\bar{\nu})$, and the emission spectrum of the donor, given as a fraction, f_D, normalized so that

$$\int f_D(\bar{\nu}) \, d\bar{\nu} = 1 \tag{2-52}$$

The donor emission may be either fluorescence or phosphorescence, but because the extinction coefficient of the acceptor enters directly, the acceptor transition should have a relatively large oscillator strength. It is unlikely that this mechanism will be of major importance for inorganic complexes, certainly as acceptors. It might interfere, however, with collisional energy transfer, especially if the latter is particularly slow because of viscosity or of some other effects.

B. Collisional Energy Transfer

Energy transfer is the process by which one excited molecule donor, D, is quenched and simultaneously another, the acceptor, A, is raised to an excited level

$$D^* + A \rightarrow D + A^* \tag{2-53}$$

Sensitization occurs by energy transfer, as does quenching in some cases. Quenching, however, can be the result of intermolecular electronic to vibrational energy conversion

$$D^* + Q \rightarrow D + Q \tag{2-54}$$

Of course, if the species, A^*, is immediately deactivated to A, then Eq. (2-53) becomes the equivalent of Eq. (2-54).

As is implicit in the name, collisional-energy transfer, a bimolecular encounter must occur between donor and acceptor molecules. It is often found that, effectively, every encounter results in energy transfer. Then the diffusion-controlled rate constant is given by the modified Stokes–Einstein equation

$$k_d = \frac{8kT}{3000\eta} \tag{2-55}$$

The energy-transfer rate is an inverse function of the viscosity, η. For a solvent such as water at room temperature, k_d is approximately $5 \times 10^9 \, M^{-1} \, sec^{-1}$, while for glycerine it is about $5 \times 10^6 \, M^{-1} \, sec^{-1}$. The effective activation energies are 4 to 6 kcal mole^{-1}. Thus, cooling a solvent from room temperature to $-100°C$ decreases k_d by almost four orders of magnitude.

In systems of low viscosity where the diffusion rate is rapid, the energy-transfer rate constants, k_{et}, may be controlled by other factors, and thus have a value appreciably less than that given by Eq. (2-55). One of the more obvious factors is energy. If the donor excited level lies below that of the acceptor there will be a normal Boltzmann factor on the energy-transfer rate constant.[78]

1. Spin Correlation. It is normally considered that spin correlation, as suggested by Wigner,[79] applies to these processes, although there has been no real experimental confirmation in these systems. For a reaction such as Eq. (2-53), the spin-correlation rule requires that

$$S_{D*} + S_A, \qquad S_{D*} + S_A - 1, \ldots, |S_{D*} - S_A| \qquad (2\text{-}56)$$

contain at least one value in common with

$$S_D + S_{A*}, \qquad S_D + S_{A*} - 1, \ldots, |S_D - S_{A*}| \qquad (2\text{-}57)$$

Otherwise there will be a prohibition factor that depends on the extent of spin-orbit coupling in donor and acceptor. For organic molecules having singlet ground states and light atoms, the restriction is more severe than for paramagnetic transition-metal complexes. For example, while the energy transfer

$$D*(S = \tfrac{1}{2}) + A(S = \tfrac{3}{2}) \rightarrow D(S = \tfrac{3}{2}) + A*(S = \tfrac{1}{2}) \qquad (2\text{-}58)$$

is allowed, so also is

$$D*(S = \tfrac{3}{2}) + A(S = \tfrac{3}{2}) \rightarrow D(S = \tfrac{3}{2}) + A*(S = \tfrac{1}{2}) \qquad (2\text{-}59)$$

In any case, even for a nominally spin-forbidden process such as

$$D*(S = \tfrac{1}{2}) + A(S = 0) \rightarrow D(S = \tfrac{3}{2}) + A*(S = 0) \qquad (2\text{-}60)$$

it is not known just how much of a prohibition factor need be applied. If that factor is small by reason of spin-orbit coupling, such energy transfer might still be diffusion controlled, especially if the viscosity of the medium is high.

Both the spin-selection rule and the energy requirements of energy transfer restrict the possible electronic states of donor and acceptor involved. However, even then it is often difficult to specify exactly and unambiguously the state of the energy transfer partners. To demonstrate the role of a specific excited state of a molecule in a particular process, either photophysical or photochemical, the molecule in question may have to be used in turn as donor and as acceptor, that is, as sensitizer and as quencher.

2. Experimental Aspects of Energy Transfer. From the pattern of excited states of $[Cr(CN)_6]^{3-}$ shown in Fig. 2-19, it is apparent that a donor having an excited state above $12{,}500 \text{ cm}^{-1}$, but below about $20{,}000 \text{ cm}^{-1}$ can, if it has a sufficiently long lifetime, populate the ^2E, $^2T_1$, and $^2T_2$ states on an

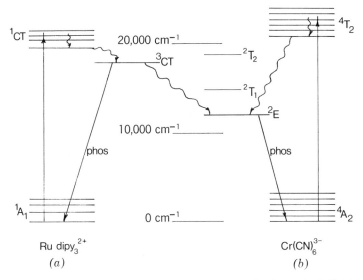

Fig. 2-19. Energy levels in the energy-transfer system: $Ru(bipy)_3^{2+}$–$Cr(CN)_6^{3-}$.[82] (a) indirect excitation via the ^3CT state of $Ru(bipy)_3^{2+}$, and (b) direct excitation of the $^4T_2$ state of $Cr(CN)_6^{3-}$.

energetically favorable basis, but not the $^4T_2$ or $^4T_1$ states. Among the donors used in this specific case are benzil,[80] with a long-lived triplet state at $18,800 \text{ cm}^{-1}$, *trans*-$[Cr(NH_3)_2(NCS)_4]^-$,[81] which has its ^2E state at $13,300 \text{ cm}^{-1}$, and $[Ru(bipy)_3]^{2+}$, for which the low-lying charge-transfer triplet state is about $18,000 \text{ cm}^{-1}$ above the ground singlet state.[82] In each case, the energy transfer is also fully allowed according to the spin-correlation rule. Each of the three donors sensitizes the phosphorescence of $[Cr(CN)_6]^{3-}$ in fluid medium and, in turn, has its own phosphorescence quenched by $[Cr(CN)_6]^{3-}$. As to which of the three doublet states of the acceptor is reached, or in what proportion they are reached by energy transfer, the rapid internal conversion of $^2T_1$ and $^2T_2$ to the ^2E state precludes a kinetic distinction among the three.

Inclusion of energy transfer requires modification of the mechanism of the primary steps involved in the photoprocesses. For quenching we must include an additional step such as

$$N_1 + Q \rightarrow M + Q_1; \qquad k_q[Q][N_1] \qquad (2\text{-}61)$$

in competition with phosphorescence and the other processes that degrade the emitting state, N_1. The quantum yield of emission becomes

$$\phi_{\text{phos}} = \frac{k_4}{k_3 + k_4} \frac{k_5}{k_5 + k_6 + k_q[Q]} \qquad (2\text{-}62)$$

which simplifies to the *Stern–Volmer equation:*

$$\frac{\phi^0_{phos}}{\phi_{phos}} = 1 + \frac{k_q}{k_5 + k_6}[Q] \tag{2-63}$$

in which ϕ^0_{phos} is the quantum yield in the absence of quencher. Absolute quantum yields are not required for the determination of the quenching constant, $k_q/(k_5 + k_6) = K_{SV}$. In practice, the ratio of intensity without to that with quencher is plotted against [Q], and the slope is the Stern–Volmer quenching constant. The value of K_{SV} also equals the quantity $k_q\tau$.

The lifetime of the donor can also be monitored as a function of quencher concentration. The Stern–Volmer equation becomes

$$\frac{\tau^0}{\tau} = 1 + k_q\tau^0[Q] \tag{2-64}$$

where τ^0 and τ are the luminescence lifetimes in the absence and presence, respectively, of quencher. Lifetime measurements have the advantage of giving directly the rate constant, k_q.

Regardless of the property that is sensitized by energy transfer to the acceptor molecule, the kinetics should also obey the Stern–Volmer law, but in the form:

$$\frac{\phi^\infty_{sens}}{\phi_{sens}} = 1 + \frac{(k_5 + k_6)}{k_q[Q]} \tag{2-65}$$

in which ϕ^∞ is the limiting quantum yield at very high acceptor concentrations. The slope of this plot is just the reciprocal of that relating to Eq. (2-63), so that the quenching-rate constant can be obtained from the slope and the value of the lifetime of the *donor* excited state.

A very recent example of the use of energy transfer is a measurement made of the quantum yield of intersystem crossing in $Cr(CN)_6^{3-}$.[82] The donor was the charge-transfer triplet state of $Ru(bipy)_3^{2+}$; both donor and acceptor phosphorescence could be followed at room temperature in the solvent, DMF. In the experiment, the yield of phosphorescence of the hexacyanochromate ion was compared under two conditions of excitation: direct absorption to the $^4T_2$ state of $Cr(CN)_6^{3-}$, and indirect excitation to the ^2E state by energy transfer. Under comparable rates of excitation, ^2E excitation by energy transfer produced twice the intensity of phosphorescence as direct excitation via the $^4T_2$ state. Therefore, it was concluded that the product of the efficiency with which the ^2E state is populated (at the limiting high concentration of acceptor) and the intersystem-crossing yield in $Cr(CN)_6^{3-}$ is 0.5. The efficiency of energy transfer would be unity in the absence of any quenching step that leads to other than the ^2E state of the Cr(III) complex. Since concentration quenching is usually not observed in transition-metal-complex ion emission, there is no reason to expect quenching here other than energy-transfer quenching. This result also depends on a value of the quantum yield of formation of the charge-transfer triplet state of the Ru(II) complex of one. Hence the conclusion that $\phi_{isc} \leq 0.5$.

Energy transfer has been used to investigate the mechanism of photochemical reactions. The principles are the same as those discussed here concerning quenching and sensitization. These aspects are however deferred to the appropriate later chapters.

REFERENCES

1. H. Kallman and G. M. Spruch, *Luminescence of Organic and Inorganic Materials*, Wiley, New York, 1962.
2. J. A. Khvostikov, *Tr. Gos. Optich. Inst. Leningrad* **12**, 3 (1937).
3. C. A. Parker, *Photoluminescence of Solutions*, Elsevier, Amsterdam, 1968.
4. J. B. Birks, *Photophysics of Aromatic Molecules*, Wiley-Interscience, New York, 1970.
5. A. Einstein, *Phys. Z.*, **18**, 121 (1917).
6. M. Kasha, *Discuss. Faraday Soc.* **9**, 14 (1950).
7. G. Viswanath and M. Kasha, *J. Chem. Phys.* **24**, 574 (1956).
8. E. Drent, G. Makkes van der Deijl, and P. J. Zanstra, *Chem. Phys. Letters* **2**, 526 (1968).
9. P. M. Rentzepis, *Chem. Phys. Letters* **3**, 717 (1969).
10. S. I. Vavilov, *Phil. Mag.* **43**, 307 (1922).
11. D. Ricard, W. H. Lowdermilk, and J. Ducuing, *Chem. Phys. Letters* **16**, 617 (1972).
12. D. S. McClure, *J. Chem. Phys.* **17**, 665 (1949).
13. G. B. Kistiakowsky and C. S. Parmenter, *J. Chem. Phys.* **42**, 2942 (1965).
14. B. Henry and M. Kasha, in *Annual Review of Physical Chemistry* (H. Eyring, C. J. Christensen, and H. S. Johnston, Eds.), Vol. 19, p. 161, Annual Reviews, Palo Alto, 1968.
15. C. E. White, in *Fluorescence; Theory, Instrumentation, and Practice* (G. G. Guilbault, Ed.), p. 275, Marcel Dekker, New York, 1967.
16. A. Reinberg, *Phys. Rev. B* **3**, 41 (1971).
17. J. E. Ralph and M. G. Townsend, *J. Chem. Phys.* **48**, 149 (1968).
18. J. N. Demas and G. A. Crosby, *J. Amer. Chem. Soc.* **93**, 2841 (1971).
19. D. F. Nelson and M. D. Sturge, *Phys. Rev. A* **137**, 1117 (1965).
20. L. S. Forster, in *Transition Metal Chemistry* (R. L. Carlin, Ed.), Vol. V, p. 1, Marcel Dekker, New York, 1969.
21. P. D. Fleischauer and P. Fleischauer, *Chem. Rev.* **70**, 199 (1970).

22. H. F. Wasgestian, *J. Phys. Chem.* **76**, 1947 (1972).
23. C. D. Flint and P. Greenough, *Chem. Phys. Letters* **16**, 369 (1972).
24. H. P. de la Garanderie, *C.R. Acad. Sci., Paris* **245**, 2047 (1957).
25. A. Pfeil, *Spectrochim. Acta A* **26**, 1341 (1970).
26. R. A. Condrate and L. S. Forster, *J. Chem. Phys.* **48**, 1514 (1968).
27. H. Witzke, *Theor. Chim. Acta (Berlin)* **20**, 171 (1971); and Ph.D. Thesis, University of British Columbia, 1969.
28. K. K. Chatterjee and L. S. Forster, *Spectrochim. Acta* **20**, 1603 (1964).
29. A. Pfeil, *J. Amer. Chem. Soc.* **93**, 5395 (1971).
30. H. L. Schläfer, *J. Phys. Chem.* **69**, 2201 (1965).
31. D. S. McClure, in *Solid State Physics* (F. Seitz and D. Turnbull, Eds.), Vol. 9, p. 399, Academic, New York, 1959.
32. G. B. Porter and H. L. Schläfer, *Z. Physik. Chem. (Frankfurt)* **37**, 109 (1963).
33. H. L. Schläfer, H. Gausmann, and H. Witzke, *J. Chem. Phys.* **46**, 1423 (1967).
34. S. N. Chen and G. B. Porter, *J. Amer. Chem. Soc.* **92**, 2189 (1970).
35. F. Castelli and L. S. Forster, *J. Amer. Chem. Soc.* **95**, 7223 (1973).
36. H. Zander, Ph.D. Thesis, University of Frankfurt, 1969.
37. J. L. Laver and P. W. Smith, *Aust. J. Chem.* **24**, 1807 (1971).
38. R. Dingle, *J. Chem. Phys.* **50**, 1952 (1969).
39. F. Castelli and L. S. Forster, private communication.
40. S. N. Chen and G. B. Porter, *Chem. Phys. Letters* **6**, 41 (1970).
41. N. Sabbatini and V. Balzani, *J. Amer. Chem. Soc.* **94**, 7587 (1972).
42. P. N. Everett, *J. Appl. Phys.* **41**, 3193 (1970).
43. M. Mingardi and G. B. Porter, *J. Chem. Phys.* **44**, 4354 (1966).
44. M. K. deArmond and J. E. Hillis, *J. Chem. Phys.* **54**, 2247 (1971).
45. D. H. W. Carstens and G. A. Crosby, *J. Mol. Spectrosc.* **34**, 113 (1970).
46. J. N. Demas and G. A. Crosby, *J. Amer. Chem. Soc.* **92**, 7262 (1970).
47. S. J. Strickler and R. A. Berg, *J. Chem. Phys.* **37**, 814 (1962).
48. D. M. Klassen and G. A. Crosby, *J. Chem. Phys.* **48**, 1853 (1968).
49. A. A. Schilt and R. C. Taylor, *J. Inorg. Nucl. Chem.* **9**, 211 (1959).
50. R. J. Watts and G. A. Crosby, *J. Amer. Chem. Soc.* **93**, 3184 (1971).
51. L. A. Rosiello, *Ric. Sci. Parte 2, Sez. A* **6**, 438 (1964).
52. F. A. Kröger, H. J. Vink, and J. van der Boomgaard, *Physica* **18**, 77 (1952).
53. W. Low, *Phys. Rev.* **109**, 247 (1958).
54. N. B. Manson, *Phys. Rev. B* **4**, 2645, 2656 (1971).
55. P. B. Dorain, H. H. Patterson, and P. C. Jordan, *J. Chem. Phys.* **49**, 3845 (1968).
56. S. N. Chen, Ph.D. Thesis, University of British Columbia, 1970.
57. W. Targos and L. S. Forster, *J. Chem. Phys.* **44**, 4342 (1966).
58. F. D. Camassei and L. S. Forster, *J. Chem. Phys.* **50**, 2603 (1969).
59. R. W. Harrigan, G. D. Hager, and G. A. Crosby, *Chem. Phys. Letters* **21**, 487 (1973).
60. F. Castelli and L. S. Forster, private communication.
61. A. Pfeil, Ph.D. Thesis, University of British Columbia, 1971.

62. J. Jortner, S. A. Rice, and R. M. Hochstrasser, in *Advances in Photochemistry* (J. N. Pitts, Jr., G. S. Hammond, and W. A. Noyes, Jr., Eds.), Vol. 7, p. 149, Wiley-Interscience, New York, 1969.

63. J. Jortner and R. S. Berry, *J. Chem. Phys.* **47**, 2757 (1968).

64. R. Englman and J. Jortner, *Mol. Phys.* **18**, 145 (1970).

65. B. Barnett and R. Englman, *J. Luminescence* **3**, 55 (1970).

66. G. W. Robinson and R. P. Frosch, *J. Chem. Phys.* **37**, 1962 (1962).

67. J. P. Byrne, E. F. McCoy, and I. G. Ross, *Aust. J. Chem.* **18**, 1589 (1965).

68. W. Siebrand, in *The Triplet State, Beirut Symposium* (A. B. Zahlan, Ed.), p. 31, Cambridge University Press, London, 1967.

69. G. W. Robinson and R. P. Frosch, *J. Chem. Phys.* **38**, 1187 (1963).

70. T. R. Thomas, R. J. Watts, and G. A. Crosby, *J. Chem. Phys.* **59**, 2123 (1973).

71. F. Castelli and L. S. Forster, *J. Amer. Chem. Soc.* **95**, 7223 (1973).

72. P. Kisliuk and C. A. Moore, *Phys. Rev.* **160**, 307 (1967).

73. M. Anson and R. C. Smith, *IEEE J. Quantum Electronics* **QE-6**, 268 (1970).

74. A. M. Glass, *J. Chem. Phys.* **50**, 1501 (1969).

75. C. D. Flint and P. Greenough, *J. Chem. Soc. Faraday II* **1972**, 897 (1972).

76. G. B. Porter, S. N. Chen, H. L. Schläfer, and H. Gausmann, *Theor. Chim. Acta (Berlin)* **20**, 81 (1971).

77. F. Wilkinson, in *Advances in Photochemistry* (W. A. Noyes, Jr., G. S. Hammond, and J. N. Pitts, Jr., Eds.), Vol. 3, p. 241, Wiley-Interscience, New York, 1964.

78. K. Sandros, *Acta Chem. Scand.* **18**, 2355 (1964).

79. E. Wigner, *Nachr. Ges. Wiss. Göttingen Math. Physik Kl.* **1927**, 375 (1927).

80. D. J. Binet, E. L. Goldberg, and L. S. Forster, *J. Phys. Chem.* **72**, 3017 (1968).

81. S. N. Chen and G. B. Porter, *J. Amer. Chem. Soc.* **92**, 3196 (1970).

82. N. Sabbatini, M. A. Scandola, and V. Balzani, *J. Phys. Chem.* **78**, 541 (1974).

3

CHARGE-TRANSFER PHOTOCHEMISTRY

John F. Endicott

Department of Chemistry
Wayne State University
Detroit, Michigan

I. INTRODUCTION

The stimulation of radial redistribution of electron density in transition-metal complexes can result in a remarkable variety of chemical reactions. The most characteristic of these is the redox decomposition of the transition metal substrate. The charge-transfer, the metal-centered (ligand field), and the ligand-centered absorption bands are not always well separated or pure in character in any given complex. Furthermore, in many extensively studied systems, as, for example, complexes of Co(III), the spectroscopic charge-transfer states are higher in energy than at least some of the ligand-field states (or sometimes ligand-centered states). Thus, in principle, these lower energy states may be populated by means of electronic relaxation from the higher energy charge-transfer states, and the chemical reactions stimulated by irradiation of charge-transfer absorption bands need not always be due to reactions of charge-transfer excited states. Designing experiments and developing comparisons that discriminate between the reaction patterns characteristic of different excited states is not trivial. In this chapter the reaction modes characteristic of other types of excited states are examined, at least to the extent permitted by available information while being limited by the necessity of minimizing duplication of material presented elsewhere in this book.

Material is arranged in this chapter according to an operational logic: (1) How does one identify charge-transfer absorption bands? (2) What kind of species should one expect to observe as primary photoredox products? (3) What are the chemical properties of such species? (4) What intermediates (excited states, radical pairs, etc.) are expected as precursors to the primary photoredox products? (5) How does one obtain information about these intermediates? (6) What models might one consider in attempting to understand the photoreaction pathways?

It is the intent of this chapter to examine the behavior of classes of complexes, rather than comprehensively document all the relevant literature. The material reviewed will therefore be largely restricted to monomeric complexes containing σ-bonded ligands. Of necessity, additional restrictions are imposed by the author's interests and the availability of information pertinent to the specific discussions below.

The lifetimes of transition-metal excited-state species are commonly less than 10^{-6} sec (see Chapters 1 and 2). Thus, this chapter is mostly concerned with the chemical reactions occurring less than 1 μsec after electronic excitation. In the majority of product and quantum-yield determinations, analytical measurements are performed many minutes after the irradiation has been stopped. Since photoredox decompositions invariably involve the production of reactive radicals, the ultimate reaction products are often unreliable indicators of the primary processes or their yields. The processes occurring in the time interval between the extinguishing of the excitation source and the formation of final products have been carefully explored only in a few laboratories, generally using flash-photolysis techniques. These studies of transient species have almost exclusively examined the chemical behavior of the products of excited-state reactions. To the present there has been very little direct observation of charge-transfer excited-state processes in transition-metal complexes (see also Chapters 1 and 2).

The reader wishing a more comprehensive survey of charge-transfer photochemistry than contained in this chapter is referred to the excellent monograph of Balzani and Carassiti[1] and several recent reviews.[2-6]

II. CHARGE-TRANSFER ABSORPTION SPECTRA

Most charge-transfer photochemistry is initiated by the irradiation of some kind of charge-transfer absorption band. Although the theoretical description of charge-transfer excited states is still relatively primitive, the correlation and assignment of charge-transfer spectra in the simple complexes considered here is quite well established.

The various types of charge transfer absorption spectra have been discussed by Dunn,[7] Jorgensen,[8] Schmidtke,[9] and Schläfer and Glieman.[10] The transitions in question all involve a radial redistribution of electron density within the molecule. This chapter is mostly concerned with systems in which excitation results in a transfer of electronic charge away from the ligands toward the metal center, charge transfer to metal (CTTM) transitions. Excitations which result in an outward flow of electronic charge, charge transfer to ligand (CTTL) and charge transfer to solvent (CTTS), are discussed extensively in Chapters 5 and 8.

Jorgensen[8] has proposed a semiempirical correlation of the energies of CTTM transition maxima to the electronegativity difference between metal and ligand, $\chi_L - \chi_M$, the d-orbital energy difference caused by ligand-field splitting, 10 Dq, the difference in the spin-pairing energy of the central metal in the ground and excited states, δSP, and the Racah parameter, B. Thus, the maximum of a CTTM transition is expected to occur for[9]

$$\bar{\nu}_{max} = 30(\chi_L - \chi_M) + 10\ Dq + \delta SP + aB \qquad (3\text{-}1)$$

In many complexes one would expect $X^- \to M$, CTTM transitions at different energies, depending on (1) whether the ligand electrons involved in the transition have σ- or π-symmetry with respect to the metal–ligand bond (hence, one defines π and σ electronegativities for the ligand[8]); (2) whether the acceptor orbital is antibonding ($^*e$) or nonbonding [t_2; e.g., both possibilities exist in Cr(III) and Ru(III) complexes]; and (3) the values, Dq and χ_M, appropriate to the specific metal-centered acceptor orbital involved in the transition. Although one would generally expect $d_{x^2-y^2}$ and d_{z^2} orbitals to differ in energy in $M^{III}L_5X$ complexes, the last point has not been carefully explored, and one generally uses an average optical electronegativity and an average value of 10 Dq if the transition involves an antibonding orbital. If nonbonding metal orbitals are involved, the 10 Dq term does not appear in the correlative equation.

Equation (3-1) is a very useful basis for correlating and assigning CTTM absorption spectra. A few examples of such correlations are presented in Table 3-1. The CTTM spectra of $Co^{III}(NH_3)_5X$ complexes have been used as a basis for obtaining "σ" and "π" electronegativities of the ligand, X, and these parameters have been used to predict the CTTM spectra of several $Ru^{III}(NH_3)_5X$, $Rh^{III}(NH_3)_5X$, and $Co^{III}(HED-TA)X$ complexes. The average difference in values of ν_{max}(obsd) and ν_{max}(calcd) for the entries in Table 3-1 is 2.5 kK, about as expected.[9] It is not clear whether the agreement could be improved through a more judicious selection of the values of χ_M, 10 Dq, and δSP and inclusion of the last term in Eq. 3-1. In fact, there is some evidence that the energy and absorptivities of CTTM transition maxima vary somewhat with the composition of the medium, so no very precise correlation of the type given by Eq. 3-1 would be expected. It remains to be seen whether the solvent contributes significantly to the binding energy of CTTM excited

Table 3-1. Analysis of CTTM Spectra of Some $M^{III}L_5X$ Complexes

M	$\chi_M{}^a$	L	X	$\chi_X{}^b$	$30(\chi_X - \chi_M)$, kK	$10\,Dq$, kK	δSP, kK^c	ν_{max}(calcd), kK	ν_{max}(obsd), kK
Co	2.3	NH$_3$	NH$_3$	3.28	—	24.9	−4.4	—	50
			F	(3.9)	48	23.8	—	67.4	>50
			Cl	3.12, σ		23.3			43.5
				2.89, π					36.5
				(3.0)					
			Br	2.99, σ		23.0			39.2
				2.76, π					32.3
				(2.8)					
			I	2.87, σ		22.4			35.1
				2.6, π					27.0
				(2.5)					
			NO$_2$	2.98, σ		25.2			41.3
				2.62, π					30.5
		EDTA	^{e}	3.12	—	22.4^f			42.5
		HEDTA^g	Cl	3.12, σ	24.6	22.0^f		42.2	38
				2.89, π	17.7			35.3	~30(?)
			Br	2.99, σ	20.7	21.8^f		38.1	36
				2.76, π	13.8			31.2	29
			NO$_2$	2.98, σ	20.4	23.8^f		39.8	38.3
				2.62, π	9.6			29.0	28.5

Metal		X						
Rh	2.3	NH_3			34.6[h]	-3.0	56.2	>52.6
		Cl	3.12, σ	24.6			49.3	51.8
			2.89, π	17.7			51.8	>50
		Br	2.99, σ	20.7	34.1[h]		44.9	48.8
			2.76, π	13.8			47.3	44.2
		I	2.87, σ	17.1	33.2[h]		39.2	36.0
			2.6, π	9				
Ru	2.05	NH_3	3.28	36.9	[i]	2.4	39.3	38.9
		Cl	3.12, σ	32.1	[i]		34.5	~40(?)
			2.89, π	25.2			27.6	30.5
		Br	2.99, σ	28.2	[i]		30.6	(?)
			2.76, π	21.3			23.7	25.1
		I	2.87, σ	24.6	[i]		27	(?)
			2.6, π	16.5			18.9	18.5

[a] Ref. 8.

[b] Calculated values based on $X^- \rightarrow Co^{III}$, σ and $X^- \rightarrow Co^{III}$, π transitions in $Co^{III}(NH_3)_5X$ complexes (Ref. 8).

[c] Based on $E(^1A_1 \rightarrow {}^1A_2) = 10\,Dq - C$ and assuming an average field substituting X for NH_3; values of $Dq(X)$ and $C = 3.7\,kK$ from Ref. 11.

[d] See Ref. 9.

[e] Six-coordinate EDTA.

[f] Based on observed transition in $Co(EDTA)^-$ and assumptions analogous to those noted in c.

[g] Pentacoordinate HEDTA.

[h] Estimated assuming $10\,Dq = g_M \times f_L$ and that $g_M = 18.2$ for Co(III) and 27.0 for Rh(III) (Ref. 8).

[i] Lowest-energy CTTM excited state has a formal t_{2g}^6 configuration of the metal, so the $10\,Dq$ term does not enter into the calculation.

states, analogous to the case for the CTTS excited states of many anions (see Chapter 8 and Section IV.D. below).

III. SURVEY OF CHARACTERISTIC REACTION PATHWAYS

A. Products Resulting from the Excitation of Electron Density Toward the Metal Center (Charge Transfer to Metal Excited States)

1. *Oxidation-Reduction Processes.* The formation of a one-electron reduced metal species and a radical fragment from oxidation of a coordinated ligand is the type of reaction most characteristic of charge transfer to metal (CTTM) excitation. These species are the primary photoredox products, and one should know their identity and yields before attempting to discuss the mechanism of the photochemical reaction.

a. *The Detection and Properties of Simple Radicals in Aqueous Systems.* The one-equivalent oxidation of almost any Lewis base produces a radical. Some of these radicals are reasonably stable (e.g., NO_2), and some are extremely reactive (e.g., NH_3^+ or $\cdot Cl$) in aqueous solution. In this regard, it is an important property of water as a solvent system that it does not efficiently oxidize or reduce most simple radicals; that is, the radicals obtained from water are among the most reactive radicals known (standard reduction potentials have been estimated as -2.75 V,[12] -2.1 V,[13] and $+2.8$ V[13] for the aqueous electron, hydrogen atom, and hydroxyl radical, respectively). In acidic aqueous solution the formation energetics for the hydrogen atom, $\cdot H$, and the hydroxyl radical, $\cdot OH$, provide practical reactivity limits for any observable radical. Bounded by these limits there are many very reactive and detectable radicals. Table 3-2 summarizes the properties of some radicals characteristic of those obtainable in the photoredox decompositions of various coordination complexes.

For those simple radicals that do not react in some way with the solvent, the natural radical lifetimes most often depend on the rates of radical–radical coupling reactions (e.g., $2 \cdot Br \rightarrow Br_2$) or bimolecular disproportionation reactions (e.g., $2Br_2^- \rightarrow Br_2 + 2Br^-$). Complex (or polyatomic) radicals may undergo a variety of cleavage reactions yielding simpler, more stable radical fragments. Finally, many of the more reactive radicals are transformed into less reactive radical species by means of hydrogen atom abstraction reactions; in some natural radical-decay modes in aqueous solution, hydrogen-atom abstraction reactions amount to disproportionation reactions (e.g., $2 \cdot C_2H_5 \rightarrow C_2H_4 + C_2H_6$). These various decay modes of the primary radical species are often found to compete with reactions with radical scavengers (including some substrate species).

A variety of techniques have been used to characterize primary radicals. Pulse techniques (flash photolysis, laser photolysis, or pulse radiolysis) are probably the most dependable, since these techniques can permit the analytical observations to be made within the radical lifetime. However, it is generally judicious to combine pulse and chemical-scavenging techniques to facilitate identification of transient radical species. One limitation of pulse techniques has been that transients are usually detected by means of changes in the optical spectrum, and not all radicals absorb more strongly than the substrates irradiated. In many cases, weakly absorbing radicals can be readily detected and character-ized by means of their reactions with simple scavenging substrates; for example, the chlorine radical, $\cdot Cl$, absorbs significantly only in the deep ultraviolet, but is very readily detected by association with Cl^- to form Cl_2^-, which absorbs strongly in the near ultraviolet. Most of the informa-tion in Table 3-1 has been obtained from pulse studies of one sort or another. Recently, electrochemical[31,51,57,58] and conductivity[59] detection methods have been used in pulse studies.

It is also possible to infer a great deal about the nature and identity of primary radicals from chemical-scavenging techniques. The basic ap-proach here is to examine the variation of final product yields (or transient yields and lifetimes) in the presence of various types of known radical scavengers. Although this can be a very powerful approach, some caution must be exercised for at least the following reasons: (1) the final reaction products are often many steps removed from the primary radicals of interest (e.g., in the case of NH_3^+ the final product is N_2); (2) reaction of a primary radical with a chemical scavenger most often results in the formation of a new radical species that may exhibit some unique form of reactivity not present in the primary radical (e.g., 2-propanol scavenges efficiently for the chlorine atom, a very powerful oxidizing radical, to form the 2-hydroxy-2-propyl radical, a powerful reducing radical); and (3) when very reactive primary radicals are produced, the final products are often formed in nonstoichiometric yields owing to a variety of experimen-tal problems. Irreproducible or nonintegral product yields have been found especially troublesome for photoreactions producing radicals such as $\cdot Cl$, NH_3^+ or $\cdot N_3$. One source of difficulty when these radicals are produced from highly absorbing transition-metal substrates is that the radical densities may tend to be extremely high in a very small region of the photolysis cell nearest the source of radiation; this can lead to such difficult problems as secondary photolysis, exceptionally efficient radical-radical (compared to radical–scavenger) reactions, and efficient reactions

Table 3-2. Selected Properties of Aqueous Solutions of Some Characteristic Radicals[a].

Radical	$k_D \times 10^9$, $M^{-1}\,sec^{-1}$ [b]	pK_a	E^0, V ($\cdot R + e^- \xrightarrow{H^+} RH$)[c]	E^0, V ($R^+ + e^- \xrightarrow{H^+} \cdot R$)[c]	λ_{max}, nm	ε_{max}, $M^{-1}\,cm^{-1}$	Other Important Properties
$\cdot OH$	5±1 (14, 15)	11.9±0.2 (13, 14)	2.8 (13)	—	260 (18)[d]	320 (18)	—
$\cdot H$	20 (19)	9.7 (12)	—	-2.1 (13)	—	—	—
e^-	10 (20, 21)[c]	—	—	-2.75 (15)	700 (22)	1.8×10^4 (22)	—
$\cdot Br$	10 (22)	—	2.07 (24)	—	—	—	—
Br_2^-	2–6 (23, 25–27)[e]	—	1.69 (24)	0.51 (24)	364 (22)	7.8×10^3 (27)	$K_{diss} = 4.6 \times 10^{-6}$ M (27)[f]
$(BrSCN)^-$	—	—	—	—	400 (28)	2.3×10^3 (28)	$K_{diss} = 6.0 \times 10^{-4}$ M (28)[g] ; $K_{diss} = 4.6 \times 10^{-9}$ M (28)[h]
$\cdot CH_2OH$	2.4 (29)	10.7 (30)	1.29	-0.92	<210 (29)	—	—
$\cdot CH_2O^-$	0.9 (29)[e]	—	—	-0.86 (31)[i]	<220 (29)	—	—
$CH_3\dot{C}HOH$	2.3 (29)	11.6 (30)	~0.95	-1.49 (31)[i]	<210 (30)	—	—
$CH_3\dot{C}HO^-$	0.5 (29)[e]	—	—	-0.61	<220 (30)	—	—
$CH_3\dot{C}(OH)CH_3$	1.4 (29)	12.2 (30)	—	-0.97 (31)[i]	<230 (30)	—	—
$CH_3\dot{C}(O^-)CH_3$	0.4 (29)[e]	—	—	-1.65 (31)[i]	<230 (29)	—	—
$CH_3-\overset{CH_3}{\underset{\cdot CH_2}{C}}-OH$	1.4 (29)	—	0.6 (31)[i]	-0.1 (31)[i]	—	—	—
$\cdot CH_2CO_2H$	1.8 (32)	4.5 (32)	~1.7	—	320 (32)	650 (29)	—
$\cdot CH_2CO_2^-$	1.0 (32)[e]	—	~2.1	—	350 (32)	800 (32)	—
$CH_3CO_2\cdot$	1.1 (32)	—	—	—	—	—	—
$\cdot CH(OH)CO_2H$	0.85 (32)[e,k]	8.8 (32)	—	—	<245 (32)	—	—
$\cdot CH(OH)CO_2^-$	0.015 (32)[e]	—	—	-1.2 (31)[i]	245 (32)	5700 (32)	—
$\cdot CH(O^-)CO_2^-$	—	—	—	-1.96 (31)[i]	255 (32)	5400 (32)	—
$\cdot CH_3$	~10 (23, 34)	—	2.25; ≥2[i]	—	—	—	—
$\cdot Cl$	4.0 (23, 26, 34)[e]	—	2.6 (26)	0.6 (26)	—	—	—
Cl_2^-	—	—	2.3 (26)	—	340 (34)	1.2×10^4 (34)	$K_{diss} < 10^{-2}$ M (34)[f]
$\cdot CN$	—	—	~2.8	—	—	—	—
$\cdot COH$	—	9.5 (36)	—	-0.97	235 (32)	3×10^3 (32)	—
$\cdot CO_2H$	1.5±0.2 (32, 37)	1.4 (38)	0.65	-0.6±0.1 (31)[i]	—	—	—
CO_2^-	1.5±0.2[e]	—	—	-1.2±0.1 (31)	260 (37)	2.2×10^3 (37)	—

·CO₃H	$(39–41)^m$	9.5 (43)	~2.1	—	600 (39, 43)	—	—	
CO₃⁻	$0.2\ (39–41)^m$	—	—	—	600 (39)	$(1.8\pm0.6)\times10^3$ (40, 41)	Reacts with O_2^- (42)	
·C₂O₄H	—	$1.4\ (44)^n$	~2.1	~1.3^o	<250	—	—	
C₂O₄⁻	$0.96\ (45)^e$	—	—	—	<250	—	—	
·I	10 (23)	—	—	—	<250	—	—	
I₂⁻	$9\ (23, 46, 47)^e$	—	1.42 (24)	0.11 (24)	380 (23, 46, 47)	1.4×10^4 (46, 47)	$K_{diss}=8.8\times10^{-6}$ M $(46)^f$	
(ISCN)⁻	—	—	1.13 (24)	—	420 (48)	9.2×10^3 (48)	$K_{diss}=1.3\times10^{-8}$ M^g $K_{diss}=4.8\times10^{-4}$ M^h	
(SCN)₂⁻	2.1 (46, 48)	—	1.5 (103c)	—	475 (48)	7.3×10^3 (48)	$K_{diss}=5.0\times10^{-6}$ $(48)^f$	
·N₃	6.8 (49, 50)	—	~1.9^p	—	275 (49, 50)	1.1×10^3 (49, 50)	$K_{diss}=1.5\times10^{-5}$ M $(51)^q$	
·NO₂	0.90 (51)	—	1.13 (13)	0.74 (13)	400 (51)	201 (51)	—	
·NO₃	$0.0108\ (52)^r$	—	~1.9	—	635 (52, 53)	250 ± 90 (52)	—	
NH₃⁺	—	6.7 (54)	2.7^s	~-0.8^t	—	—	—	
·NH₂	—	—	—	—	—	—	—	

a Properties are selected from references cited in parenthesis.

b For radicals exhibiting radical–radical coupling or disproportionation reactions.

c Estimated from tabulated (e.g., as in Ref. 13) thermodynamic parameters except as indicated; conventions as in Ref. 13: versus S.N.E. In cases that no reference is cited estimates have been made for this tabulation by the author using data from Refs. 13, 55, and 56.

d Not a distinct maximum.

e k_D varies with ionic strength, and values cited have not been extrapolated to zero ionic strength.

f $K_{diss}=$ dissociation constant for $X_2^- \underset{k_f}{\overset{k_d}{\rightleftharpoons}} \cdot X + X^-$; k_f is generally diffusion controlled (~10^{10} M⁻¹ sec⁻¹); see Ref. 21.

g $K_{diss}=$ dissociation constant for $(XSCN)^- \rightleftharpoons X^- + \cdot SCN$.

h $K_{diss}=$ dissociation constant for $(XSCN)^- \rightleftharpoons \cdot X + SCN^-$.

i Based on the reversible polarographic half-wave potential at high pH for RCHO⁻; potentials have been referred to standard hydrogen electrode, $[H^+]=1.0$ M using the pK_a value cited. Measured potentials in acidic solution are often irreversible and pH insensitive (see also Ref. 57).

j Based on determination of polarographic half-wave potential at pH = 13.5 and corrected for standard calomel eletrode potential; no Nernst correction for [H⁺] applied.

k At pH 7.2.

l Efficient oxidation of Br⁻ observed but not Cl⁻.[33]

m Initial studies[39–41] did not recognize the complications which arise due to the reaction of CO_3^- with O_2^{-}[42] and/or H⁺.[43]

n Based on the mechanism in Ref. 44 and presumed similarity to $H_2C_2O_4$ (M. Z. Hoffman, private communication).

o For the couple, $2CO_2 + H^+ + 2e^- \rightleftharpoons HC_2O_4^-$.

p Based on the formation of $(BrN_3)^-$.[26]

q $K_{diss}=$ dissociation constant for $N_2O_4 \rightleftharpoons 2NO_2$.

r A first-order decay process has been reported for this radical.[53]

s Estimated for $NH_3^+ + e^- \rightleftharpoons NH_3$; for $NH_3^+ + e^- + H^+ \rightleftharpoons NH_4^+$; standard reduction potential ~3.3 V.

t For the couple $\frac{1}{2}N_2 + 2H^+ + 2e^- \rightarrow \cdot NH_2$.

with contaminants on the surface of the photolysis vessel.

The actual reaction modes exhibited by aqueous radicals are dependent on a combination of thermodynamic feasibilities and mechanistic propensities. The thermodynamic possibilities in aqueous systems are most conveniently summarized in terms of standard reduction potentials (SRP). Reasonable estimates of SRPs have been made in a large number of cases,[12,13,24,26,31] and these are included in Table 3-1. In some cases such as the halogen atoms and the dihalogen radical anions, the thermochemical estimates[24,26] have to be very good, and thermochemical arguments may confidently be applied to assess the validity of attempts to determine potentials experimentally (see footnote 19 in Ref. 24).

Reasonable thermochemical estimates can often be made, even when precise thermochemical data in aqueous solution is not available for the parents or products of radical species. Several such estimates[55,56] have been included in Table 3-2. One case, $\cdot CH_2OH$, will be considered in detail to illustrate the approach.

In order to estimate an SRP for the couple

$$\cdot CH_2OH + e^- + H^+ \rightleftharpoons CH_3OH \tag{3-2}$$

one needs to know the standard free energies of formation of $\cdot CH_2OH$ and CH_3OH in aqueous solution. Such information is available for CH_3OH but not for $\cdot CH_2OH$.[13] However, good estimates of the enthalpy of formation in the gas phase are available for both species, -162 kJ/mole[13] and -33 kJ/mole,[55] respectively.

$$CH_3OH(g) \rightarrow \cdot CH_2OH(g) + \cdot H(g) \tag{3-3}$$

If we assume that the entropy change in reaction (3-3) is small, and that the free energies of solution of CH_3OH and $\cdot CH_2OH$ are nearly equal, then the SRP for $\cdot CH_2OH$ acting as an oxidant, Eq. (3-2), is 1.29 V. In a similar manner, using the above-estimated value of ΔG_f^0 for $\cdot CH_2OH$ and tabulated data for CH_2O,[13] one finds SRP $= -0.92$ V for $\cdot CH_2OH$ acting as a reductant.

$$CH_2O + e^- + H^+ \rightleftharpoons \cdot CH_2OH \tag{3-4}$$

In a very elegant experiment Lilie, Beck, and Henglein[31] determined that the half-wave potential for the reversible oxidation of $\cdot CH_2O^-$ was -1.73 V versus S.C.E. at pH $= 13.5$; referred to the standard hydrogen electrode, at $[H^+] = 1$ M (p$K_a = 10.7$ for $\cdot CH_2OH$[30]), this would be -0.86 V, in excellent agreement with the estimate. The polarographic oxidation of $\cdot CH_2OH$ was irreversible, with a weaker than predicted dependence on $[H^+]$, and occurred at more positive potentials than predicted; since the polarographic wave is irreversible it cannot be readily related to an SRP (see also Ref. 57).

The redox thermodynamics for $\cdot CH_2OH$ may be summarized in the reduction-potential diagram.

$$CH_2O \xrightarrow{\ -0.92\ } \cdot CH_2OH \xrightarrow{\ 1.29\ } CH_3OH$$

$$0.19$$

(3-5)

Similar approaches have been used for several other radicals listed in Table 3-1. In several cases, enthalpies of formation of radicals had to be calculated from bond-energy estimates.

Estimates of ΔH_f^0 for two of the radicals in Table 3-2, $CH_3CO_2\cdot$ and $HC_2O_4\cdot$, indicate that these radicals are unstable with respect to cleavage to form CO_2 and $\cdot CH_3$ (by 62.7 kJ mole^{-1}) or $HCO_2\cdot$ (by 28 kJ mole^{-1}), respectively. This certainly seems consistent with observations in the case of $CH_3CO_2\cdot$,[60,61] but there is some evidence that HC_2O_4 or $C_2O_4^-$ persists in aqueous solution.[44,45,62] It should be observed that a very large number of radicals can undergo irreversible redox disproportionation reactions, such as

$$2\cdot CH_2OH \rightarrow CH_2O + CH_3OH \tag{3-6}$$

and are, therefore, in principle, capable of functioning as either oxidants or reductants.

b. *Metallo Fragments Produced in Redox Decompositions.* Few primary metallo fragments of redox decompositions have been observed in flash-photolysis studies. However, a great deal is known about the chemical behavior of most transition metals in most accessible oxidation states. It is, therefore, generally possible to infer the nature of the primary metallo products from strictly chemical information such as product yields or the variations of products and yields in the presence of chemical scavengers.

In the much-studied cases of photoreduction of Co(III) complexes, the substrates are inert, and the product metallo fragments [Co(II) complexes] are generally labile to substitution. This happens to be a particularly fortunate combination of circumstances since the Co(II) products generally equilibrate rapidly[63] and the resulting $Co(OH_2)_6^{2+}$ is very difficult to oxidize.[13] As a result, bulk solution recombination reactions between oxidized ligand radicals and Co(II) species have rarely been observed.[64]

Table 3-3. Radical and Metallo Fragments Characteristic of Redox Reactions Following Direct CTTM Excitation of Transition-Metal Complexes[a]

Substrate Irradiated	Primary Radical	Identification Based on[b]	Probable Metallo Fragment	Estimated Lifetime of Metallo Fragment, sec[c]	Observed Radical Reactions with Substrate (P = A) or Metallo Fragment (P = R) $k_p \times 10^7$, M^{-1} sec^{-1}	Other Complicating Features
$Co(NH_3)_6^{3+}$	NH$_3^+$ (73)[d]	CI	$Co(NH_3)_5^{2+}$	<10^{-6} (67, 68, 74)	—	—
$[Co(NH_3)_5OH_2]^{3+}$	NH$_3^+$ (73)[d]	CI	$Co(NH_3)_5^{2+}$	—	—	Photoaquation of NH$_3$ leads to photosensitive products (66, 76)
$[Co(NH_3)_5NCS]^{2+}$	·NCS (71, 75)	FP[(NCS)$_2$]	$Co(NH_3)_5^{2+}$	—	—	—
$[Co(NH_3)_5N_3]^{2+}$	·N$_3$ (66, 71, 76, 77)	CI	$Co(NH_3)_4^{2+}$[e]	—	—	—
$[Co(NH_3)_5O_2R]^{2+}$[f]	·R (62, 78)[f]	CI	$Co(NH_3)_5^{2+}$	—	[f]	—
$[Co(NH_3)_5O_2CCO_2]^{+}$	C$_2$O$_4^-$/CO$_2^-$ (44, 79, 80)	FP[C$_2$O$_4^-$/CO$_2^-$]	$Co(NH_3)_5CO_2H^{2+}$[h]	~10^{-3} (44, 79, 80)	~10; P = A (79)	—
$[Co(NH_3)_5OCO_2]^{+}$	CO$_3^-$ (81–83)	FP[CO$_3$]	$Co(NH_3)_5^{3+}$		~1; P = A (84)	—
	·H (84)	CI	$Co(NH_3)_5^{3+}$	~10^{-3} (84)	<1; P = A (84)	—
$[Co(NH_3)_5OCHO]^{2+}$	CO$_2^-$ (84)[i]	FP[CO$_2$]	$Co(NH_3)_5CO_2H^{2+}$[h]			
$Co(NH_3)_5Cl^{2+}$	NH$_3^+$ or ·Cl (35)	CI, FP[NH$_2$Cl$^-$?]	$Co(NH_3)_5^{2+}$		<10^{-5}; P = A (26, 86)	—
$Co(NH_3)_5Br^{2+}$	·Br (26, 85)	FP[Br$_2$]	$Co(NH_3)_5^{3+}$			
$Co(NH_3)_3I^{2+}$	·I (26, 85)	FP[I$_2$]	$Co(NH_3)_5^{2+}$		~2×10^{-3}; P = A (26)	
$[Co(NH_3)_5NO_2]^{2+}$	NO$_2$ (87, 88)	CI	$Co(NH_3)_5^{3+}$			
			$Co(NH_3)_5$——ONO^{2+}[j]	>10^2		
$Co(en)_3^{3+}$	en$^+$ (89)	CI	$Co(en)_3^{2+}$	10^{-2} (67)[k]	—	—
$[Co(CN)_5NCS]^{3-}$	·NCS (66, 75)	FP[(NCS)$_2$]	$Co(CN)_5^{3-}$	10 (90)[l]	—	Facile reactions of Co(CN)$_5^{3-}$ with O$_2$
$[Co(CN)_5SCN]^{3-}$	·NCS (66, 75)	FP[(NCS)$_2$]	$Co(CN)_5^{3-}$		>1; P = R (66, 77)	
$[Co(CN)_5N_3]^{3-}$	·N$_3$ (66, 77)	CI, FP[I$_2$][m]	$Co(CN)_5^{3-}$		~10^2; P = R (66)	
$Co(CN)_5I^{3-}$	·I (66, 75)	CI, FP[I$_2$]	$Co(CN)_5^{3-}$			
$[Co(CN)_5SO_3]^{4-}$	SO$_3^-$ (91)	CI	$Co(CN)_5^{3-}$			
$Co(CN)_5R^{3-}$[o]	·R (91)	CI	$Co(CN)_5^{3-}$		[n]	
$Co(EDTA)^{-}$	·Y (92, 93)[p]	CI	Co(·Y)$^-$		>10^4; P = A (93)	Primary or secondary radical scavenged by substrate
$[Co(HEDTA)Cl]^{-}$	·Y (92, 93)[p]	CI, FP	Co(·Y)$^-$		>10^4; P = A (93)	
$[Co(HEDTA)Br]^{-}$	·Y (92, 93)[p]	CI, FP	Co(·Y)$^-$		>10^4; P = A (93)	
$[Co(HEDTA)NO_2]^{-}$	·Y (92, 93)	CI, FP	Co(·Y)$^-$		>10^4; P = A (93)	
$[Co(N_4)Cl_2]^{+}$[q]	·Cl (26)	FP[Cl$_2$]	$Co(N_4)^{2+}$	>10^5	1.0×10^2; P = R (26)	
$[Co(N_4)Br_2]^{+}$[q]	·Br (26)	FP[Br$_2$]	$Co(N_4)^{2+}$		1.4×10^2; P = R (26)	
$[Co(N_4)(OH_2)CH_3]^{2+}$[q,r]	·CH$_3$ (33, 94–97)	CI	$Co(N_4)$		2×10; P = R (94–97)	
$[Rh(NH_3)_5NCS]^{2+}$	·NCS (75)	FP[(NCS)$_2$]	$Rh(NH_3)_4^{2+}$	>10^{-3} (67, 98)	—	Linkage isomerization of NCS

Complex[a]	Radical[b]	Method[b]	Metallo fragment[c]		
[Rh(NH$_3$)$_5$SCN]$^{2+}$	·NCS (75)	FP[(NCS)$_2$]	Rh(NH$_3$)$_4$$^{2+}$	10^{-6}–10^{-3} (77)^s	—
[Rh(NH$_3$)$_5$N$_3$]$^{2+}$	NH (66, 77, 99)	Cl, FP	Rh(NH$_3$)$_5$NH^{3+}	—	—
Rh(NH$_3$)$_5$Cl^{2+}	·Cl (100)	FP[Cl$_2$]^t	Rh(NH$_3$)$_4$$^{2+}$	—	—
Rh(NH$_3$)$_5$Br^{2+}	·Br (100)	FP[Br$_2$]^t	Rh(NH$_3$)$_4$$^{2+}$	—	—
Rh(NH$_3$)$_5$I^{2+}	·I (98)	FP[I$_2$]^t	Rh(NH$_3$)$_5$$^{2+}$	~10^2; P = R (98)	—
[Ir(NH$_3$)$_5$N$_3$]$^{2+}$	·NH (99, 101)	Cl, FP	Ir(NH$_3$)$_5$NH^{3+}	—	—
PtCl$_6$$^{2-}$	·Cl (102)	FP[Cl$_2$]	PtCl$_4$	10^{-1} (102)	—
Fe(OH$_2$)$_5$Br^{2+}	·Br (103)	FP[Br$_2$]	Fe(OH$_2$)$_6$$^{2+}$	—	—
[Fe(OH$_2$)$_4$C$_2$O$_4$]$^+$	C$_2$O$_4^-$ (62)	FP[C$_2$O$_4^-$]	Fe(OH$_2$)$_6$$^{2+}$	—	—
Fe(C$_2$O$_4$)$_3$$^{3-}$·^{u}	C$_2$O$_4^-$ (58b 104, 105)	Cl, FP		—	—
[Fe(HEDTA)OH$_2$]	·Y (93b)^p	Cl	Fe(·Y)$^-$	1.4×10^{-2}; P = A (93b)	—
Ru(NH$_3$)$_5$X^{2+}·^{v}	None (106)	Cl, FP		Predominately photo- aquation observed	—

[a] In acidic aqueous solution at 25°C, except as indicated. References in parentheses.

[b] Abbreviations: CI, chemical inference, based on product analyses and scavenging studies; FP, flash photolysis (species observed indicated in brackets). Radical decay modes in the absence of substrate or metallo fragments are indicated in Table 3.2.

[c] Normal lifetime for species indicated in the absence of radicals. The lifetime of [Co(NH$_3$)$_5$]$^{2+}$, with respect to formation of Co^{2+}, is less than 10^{-6} sec.[67,68,74] Exceptions as noted.

[d] Final product in this case is N$_2$.[35]

[e] Photoaquation of ammonia occurs in this system with very high yield,[66,76] and the evidence[66] indicates the occurrence of [Co(NH$_3$)$_4$]$^{2+}$ as the primary metallo fragment.

[f] For R = —CH$_3$, —CH$_2$Cl, —CH$_2$Br, —CCl$_3$, —CF$_3$, —CH$_2$OH, —CH$_2$CN, —CH$_2$CH$_3$, —C(CH$_3$)$_2$OH, —CH=CH$_2$, —C≡CH, —CH—CH$_2$, —C$_6$H$_{11}$, —CH$_2$Ph.
                           CH$_2$

[g] Product yield indicates that less than 10% of Co^{2+} is produced in radical reduction of substrate.[61]

[h] Heterolytic bond cleavage in ligand followed by linkage isomerization to produce a metastable Co—C bonded formate.

[i] Secondary radical formed in decay of linkage isomer.

[j] Linkage isomer.

[k] At pH = 5.

[l] At pH = 1.0.

[m] For reactions run in dilute I$^-$; ·N$_3$ + I$^-$ → N$_3^-$ + ·I.

[n] Some evidence for ([Co(CN)$_5$]$^{3-}$ + ·R) recombination is presented in Ref. 91.

[o] R = ·CH$_2$—Ph.

[p] ·Y = N-methyleneethylenediaminetriacetate. It is presumed that this radical remains coordinated to the reduced metal after the initial photoreaction.

[q] Y (N$_4$), a cyclic tetradentate nitrogen donor ligand; ligands used include 5,7,7,12,14,14-hexamethyl-1,4,8,11-tetraazacyclotetradeca-4,11-diene and ammine analogs (Ref. 26 and S. D. Malone and J. F. Endicott, unpublished observations).

[r] Similar behavior observed in cases that (N$_4$) = bis(dimethyl)glyoximate and corrinoids.[95-97]

[s] pH- and medium-dependent lifetime.

[t] May have resulted from irradiation of ion pair.

[u] Actually a mixture of bis and tris oxalato complexes.

[v] For X = Cl, Br, I: Siegel and Armor[107] have found no evidence for significant photoredox reactions of Ru(NH$_3$)$_5$Cl^{2+} even on irradiation at 185 nm; no radicals were detected on irradiation of the lowest energy CTTM band.[106]

Exceptions to this occur when the product Co(II) species is unusually stable (e.g., $[Co(CN)_5]^{3-}$) and when the product species contains a cyclic tetradentate ligand so that the formation of $Co(OH_2)_6^{2+}$ occurs very slowly.[26,65] In such cases, recombination reactions of the primary photoredox products have been observed,[26,66] and primary redox quantum yields are difficult to obtain.

For several known photoredox reactions the likely immediate, or primary, metal fragments have been tentatively identified in Table 3-3. To a significant extent, this identification is an intelligent guess based on the identification of some primary radical derived from oxidation of a ligand species and known aspects of the chemistry of the reduced metal. Thus for most of the early entries the metal fragment is identified as $[Co(NH_3)_5]^{2+}$, or the cationic residue remaining when the oxidized ligand is removed from the coordination sphere. The rate at which a $[Co(NH_3)_5]^{2+}$ species would aquate is not known with certainty. However, pulse-radiolysis studies indicate that the first three ammonias of $Co(NH_3)_6^{2+}$ are replaced by water in less than 10^{-6} sec,[67] and this complex exchanges ligands with liquid ammonia with a rate constant of 7.2×10^6 sec^{-1},[68] so the lifetime of $[Co(NH_3)_5]^{2+}$ is certainly less than 10^{-6} sec, but probably greater than 10^{-8} sec for high-spin Co(II).

Since spin relaxation is no doubt as rapid in complexes of Co(II) as those of Cr(III)[69] or Fe(III),[70] the Co(II) ammine species may be presumed to be high spin after a very short time (less than 10^{-6} sec). Whether the CTTM excited states fragment into high- or low-spin Co(II) is an important question, but little available experimental information bears directly on this point. Since this problem has a bearing on the cage recombination mechanism,[3,71,72] it will be considered more critically in Section V below. The Co(II) ammine complexes have not been detected in either flash-photolysis or chemical-scavenging studies, presumably due to the very rapid equilibration of Co(II) in acidic aqueous solution.[63,71] Clearly, the structure and composition of the metal fragment is ambiguous in these cases. Similar reservations may be raised with regard to the species identified as $[Co(en)_2]^{2+}$.

The mechanistic importance of the structure of the metallo fragment is illustrated in contrasting features of the chemistry of cobalt-ammine complexes and cobalt-cyclic amine complexes. Co(II) ammine complexes are high spin and six coordinate;[63] thus equilibrated CTTM excited states that produce high-spin Co(II) products would be expected to result in the formation of $[Co(NH_3)_5]^{2+}$ [or $Co(NH_3)_5OH_2^{2+}$ after collapse of the solvent cage] species. However, the Co(II) complexes with tetradentate, planar macrocyclic ligands tend to be low spin,[108,109] and recent structural work has indicated a large axial distortion in the $[Co(Me_6[14]dieneN_4)(OH_2)_2]^{2+}$

complex (Co—OH_2 distances are 2.48 Å, compared to Co—N distances of about 1.9 Å).[110] If this is taken as a model for low-spin Co(II), then equilibrated CTTM excited states of Co(II) intermediates with this metal-ion electronic configuration would be expected to suffer a similar axial distortion, and this should labilize two-coordination positions (presumably, the oxidized ligand and the ligand *trans* to it). Thus the observation of a $[Co(NH_3)_4(OH_2)_2]^{2+}$ photochemical product would be a clue that a low-spin species was a key reaction intermediate. Such observations have not been made for cobalt complexes owing to the lability of the ammine ligands, but there is strong evidence for the intermediacy of axially distorted Rh(II) in the CTTM photochemistry of $Rh^{III}(NH_3)_5X$ (X = Cl, Br, I).[67,98,100]

When the ligand oxidized is a polydentate chelate, the "primary radical" may be incorporated into the "metallo fragment." This is necessarily the case with complexes of ethylenediaminetetraacetate (EDTA). For example, CTTM excitation of Co(EDTA)$^-$ in acidic solution results in the formation of Co^{2+}, CO_2, CH_2O, and other fragments of the EDTA ligand.[92,93] Since Co^{2+} and CO_2 appear to be primary products,[93a] one would postulate that the overall photoreduction proceeds stepwise through a reaction sequence such as in Eqs. (3-7) to (3-11), where $\cdot Y^{3-}$ is N-methylethylenediaminetriacetate and $\cdot S$ represents radical species (one or more) obtained from the fragmentation of the radical-EDTA species.

$$Co(EDTA)^- + h\nu \longrightarrow Co(\cdot Y)^- + CO_2 \qquad (3\text{-}7)$$

$$Co(\cdot Y)^- + nH^+ \longrightarrow H_nY\cdot + Co^{2+} \qquad (3\text{-}8)$$

$$H_nY\cdot \longrightarrow \cdot S + \cdots \qquad (3\text{-}9)$$

$$Co(\cdot Y)^- \xrightarrow{+H^+} \cdot S + Co^{2+} + \cdots \qquad (3\text{-}10)$$

$$(\cdot S \text{ or } H_nY\cdot) + Co(EDTA)^- \longrightarrow Co^{2+} + CH_2O + CO_2 + \cdots \quad (3\text{-}11)$$

Strongly "reducing" radicals are produced in the photoredox decomposition of EDTA complexes, and primary and secondary radicals very efficiently reduce the M^{III} EDTA substrates. Despite the stability of metallo EDTA complexes, it has not been possible to intercept the $M^{II}(\cdot Y)$ species, quite possibly due to rapid intramolecular fragmentation of the coordinated radical ligand.[93a]

A few of the metallo fragments listed in Table 3-2 have been observed as transient species in flash-photolysis studies. Thus the near-ultraviolet or visible charge transfer to ligand (CTTL) transitions have been used to identify $[Co(N_4)]^{2+}$ species in two cases where (N_4) was a partially unsaturated, tetradentate macrocyclic ligand. In these cases (i.e., for

$Me_6[14]dieneN_4$ = 5,7,7,12,14,14- = hexamethyl-1,4,7,11-tetraazacyclotet-radeca-4,11-diene   or $Me_4[14]tetraeneN_4$ = 2,3,9,10-tetramethyl-1,4,8,-11-tetraazacyclotetradeca-1,3,8,10-tetraene), the Co(II) complexes, $[Co(N_4)]^{2+}$, are low spin[108,109] and axially distorted,[110] and the bulk solution recombination reactions of $[Co(N_4)]^{2+}$ with X_2^- [26] or $\cdot CH_3$[33,94] are nearly diffusion controlled. Although Co(II) complexes of cyclic tetraamines ($[14]aneN_4$ = 1,4,7,11-tetraazacyclotetradecane, or $Me_6[14]aneN_4$ = 5,7,-7,12,14,14-hexamethyl-1,4,7,11-tetraazacyclotetradecane) are not as readily identified directly in dilute solution, they are similar to the imine complexes in lability, in standard reduction potentials of the Co(III)/Co(II) couples,[65,111,112] in the efficiency of $[Co(N_4)]^{2+}/\cdot CH_3$ reactions,[33,94] and in magnetic properties,[108,109] so their general patterns of redox behavior are expected to be similar. In these cases, as in all cases where there are efficient back reactions in bulk solution, the determination of primary photoredox quantum yields is extremely difficult.

Pentacyanocobaltate(II) is sufficiently stable that it can be detected following photoredox reactions of $Co^{III}(CN)_5X$ complexes in neutral or basic solution.[66] This Co(II) species is an extremely powerful reductant,[13] and it is sufficiently long lived ($t_{1/2} \simeq 10$ sec in 0.1 M H^+ [90]) that facile back reactions occur between $\dot{C}o(CN)_5^{3-}$ and primarily radicals (such as $\cdot I$ or $\cdot N_3$) or oxidized products (e.g., I_2).[66,75]

One particularly clear instance of metallo-fragment identification appears to be in the flash-photolysis study of $PtCl_6^{2-}$ by Wright and Laurence.[102] These authors have identified primary product species as $[PtCl_4]^-$ and Cl_2^-, indicating that

$$PtCl_6^{2-} + h\nu \rightarrow [PtCl_4^-]^- + Cl^- + \cdot Cl \qquad (3\text{-}12)$$

$$\cdot Cl + Cl^- \rightleftharpoons Cl_2^- \qquad (3\text{-}13)$$

describe the photoredox process in this case, analogous to the behavior of Rh(III) ammines and $Co^{III}(N_4)X_2$ complexes. The photochemistry of this and related Pt(IV) complexes is discussed in Chapter 5.

In several cases of Rh(III) ammine complexes, CTTM excitation leads to production of oxidized ligand radicals and a Rh(II) species.[75,98,100] The net products of such reactions were always found to be trans-$[Rh^{III}(NH_4)(OH_2)X]$ species, the product yields were pH independent ($1 \leq pH \leq 3$), and Rh(II)-radical back reactions were found to occur at nearly diffusion limited rates. Therefore, it has been inferred that the Rh(II) species must be essentially four coordinate, that is, a tetraammine.[98,100] This inference has been supported by the induced anation of (relatively nonabsorbing amounts of) $[Rh(NH_3)_5(OH_2)]^{3+}$ observed when

I^- is irradiated in acidic solution[98]

$$I^- + h\nu \rightarrow \cdot I + e^- \tag{3-14}$$

$$\cdot I + I^- \rightleftharpoons I_2^- \tag{3-15}$$

$$2I_2^- \rightarrow I_3^- + I^- \tag{3-16}$$

$$e^- + [Rh(NH_3)_5(OH_2)]^{3+} \rightarrow Rh(II) \tag{3-17}$$

$$Rh(II) + I_2^- (\text{or } I_3^-) \rightarrow trans\text{-}[Rh(NH_3)_4(OH_2)I]^{2+} \tag{3-18}$$

$$trans\text{-}[Rh(NH_3)_4(OH_2)I]^{2+} + I^- \rightarrow trans\text{-}[Rh(NH_3)_4I_2]^+ \tag{3-19}$$

and by the rapid ($t_{1/2} < 2 \times 10^{-6}$ sec) aquation of only two ligands in the products of electron capture by $Rh(NH_3)_5X^{2+}$ complexes.[67b]

For several of the entries in Table 3-2, CTTM excitation has been observed to lead to some reaction centered at the ligand rather than, or in many instances along with, the kinds of photoredox behavior discussed above. Thus irradiation of $[Co(NH_3)_5O_2CCO_2H]^{2+}$ leads to the formation of Co^{2+} as well as the formation of a C-bonded formatopentaammine Co(III) intermediate,[44,78] and CTTM excitation of $[Rh(NH_3)_5N_3]^{2+}$ and $[Ir(NH_3)_5N_3]^{2+}$ leads to the predominant formation of nitrene intermediates with no detectable azide radical formation.[66,77,99,101]

A few complexes of labile metals have been included in Table 3-2. A detailed discussion of the photochemistry of labile metal ions may be found in Chapter 8. Particularly worthy of note in the present section are the thorough flash-photolysis studies by Cooper and DeGraff[62,105] of the oxalato complexes of Fe(III). These studies indicate a combination of primary photoredox and photosubstitution processes, complicated by oxalate radical ($C_2O_4^-$ or $[Fe(C_2O_4)]^-$) reduction of the Fe(III) substrate in competition with a radical-radical redox disproportionation reaction. Despite the apparent care with which these studies were executed, more recent work employing both spectrophotometric and electrochemical detection methods[58b] has provided evidence for previously unsuspected photoprocesses in ferrioxalate. Whereas the Cooper–DeGraff work provides strong evidence[62] for the formation of some $C_2O_4^-$ radical, Patterson and Perone[58b] propose a homolytic C—C cleavage of coordinated oxalate. Clearly, oxalato complexes must rank among the most challenging and difficult systems studied. As yet, no one has attempted to reconcile the differences in photoreaction mode observed for oxalato Fe(III)[58b,62,105,114] and oxalato Co(III)[44,79,80,115] complexes.

2. Identification of Species Preceding Bulk Radical Formation.

Identification of the primary radical fragments released into the bulk solution has proved difficult and sometimes ambiguous (see preceding section).

The identification of the precursors of these primary radical species has been far more difficult, and attempted identifications have almost always been found to be equivocal. Very little direct experimental evidence bearing on this fundamental mechanistic problem has been accumulated.

One would like to be able to identify the excited states responsible for the photochemical act. This identification would ideally include the molecular geometry as well as the electronic configurations and distribution. Unfortunately, excited-state lifetimes are extremely short, and direct observation of the relevant excited-state processes has not been successful as of this writing. On the other hand, a great deal can be inferred about these excited states from a careful consideration of the hierarchy of energy states for a given molecule, the nature of the primary process, and possibly from sensitized excitation and excited-state quenching studies. In contrast to the paucity of relevant experimental data, there is an abundance of hypothetical models proposed to describe various aspects of CTTM photochemistry; some of these models will be discussed in Sections IV and V below. Most of the limited available information deals with Co(III) complexes.

There have been several attempts to use chemical scavengers (e.g., alcohols or halides) to intercept intermediate species.[35,100,106,116] However, to the present, these attempts have either failed[100,106] or resulted only in the scavenging of radical species in the bulk of solution.[61,117] Since relatively large scavenger concentrations are usually required in such studies, quantitative interpretations must depend on a precise knowledge of the variations of primary quantum yields with variations in solvent compositions.

There have been a number of successful intermolecular, sensitized redox decompositions of Co(III) and other substrates.[1,3,92,93,118–123] The most interesting of these have employed donor molecules, D, with relatively long-lived, emitting triplet states. In the presence of Co(III) substrates one may observe the concomitant quenching of the donor emission and the formation of Co^{2+}. The inference often drawn from such studies is that the Co(II) products are formed in a triplet-to-triplet energy-transfer process (see also Chapter 1) and that the charge-transfer triplet state of the Co(III) complex is the photoactive state in the photosensitized and directly excited redox decompositions. This mechanism for the sensitization process is represented in general terms by Eqs. (3-20) to (3-25), and the corresponding direct excitation processes are described by Eqs. (3-26) to (3-28) followed by (3-24) and (3-25).

$$^1D_0 + h\nu \rightarrow {}^1D_1 \tag{3-20}$$

$$^1D_1 \rightarrow {}^1D_0 + \text{heat} \tag{3-21}$$

$$^1D_1 \rightarrow {}^3D_1 \tag{3-22}$$

$$^3D_1 + Co^{III}L_5(X^-)(^1A_1) \rightarrow {}^1D_0 + CoL_5(X)(^3CT) \tag{3-23}$$

$$CoL_5(X)(^3CT) \rightarrow Co^{2+} + 5L + \cdot X \tag{3-24}$$

$$CoL_5(X)(^3CT) \rightarrow Co^{III}L_5(X^-)(^1A_1) + heat \tag{3-25}$$

$$Co^{III}L_5(X^-)(^1A_1) + h\nu \rightarrow CoL_5X(^1CT) \tag{3-26}$$

$$CoL_5X(^1CT) \rightarrow Co^{III}L_5(X^-)(^1A_1) + heat \tag{3-27}$$

$$CoL_5X(^1CT) \rightarrow CoL_5X(^3CT) \tag{3-28}$$

Unfortunately, these are very complex systems to investigate, and the literature on the subject contains many errors. For example, in studies using biacetyl as a sensitizer, the "sensitization" reaction was found to proceed by means of a chemical mechanism[119,120] rather than triplet-to-triplet energy transfer as originally claimed[118]; the chemical reactivity of the triplet excited states of ketones is well documented and reviewed elsewhere.[122]

Recently, Gafney and Adamson[123] have hypothesized that the $Ru(bipy)_3^{2+}$-sensitized redox decompositions of Co(III) complexes may occur by means of an electron-transfer reaction from the charge-transfer triplet excited state of $Ru(bipy)_3^{2+}$; thus these authors would visualize the sensitization process as one in which reactions corresponding to Eqs. (3-20) to (3-22) are followed by (3-29). Such an electron transfer process, possibly complicated by electronic energy transfer, has been found to result from the $Ru(NH_3)_6^{3+}$ quenching of the CTTL excited state of $Ru(bipy)_3^{2+}$,[124] and the one-to-one stoichiometric correspondence of the quenching of $Ru(bipy)_3^{2+}(^3CT)$ and the generation of $Ru(bipy)_3^{3+}$ has been cited as evidence for similar electron-transfer quenching processes in $Ru(NH_3)_6^{3+}$ and $Co^{III}(NH_3)_5X$ ($X = NH_3$, H_2O, Cl^-, Br^-) acceptors.[125] The excited-state electron transfer process can, in principle, be readily distinguished from Eqs. (3-20) to (3-25), since free radicals are formed in reaction (3-24), while $Ru(bipy)_3^{3+}$, but no free radical, is a primary product of

$$Ru(bipy)_3^{2+}(^3CT) + Co^{III}L_5X \rightarrow Ru(bipy)_3^{3+} + Co(II) + \cdots \tag{3-29}$$

Furthermore, $Ru(bipy)_3^{3+}$ is efficiently oxidized by several likely primary radicals,[86,93a] and the SRP for the $Ru(bipy)_3^{3+,2+}$ couple is 1.26 V,[126] compared to the many larger values in column 4 of Table 3-2. Studies in 50% 2-propanol have suggested the formation of radicals as primary products in the $Ru(bipy)_3^{2+}$-sensitized redox decomposition of $Co(NH_3)_5Br^{2+}$, $Co(EDTA)^-$, and $[Co(HEDTA)X]^-$ ($X = Cl$, Br, NO_2).[93]

Evidence has also been cited for electron-transfer quenching of

$Ru(bipy)_3^{2+}$ (^3CT) by $Co(C_2O_4)_3^{3-}$,[127] Fe^{3+},[124] *trans*-1,2,-bis(N-methyl-4-pyridyl)ethylene[124] and 1,1'-dimethyl-4,4'-bipyridine.[124]

Crosby and co-workers[128] have introduced yet another complication into interpretation of sensitization studies using Ru(II) donors. These workers claim that $Ru(bipy)_3^{2+}$ excited states do not have well-defined singlet or triplet character and so cannot be regarded as well-defined triplet donors. Although the $Ru(bipy)_3^{2+}$ sensitization of well-defined triplet-state reactions has been observed,[129] if the donor spin state is not well defined (i.e., if spin is not a "good quantum number") then quenching of the donor excited states cannot be assumed to result in population of acceptor excited states of triplet spin multiplicity.

Unfortunately, the biphenyl[119] and the quinoline[120] sensitized redox decompositions of $Co(NH_3)_6^{3+}$ have not been examined critically for contributions of electron transfer steps analogous to (3-29). These reactions are qualitatively similar to the $Ru(bipy)_3^{2+}$-sensitized reactions discussed above, but were carried out in alcohol-containing media and may be complicated by reactions of the strongly reducing $R\dot{C}OH$ radicals. For example, the biphenyl sensitization[119] was carried out in ethanol, so if the mechanism were triplet-to-triplet sensitization, the limiting yield could be as low as 0.4 when corrected for $CH_3\dot{C}OH/Co(NH_3)_6^{3+}$ reactions: some of the medium dependences observed by Gafney and Adamson[120] could arise from the competition between normal decay modes of the primary radicals and the scavenging of radicals by alcohols. The lower energy triplet donors, $Ru(bipy)_3^{2+}$ and biacetyl, do not sensitize the redox decomposition of $Co(NH_3)_6^{3+}$, although electron-transfer quenching of $Ru(bipy)_3^{2+}$ (^3CT) by $Co(NH_3)_6^{3+}$ has been reported.[125]

Finally, one may question whether the approach of using triplet (or other excited state) sensitizers can even in principle be a mechanistically useful probe of the nature and reactivity of CTTM excited states. To the extent that solvent polarization contributes to the binding energy of CTTM excited states, the CTTM excited state produced in any process such as (3-23) must be different from the corresponding state produced on direct excitation since the solvation of initial collision (or "precursor") complex, $\{D_1, {}^1A_1\}$, must differ from solvation of the acceptor ground state, $^1A_1$. Thus, within the context of models discussed in Sections IV and V below, one could be justified in arguing that the excited-state electronic energy manifolds differ in the successor complex, $\{^1D_0, {}^*A\}$, for energy transfer and in the photoactive CTTM state generated following direct excitation of the acceptor. Such differences in their electronic manifolds would be expected to produce different values for the rate constants describing excited-state reactions, relaxation, and so forth. On the basis of such considerations, one might expect that quantum yields obtained from sensitized and direct excitation studies should be different, and one would expect the limiting sensitized yields to vary strikingly with the interactions between donor and solvent. At the present time, the mechanistic details of neither the direct nor sensitized photoredox reactions are well enough understood for one to make confident predictions of relative quantum yields.

The preceding paragraphs summarize attempts to explore the possibility of the intermediacy of reactive excited states in photoredox reactions. Regardless of the identity of the excited state responsible for the photoredox reaction pathway, the primary redox products must be a metallo fragment and an oxidized ligand radical formed in close proximity to one another. Such "radical pairs" may either recombine to regenerate substrate species (sometimes with change of ligation) or diffuse apart to form the observed redox products. It seems evident that radical pair species should be regarded as the primary products of photoredox reactions, and that the quantum yields for formation of radical pairs should be regarded as the mechanistically significant photochemical information. Unfortunately, the yields of radicals or reduced-metal species in bulk solution are not reliable indices of the primary yields of radical pairs. Until very recenly, few investigators had systematically sought out evidence for the intermediacy and the chemical behavior of radical pair species. The importance of radical-pair recombination reactions has now been elegantly established by Scandola.[130] In this further investigation[87,88] of the photochemistry of $[Co(NH_3)_5NO_2]^{2+}$, Scandola and co-workers found that the yields of the Co^{2+} and $[Co(NH_3)_5ONO]^{2+}$ products are inversely related, the former decreasing and the latter increasing, through increases in the medium viscosity. Although the local viscosity near the complex ion is unknown in solutions of mixed solvents (Scandola used water/glycerol mixtures), the qualitative dependence of yields on viscosity and the inverse coupling of diffusive separation (to form NO_2 and Co^{2+}) and cage recombination (to form $[Co(NH_3)_5ONO]^{2+}$) processes are as one would expect for a system exhibiting significant radical-pair recombination.[60,131]

Although aspects of the photochemistry of $[Co(NH_3)_5NO_2]^{2+}$ approach the behavior predicted from a limiting radical-pair model (Section V.A.1 below), the other complexes examined to date behave quite differently. For example, for $Co(NH_3)_5Cl^{2+}$, $Co(NH_3)_5Br^{2+}$, and $Co(NH_3)_5NCS^{2+}$, the photoredox quantum yields are strongly wavelength dependent and the qualitative kind of wavelength dependence is very different in aqueous, in glycerol/water, and in phosphoric acid/water media.[132] In contrast, $\phi_{Co^{2+}}$ is nearly independent of wavelength for CTTM irradiations (300 nm $\geq \lambda \geq$ 254 nm) of $[Co(NH_3)_5N_3]^{2+}$, but the nearly constant values of $\phi_{Co^{2+}}$

obtained from irradiations over this wavelength range do decrease systematically, and the quantum yields of ammonia aquation do increase systematically with increasing viscosity.[66b,132a] This behavior of the azido complex suggests that the excited-state precursors to the formation of radical pairs may have been thermally equilibrated. The total (350 nm $\geq$ $\lambda \geq 214$ mm) medium dependence of the yields from $Co(NH_3)_5X^{2+}$ (X = Cl, Br, N_3, NCS) complexes cannot be accounted for only in terms of a competition between the diffusive separation and recombination reactions of radical pairs.

Investigations of the medium dependencies of photoredox processes are still evolving as this chapter is being written. Although this approach is providing much new insight into photoredox processes, many new problems have been raised. Since these current investigations have been designed specifically to investigate proposed photoredox models, they will be discussed further in Section V below.

The studies summarized above represent what seem to be the most promising current approaches to obtaining information about the nature of photochemically significant intermediates. Although each of these approaches has serious limitations, as noted, the more numerous attempts to deduce the nature of these intermediates from the nature of the excited state populated by initial absorption of radiation or from the variations of quantum yields with wavelength have been far less instructive and are intrinsically much more equivocal. Aspects of these approaches play a role in the discussion below, but there are additional points which should be mentioned here. For several types of complexes, especially those of Co(III) and Rh(III), CTTM excitation very often results in a combination of redox and ligand substitution reactions. Some, but not all, of the ammonia aquation produced on CTTM excitation of $[Co(NH_3)_5N_3]^{2+}$ and apparently all the linkage isomerization from irradiations of $[Co(NH_3)_5NO_2]^{2+}$ is the result of reactions involving the primary, solvent-cage trapped, redox products. In the remaining cases, the photoredox and photoaquation products may or may not have common precursors. For the example of $[Co(NH_3)_5N_3]^{2+}$, the fraction (about 50%) of ammonia aquation following CTTM excitation, but that does not appear to be coupled to Co^{2+} formation, may result alternatively from (1) a reaction of the CTTM excited state before it decomposes into radical pairs or (2) from a reaction of a ligand-field excited state populated by electronic relaxation of the CTTM excited state. In either case, the communication between charge-transfer and ligand-field excited states would appear to be inefficient; that is, in neither case could one claim rapid electronic relaxation to the lowest energy state of a given spin multiplicity and that the photochemical products were obtained only from that lowest energy excited state. For most of the Co(III) ammine and EDTA complexes studied to date, the observed photoaquation processes apppear to reach a maximum value in the near-ultraviolet region of observed overlap of the CTTM and ligand-field absorption bands.

3. *Alteration of Product Yields Due to Radical Substrate Reactions.* Radical species formed in redox decompositions of transition-metal complexes are most often powerful oxidants and/or powerful reductants (Table 3-2). As a result, radical-substrate or radical-product reactions occur in many systems and frequently complicate the net product stoichiometries. Neglect of such reactions has often led to erroneous mechanistic inferences. However, the detection and evaluation of such reactions has generally proved difficult. A few examples are discussed here to illustrate the range of reactions that have now been established; these and other specific cases that have been investigated are noted in Column 6 of Table 3-3.

In one of the earliest simple cases of radical-substrate reactions, Haim and Taube[133] postulated an efficient iodine atom reduction of $Co(NH_3)_5I^{2+}$. It later developed that the variation of quantum yield with radiation intensity was more consistent with an I_2^- reduction of $Co(NH_3)_5I^{2+}$ with a rate constant of the order of $10^5 M^{-1} sec^{-1}$.[26,134] Since this reaction proceeds with a much larger overall free-energy change than the Br_2^- reduction of $Co(NH_3)_5Br^{2+}$, the latter would be expected to have a rate constant of less than $10^2 M^{-1} sec^{-1}$.[26,135] Radical reductions of the substrate have been demonstrated to be very efficient for the oxalatoferrates,[62,58b,105,114] $[Fe(HEDTA)OH_2]$,[93] $Co(EDTA)^-$,[92,93] and $[Co(HEDTA)X]^-$ $(X = Cl, Br, NO_2)$.[92,93] The $C_2O_4^-$ reduction of monooxalato Co(III) complexes has been demonstrated, but, in some instances, it occurs in competition with redox disproportionation of the oxalate radical anion.[44,80] The formation of a metastable Co(III) intermediate complicates these systems further. Although oxalate radical reduction of $Co(C_2O_4)_3^{3-}$ is very efficient,[136] as has been traditionally assumed,[137,138] the photoreduction of this complex has been found to proceed through a long-lived (lifetime $\sim 10^{-2}$ sec), metastable intermediate that decays relatively slowly into Co^{2+} and some oxalate radical.[115]

Radical-metallo-fragment recombination reactions are important in cases where reactive, reasonably stable metallo fragments are formed. For example, the macrocyclic ligand complexes of Co(II), $[Co(N_4)]^{2+}$ $(N_4 = Me_6[14]aneN_4, Me_6[14]dieneN_4, Me_4[14]tetraeneN_4,$ etc.), mentioned above and in Table 3-3, have been independently prepared and characterized[26,65,109–112,139] and are very stable to substitution of the macrocycle (N_4) in acidic aqueous solution. In these cases, the $[Co(N_4)(OH_2)_2]^{3+,2+}$ SRPs are between 0.53 V and 0.59 V,[65,112,139] and the $[Co(N_4)]^{2+}$ species have been observed to be efficiently and rapidly oxidized by $\cdot CH_3$[93] and X_2^- (X = Cl, Br, I).[26] Although $Co(CN)_5^{3-}$ is a much stronger reducing agent[13] than $[Co(N_4)]^{2+}$, metallo-fragment-radical recombination reactions have been less well documented for the cyano systems. Very recent work[66,75,77] has documented the efficiency of the recombination reactions of several radicals with $Co(CN)_5^{3-}$; the efficiency of these recombination reactions is probably one reason the photoredox chemistry of pentacyanocobaltates has been neglected for so long.

In contrast, the monomeric Rh(II) species are not at all well characterized, well known, or even very stable. In a limited sense, these species might well be regarded a radicals themselves, and recombination with other radical species seems generally very rapid.[98,100]

A few additional examples of radical-substrate or radical-product reactions are noted in Table 3-3. The possibility of such reactions has been neglected in much of the literature on charge-transfer photochemistry, and mechanistic hypotheses have often been based on final-product yields and ratios. Clearly, the reactions discussed in this section alter the product yields: the net yields of reduced metallo species are generally increased by radical attack on the substrate, while the net yields of reduced metallo species are decreased (often to the vanishing point, as for $Co(CN)_5I^{3-}$ and $Rh(NH_3)_5I^{2+}$) by radical-metallo-fragment recombination reactions. Net product ratios can be expected to vary even more wildly, depending, as they often do, on competitive scavenging for radical species. The potential radical reactions in each system investigated should be considered with care; a careful consideration of the energetics of likely radical reactions, using data such as that in Table 3-2, should help systematize the search.

B. Classification of Reactions Observed to Follow CTTM Excitation

Many different reactions have now been found to result from CTTM excitation of coordination complexes. If we restrict discussion to complexes of the type, $M^{III}L_5X$, then these reactions may be systematically classified as below; known examples are cited (in parentheses) where possible.

Class 1 photoredox reactions: oxidation of X, coupled with
  a. no aquation reactions ($Co(NH_3)_5Br^{2+}$ irradiated at 254 nm).[73,132]
  b. aquation of X ($Co(NH_3)_5Br^{2+}$ irradiated at 360 nm).[71,72,132]
  c. aquation of L ($[Co(NH_3)_5N_3]^{2+}$).[66b,76]

Class 2 photoredox reactions: oxidation of L, coupled with
  a. no aquation reactions.

b. aquation of L.

c. aquation of X ([Co(HEDTA)Br]$^-$).[92]

Class 3 photoredox reactions: ligand-centered reactions

a. heterolytic cleavage of intraligand bonds in X ([Rh(NH$_3$)$_5$-N$_3$]$^{2+}$,[66,77,99] [Co(en)$_2$C$_2$O$_4$]$^+$ [44,80]).

b. Redox and linkage isomerization ([Co(NH$_3$)$_5$NO$_2$]$^{2+}$).[87,88]

Even from the citation of characteristic examples, it is clear that a variety of behavior has been observed among Co(II) complexes. For many coordination complexes, the class of photochemical behavior observed depends on the absorption feature that is irradiated. In fact, this would be the expected result of excitation of different, poorly communicating CTTM absorption bands; for example, irradiation of an absorption feature which is largely X → M(III) in character might well lead to different products than irradiation of an absorption feature that is largely L → M(III) in character. This possibility has not been explored to the extent that generalizations are warranted as of this writing. There is one class of compounds in which the variation of reaction class with CTTM absorption feature irradiated has been examined systematically: the [Co(HEDTA)X]$^-$ (X = Cl, Br, NO$_2$) complexes. In these investigations it was found that neither the quantum yield nor the reaction class was very sensitive to changes in the ligand, X, and that the reaction class was independent of the kind of absorption band irradiated.[92]

An extensive classification of photoredox reactions using the descriptive scheme outlined above seems inappropriate here; however, a few comments on some specific examples are in order.

Class 1a. This seems to be the most characteristic, but not exclusive, reaction mode for acidopentamminecobalt(III) complexes irradiated in high-energy CTTM absorption bands. However, it should be recognized that minor aquation pathways, particularly of NH$_3$, are very difficult to detect in these systems. For example, the small yield for ammonia aquation from the 365 nm irradiation of [Co(NH$_3$)$_5$O$_2$CCH$_3$]$^{2+}$ would necessarily have gone undetected in the 254 nm irradiations of this complex[61]; however, the ratio of redox and aquation processes for this complex does change dramatically.

Class 1b. This has long been regarded to be the type of behavior most characteristic of Co(III) complexes.[1–6,71,72] As it turns out, there are relatively few Co(III) complexes that fall unequivocally into this class. For example, Co(NH$_3$)$_5$Cl^{2+} and [Co(NH$_3$)$_5$NO$_2$]$^{2+}$ were originally placed in this class; however, recent studies have shown that at least some ammonia ligands are oxidized in the primary photoprocess in the former case[35] and that linkage isomerization, but no detectable NO$_2^-$ aquation, occurs in the latter.[87,88] A unique and interesting example of this class of behavior is that of $[Co(NH_3)_4CO_3]^+$, in which CTTM irradiation produces a combination of photoredox (forming Co^{2+} and CO_3^-)[81] and carbonate-ring opening (forming $[Co(NH_3)_4(OH_2)CO_3]^+$).[82].

Classes 1c and 2c. Work in recent years has demonstrated that the ligand oxidized in the photoredox process is not always the ligand photoaquated. The contrary has been demonstrated to be the case for excitations in several absorption bands of $Rh(NH_3)_5I^{2+}$ [98] and $[Co(HED-TA)X]^-$ (X = Br, Cl).[92,93] The difficulty in obtaining reliable yields for photoaquation of ammonia in the presence of substantial photoredox of $Co^{III}(NH_3)_5X$ complexes is probably one reason this class of behavior has not been more widely recognized. For $[Co(NH_3)_5N_3]^{2+}$, the quantum efficiency of ammonia aquation is high relative to photoredox, and it has been possible to achieve cation-exchange separations of NH_4^+, Co^{2+}, and $[Co(NH_3)_4(OH_2)N_3]^{2+}$ from unphotolyzed $[Co(NH_3)_5N_3]^{2+}$.[66]

Class 3a. This is a relatively recently discovered photoreaction class that is best exemplified in the behavior of $[Ir(NH_3)_5N_3]^{2+}$ [99,101] and $[Rh(NH_3)_5N_3]^{2+}$.[66,77,99] Several oxalato complexes of Co(III) also appear to fall within this class. Thus ultraviolet excitation of monooxalato ammine (or amine, monodentate or bidentate oxalate) complexes of Co(III) leads to photoredox (forming $C_2O_4^-$) and heterolytic cleavage of the oxalate ligand (forming $CO_2 + Co^{III}$—CO_2H^-).[44,80]

Class 3b. This type of behavior has been established to result from excitation of a variety of absorption bands in $Co^{III}(L_5)NO_2$ (L = NH_3 or HEDTA/5) complexes.[87,88,92] It has also been found to occur for

$$[Co(NH_3)_5O\overset{\overset{\displaystyle O}{\|}}{C}H]^{2+}\ [84]$$

(note that HCO_2^- and NO_2^- are isoelectronic) and $[Rh(NH_3)_5NCS]^{2+}$.

C.   Reaction Modes Characteristic of Irradiation of Other Types of Charge-Transfer Absorption Bands

There have been some studies of the photochemical processes following irradiation of charge transfer to ligand (CTTL) absorption bands and many studies of charge transfer to solvent excitations; the former studies have been mostly confined to heavy metals and are reviewed in Chapter 5, while the latter are discussed in Chapter 8.

For some complexes discussed in the present chapter there is considerable ambiguity about the assignment of their ultraviolet absorption bands. Notable examples may be found among the acidopentacyano-cobaltate(III) complexes, in which the deep ultraviolet absorption bands

($\lambda < 225$ nm) have been assigned as (CTTL)[Co(III)$\rightarrow$CN$^-$],[140] whereas the photochemistry induced by irradiation of these bands produces cyanide oxidation products,[66b] as might be expected of a state of the CTTM[CN$^-$ $\rightarrow$ Co(III)] type. Since these are anionic complexes they might be expected to exhibit CTTS behavior at very high energy; however, there is no evidence that this is a significant process.[66b]

The photochemical behavior of ion-pair or "outer-sphere" complexes has been relatively neglected. This is unfortunate since such species could usefully "model" some aspects of CTTM photochemistry. For example, the (Ru(NH$_3$)$_6^{3+}$, X$^-$) ion pairs (X = I, Br, etc.) exhibit distinct CTTM absorption bands,[141] and flash photolysis of these species produces Ru(II) and halogen radical species.[106] Qualitatively similar behavior has been found for (Co(NH$_3$)$_6^{3+}$, X$^-$) ion-pair species.[26,73] Certain features of such systems are discussed further in Sections IV and V below.

Electronic transitions localized within unsaturated ligands are not charge-transfer transitions, but they do frequently occur at energies overlapping appreciably with CTTM and other molecular transitions. Such transitions are often suspected to be present only because the uncoordinated ligands absorb in the ultraviolet region; for example, ligand-centered transitions have not been identified in [Co(en)$_2$C$_2$O$_4$]$^+$, but some aspects of the photochemistry have been attributed to ligand-centered excited states.[44] Somewhat less equivocal were studies of stilbenecarboxalatopentaamminecobalt(III), in which it was found that excitation of ligand-centered transitions resulted in efficient transfer of excitation energy from the ligand-centered excited state to a CTTM state of the complex.[142] In contrast, irradiation of the 305 nm, possibly ligand-centered,[143] absorption band of [Co(NH$_3$)$_5$NCS]$^{2+}$ was found to result in a small (0.028), temperature-dependent, wavelength-independent quantum yield of Co^{2+}; the apparent activation enthalpy for Co^{2+} formation throughout the 305-nm band was 9 kJ/mole, in contrast to the temperature-independent Co^{2+} yield obtained for excitation of higher energy CTTM bands in the same complex.[75] The observations on [Co(NH$_3$)$_5$NCS]$^{2+}$ have been interpreted as evidence for vibrational equilibration within the excited state associated with the 305-nm band, and inefficient transfer of excitation energy into the photoactive CTTM state.[75] It seems evident that the photochemistry of complexes containing ligand-centered excited states deserves more careful study.

IV. APPROACHES TO MECHANISTIC CHARGE-TRANSFER PHOTOCHEMISTRY

The intent of the present section is to sketch some approaches to designing experiments and to the organization of observations that the author feels are useful in evolving general patterns of photochemical behavior. This section has some mechanistic biases; some alternative mechanistic models are reviewed critically in Section V below.

A. Absolute Quantum Yields

There has been some tendency in the past to draw mechanistic inferences from variations in values of absolute or relative quantum yields for different compounds.[1-6,71-73,144] Of course, such comparisons must be made with great caution since quantum yields are actually complex ratios of rate constants, and since variations in quantum yield cannot be assumed to be directly proportional to variations in the rate constant for the process of interest (e.g., k_p in Fig. 3-1). An example of a d^6 metal complex of the type, ML_6, one of the simplest energy-level diagrams pertinent to a redox process derived from a charge-transfer excited state, is illustrated in Fig. 3-1. Given the possibility of population of a photoactive state of triplet-spin multiplicity from either a ligand-field singlet or charge-transfer singlet excited state, the overall quantum yield for redox products (following CTTM excitation) would be given by

$$\phi_p = \frac{k_p}{k_p + k_4}\left\{\frac{k_1}{k_1+k_2+k_3} + \left(\frac{k_5}{k_5+k_6+k_7}\right)\left(\frac{k_3}{k_1+k_2+k_3}\right)\right\} \qquad (3\text{-}30)$$

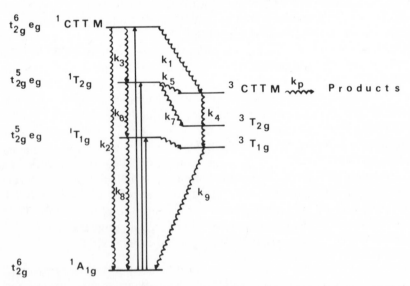

Fig. 3-1. Hypothetical energy-level scheme for a d^6 complex of O_h symmetry. Radiative processes have been omitted, and product formation has been assumed to occur from a state of triplet spin multiplicity.

A simpler expression, $\phi'_p = k'_p/(k'_p + k_1 + k_2 + k_3)$, would be obtained if the photoproducts were derived only from the CTTM singlet state. In this latter instance, an obvious limiting case is that in which $k'_p \gg (k_1 + k_2 + k_3)$ and $\phi'_p \rightarrow 1$; this corresponds to an important limiting case in which excited states play no significant chemical role in determining photochemical products or yields;[1-4,6,71,72,145] this limit is discussed in Section V.A below. In the absence of some demonstration that the various nonreactive, nonradiative electronic relaxation modes are either the same or unimportant in each of the molecules being compared, there is no reason to believe that variations in values of absolute quantum yields can be used as the major (or only) mechanistic criterion. The situation is further complicated by a tendency to compare quantum yields at a single wavelength[71,72,146] rather than quantum yields from similar absorption bands or, better yet, quantum-yield profiles as a function of wavelength. As noted above, few coordination complexes have been irradiated in all of their ultraviolet absorption bands.

Despite the pitfalls, cautious comparisons of quantum yields through homologous series of compounds can be quite instructive. The most useful quantum-yield comparisons, for purposes here, would be comparisons of the variation of quantum yields as a function of wavelength through all the charge-transfer absorption bands of interest. Such a comparison of Co(II) quantum yields for $Co^{III}(NH_3)_5X$, $[Co(HEDTA)X]^-$, and $Co(EDTA)^-$ complexes is shown in Fig. 3-2. For the acidopentaamminecobalt(III) complexes, the quantum-yield profile depends profoundly on the ligand, X, while for the $[Co(HEDTA)X]^-$ analogs, $\phi_{Co^{2+}}$ is nearly independent of X. This contrast in behavior may be associated with the fact that the pentaammines generally exhibit class 1a or class 1b behavior, while the EDTA complexes all fall into class 2c.

The $[Co(HEDTA)X]^-$ (X = Cl, Br, NO$_2$) complexes are a very instructive series of closely related compounds. Although the quantum yields for formation of Co(II) are very nearly independent of X in the $[Co(HEDTA)X]^-$ complexes (for excitation at $\lambda \leq 229$ nm), the quantum yields are much larger for these complexes than for $Co(EDTA)^-$. That limiting quantum yields ($\phi^l_{Co^{2+}} = 0.14$ and 0.025, respectively) are obtained for short-wavelength irradiation of $[Co(HEDTA)Cl]^-$ and $Co(EDTA)^-$, may indicate a rapid thermal equilibration among the high-energy CTTM excited states of singlet-spin multiplicity, which is followed by a competition between the formation of products (or the photoactive state) and the nonradiative relaxation to unreactive lower energy states. It is to be observed that the differences in limiting quantum yields of $[Co(HEDTA)Cl]^-$ and $Co(EDTA)^-$ may only indicate a relatively small difference in one of the rate constants (i.e., k_p, k_1, k_4, or $k_2 + k_3$, see Fig. 3-1) for reaction

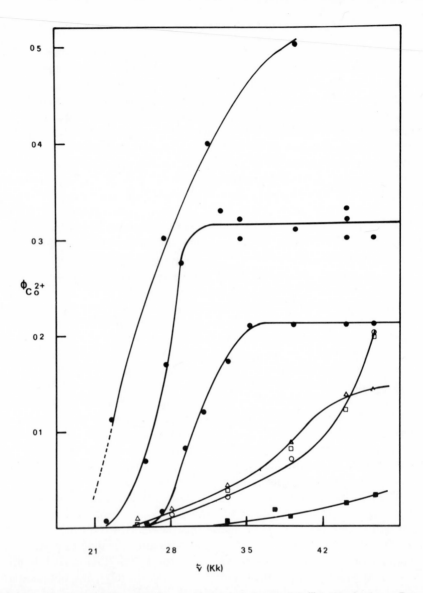

Fig. 3-2. Quantum-yield profiles for photoreduction of several $Co^{III}(L_5)X$ complexes. Data are taken from Refs. 1, 92, 93b, and 132. Acidopentammine complexes (solid circles) are: upper curve, $[Co(NH_3)_5NO_2]^{2+}$; middle curve, $[Co(NH_3)_5Br]^{2+}$; lower curve, $[Co(NH_3)_5Cl]^{2+}$. Data for EDTA complexes: $[Co(HEDTA)Cl]^-$ (open triangles); $[Co(HEDTA)NO_2]^-$ (open squares); $[Co(HEDTA)Br]^-$ (open circles); $[Co(EDTA)]^-$ (solid squares).[132b] (Reproduced from ref. 132b with permission of the copyright holder.)

or relaxation of the excited states of these complexes.

The photoredox behavior of Co(EDTA)⁻ serves as an example of the expected quantum-yield profile for a system with a single photoactive excited state (Fig. 3-3).[93b] There is a reasonably definite threshold energy, $E_{th} \simeq 27$ kK, below which $\phi_{Co^{2+}}$ is very small ($<10^{-3}$). This threshold energy corresponds roughly to the energy threshold for absorption in the

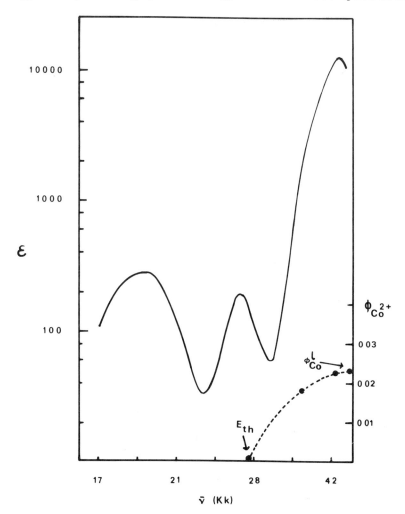

Fig. 3-3. Absorption spectrum (solid line) and redox quantum yields (dashed line) for [Co(EDTA)]⁻. For irradiations at 380 nm, $\phi_{Co^{2+}} = (3.0 \pm 0.3) \times 10^{-4}$. Approximate values for the threshold energy for the appearance of photoredox behavior, E_{th}, and the high energy-limiting quantum yield, $\phi^{l}_{Co^{2+}}$, are indicated by arrows.[93b]

spectroscopic CTTM band. Finally, $\phi_{Co^{2+}}$ reaches a limiting value, much less than unity ($\phi^l_{Co^{2+}} \simeq 0.025$) at the higher excitation energies within the CTTM band; this limiting yield implies rapid vibrational relaxation within the spectroscopic excited state, followed by the several competitive processes characteristic of the equilibrated excited state (e.g., electronic relaxation, intersystem crossing, chemical reaction, etc.). The quantum-yield profiles for $Co(NH_3)_5Cl^{2+}$ and $Co(NH_3)_5Br^{2+}$ are qualitatively similar, with E_{th} being smaller and $\phi^l_{Co^{2+}}$ being larger for the bromo complex than for the chloro complex.

The similarity of yields for $[Co(HEDTA)X]^-$ complexes, independent of absorption spectrum or X over the excitation range, $300 \, nm \geq \lambda \geq 229 \, nm$, suggests a common photoactive excited state and rapid thermal equilibration among excited states populated by absorption of radiation in this region (Fig. 3-2). This similarity of yields stands in marked contrast to the analogous $Co(NH_3)_5X^{2+}$ complexes in which the product yields from CTTM excitation are strongly dependent on X. This contrasting behavior of the HEDTA and the pentammine complexes suggests that the kinetic parameters for the state (or species), giving rise to the redox reaction mode, depend very strongly on the nature of the ligand photooxidized; that other factors can also be of major significance, however, is indicated by the large differences in photoredox yields between the $[Co(HED-TA)X]^-$ complexes and $Co(EDTA)^-$. It appears that the high-energy limiting yields do depend on X. This may indicate any or all of the following: (1) that thermal equilibration may occur in competition with other processes (e.g., intersystem crossing) in some of the complexes; (2) that the binding energy of the photoactive excited state may vary with X; or (3) that there is some spectroscopic complexity in one of the higher energy states. In terms of the primitive excited-state model implied by Fig. 3-1, and assuming very poor communication between CT and LF excited states ($k_p > k_4$, $k_1 + k_2 > k_3$), then for CTTM excitation of Co(III) complexes, Eq. (3-30) becomes

$$\phi_p \simeq \frac{k_1}{k_1 + k_2} \tag{3-30'}$$

and the variations in yield for $Co(NH_3)_5X^{2+}$ may be attributed to variation in the rates of intersystem crossing with variations in X, perhaps through spin-orbit coupling.

An interesting contrast in behavior occurs with $[Fe(HEDTA)OH_2]$. For this complex the redox yield goes through a maximum approximately parallel to the absorption profile[93b] (Fig. 3-4). It was suggested that this maximum in the yield is the result of a high-energy spectroscopic transition that populates largely a carbonyl-centered excited state, and

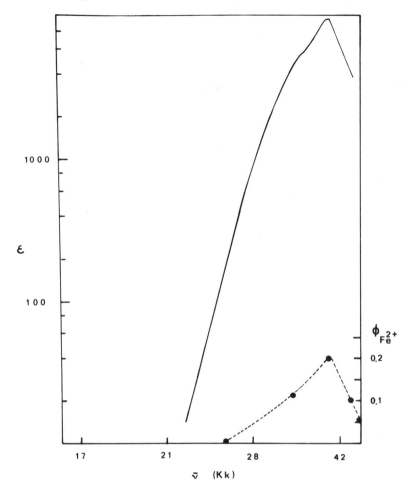

Fig. 3-4. Absorption spectrum (solid line) and redox quantum yields (dashed line) for [Fe(HEDTA)OH$_2$]. For irradiation at 405 nm, $\phi_{Fe^{2+}} = (1.0 \pm 0.2) \times 10^{-2}$.[93b]

this state efficiently intersystem crosses to a lower energy state with the carbonyl having formally triplet-spin multiplicity. Thus, if the photochemistry derives from a spectroscopic state in [Fe(HEDTA)OH$_2$] and from nonspectroscopic states in the cobalt analogs, then the contrasting behavior at very short excitation wavelengths can be associated with the efficiency of "triplet"-sensitized decomposition of the latter, but not the former (see also Ref. 142). In any case, this behavior implies some lack of communication between high-energy excited states in the iron, similar perhaps to some cobalt complexes, but in contrast to [Co(HEDTA)Cl]⁻ and Co(EDTA)⁻.

Clearly, the analysis of quantum-yield profiles such as in Figs. (3-2) to (3-4) is instructive. Unfortunately, there have not been a sufficient number of such studies performed at this writing to permit an entirely satisfying analysis of patterns of reactivity. It is to be hoped that future studies will explore variations in photochemical behavior at least in all the principal CTTM absorption bands. Eventually, one would like to make comparisons of homologous series of compounds along the lines of Fig. 3-2, perhaps determining the dependence of the threshold energies and high-energy limiting yields on structural parameters of the coordination-complex substrates.

B. Communication Between Excited States

The related questions: (1) whether the population of two excited states of comparable energies in a large collection of molecules do approach an equilibrium distribution, and (2) whether the relaxation of high-energy excited states in a molecule always results in the transient population of all lower-energy excited states of the same spin multiplicity, are of major mechanistic concern. Aspects of these questions are essential components of the preceding and following discussions, so the present section is limited to general features and a few illustrative examples.

The focus in the present section is mechanistic photochemistry; however, the questions posed are much more general, with spectroscopic as well as photochemical implications. Spectroscopic aspects of the problems of communication between excited states are discussed in Chapter 1.

1. Communication Between Charge-Transfer Excited States. Of the few Co(III) complexes that have been irradiated through their several charge-transfer bands, most indicate a single characteristic reaction mode and possibly a single photoactive state. Exceptions have been found for $[Co(CN)_5N_3]^{3-}$ and $Co(CN)_5I^{3-}$, where excitation of low-energy CTTM bands produces $[Co(CN)_5]^{3-}$ and $\cdot X$, while higher energy excitations result in several processes among which are oxidation of cyanide.[66,75] If the observed photochemical reactivity were characteristic of a single, probably lowest energy state, then one would expect a single photochemical process for all CTTM excitations, and one would expect the yield for this process to reach its limiting value for relatively low-energy CTTM excitations (see Section IV.A). It is not yet clear whether this will prove to be the prevailing pattern.

For most of the $Rh^{III}(NH_3)_5X$ complexes that have been examined, it appears that a single CTTM state is responsible for the redox photoreactivity. In general, such states appear to be efficiently populated by irradiation of absorption bands well above their threshold.[98,100] A notable exception is $[Rh(NH_3)_5N_3]^{2+}$. For this complex, the most important redox process—the primary formation of $[Rh(NH_3)_5N]^{2+}$ and N_2—occurs over a relatively narrow range of excitation energies.[66] Such behavior indicates that relaxation of higher energy charge-transfer states in this complex can occur without population of the state responsible for nitrene formation; in fact, the quantum-yield profile[66] indicates that the nitrene intermediate is produced from an exceptionally isolated excited state.

2. *Communication Between Charge-Transfer and Ligand-Field Excited States.* It has been often argued that communication between CTTM and ligand-field excited states is inefficient in cobalt complexes (e.g., see Refs. 71, 72, and 145); however, since the quantum yields obtained from CTTM excitation have often been less than 0.2 and since the ligand-field states are generally photoinsensitive ($\phi < 0.01$), the possibility of appreciable population of ligand-field excited states by means of the nonradiative relaxation of CTTM excited states can rarely be ruled out. Probably the least equivocal example of inefficient communication between CTTM and ligand-field states occurs for $Rh(NH_3)_5I^{2+}$.[98] Ligand-field excitation of this complex results in *trans*-ammonia aquation with a quantum yield of about 0.9, while CTTM excitation ($\lambda \leq 254$ nm) results in both ammonia aquation ($\phi \approx 0.2$) and redox ($\phi \approx 0.2$) processes, so that at least 50% of the time the CTTM excited states relax directly to the ground state without populating the lower energy ligand-field excited states.

In contrast to the above, there is appreciable spectroscopic[147,148] and some photochemical evidence[149] for efficient communication of CTTL and ligand-field states in complexes of second- and third-row transition metals (see also Chapters 1 and 5). The population of CTTM states following ligand-field excitation has not been specifically investigated in many systems. However, for most Co(III) complexes, small yields of Co(II), $\phi_{Co^{2+}} \leq 10^{-3}$, have been reported for excitation energies well below the "threshold" energy defined in Section III.A above. If these photoredox yields result from primary, monophotonic photochemical processes, then they may arise from a very small yield for intersystem crossing or internal conversion from ligand-field excited state to a reactive charge-transfer state [e.g., a small contribution from the second term in the brackets in Eq. (3-30)].

3. *Communication Between Ligand-Centered and Charge-Transfer or Ligand-Field States.* The possible isolation of ligand-centered excited

states has received relatively little attention. As of this writing, information from a few systems indicate that in some complexes ligand-centered states can be relatively isolated from states involving the metal ion.[75,150]

C. Dependence of Yield and Class of Behavior on the Electronic Structure of the Central Metal

The probing of the relationships between the electronic structure of the central metal and the photochemical reactivity of the complex is one of the great problems, and opportunities, of the photochemistry of coordination complexes. In principle, one may examine variations in photochemical reaction mode and quantum yield on direct or sensitized excitation of similar CTTM states in series of complexes, such as $M^{III}(NH_3)_5X$ or $M^{III}(HEDTA)X$, with similar ligands but different central metals. In practice, such studies have not progressed far. For example, although Cr(III) complexes have a reputation of not exhibiting photoredox behavior,[1-6,151] very few Cr(III) complexes have been irradiated in their CTTM absorption bands. The acquisition of useful information is further complicated by facile radical-metallo fragment recombination reactions in many systems of interest; thus "net" photoredox is rarely observed in Rh(III) or acidopentacyanocobaltate(III) complexes, owing to the very rapid reoxidation of metallo fragments by radicals and other photooxidized ligand species.

Given these restrictions of the information now available, this section is necessarily speculative or (hopefully) anticipatory.

1. Correlations of Photochemical Reactivity with the Number of Metal-Centered d Electrons. Among substitution inert complexes of the type, $M^{III}(NH_3)_5X$, the relative photoinertness of the low-energy CTTM bands of $Ru^{III}(NH_3)_5X$ contrasts to the photosensitivity of these same bands in $Co^{III}(NH_3)_5X$ and $Rh^{III}(NH_3)_5X$ complexes. Since the ground-state species have low-spin $4d^5$. $3d^6$, and $4d^6$ metal centers, respectively, this suggests that excitations populating σ-antibonding metal orbitals ($*e$) are more likely to produce redox products than excitations populating π-nonbonding metal orbitals (t_2). Thus spin-allowed CTTM excitations of low-spin nd^6 complexes must always produce a metal-centered electronic configuration such as t_2^6*e, while the lowest-energy CTTM transitions in nd^3 or nd^5 complexes are expected to result in metal-centered electronic configurations of the types, t_2^4 and t_2^6, respectively. Correlation of photoredox reactivity with these changes in electronic configuration of the central metal presumes that in a complex with an excited-state electronic configuration of the type, t_2^6*e the vibrationally relaxed excited state is labilized, by means of a Jahn–Teller distortion, along the metal-radical ligand axis. Transition-metal excited-state lifetimes are generally very short, less than 10^{-6} sec. Consequently, one must be careful in the use of ground-state labilities as models for excited-state lability; that is, for ligands to be labile within the lifetime of an excited state the ligand substitution rate must be greater than or equal to the rate of nonreactive decay of the excited state. On the basis of the $\sim 10^8\,M^{-1}\,sec^{-1}$ rate constant for the oxidative addition[94] of $\cdot CH_3$ to the Jahn–Teller distorted $3d^7$ complex,[110] $[Co([14]dieneN_4)(OH_2)_2]^{2+}$, one may infer that the lifetime of the axial water molecules in this complex is less than 10^{-9} sec; thus a Jahn–Teller distorted, low-spin nd^7 metal might well be labile even on the time scale of excited-state lifetimes.

The low-spin nd^5 and nd^3 complexes are particularly interesting in the present context since spin-allowed CTTM transitions are expected to populate nonbonding orbitals (t_2) at low energy and antibonding orbitals $(*e)$ at higher energy; the energy difference between such transitions should be about the magnitude of $10\,Dq$ for the complex. The excited states with electron density in antibonding orbitals should resemble nd^7 systems in lability, and excitations producing excited states with such metal-centered electronic configurations would be expected to produce more photoreduction than the lower energy CTTM excitations, provided there is not efficient communication between CTTM excited states of such different electronic configurations as (t_2^6) and $(t_2^5 e_g)$. This possibility has received very little attention, but Siegel and Armor's results with Ru(III) complexes[107] may be consistent with this hypothesis.

In principle, excited-state distortions that may labilize the complex leading to photoredox reactivity, as hypothesized above, may also provide a mechanism for nonreactive electronic relaxation of the excited state through strong coupling (or intersection) of excited and ground electronic-state potential-energy surfaces. As noted above, the yield of products resulting from the irradiation of $Rh(NH_3)_5I^{2+}$ actually decreases as the excitation energy increases. For this complex about 50% of the CTTM excited states initially populated appear to decay directly to the ground state, and it has been proposed that the efficiency of this process is due to a strong coupling mechanism.[98] Thus the effects of axial labilization and strong excited-state to ground-state coupling, both results of excited-state distortion, might well be opposed to one another. This competition between nonreactive electronic relaxation and reactive radical dissociation in the excited state might be biased by suitable choice of ligands; for example, if the ligand photooxidized were a polydentate chelate, the rate of dissociation should be greatly restricted and excited-state distortions should favor nonradiative relaxation to the ground state. This argument provides a plausible explanation of the greater photosensitivity observed for [Fe(HEDTA)OH₂] than for Co(EDTA)⁻.[93b]

2. Correlations of Orbital Extension and Excited-State π-Bonding Interactions.

In a number of recent studies, coordination complexes have been found to undergo some sort of bond-cleavage reaction centered on the ligand as well as, or sometimes rather than, cleavage of a metal-ligand bond following CTTM excitation. In many of these cases the ligands involved in the photochemical reactions contain low-energy antibonding orbitals, raising the possibility that the photochemical reaction mode is influenced by $d \to \pi$ back bonding in the excited state. The only photochemical reactions considered in this section are those that have been observed to follow CTTM excitation of azido complexes. This is a speculative categorization since these reactions are complex and of recent vintage; however, they do raise some interesting mechanistic possibilities.

The photosensitivity of azido complexes of transition metals has been well known for a very long time (see Ref. 1). Recent systematic studies have led to a number of surprising features of these reactions:

1. Photo aquation of ammonia is an efficient process[66b,76] for all excitations of $[Co(NH_3)_5N_3]^{2+}$; ligand-field excitation of this complex does not lead to significant photoredox (early reports[71,72] apparently did not take account of secondary photolysis); ligand-field excitation of $[Rh(NH_3)_5N_3]^{2+}$ leads largely to N_3^- aquation.[66]

2. Basolo and co-workers[99,101] have discovered that CTTM excitation of $[Rh(NH_3)_5N_3]^{2+}$ and $[Ir(NH_3)_5N_3]^{2+}$ leads to N^-—N_2 cleavage of the coordinated azide, resulting in the formation of nitrene (M—NH) intermediates; CTTM excitation of the $[Co(NH_3)_5N_3]^{2+}$ and $[Co(CN)_5N_3]^{3-}$ lead predominately to Co(II) complexes and azide radicals.[66,77]

3. Nitrene formation from $[Rh(NH_3)_5N_3]^{2+}$ occurs over a narrow range of excitation energies.[66]

The first feature is somewhat outside the province of this chapter, but does suggest appreciably different geometries for the equilibrated excited states of $[Co(NH_3)_5N_3]^{2+}$ and $[Rh(NH_3)_5N_3]^{2+}$.

The comparison between $[Rh(NH_3)_5N_3]^{2+}$ and $[Co(CN)_5N_3]^{3-}$ is particularly instructive since the values of 10 Dq and optical electronegativity are about the same for these complexes, with a resulting similarity in the sequence of energy levels. There may be a small contribution from a nitrene pathway following CTTM $[N_3^- \to Co(III)]$ excitation of $[Co(CN)_5N_3]^{3-}$, but over 90% of the redox decomposition of coordinated azide results in formation of azide radicals; nearly the opposite ratio of pathways is exhibited on irradiation of the same absorption band of $[Rh(NH_3)_5N_3]^{2+}$.[66] Clearly, the dramatic difference in photochemical behavior must be dictated by some electronic feature of these complexes other than the sequence of accessible energy levels. Some possibilities are that: (1) N—N_2 cleavage to form the nitrene intermediates is promoted by $4d$ (or $5d$) to $*\pi$ back bonding to stabilize the $M(II) \langle N \cdots N_2$ configuration in the CTTM excited state; (2) a part of the photochemistry derives from a ligand-centered excited state; or (3) some aspects of the photochemistry originate in metal-to-ligand excited states in the heavy metals. The radial extension of the $3d$ orbitals is smaller, apparently not sufficient for back bonding in the ground state of $[Co(CN)_5N_3]^{3-}$,[140] and π-back bonding may not be adequate to stabilize such a nuclear arrangement in the excited state.

D. Some Correlations of Photoredox and Radical Reaction Energetics

The Jørgensen model for correlation of CTTM transition energies is based on a model involving fractional oxidation in the excited state of each of the ligands, X^-, in $M^{n+}(X^-)_6$ complexes. Thus both the excited and ground states may be represented by distributions of positive (at the metal) and negative (averaged over the ligands) charge in concentric, nearly spherical shells, with a smaller average density of charge in each concentric shell in the excited state than in the ground state. Although it is conventional to visualize CTTM transitions as corresponding to an inward flow of negative charge, they are equivalent to an outward flow of positive charge. One might, therefore, expect some formal similarity between models for CTTM transitions and the models of CTTS transitions discussed in Chapter 8, with proper compensation for the different senses of charge flow. The striking difference between these models is that the latter allow for specific involvement of the solvent, while the former do not. Although the Jørgensen approach provides powerful and useful tools for analyzing CTTM spectra, it is likely that any model completely neglecting the solvent environment cannot be entirely satisfactory or provide much insight into the subtleties of the associated photochemical and spectroscopic processes. This viewpoint seems particularly appropriate when complexes of the ML_5X type are considered. It is readily demonstrated that the energies of CTTM transitions in such complexes do vary with the solvent environment,[132] and a careful analysis of the energies of CTTM transitions and radical recombination reactions strongly suggests significant contributions of the solvent to the energy of the electronic transition.

Rather than beginning with an approximate discussion of some typical

ML_5X complex, let us first consider a more primitive kind of complex in which the energy terms are more clearly defined. The outer-sphere or ion-pair complexes, $\{Co(NH_3)_6^{3+}, I^-\}$ and $\{Ru(NH_3)_6^{3+}, I^-\}$, are known species and have characteristic absorption spectra;[141,152,153] in fact, the latter exhibits an absorption band at 390 nm, where both of its constituent species are transparent.[141] The SRP of the $Ru(NH_3)_6^{3+}/Ru(NH_3)_6^{2+}$ couple is 0.1 V;[154] the SRP for the $\cdot I/I^-$ couple is 1.42 V,[24] and the entropies for the couples are estimated to be $\Delta S^0 = 13.6$[155] and -2.0[24] kK degree^{-1} mole^{-1}, respectively. From these parameters one finds that $\Delta G^0 = 10.7$ kK mole^{-1} and $\Delta H^0 = 15.2$ kK mole^{-1} for

$$Ru(NH_3)_6^{3+} + I^- \rightarrow Ru(NH_3)_6^{2+} + \cdot I \qquad (3\text{-}31)$$

Since the formation constants for the ion pair and the "radical pair" would be expected to be of the order of 10 and 0.1, respectively,[74,152,153] the enthalpy change for

$$\{Ru(NH_3)_6^{3+}, I^-\} \rightarrow \{Ru(NH_3)_6^{2+}, \cdot I\} \qquad (3\text{-}32)$$

would be $\Delta H_{IP}^0 = 17$ kK mole^{-1}. This may be compared to an energy of approximately 18 kK mole^{-1} for the onset of CTTM absorption (E'_{th}) of the ion pair.[132b,141] For $\{Co(NH_3)_6^{3+}, I^-\}$, the thermodynamic parameters for formation of $\{Co(NH_3)_6^{2+}, I\}$ containing high-spin Co(II) are nearly identical,[74,152,153,155] and the onset for CTTM absorption occurs at approximately 26 kK mole^{-1}.[132b,150]

The near identity of ΔH_{IP}^0 and E'_{th} for the ruthenium-ion pair is likely to be somewhat fortuitous since ΔH_{IP}^0 must have an uncertainty of the order of 3 kK and E'_{th} of at least 1 kK. The solvent environments of $\{Ru(NH_3)_6^{3+}, I^-\}$ and $\{Ru(NH_3)_6^{2+}, \cdot I\}$ must be considerably different. It seems quite likely that the breadth of the CTTM transition is a manifestation of the statistics of ion-pair solvation and the Franck–Condon constraints on the electronic transition. Since there is little difference in bond length between $Ru(NH_3)_6^{3+}$ and $Ru(NH_3)_6^{2+}$,[158] the Franck–Condon contributions to the transition must be largely due to the orientations of solvent molecules.[156] The associated reorganization of solvent molecules is very similar to processes occurring for thermal electron transfer reactions[135] and would give rise to an activation barrier for radical recombination. The photochemistry of $\{Ru(NH_3)_6^{3+}, X^-\}$ ion pairs has not been explored, but there is no obvious reason that the quantum yield for formation of primary products should be unity.

Three factors are likely to contribute to the 8 kK mole^{-1} greater value of E'_{th} than ΔH_{IP}^0 for the $\{Co(NH_3)_6^{3+}, I^-\}$ ion pair: (1) low-spin $Co(NH_3)_6^{2+}$ is higher in energy than the high-spin ground state, (2) vibrationally relaxed low-spin complexes of Co(II) can suffer Jahn–Teller distortions of as much as 0.5 Å in each of two Co—NH$_3$ bonds,[108] (3) the Franck–Condon excited state is generated with the ground state solvent environment rather than a solvent environment appropriate to {Co(NH$_3$)$_6^{2+}$, ·I}. Compression to the Co(III) bond length in such an equilibrated Co(II) complex would result in a vibrational excitation of about 8 kK[157] in the *CTTM excited state reached in a Franck–Condon-allowed transition; the vibrational excitation energy contained in E'_{th}, ΔH^0_{FC}, should be much less than this. The {Co(NH$_3$)$_6^{3+}$, I$^-$} ion pair has its lowest energy absorbance maximum about 12 kK higher than the lowest energy absorbance maximum of the ruthenium analog. Estimates of the energy difference (ΔH^0_{spin}) between high-spin and low-spin cobalt-ammine complexes are in the range of 5 kK mole^{-1} [155] to 9 kK mole^{-1}.[158] The difference of 8 kK mole^{-1} found between ΔH^0_{IP} and E'_{th} for {Co(NH$_3$)$_6^{3+}$, I$^-$} and between E'_{th} for the cobalt and ruthenium ion pairs has to be an upper limit on the difference in energy between the high- and low-spin Co(II) species. On the other hand, the effect of differences between Co^{III}—NH$_3$ and Co^{II}—NH$_3$ bond lengths in the ground state and equilibrated low-spin product would be to shift the potential-energy surface describing the *CTTM excited state of the cobalt system along the reaction coordinate with respect to the similar surface of the ruthenium system. Since a relatively small difference in bond lengths is observed between Ru(NH$_3$)$_6^{3+}$, and Ru(NH$_3$)$_6^{2+}$,[158] the observed value of E'_{th} for {Ru(NH$_3$)$_6^{3+}$, I$^-$} may correspond to population of the lowest allowed vibrational states of *CTTM. In contrast, very small values of Franck–Condon overlap factors probably prohibit the population of the lowest vibrational levels in the *CTTM excited state of {Co(NH$_3$)$_6^{3+}$, I$^-$}.

In summary, one would expect the potential energy surface for the *CTTM state of cobalt to be both shifted along the photoreaction coordinate and raised in energy with respect to the potential surface for the *CTTM state of ruthenium; these two effects imply that $E'_{th}(Co) - E'_{th}(Ru) = 8$ kK $\approx (\Delta H^0_{spin} + \Delta H^0_{FC})$.

Now let us consider Co(NH$_3$)$_5$Br^{2+} as a prototypical coordination complex. In order to calculate the thermodynamic enthalpy of forming a {[Co(NH$_3$)$_5$]$^{2+}$, ·Br} radical pair, we must find some means of accounting for the cleavage of a cobalt–bromine bond. An approximate treatment may be formulated as follows. Consider the following scheme of reaction:

$$\{Co(NH_3)_5^{2+}, \cdot Br\}$$

$$\Big\uparrow \Delta H^0_a \qquad \overset{\Delta H^0_c}{\searrow} \{Co(NH_3)_5^{3+}, Br^-\} \qquad (3\text{-}33)$$

$$Co(NH_3)_5Br^{2+} \quad \overset{(\Delta H^0_h + \Delta H^0_{solv})}{\nearrow}$$

The quantity to be estimated is $\Delta H_a^0 = \Delta H_b^0 + \Delta H_{solv}^0 + \Delta H_c^0$. As an approximation to $\Delta H_b^0 + \Delta H_{solv}^0$ we may use the activation enthalpy found for the thermal hydrolysis of $Co(NH_3)_5Br^{2+}$; that is, assume $\Delta H^{\ddagger} = 7.5 \text{ kK}$ mole^{-1} [63] $\simeq (\Delta H_b^0 + \Delta H_{solv}^0)$. Analogous to calculations in the preceding paragraphs, it is easily shown that $\Delta H_{IP}^0 \simeq 23 \text{ kK} \text{ mole}^{-1}$ for $\{Co(NH_3)_6^{3+}, Br^-\} \rightarrow \{Co(NH_3)_6^{2+}, \cdot Br\}$. For low-spin Co(II) complexes, it has been found that the difference in reduction potentials is about one volt for the couples, $Co^{III}(N_4)(NH_3)_2 | Co^{II}(N_4)(NH_3)_2$ and $Co^{III}(N_4)X_2 | Co^{II}(N_4)X_2$, when X has about one half the crystal-field strength of NH_3.[111,155] If the contribution of $\Delta H_{spin}^0 \simeq 4 \text{ kK} \text{ mole}^{-1}$ then $\Delta H_c^0 \sim 19 \text{ kK} \text{ mole}^{-1}$, and $\Delta H_a^0 \simeq 26 \text{ kK} \text{ mole}^{-1}$. For this complex the onset of CTTM absorbance and E_{th}, as defined previously, are both about 22 kK mole^{-1}.

Acidopentaammineruthenium(III) | ruthenium(II) redox couples are less sensitive to the nature of the sixth ligand than are their cobalt analogs, presumably because the crystal-field stabilization energies of Ru(II) complexes are larger than those of Co(II) complexes. Either on the basis of only taking account of the differences in crystal-field stabilization energies, or simply by guessing at a difference of 0.2 V in the pentaammine and hexaammine ruthenium couples, one may estimate that $\Delta H_c^0 \sim 29 \text{ kK}$ for the $Ru(NH_3)_5Br^{2+}$ system. From kinetic parameters for the acid hydrolysis of this complex, $(\Delta H_b^0 + \Delta H_{solv}^0) \simeq \Delta H^{\ddagger} = 7.5 \text{ kK},$[159] so that $\Delta H_c^0 \simeq 36 \text{ kK}$. For this complex $E_{th}' \simeq 19 \text{ kK}.$[160] The estimate of ΔH_c^0 is no doubt appreciably in error; however, for this complex one expects that ΔH_c^0 should be considerably greater than E_{th}' since the formation of the photoredox product, $\{[Ru(NH_3)_5]^{2+}, \cdot Br\}$, involves creating a five-coordinate Ru(II) species and since Ru(II) has a relatively large crystal-field stabilization energy. The observation that irradiation of the lowest-energy CTTM band of $Ru(NH_3)_5X^{2+}$ complexes does not produce significant photoreduction products[106] is also consistent with a large difference between ΔH_c^0 and E_{th}'.

There should be several qualitative similarities between the CTTM excited states of $Ru(NH_3)_5Br^{2+}$ and $\{Ru(NH_3)_6^{3+}, I^-\}$; for example, in each case the excited- and ground-state potential energy surfaces would be approximately nested were it not for contributions of the solvent to the excited-state potential energy.

For the same reasons as entered into the earlier discussion of the ruthenium- and cobalt-ion pairs, one would expect the *CTTM surface for $Co(NH_3)_5Br^{2+}$ to be shifted with respect to the ground state along the reaction coordinate. The amount of molecular vibrational excitation associated with E_{th}' should be less for the spin-allowed transition in $Co(NH_3)_5Br^{2+}$ than in $\{Co(NH_3)_6^{3+}, I^-\}$ since in the former instance one would presume the transition leaves the vibrational energy localized along a NH_3—Co^{II}—Br (radical) axis, and in the latter case the axis must be NH_3—Co^{II}—NH_3. Nevertheless, one would expect the vibrationally equilibrated CTTM excited state for this complex must have an energy $E_0 < E'_{th} \simeq 22$ kK. Owing to the differences in loss of crystal-field stabilization energy on generating a five-coordinate product species,[63] ΔH_c^0 is expected to be much smaller for $Co(NH_3)_5Br^{2+}$ than for $Ru(NH_3)_5Br^{2+}$. For $Co(NH_3)_5Br^{2+}$, E_0 is probably less than ΔH_c^0, although the difference is expected to be far less than for $Ru(NH_3)_5Br^{2+}$.

The above estimate of ΔH_c^0 is for a low-spin Co(II) product. Such a low-spin Co(II) species may be presumed to correlate with the spectroscopic, spin-allowed CTTM transition, whereas ground-state, high-spin Co(II) species cannot be readily correlated with the spectroscopic transition. On the other hand, a Co(II) fragment with a high-spin electronic configuration could be correlated, through a radical pair of net triplet-spin multiplicity, $^3\{[Co(NH_3)_5]^{2+}, \cdot Br\}$, to a CTTM triplet state with a thermally

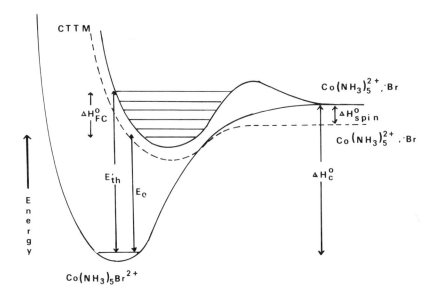

Fig. 3-5. Hypothetical potential-energy surfaces for $[Co(NH_3)_5Br]^{2+}$ and for two of the $\{Co(NH_3)_5^{2+}, \cdot Br\}$ radical pairs. The upper CTTM curve represents a potential-energy surface for a CTTM excited state and correlated radical pair of singlet spin multiplicity; the lower, dashed CTTM curve represents a potential energy surface for a CTTM excited state of triplet spin multiplicity that is correlated to a radical-pair product containing a high-spin Co(II) fragment.

equilibrated energy, $E_0' < E_0$; that is, $E_0 - E_0' \leq \Delta H_{spin}^0$.

The major features of the above discussion are represented in Fig. 3-5. The application of the approaches outlined above to other $M^{III}L_5X$ complexes is under investigation.[132b] The models sketched in this section are discussed further below.

E. Medium Effects

There have been relatively few systematic studies of the variations in photoredox processes in different solvent media. In principle, the "medium" might be generally defined to include all species except the substrate of photochemical interest. Thus "medium" effects might encompass the sensitization and quenching of photochemical processes by various solutes, self-quenching phenomena, and so forth. For purposes of the present section the "medium" is defined as solvent and solute species providing the microscopic environment of the photochemical substrate— the effects considered are environmental effects, exclusive of bimolecular electronic energy-transfer processes or specific chemical reactions.

At the present time, there appear to be two general categories of environmental effects affecting photochemical processes: the effects of medium viscosity and the contribution of substrate solvation to the excited-state potential-energy manifold.

1. *Viscosity Effects.* The most predictable effect of the solvent environment on the yields of photoredox products is a result of the variation with solvent viscosity of the distribution of geminate and secondary radical pairs and/or alteration of the probabilities of recombination and dissociation of such radical pairs. Medium effects of this kind are most readily discussed in terms of solvent viscosity and have been the subject of extensive theoretical and experimental investigation.[60,131] Very recently, medium effects of this kind have been found to be an important feature of the photoredox chemistry of Co(III) complexes.[66b,130,132] From the studies that have been performed it is evident that the primary products of CTTM excitation are geminate and/or secondary radical-pair species, the distribution of the types of radical pair varying with excitation energy and solvent viscosity in a somewhat predictable manner.[66b,130,132] The chemistry of these radical pair species is just beginning to be investigated.

2. *Effects of Solvent Repolarization on CTTM Excited States.* As the viscosity of the medium is increased, one would expect to find radical recombination processes enhanced, and, for a given excitation wavelength, one would expect the ratio of the yields of secondary to

primary radical pairs to be decreased. While these qualitative predictions have been found to be the case for 254 nm excitations of $[Co(NH_3)_5NO_2]^{2+}$ [130] and for ultraviolet excitations (in the range, 300 nm $\geq \lambda \geq 254$ nm) of $Co(NH_3)_5Br^{2+}$ [132] and $[Co(NH_3)_5N_3]^{2+}$,[66b,132] contrasting behavior has been observed for higher energy excitations of $Co(NH_3)_5X^{2+}$ (X = Br, N₃, NCS)[75,132] and all excitations ($300 \geq \lambda \geq 214$ nm) of $Co(NH_3)_5Cl^{2+}$.[132]

The behavior of $Co(NH_3)_5Br^{2+}$ is instructive in the present context. In water a limiting photoredox yield, $\phi^l_{Co^{2+}} = 0.32 \pm 0.02$, is obtained for excitation at $\lambda \leq 300$ nm (Fig. 3-2). This is consistent with the formation of only secondary radical pairs from high-energy excitations, provided such secondary radical pairs do not recombine (see Sections VI.A.1 to VI.A.3 below); however, such an explanation requires that two thirds of the species excited relax to the ground state without formation of radical-pair products. More appropriate mechanistic hypotheses would be that either (1) vibrational relaxation is more rapid than bond fragmentation in the CTTM excited state, and that a constant ratio of geminate to secondary radical pairs is produced from the thermally equilibrated excited state, or (2) that generation of secondary radical pairs, geminate radical pairs, and ground-state species (the latter by means of electronic relaxation) are competitive processes in the initially formed Franck–Condon excited state. Either hypothesis requires some binding energy in the initial CTTM excited state.

In glycerol (50% or 70%)/water mixtures, no limiting quantum yield has been observed; for high-excitation energies, the photoredox yield in such mixtures exceeds (by about a factor of 2 for $\lambda = 229$ nm) the yields observed in water; that is, in the mixed-solvent system $Co(NH_3)_5Br^{2+}$ approaches photodissociative behavior in which the net redox yield depends only on the ratio of secondary to geminate radical pairs.

The Franck–Condon-allowed, spin-allowed, CTTM transitions leave the initial excited state with a solvation environment similar to that of the ground state. Since such transitions involve a large decrease of the substrate electric dipole moment, the thermal equilibration of CTTM excited states would involve a compensatory adjustment of this solvation environment. Thus it seems likely that movement of the excited-state system along a reaction coordinate toward a geminate radical pair product should involve some activation barrier for solvent reorganization. These considerations and the above photochemical observations have been major factors in the evolution of ideas represented in the preceding section: both lines of thought point to the possibility that the solvent can make substantial contributions to the binding energies of CTTM excited states.

Although the points discussed above have precedent in related areas of photochemistry (e.g., see Ref. 156), they have not been subjected to examination or vigorous investigation in studies of the photochemistry of transition-metal complexes.

Additional effects of the solvent environment are very likely, but have less experimental support. Among these other environmental effects, two seem particularly worthy of future attention. A change in solvation energy is a major component of ΔH_a^0, calculated in the preceding section. Thus the energetics of excited state dissociation might be expected to vary appreciably from solvent to solvent; for example, owing to selective solvation effects in mixed-solvent media one might expect ΔH_a^0 to vary more strongly with solvent composition than E_{th}, and this effect should contribute to the binding of the CTTM excited state.

Finally, since nonradiative relaxation involves the "transfer" of substrate electronic excitation energy into solvent vibrational energy, any strong coupling (e.g., hydrogen bonding) of solvent and substrate should affect the rate constant for nonradiative relaxation and, therefore, quantum yields for products. Although evidence for the importance of such coupling has been found in the variations of the lifetimes of ligand-field triplet states[161] and in some ligand-field photochemistry[162,163] (see also Chapter 5), there is little such information available pertinent to CTTM photochemistry.

It seems clear that further studies of the charge-transfer photochemistry of transition-metal complexes in other than aqueous media can be expected to provide much new and mechanistically crucial information.

V. MODELS FOR PHOTOREDOX SYSTEMS

The theoretical basis for charge-transfer photochemistry is still pretty primitive. However, several mechanistic models have been proposed, and these provide useful bases for organizing observations (as in the preceding section of this chapter) and for designing new experimental studies. As these or other models evolve they should provide an eventual basis for a critical theoretical approach. The formulation, assumptions, and applicability of several important models are discussed below.

A.   Radical-Pair Models

Several models have attempted to attribute all or some of the features of the photochemistry of Co(III) complexes to the primary generation of a radical pair followed by geminate recombination or diffusive separation of the radical species. The early model of Sporer and Adamson[71,72] has been successively modified;[3,6,142] only the most recent version of this model is treated in Section V.A.1 below.

1. The Limiting Radical-Pair Model of Adamson.

This model was proposed some 15 years ago[71,72] and continues to have its strong advocates.[3,6,117,145] The basic radical-pair model is relatively simple and has been reviewed in detail elsewhere.[60,131] The application of this model to the photoredox chemistry of Co(III) complexes has only recently been subjected to very critical tests despite reasonably well established predictive aspects of the radical-pair model and despite the interest of several laboratories in the mechanistic aspects of the photochemistry of coordination compounds. A useful and concise statement of the limiting model has been made in a recent review[3]:

> The photochemistry of d^6 complexes conforms to certain predictions of excited state reactivity made in the early literature .... All these results can be accounted for by assuming that the primary reaction for absorption in a CT band is the formation of a cage species via homolytic fission.

$$Co\!-\!X^{2+} \xrightarrow{\ h\nu\ } \{Co^{2+}, X\} + \Delta \tag{3-34}$$

> The quantity Δ represents the amount of light energy absorbed in excess of that necessary for the electron transfer, and it determines to some extent the magnitude of the quantum yields for either aquation or redox. If Δ is small (i.e., the kinetic energies of X and the cobalt entity are low), then primary recombination of the cage partners is favorable, while, for large Δ, X may diffuse far enough from the cobalt entity for a solvent (S) molecule to enter the cage.

$$\{Co^{2+}, X\} \to CoX^{2+} \qquad \Delta \ small \tag{3-35}$$

$$\{Co^{2+}, X\} \to Co^{2+}(S)X \qquad \Delta \ large \tag{3-36}$$

> Following reaction [(3-36)] the original reaction partners may then diffuse apart with or without transfer of the electron back to X.

$$Co^{2+}(S)X \to Co^{2+} + S + X \tag{3-37}$$

$$Co^{2+}(S)X \to CoS^{3+} + X^- \tag{3-38}$$

This model allows no specific role for the CTTM excited states[6,71,72] and describes the limiting case of a photodissociative excited state (see Section IV.A). Thus the limiting model predicts a unitary yield for radical pairs (geminate plus secondary). In addition, the Adamson model proposes that recombination to give the original complex can only occur for the geminate pair and that the only secondary recombination process is back-electron transfer resulting in aquation of the complex.

Several obvious predictions may be made from this formulation of the radical-pair model.

i. $\phi_{Co^{2+}}$ *should be wavelength dependent*. This is almost always observed (see Figs. 3-2 and 3-3 and Table 3-4).

ii. *The sum* $(\phi_{aq} + \phi_{Co^{2+}})$ *should approach unity at very high photon energies*. This is almost never observed (e.g., see Table 3-4) in aqueous solutions. Exceptions may be $[Co(NH_3)_5NO_2]^{2+}$ in water and $Co(NH_3)_5X^{3+}$ (X = Cl, Br) in glycerol/water solutions.

iii. *The ratio of yields of aquation and redox processes should be independent of wavelength*. In a restricted sense this prediction is approximately true for the five pentaammine complexes in Table 3-4; however, of those complexes, three do not produce significant photoaquation on irradiation of pure CTTM bands, one results in ammonia aquation (see Section V.A.1.iv), and one results in linkage isomerization rather than aquation. This prediction has been discussed previously by Balzani and Carassiti (Ref. 1, p. 213) largely in connection with the photochemistry of $[Co(NH_3)_5NO_2]^{2+}$; for this complex the ratio of linkage isomerization to redox is a constant.[87,88] If consideration is extended over the range of excitation energies, the ratio of halide or ammonia aquation to redox is not a constant for the first four complexes in Table 3-4; over such an extended range of excitations this ratio generally decreases with increasing energy of irradiation of these complexes.

iv. *The ligand aquated is the ligand photooxidized in the primary step*. This is not strictly true for any of the complexes cited in Table 3-4. Some data for ligand-field excitations are included in Table 3-4 since advocates of this model are inclined to cite data for the region of overlap of CTTM and ligand-field absorption bands; some aquation is generally associated with ligand-field excitation.

v. *The quantum yield for photoredox should decrease in a simple manner with increasing solvent viscosity*. Observed for 254 nm excitations of $[Co(NH_3)_5NO_2]^{2+}$ and $Co(NH_3)_5Br^{2+}$, but the variations of quantum yield with medium are far too complex to be accommodated by the simple models.

Certainly, this specific radical pair model lacks the generality which has been claimed for it, if only because it predicts that all photoredox reactions should fall into class 1b, whereas class 1a seems to be more common while classes 1c and 2c are also important. However, a limiting model, along the lines proposed by Adamson, would be very useful from a mechanistic point of view since some cobalt complexes do seem to approach limiting photodissociative behavior under some circumstances. Viewed as a description for limiting photodissociative behavior, the major deficiency of the Adamson model seems to lie in Eq. (3-38). This suggests the following modification of the Adamson model.

2. A Limiting Radical-Pair Model for the Photodissociative Case.
Equations (3-34) to (3-37) describe a limiting type of photoredox behavior which many coordination complexes of cobalt appear to approach under some circumstances. For example, this seems an appropriate model to describe the photoredox behavior approached by $Co(NH_3)_5Br^{2+}$, in water/glycerol mixtures.

3. Models Allowing for Secondary Recombination. In connection with their study of the photochemistry of $[Co(NH_3)_5NO_2]^{2+}$, Balzani and co-workers[87] proposed that the linkage isomerization arose through secondary recombination of solvent separated $[Co(NH_3)_5]^{2+}$ and NO_2 radicals. This mechanism has been elaborated further by Scandola and co-workers.[88] This approach suggests some alternatives to the limiting models discussed in the preceding sections. Thus the possibility of regenerating a "Co(III)—X$^-$" substrate species by means of secondary recombination also may be considered, and would be described by reactions (3-34) to (3-37) and

$$\{Co^{2+}(S)X\} \xrightarrow{k_{39}} Co(III){-}X^- + S \qquad (3\text{-}39)$$

Such a model does embody many features of the photoredox chemistry of Co(III). If we associate rate constants, k_{35}, k_{36}, and k_{37} with reactions (3-35), (3-36), and (3-37), respectively, the following predictions may be made on the basis of this model:

i. $\phi_{Co^{2+}}$ *should be wavelength dependent* as observed, since the ratio of secondary and geminate radical pairs will depend on the amount of excess photon energy.

ii. *The sum of observed photochemical reactions may approach a limit less than unity.* For example,

$$\phi_{Co^{2+}} \rightarrow \left\{\frac{k_{37}}{k_{37}+k_{39}}\right\} = \phi^l_{Co^{2+}}$$

at very high photon energies. However, in order for step (3-39) to occur on the time scale of radical-pair lifetimes the water molecule in a species resembling $[Co(NH_3)_5OH_2]^{2+}$ must be very labile ($t_{1/2} \simeq 10^{-10}$ sec); this would be reasonable for a low-spin, axially distorted Co(II) species. Thus the model seems to require that spin relaxation is slow compared to the diffusive collapse of the secondary solvent cage. This inference from the model suggests that ammonia aquation could be diagnostic of the significance of this pathway.

Table 3-4. Photoredox Behavior of Some $Co^{III}L_5X$ Complexes in Aqueous Solutions[a]

L	X	Excitation Energy, kK	Band Assignment[b]	Redox Yields ϕ_L	Redox Yields ϕ_x	$\phi_{aq} + \phi_{redox}$	ϕ_{aq}/ϕ_{redox}	Ligand Labilized
NH$_3$	Cl	20.5[c]	LF	1.2×10^{-4} [d]	—	7×10^{-3}	57	NH$_3$, Cl$^-$
		39.4	Cl$^- \to$ Co, π	0.2[d]	—	0.2	~0	—
		43.7	Cl$^- \to$ Co	0.2[d]	—	0.2	~0	—
		46.7	Cl$^- \to$ Co	0.2[d]	—	0.2	~0	—
	Br	19.2	LF	—	10^{-4}		20	Br$^-$
		22.2	LF	—	8×10^{-3}	0.03	3.4	Br$^-$
		24.4	LF	—	0.071	0.14	0.97	Br$^-$
		27.0	Br$^- \to$ Co, π + LF	—	0.15	0.2	0.45	Br$^-$
		39.4	Br$^- \to$ Co, σ	—	0.33	0.3	~0	
		43.4	Br$^- \to$ Co, σ	—	0.3	0.3	~0	
		46.7	NH$_3 \to$ Co	—	0.3	0.3	~0	none
	N$_3$	19.2	LF	—	$<10^{-3}$	0.15	>160	
		22.2	LF	—	0.04	0.5	11	NH$_3$
		24.1	N$_3^- \to$ Co	—	0.14	0.7	4.3	NH$_3$
		30.3	N$_3^- \to$ Co	—	0.19	0.8	3	NH$_3$
		39.4	N$_3^- \to$ Co	—	0.20	0.8	3	NH$_3$
		46.7	NH$_3 \to$ Co(?)	—	0.20	0.8	3	NH$_3$
	NCS	19.2	LF	—	3×10^{-4}	8×10^{-4}	1.67	NCS$^-$
		23.9	LF + L	—	9×10^{-3}	0.013	0.45	NCS$^-$
		27.0	L + NCS$^- \to$ Co[e]	—	0.025	0.035	0.40	NCS$^-$
		28.5	L + NCS$^- \to$ Co[e]	—	0.026	0.031	0.19	NCS$^-$
		30.2	L + NCS$^- \to$ Co[e]	—	0.028	0.030	<0.07	—

	35.2	L+NCS→Co[e]	—	0.039	0.039	<0.02	—
	39.4	NCS→Co	—	0.13	0.13	<0.007	—
	46.7	NCS→Co	—	0.17	0.17	<0.006	—
—NO₂	22.6		—	0.12	0.16	0.29	NO₂⁻ (linkage isom.)
	27.4	NO₂⁻→Co	—	0.31	0.4	0.18	NO₂⁻ (linkage isom.)
	31.9	NO₂⁻→Co	—	0.41	0.5	0.27	NO₂⁻ (linkage isom.)
	39.3	NO₂⁻→Co	—	0.51	0.6	0.25	NO₂⁻ (linkage isom.)
HEDTA/5 Cl	22.2	LF	0.012	—	0.02	0.8	Cl⁻
	25.0	LF	0.015	—	0.02	0.7	Cl⁻
	33.3	LF+CT	0.040	—	0.08	1.0	Cl⁻
	39.3	CT	0.09	—	0.14	0.55	Cl⁻
	43.7	CT	0.14	—	0.19	0.36	Cl⁻
	46.7	CT	0.14	—	0.19	0.36	Cl⁻
Br	22.2	LF	0.004	—	0.03	7.5	Br⁻
	28.6		0.018	—	0.10	5	Br⁻
	33.3	Br⁻→Co	0.028	—	0.06	1.1	Br⁻
	39.3	Br⁻→Co	0.07	—	0.11	0.6	Br⁻
	46.7	CT	0.22	—	0.25	0.14	Br⁻

[a] Quantum yields taken from Refs. 1, 91, and 133, except as indicated.

[b] Abbreviations: LF = ligand-field transition, X⁻ → Co and CT = CTTM transitions, L = ligand-centered transition.

[c] Data from R. A. Pribush, C. K. Poon, C. M. Bruce, and A. W. Adamson, J. Amer. Chem. Soc., **96**, 3027 (1974).

[d] There is some ambiguity as to whether NH₃ or Cl⁻ is oxidized in the primary step (Ref. 35).

[e] Both ligand-centered and CTTM transitions are expected in this region.

iii. *The ratio of yields of secondary recombination reactions and redox is constant.* This model does allow for labilization reactions of other than the photooxidized ligand, X. Such a constancy of the ratio of yields of labilization to redox have been found for all irradiations of $[Co(NH_3)_5NO_2]^{2+}$ and for pure CTTM irradiations of $[Co(NH_3)_5N_3]^{2+}$.

iv. *The secondary recombination yields should increase with the SRP of the radical formed.* In principle, k_{39} should increase as the free energy of the reaction described becomes more negative[135]; therefore, recombination efficiency should increase in the same manner. In practice, radical reactions with coordination complexes often do not follow predictions of simple free-energy relations.[166] The recombination efficiencies (within the context of this model) of four of the pentaammines in Table 3-4 may be estimated: $Co(NH_3)_5Cl^{2+}$ ($\phi_{rec} = 0.77$; SRP = 2.6 V), $Co(NH_3)_5Br^{2+}$ ($\phi_{rec} = 0.67$; SRP = 1.96 V), $[Co(NH_3)_5N_3]^{2+}$ ($\phi_{rec} = 0.8$; SRP = 2 V), $[Co(NH_3)_5NO_2]^{2+}$ ($\phi_{rec} = 0.5$; SRP = 1.13 V). There may be a weak correlation.

v. $\phi^l_{Co^{2+}}$ *should decrease with solvent viscosity, and in a plot of* $\phi_{Co^{2+}}$ *versus excitation energy, the approach to* $\phi^l_{Co^{2+}}$ *should be more gradual in solutions of higher viscosity.* According to this model,

$$\phi_{Co^{2+}} = \frac{k_{36}}{k_{35} + k_{36}}\left(\frac{k_{37}}{k_{39} + k_{37}}\right)$$

where both k_{36} and k_{37} should decrease as solvent viscosity increases, while k_{36} should increase as excitation energy increases. The experimental information presently at hand indicates a far more complex variation of yields with medium than can be inferred from this model. Of the $Co(NH_3)_5X^{2+}$ (X = Cl, Br, N$_3$, NCS) complexes, only $[Co(NH_3)_5N_3]^{2+}$ and $Co(NH_3)_5Br^{2+}$ might be presumed to approach this prediction, but only if one were to assume that the similar approaches of the azido complex to different values of $\phi^l_{Co^{2+}}$ in water and in water/glycerol were a manifestation of selective solvation of the substrate in the mixed solvent medium; while for the bromo complex it would be necessary to assume that the values of $\phi^l_{Co^{2+}}$ are obscured by some new photoredox process at high energy, but that approach to these values is qualitatively as predicted.

4. *Further Variation in Radical-Pair Models.* Scandola[130] has remarked that some confusion has resulted from a prevailing attitude in the literature that there is but one radical-pair model. Of course, there are some additional variations in radical-pair models that seem to have escaped attention in the literature on the charge-transfer photochemistry of coordination complexes. The most important of variations that should be specifically considered is the possibility of differences in electronic-spin multiplicities of radical-pair species. The preceding discussions have already taken note of the possibilities of formation of high- or low-spin Co(II) fragments in the photoredox decompositions of Co(III) complexes. The radical pairs for either kind of Co(II) fragment may have the spin moments of the cobalt and the radical fragments parallel or opposed. Thus, on the basis of possible spin statistics alone, one should consider that any or all of four different geminate and four different secondary radical pairs may be formed in the photoredox decompositions of Co(III) complexes; that is, for both the geminate and secondary radical pairs, the pairs may have net singlet, triplet (two kinds), and quintet spin multiplicities. The radical recombination reactions, (3-35) and (3-39), are spin forbidden in all except the singlet radical pairs. Since radical-pair lifetimes are very short (less than 10^{-9} sec), spin forbiddance may have a significant effect on the probabilities of radical recombination; furthermore, these considerations present yet another ambiguity in the interpretation of triplet-to-triplet sensitized photoredox reactions. At the present writing, no study has been reported that examines the question of the spin multiplicity of the radical-pair species obtained from the photoredox decompositions of transition-metal complexes. Such studies seem likely in the near future.

Finally, the possibility of ligand-exchange reactions occurring in radical-pair species should always be considered when the metallo fragment contains (or could contain) a low-spin d^7 metal.

The intervention of a radical-pair species at some point along the photoredox reaction coordinate is reasonable and seems well established for several complexes. There are many important unresolved questions as to how many of the photochemical details can be attributed to the reactions of such radical pairs. There are a surprising number of problems in this area that have received very little attention. For example, a radical pair containing low-spin Co(II) correlates well with the excited state generated from a spin-allowed CTTM transition, whereas a radical pair containing high-spin Co(II) does not. Since the recombination efficiencies may well be smaller for the spin-forbidden reaction of $\cdot X$ with high-spin (i.e., ground-state) $[Co(NH_3)_5]^{2+}$ than with the low spin analog, one might expect substantially different behavior from $Co^{III}(NH_3)_5X$ complexes and from complexes such as $Co^{III}(N_4)X_2$, $Co^{III}(CN)_5X$, or $Rh^{III}(NH_3)_5X$ in which the ground state of the reduced metallo product is low spin. There is little experimental information on this point.

B. Concerning the Generation of Models for the Contributions of Excited States

Much of the mechanistic thinking in studies of the charge-transfer photochemistry of coordination complexes has been organized along antagonist lines which claim either that all the photochemical behavior can be explained in terms of the chemistry of radical pairs or in terms of the chemistry of CTTM excited states. Thus, while there have been frequent discussions advocating one or the other of the allegedly antagonistic hypotheses, there have been few systematic formulations or critiques of either. It now seems obvious that radical pairs and excited states play complementary, rather than antagonistic roles, and that a satisfactory understanding of this subject will be achieved only when the contributions of each can be represented by a useful model. At the present time, the radical-pair component of charge-transfer photochemistry has received far more attention than has the excited-state component: specific models for the chemical behavior of radical pairs have been proposed in the literature and subjected to experimental examination (see Section V.A above). Although excited-state models appear to be evolving, specific models have not been clearly formulated in the literature nor subjected to careful scrutiny and testing. As a consequence, the present section reflects almost exclusively the viewpoints of the author and is to a significant extent an attempt to anticipate the direction in which mechanistic thinking will evolve.

At the present time, it seems evident that limiting excited-state models will not prove very useful of themselves simply because the primary photoredox products must be some kind of radical pair. However, it is useful to assume that the chemical effects attributable to radical-pair reactions can be separated from the chemically significant behavior in excited-state species that are precursors to radical pairs, and thus to separate the discussions of radical-pair and excited-state models. Many aspects of the present discussion have been anticipated in Section IV, and especially in Section IV.D.

1. *Molecular Properties Which One Would Like Treated by an Excited-State Model.* A number of questions concerning the nature and properties of CTTM excited states are posed, or alluded to, elsewhere in this chapter; the most important of these are summarized briefly here:

a. *Energies of CTTM Excited States.* Although a complete description seems impossible for the potential-energy surface for a CTTM excited state of any coordination complex, it does seem that estimates (experimental or theoretical) are possible for a number of key points on such surfaces:

i. The minimum energy difference between the thermal ground state and the primary radical-pair product (Section IV.D).

ii. The excited-state binding energy. It seems likely that there are two major components to the energy barrier to homolytic bond cleavage in CTTM excited state: (1) residual chemical bonding (σ-, π-back bonding, etc.) between the reduced metallo fragment and the oxidized ligand and (2) a solvent reorganizational energy required to change from the solvation of the ground state to the solvation of a geminate radical pair.

iii. The energy that would be characteristic of a thermally equilibrated excited state. Estimates of this quantity would depend on obtaining reasonable estimates of excited-state binding energies and excited-state distortions.

b. *The existence of CTTM Electronic Manifolds of Differing Spin Multiplicities.* The closely related questions of (1) the possible intermediacy of nonspectroscopic CTTM states in the photoredox reactions stimulated by direct excitation and (2) the chemical reactivity of such nonspectroscopic states have often been posed, but never resolved. Yet, for cobalt complexes, the thermally stable, high-spin Co(II) products do not correlate with either the spectroscopic CTTM states (of singlet-spin multiplicity) or with radical pairs of net singlet-spin multiplicity. On the other hand, high-spin Co(II) species do correlate well with CTTM states and radical pairs of triplet-spin multiplicity. Thus, on the basis of spin correlations, one would expect some CTTM state of triplet-spin multiplicity to lie lower than the spectroscopic CTTM state by the approximate energy difference of the appropriate high-spin and low-spin Co(II) species. It is to be noted that the singlet and triplet states discussed in this approach necessarily differ in their electronic configurations.

c. *The Isolation and Chemical Role of Ligand-Centered States.* Electronic states that are largely centered within a coordinated ligand have now been implicated in many photochemical processes. It is important to understand how and when one can expect such states to be isolated from other electronic states of the coordination complex.

2. *Molecular Processes Whose Significance Should be Considered Within the Context of Any Excited-State Model.*

a. *Molecular Distortion.* Depending on the excited-state electronic configuration of the central metal, the possibilities of homolytic and heterolytic bond cleavage exist in a CTTM excited state. Again for Co(III) complexes, the d^7 center of the CTTM excited state might be expected to be axially distorted with both the oxidized ligand and the ligand trans to it labilized. In addition to the obvious relevance to chemical deactivation modes of the excited state, excited-state distortions are major factors in determining the amount of vibrational excitation associated with the spectroscopic transitions and with the rates of nonreactive electronic relaxation processes.

b. *Rates of Electronic Relaxation.* Radiative relaxation of CTTM excited states has not been observed. The effects of the various nonradiative relaxation rates on observed quantum yields are discussed in Section IV.A.

3. *Toward Working Models for CTTM Excited States.* The major features in currently evolving models for CTTM excited states have been sketched in Sections IV.A and IV.D. Important assumptions of these models are (1) the spectroscopic CTTM excited states can be bound states (depending on the system) in which some vibrational and electronic relaxation may occur; (2) reactive CTTM excited states decompose into radical-pair species as the primary photoredox products; (3) the binding energy of the spectroscopic CTTM excited state is composed of some residual bonding between the incipient radical fragments, as well as a significant contribution due to the polarization of the solvent medium; (4) nonspectroscopic CTTM states are possible, but they may be less tightly bound than the spectroscopic states (some residual bonding must still be considered), since one would expect that some solvent repolarization should have occurred during the sequence of events resulting in the population of such nonspectroscopic states.

VI. FUTURE PROBLEMS

Future developments appear to depend critically on the evolution of an integrated, useful working model for the contributions of radical-pair and CTTM species. Once useful models have evolved, new model systems and experiments can be designed to probe some of the issues raised in the preceding sections. The evolution of models of at least limited usefulness seems likely in the relatively near future.

Although the evolution of useful models appears as the most important problem to the immediate future, it is obvious that certain experimental problems need more critical examination. Among these are the charge-transfer photochemistry of the homologous series of complexes: $M^{III}(NH_3)_5X$ ($M = Cr$, Rh, and Ru), $M^{III}(CN)_5X$ ($M = Fe$, Ru), and $M^{III}(HEDTA)X$ ($M = Co$, Rh, Cr, Ru). Similarly, it is obvious that further studies should be carried out with electronic sensitizers in order to determine whether useful mechanistic information about charge-transfer photochemistry may be obtained with this approach.

Finally, the application of high-intensity, pulsed-laser systems to the systems considered in this chapter should at the minimum be useful in delineating the chemical behavior of radical-pair species.

ACKNOWLEDGMENTS

The author is indebted to Dr. G. J. Ferraudi for assistance with some of the tabulations, much of the art work, for a careful reading of the manuscript, and for many stimulating and critical discussions. The author also thanks Professors V. Balzani and M. Z. Hoffman for reading and commenting on the manuscript. The author is also grateful to a large number of co-workers and colleagues for helpful discussions on parts of this manuscript and to several colleagues who made available preliminary reports of their research.

Partial support by the National Science Foundation of much of the research mentioned in this chapter is gratefully acknowledged.

REFERENCES

1. V. Balzani and V. Carassiti, *Photochemistry of Coordination Compounds*, Academic, New York, 1970.
2. W. L. Waltz and R. G. Sutherland, *Chem. Soc. Rev.* **1**, 241 (1972).
3. P. D. Fleischauer, A. W. Adamson, and G. Sartori, *Progr. Inorg. Chem.* **17**, 1 (1972).
4. D. W. Watts, *MTP Int. Rev. Sci., Inorg. Chem.* [1] **9**, 52 (1972).
5. D. R. Eaton, *Spectrosc. Inorg. Chem.* **1**, 29 (1970).
6. A. W. Adamson, W. L. Waltz, E. Zinato, D. W. Watts, P. D. Fleischauer, and R. D. Lindholm, *Chem. Rev.* **68**, 541 (1968).
7. T. M. Dunn, *Modern Coordination Chemistry* (J. Lewis and R. G. Wilkins, Eds.), Chap. 4, Wiley-Interscience, New York, 1960.
8. (a) C. K. Jørgensen, *Progr. Inorg. Chem.* **4**, 73 (1962); (b) *Orbitals in Atoms and Molecules*, Academic, New York, 1962; (c) *Absorption Spectra and Chemical Bonding in Complexes*, Oxford U.P., Oxford, England, 1962; (d) *Oxidation Numbers and Oxidation States*, Springer, New York, 1969.
9. H.-H. Schmidtke, *Physical Methods in Advanced Inorganic Chemistry*, (H. A. O. Hill and P. Day, Eds.), Chap. 4, Wiley-Interscience, New York, 1968.
10. H. L. Schläfer and G. Gleiman, *Ligand Field Theory*, Chap. 1, Wiley-Interscience, New York, 1969.
11. R. A. D. Wentworth and T. S. Piper, *Inorg. Chem.* **4**, 709 (1965).
12. J. K. Thomas, *Advan. Radiation Chem.* **1**, 103 (1969).

13. W. Latimer, *Oxidation Potentials*, 2nd ed., Prentice-Hall, Englewood Cliffs, N.J., 1952.

14. A. K. Pikaev, *Pulse Radiolysis of Water and Aqueous Solutions*, Indiana U.P., Bloomington, Indiana, 1967.

15. H. Fricke and J. K. Thomas, *Radiation Res. Suppl.* **4**, 35 (1964).

16. J. Rabani and M. S. Matheson, *J. Phys. Chem.* **70**, 761 (1966).

17. G. E. Adams, J. W. Boag, J. Currand, and B. D. Michael, in *Pulse Radiolysis* (M. Ebert, J. P. Kenne, A. J. Swallow, and J. H. Baxendale, Eds.), p. 117, Academic, New York, 1965.

18. J. K. Thomas, *Trans. Faraday Soc.* **61**, 702 (1965).

19. M. Anbar and P. Neta, *Int. J. Appl. Radiat. Isotopes* **18**, 439 (1967).

20. S. Gordon, E. J. Hart, M. S. Matheson, J. Rabani, and J. K. Thomas, *J. Amer. Chem. Soc.* **85**, 1375 (1963).

21. L. M. Dorfman and I. A. Taub, *J. Amer. Chem. Soc.* **85**, 2370 (1963).

22. (a) E. J. Hart and J. W. Boag, *J. Amer. Chem. Soc.* **84**, 4090 (1962); (b) *Nature* (*London*) **197**, 45 (1963).

23. L. I. Grossweiner and M. S. Matheson, *J. Phys. Chem.* **61**, 1089 (1957).

24. W. H. Woodruff and D. W. Margerum, *Inorg. Chem.* **12**, 962 (1973).

25. M. E. Langmuir and E. Hayon, *J. Phys. Chem.* **71**, 3808 (1967).

26. (a) S. D. Malone and J. F. Endicott, *J. Phys. Chem.* **76**, 2223 (1972); (b) unpublished observations.

27. M. S. Matheson, W. A. Mulac, J. L. Weeks, and J. Rabani, *J. Phys. Chem.* **70**, 2092 (1966).

28. M. Schöneshöfer and A. Henglein, *Ber. Bunsenges. Phys. Chem.* **73**, 289 (1969).

29. K. D. Asmus, A. Henglein, A. Wigger, and G. Beck, *Ber. Bunsenges. Phys. Chem.* **70**, 756 (1966).

30. M. Simic, P. Neta, and E. Hayon, *J. Phys. Chem.* **73**, 3794 (1969).

31. J. Lilie, G. Beck, and A. Henglein, *Ber. Bunsenges, Phys. Chem.* **75**, 458 (1971).

32. P. Neta, M. Simic, and E. Hayon, *J. Phys. Chem.* **73**, 4207 (1969).

33. T. S. Roche and J. F. Endicott, *Inorg. Chem.* **13**, 1575 (1974).

34. M. Anbar and J. K. Thomas, *J. Phys. Chem.* **68**, 3829 (1964).

35. G. Caspari, R. G. Hughes, J. F. Endicott, and M. Z. Hoffman, *J. Amer. Chem. Soc.* **92**, 6801 (1970).

36. A. Fojtik, G. Czapski, and A. Henglein, *J. Phys. Chem.* **74**, 3204 (1970).

37. J. P. Keene, J. Raef, and A. J. Swallow, *Pulse Radiolysis* (M. Ebert, J. P. Keene, A. J. Swallow, and A. J. Baxendale, Eds.), p. 99, Academic, New York, 1965.

38. G. V. Buxton and R. M. Sellers, *J. Chem. Soc. Faraday Trans.* **69**, 555 (1973).

39. E. Hayon and J. T. McGarvey, *J. Phys. Chem.* **71**, 1472 (1967).

40. G. E. Adams, J. W. Boag, and B. D. Michael, *Proc. Roy. Soc. A* **287**, 32 (1965).

41. J. L. Weeks and J. Rabani, *J. Phys. Chem.* **70**, 2100 (1966).

42. D. Bebar, G. Czapski, and I. Duchovny, *J. Phys. Chem.* **74**, 2206 (1970).

43. S.-N. Chen and M. Z. Hoffman, *J.C.S. Chem. Commun.*, 991 (1972).

44. A. F. Vaudo, E. R. Kantrowitz, M. Z. Hoffman, E. Papaconstantinou, and J. F. Endicott, *J. Amer. Chem. Soc.* **94**, 6655 (1972).

45. N. Getoff, F. Schworer, V. M. Markovic, K. Sehester, and S. O. Nielsen, *J. Phys. Chem.* **75,** 749 (1971).

46. J. H. Baxendale, P. L. Bevan, and A. D. Stott, *Trans. Faraday Soc.* **64,** 2389 (1968).

47. J. K. Thomas, *Trans. Faraday Soc.* **61,** 702 (1965).

48. M. Schöneshöfer and A. Henglein, *Ber. Bunsenges. Phys. Chem.* **74,** 393 (1970).

49. A. Treinin and E. Hayon, *J. Chem. Phys.* **50,** 538 (1969).

50. F. Barat, B. Hickel, and J. Sutton, *Chem. Commun.*, 125 (1969).

51. M. Gratzel, A. Henglein, J. Lilie, and G. Beck, *Ber. Bunsenges. Phys. Chem.* **73,** 646 (1969).

52. R. W. Glass and T. W. Martin, *J. Amer. Chem. Soc.* **92,** 5084 (1970).

53. L. Doglioti and E. Hayon, *J. Phys. Chem.* **72,** 1800 (1968).

54. E. Hayon and M. Simic, *J. Amer. Chem. Soc.* **94,** 42 (1972).

55. J. A. Kerr, *Chem. Rev.* **66,** 465 (1966).

56. For the calculations in Table 3-2, bond energies and thermodynamic data were taken from Refs. 13, 24, 54 or *Handbook of Chemistry and Physics*, 48th ed., The Chemical Rubber Co., Cleveland, Ohio, 1967–1968, and references cited therein.

57. M. Gratzel, K. M. Bansal, and A. Henglein, *Ber. Bunsenges. Phys. Chem.* **77,** 11 (1973).

58. (a) J. I. H. Patterson and S. P. Perone, *Anal. Chem.* **44,** 1978 (1972); (b) *J. Phys. Chem.* **77,** 2437 (1973).

59. J. Lilie and R. W. Fessenden, *J. Phys. Chem.* **77,** 674 (1973).

60. J. P. Lorand, *Progr. Inorg. Chem.* **17,** 207 (1972).

61. E. R. Kantrowitz, M. Z. Hoffman, and J. F. Endicott, *J. Phys. Chem.* **75,** 1914 (1971).

62. G. D. Cooper and B. A. DeGraff, *J. Phys. Chem.* **76,** 2618 (1972).

63. F. Basolo and R. G. Pearson, *Mechanisms of Inorganic Reactions*, 2nd ed., Chap. 3, Wiley, New York, 1967.

64. An exception is the Cl_2^- oxidation of Co^{2+}; A. T. Thornton and G. S. Laurence, *J. C. S. Dalton Transactions*, 1632 (1973).

65. M. P. Liteplo and J. F. Endicott, *Inorg. Chem.* **19,** 1420 (1971).

66. (a) G. Ferraudi and J. F. Endicott, *J. Amer. Chem. Soc.* **95,** 2371 (1973); (b) J. F. Endicott, G. J. Ferraudi, and J. R. Barber, *J. Amer. Chem. Soc.*, **97,** 219 (1975).

67. (a) M. Simic and J. Lilie, *J. Amer. Chem. Soc.* **96,** 291 (1974); (b) J. Lilie, M. Simic and J. F. Endicott, submitted for publication.

68. M. Grant, H. W. Dodgen, and J. P. Hunt, *J. Amer. Chem. Soc.* **91,** 1724 (1970).

69. J. T. Yardley and J. K. Beattie, *J. Amer. Chem. Soc.* **94,** 8925 (1972).

70. J. K. Beattie, N. Sutin, D. H. Turner, and G. W. Flynn, *J. Amer. Chem. Soc.* **95,** 2052 (1973).

71. A. W. Adamson and A. H. Sporer, *J. Amer. Chem. Soc.* **80,** 3865 (1958).

72. A. W. Adamson, *Discuss. Faraday Soc.* **29,** 163 (1960).

73. J. F. Endicott and M. Z. Hoffman, *J. Amer. Chem. Soc.* **87,** 3348 (1965).

74. R. G. Wilkins, *Accounts Chem. Res.* **3,** 408 (1970).

75. G. J. Ferraudi, J. F. Endicott, and J. R. Barber, *J. Amer. Chem. Soc.*, in press.

76. J. F. Endicott, M. Z. Hoffman, and L. S. Beres, *J. Phys. Chem.* **74,** 1021 (1970).

77. G. J. Ferraudi and J. F. Endicott, *Inorg. Chem.* **12,** 2389 (1973).

78. D. D. Campano, E. R. Kantrowitz, M. Z. Hoffman, and M. S. Weinberg, *J. Phys. Chem.* **78,** 686 (1974).

79. M. Z. Hoffman and M. Simic, *Inorg. Chem.* **12,** 2471 (1973).

80. A. F. Vaudo, E. R. Kantrowitz, and M. Z. Hoffman, *J. Amer. Chem. Soc.* **93,** 6698 (1971).

81. V. W. Cope and M. Z. Hoffman, *J.C.S. Chem. Commun.,* 227 (1972).

82. V. W. Cope, S.-N. Chen, and M. Z. Hoffman, *J. Amer. Chem. Soc.* **95,** 3116 (1973).

83. S.-N. Chen, V. W. Cope, and M. Z. Hoffman, *J. Phys. Chem.* **77,** 1111 (1973).

84. E. R. Kantrowitz, M. Z. Hoffman, and K. M. Schilling, *J. Phys. Chem.* **76,** 2493 (1972).

85. S. A. Penkett and A. W. Adamson, *J. Amer. Chem. Soc.* **87,** 2514 (1965).

86. P. Natarajan and J. F. Endicott, *J. Phys. Chem.* **77,** 971 (1973).

87. V. Balzani, R. Ballardini, N. Sabbatini, and L. Moggi, *Inorg. Chem.* **7,** 1398 (1968).

88. F. Scandola, C. Bartocci, and M. A. Scandola, *J. Phys. Chem.* **78,** 572 (1974).

89. D. Klein and C. W. Moeller, *Inorg. Chem.* **4,** 394 (1965).

90. J. H. Espenson and J. R. Pipal, *Inorg. Chem.* **7,** 1463 (1968).

91. A. Vogler, private communication.

92. P. Natarajan and J. F. Endicott, *J. Amer. Chem. Soc.* **95,** 2470 (1973).

93. (a) P. Natarajan and J. F. Endicott, *J. Phys. Chem.* **77,** 1823 (1973); (b) *ibid.* **77,** 2049 (1973).

94. T. S. Roche and J. F. Endicott, *J. Amer. Chem. Soc.* **94,** 8622 (1972).

95. G. N. Schrauzer, L. P. Lee, and J. W. Sibert, *J. Amer. Chem. Soc.* **92,** 2997 (1970).

96. J. M. Pratt, *Inorganic Chemistry of Vitamin B$_{12}$,* Chap. 14, Academic, New York, 1972.

97. H. A. O. Hill, in *Inorganic Biochemistry* (G. L. Eichorn, Ed.), Elsevier Sci. Publ. Co., New York Vol. 2, p. 1067, 1973.

98. T. L. Kelly and J. F. Endicott, *J. Amer. Chem. Soc.* **94,** 1797 (1972).

99. (a) J. L. Reed, F. Wang, and F. Basolo, *J. Amer. Chem. Soc.* **94,** 7173 (1972); (b) J. L. Reed, H. D. Gafney, and F. Basolo, *ibid.* **96,** 1363 (1974).

100. T. L. Kelly, Ph.D. Dissertation, Wayne State University, 1971.

101. H. D. Gafney, J. Reed, and F. Basolo, *J. Amer. Chem. Soc.* **95,** 7998 (1973).

102. R. C. Wright and G. S. Laurence, *J.C.S. Chem. Commun.,* 132 (1972).

103. (a) A. T. Thornton and G. S. Laurence, *Chem. Commun.,* 443(1970); (b) *J.C.S. Dalton transactions,* **804** (1973); (c) G. S. Laurence, private communication.

104. C. A. Parker and C. G. Hatchard, *J. Phys. Chem.* **63,** 22 (1959).

105. G. D. Cooper and B. A. DeGraff, *J. Phys. Chem.* **75,** 2897 (1971).

106. W. L. Wells and J. F. Endicott, *J. Phys. Chem.* **75,** 3075 (1971).

107. J. Siegel and J. A. Armor, *J. Amer. Chem. Soc.* **96,** 4102 (1974).

108. L. Warner, Ph.D. Dissertation, Ohio State University, 1968.

109. (a) D. P. Rillema, J. F. Endicott, and N. A. P. Kane-Maguire, *J.C.S. Chem. Commun.,* 495 (1972); (b) D. P. Rillema and J. F. Endicott, unpublished observations.

110. M. D. Glick, J. M. Kuszaj, and J. F. Endicott, *J. Amer. Chem. Soc.* **95,** 5097 (1973).

111. D. P. Rillema, J. F. Endicott, and W. Papaconstantinou, *Inorg. Chem.* **10,** 1739 (1971).

112. D. P. Rillema, J. F. Endicott, and R. C. Patel, *J. Amer. Chem. Soc.* **94,** 394 (1972).

113. J. F. Endicott and M. Z. Hoffman, *J. Phys. Chem.* **70**, 3389 (1966).

114. C. A. Parker and C. J. Hatchard, *J. Phys. Chem.* **63**, 22 (1959).

115. N. S. Rowan, M. Z. Hoffman, and R. M. Milburn, *J. Amer. Chem. Soc.*, **96**, 6060 (1974).

116. E. R. Kantrowtiz, J. F. Endicott, and M. Z. Hoffman, *J. Amer. Chem. Soc.* **92**, 1776 (1970).

117. R. D. Lindholm and T. K. Hall, *J. Amer. Chem. Soc.* **93**, 3525 (1971).

118. A. Vogler and A. W. Adamson, *J. Amer. Chem. Soc.* **90**, 5943 (1968).

119. M. A. Scandola and F. Scandola, *J. Amer. Chem. Soc.* **92**, 7278 (1970).

120. H. D. Gafney and A. W. Adamson, *J. Phys. Chem.* **78**, 1105 (1972).

121. I. Fujita and H. Kobayaski, *Ber. Bunsenges, Phys. Chem.* **76**, 115 (1972).

122. D. S. Engle and B. M. Monroe, *Advan. Photochem.* **8**, 245 (1971).

123. H. D. Gafney and A. W. Adamson, *J. Amer. Chem. Soc.* **94**, 8238 (1972).

124. C. R. Bock, T. J. Meyer, and D. G. Whitten, *J. Amer. Chem. Soc.* **96**, 6710 (1974).

125. G. Navon and N. Sutin, *Inorg. Chem.* **13**, 2159 (1974).

126. D. B. Buckingham and A. M. Sargeson, *Chelating Agents and Metal Chelates* (F. P. Dwyer and D. P. Mellor, Ed.), Chap. 8, Academic, New York, 1964.

127. J. N. Demas and A. W. Adamson, *J. Amer. Chem. Soc.* **95**, 5159 (1973).

128. (a) K. W. Hipps, G. D. Hager and G. A. Crosby, Abstracts, 166th National Meeting of the American Chemical Society, Chicago, Ill. (Aug., 1973), PHYS 81; (b) note added in proof to R. W. Harrigan and G. A. Crosby, *J. Chem. Phys.* **59**, 3468 (1973).

129. M. Wrighton and J. Markham, *J. Phys. Chem.* **77**, 3042 (1973).

130. F. Scandola, C. Bartocci, and M. A. Scandola, *J. Amer. Chem. Soc.* **95**, 7898 (1973).

131. R. M. Noyes, *Progr. Reaction Kinetics* **1**, 128 (1961).

132. (a) J. F. Endicott and G. J. Ferraudi, *J. Amer. Chem. Soc.* **96**, 3681 (1974); (b) J. F. Endicott, J. R. Barber, and G. J. Ferraudi, *J. Phys. Chem..*, **79**, 0000 (1975).

133. A. Haim and H. Taube, *J. Amer. Chem. Soc.* **85**, 495 (1963).

134. V. Balzani and V. Carassiti, *Photochemistry of Coordination Compounds*, p. 205, Academic Press, New York, 1970.

135. For a discussion of the free energy dependence of the rate constants for reduction of Co(III) complexes, see Ref. 103 and R. A. Marcus, *Annu. Rev. Phys. Chem.* **15**, 155 (1964).

136. N. S. Rowan, R. M. Milburn, and M. Z. Hoffman, *Inorg. Chem.* **11**, 2272 (1972).

137. T. B. Copestake and N. Uri, *Proc. Roy. Soc., Ser. A.* **228**, 252 (1955).

138. G. B. Porter, J. G. W. Doering, and S. Karanka, *J. Amer. Chem. Soc.* **84**, 4027 (1962).

139. D. P. Rillema and J. F. Endicott, *J. Amer. Chem. Soc.* **94**, 8711 (1972).

140. D. F. Gutterman and H. B. Gray, *J. Amer. Chem. Soc.* **93**, 3364 (1971).

141. H. Elsgernd and J. K. Beattie, *Inorg. Chem.* **7**, 2468 (1968).

142. A. W. Adamson, A. Vogler, and I. Lantzke, *J. Phys. Chem.* **73**, 4183 (1969).

143. (a) C. K. Jørgensen, *Absorption Spectra and Chemical Bonding in Complexes*, Wiley, New York, 1961; (b) H. H. Schmidtke, *Z. Physik. Chem.* **45**, 305 (1965).

144. C. Kutal and A. W. Adamson, *Inorg. Chem.* **12**, 1454 (1973).

145. A. Vogler and A. W. Adamson, *J. Phys. Chem.* **74**, 67 (1970).

146. See also footnote 7 of Ref. 98.

147. G. A. Crosby, R. J. Watts, and D. H. W. Carstens, *Science* **170**, 1195 (1970).

148. J. N. Demas and G. A. Crosby, *J. Amer. Chem. Soc.* **93**, 2841 (1971).

149. (a) D. A. Chaisson, R. E. Hintze, D. H. Stuermer, J. D. Petersen, D. P. McDonald, and P. C. Ford, *J. Amer. Chem. Soc.* **94**, 6665 (1972); (b) G. Malouf and P. C. Ford, *ibid.* **96**, 601 (1974).

150. (a) P. P. Zarnegar and D. G. Whitten, *J. Amer. Chem. Soc.* **93**, 3777 (1971); (b) P. P. Zarnegar, C. R. Bock, and D. G. Whitten, *ibid.* **95**, 4367 (1973).

151. An exception is $[Cr(NH_3)_5N_3]^{2+}$: A. Vogler, *J. Amer. Chem. Soc.* **93**, 5212 (1971).

152. M. T. Beck, *Coord. Chem. Rev.* **3**, 91 (1968).

153. M. G. Evans and G. H. Nancollas, *Trans. Faraday Soc.* **49**, 363 (1949).

154. T. J. Meyer and H. Taube, *Inorg. Chem.* **7**, 2369 (1968).

155. P. A. Rock, *Inorg. Chem.* **7**, 837 (1968).

156. For discussions of similar problems in excited states of charge-transfer complexes and other molecules involving large differences in excited-state and ground-state dipole moments see: (a) M. Ottolenghi, *Accounts Chem. Res.* **6**, 153 (1973); (b) E. Lippert, *ibid.* **3**, 74 (1970); (c) W. S. Struve, P. M. Rentzepis, and J. Jortner, *J. Chem. Phys.* **59**, 5014 (1973).

157. Estimate based on a force constant of $0.7 \times 10^5 \, dyn \, cm^{-1}$ for $Co(NH_3)_6^{2+}$ (K. Nakamoto, *Infrared Spectra of Coordination Compounds*, Wiley, New York, 1963).

158. H. L. Stynes and J. A. Ibers, *Inorg. Chem.* **10**, 2304 (1971).

159. J. A. Broomhead and N. A. P. Kane-Maguire, *Inorg. Chem.* **7**, 2519 (1968).

160. H. Hartmann and C. Bushbeck, *Z. Physik. Chem.* (Frankfurt) **11**, 120 (1957).

161. T. R. Thomas, R. J. Watts, and G. A. Crosby, *J. Chem. Phys.* **59**, 2123 (1973).

162. T. L. Kelly and J. F. Endicott, *J. Phys. Chem.* **76**, 1937 (1972).

163. J. D. Peterson and P. L. Ford, *J. Phys. Chem.* **78**, 1144 (1974).

164. (a) H. Cohen and D. Meyerstein, *J. Amer. Chem. Soc.* **94**, 6944 (1972); (b) D. Meyerstein, private communication.

4

SUBSTITUTIONAL PHOTOCHEMISTRY OF FIRST-ROW TRANSITION ELEMENTS

Edoardo Zinato

Istituto di Chimica Generale ed Inorganica
Universita di Perugia
Perugia, Italy

I. INTRODUCTION

Reactive deactivation of electronic excited states of coordination compounds may occur through three main pathways: (1) substitution and substitution-related photoreactions, (2) oxidation-reduction photoreactions, and (3) ligand photoreactions.

It is often observed that reactions of type (1) follow ligand-field (LF) band irradiation and those of type (2) are induced by charge-transfer (CT) excitation, while type (3) processes are associated with stimulation of electronic transitions between ligand-localized orbitals. This pattern, although very attractive, may represent a deceiving oversimplification for the following reasons.

First, absorption bands of different character may overlap to some extent. Then the labeling of electronically excited states according to the localization of the orbitals involved in the transition is, in many instances, an approximation (see Chapter 1). In addition, the excited state produced by light absorption may be converted to a (lower energy) state of a different type and of qualitatively different reactivity. Finally, the stabilized reaction products may differ markedly from those generated by the primary photochemical process.

For example, the stoichiometry of a photoreaction following CT band irradiation may be found to be consistent with a substitution-type process. On the other hand, it is known that photoredox reactions always produce reactive radicals (see Chapter 3). Then, if no radical can be detected, uncertainty may remain as to whether the observed behavior is due to conversion to a substitutionally reactive state or to analytical inadequacy.

In view of these complexities, "substitutional photochemistry" should not be considered as a synonym of "ligand-field photochemistry." The two definitions, however, have a great degree of overlap since, fortunately, for many families of complexes the correlations outlined above generally hold.

This chapter deals with the photosubstitutional reactivity of Werner-type complexes of the light transition elements. The role of the ligand-field (LF) excited states receives major attention, without neglecting the bordering areas delineated above. Since families of coordination compounds and patterns of behavior are focused, the literature survey is selective, rather than exhaustive. An outstanding book[1] and several reviews[2-16] provide an up-to-date literature coverage.

II. THE STUDY OF PHOTOSUBSTITUTIONS

A. Light versus Heavy Metal Complexes

As is usually done with the ground-state (or thermal) chemistry, there are sufficient reasons for considering separately the excited-state (or photo) chemistry of the first-transition-row metals and that relative to the second and third transition rows. Several differences occur in the ground-state properties between the two families of compounds, namely, in atomic and ionic radii, oxidation states, extent of π-bonding, magnetic properties, aqueous chemistry, and so forth. They are surveyed in Chapter 5.

It seems appropriate to underline those aspects having spectroscopic, and hence photochemical, implications. The splittings of the d orbitals (in a ligand field of a given symmetry) are smaller in the first transition row than in the successive ones. This feature, coupled with a higher electron-pairing energy for the $3d$ ions, makes possible the existence of high-spin and low-spin configurations, depending on the total field strength of the ligands. For example, for $3d^4$, $3d^5$, $3d^6$, $3d^7$ octahedral complexes, two sets of energy levels are possible in principle (see Chapter 1). The differences in orbital and spin multiplicity in the ground and in the excited states in the two series, give rise to completely different absorption spectra.

Due to the lower ligand-field splitting energy, the LF absorption bands of $3d$ complex ions occur at relatively high wavelengths compared to the analogous $4d$ and $5d$ systems; that is, while the former bands are observed in the visible or near-ultraviolet region (hence the variety of colors of the first-row complexes), the onset of the latter is usually in the short-wavelength side of the visible region. As a result, in the light metal complexes the range of the LF bands (at least the low-energy ones) is fairly well separated from that of the CT bands, which occur deeper in the ultraviolet and whose positions do not seem to depend on whether the metal is a light or a heavy one. A practical consequence is that $3d$ complexes may undergo more selective irradiation, leading to a more selective population of an excited state of a given type (LF or CT) and hence to a "cleaner" photochemistry.

Another major difference is that spin-orbit coupling is practically negligible in the first-row transition metals. Here the Russell–Saunders coupling approximation applies very well, that is, the orbital (L) and spin (S) numbers are "good" quantum numbers. In particular, the designation of a given excited state by its spin multiplicity is a fairly adequate one, as shown by the almost rigorous validity of the spin-selection rules. In fact, the "spin-forbidden," LF absorption bands have a much lower extinction coefficient ($\varepsilon = 0.1$–$1\ M^{-1}\ cm^{-1}$) than the "spin-allowed" ones ($\varepsilon = 10$–$100\ M^{-1}\ cm^{-1}$). Therefore, the distinction between "spin-allowed" and "spin-forbidden" energy levels appears meaningful with respect to several properties such as geometry, lifetime, reactivity, and so forth. On the contrary, such a labeling, although commonly used, becomes almost a mere formalism with the heavier metal ions, due to their high degree of spin-orbit coupling. Often "spin-forbidden" and "spin-allowed" LF bands of $4d$ and $5d$ systems have comparable intensities. An important experimental implication is that, in $3d$ metal complexes, population of "spin-forbidden" states by direct light absorption can be seldom achieved. In the first place, the optical density of solutions is extremely low in the spectral region involved; secondly, the "spin-forbidden" bands usually appear only as weak features on the tail of the much more intense "spin-allowed" bands, and light absorption populates both types of states simultaneously and nonselectively.

B. Some Properties of Ligand-Field Excited States

Several features of the LF excited states bear directly on substitutional photoreactivity. Some of the concepts of Chapter 1 are briefly recalled, taking as an example, the $3d^3$ Cr(III) complexes, whose photochemistry and luminescence are better understood than those of any other transition metal to date.

Ligand-field theory provides an adequate rationalization of the lower-energy absorption features of coordination compounds. A schematic energy-level diagram for octahedral (O_h) and tetragonally distorted (D_{4h} or C_{4v}) d^3 complexes is shown in Fig. 4-1, along with the electronic configurations, in the limiting strong-field approximation, for O_h geometry. In octahedral symmetry three Laporte-forbidden transitions

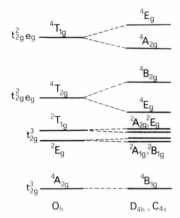

Fig. 4-1. Schematic diagram of the lowest energy levels for d^3 complexes of O_h, D_{4h}, and C_{4v} symmetries. The limiting electronic configurations refer to O_h symmetry (in C_{4v} symmetry, the g subscript should be omitted).

are possible: $^4A_{2g} \rightarrow {}^4T_{2g}$, $^4A_{2g} \rightarrow {}^4T_{1g}(F)$, $^4A_{2g} \rightarrow {}^4T_{1g}(P)$, giving rise to three absorption bands designated as L_1, L_2, L_3 in order of increasing energy.[4] A series of spin-forbidden transitions may also occur as, for example, the $^4A_{2g} \rightarrow {}^2E_g$ one. Due to the restrictive application of the selection rule, only the lowest-energy transitions are experimentally detectable as weak details on the long-wavelength side of the L_1 band. A different label is usually given to these absorptions, as D (for doublet), for Cr(III).

Figure 4-2 shows the complete electronic absorption spectrum of a typical Cr(III) complex ion. With the exception of the systems having very low-field ligands, the L_3 band is usually obscured by the more intense, fully "allowed," CT bands in the ultraviolet region. Although the chosen acidopentaammine ion has C_{4v} microsymmetry, no appreciable splitting occurs in the L_1 and L_2 bands, due to the close position of the NH_3 and NCS^- ligands in the spectrochemical series, and the compound may still be regarded as almost octahedral.

If covalent σ- and π-bonding is taken into account, a molecular orbital (MO) treatment leads to the orbital ordering illustrated in Fig. 1-7 and predicts the same number and types of transitions, which are now depicted as "metal-localized" since they involve orbitals having mainly metal character.

The principal factor determining the chemical reactivity of a molecule in the ground state is its electronic distribution. The kinetic lability or inertness of coordination compounds toward substitution is well accounted for in terms of occupancy of the various d orbitals, according to the valence bond (VB), LF, and MO theories.[17] The same considerations can be fruitfully applied in predicting, at least qualitatively, the reactivity of the excited states.

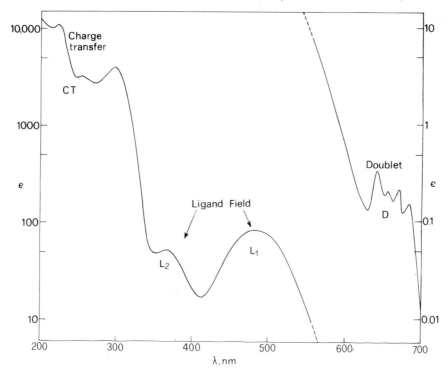

Fig. 4-2. Visible and ultraviolet absorption spectrum of aqueous $[Cr(NH_3)_5(NCS)]^{2+}$.[29]

According to LF theory, a d–d transition produces redistribution of electronic charge between orbitals belonging to the metal ion. In an octahedral complex it involves promotion of electron density from the t_{2g} orbitals, which point toward the edges of the octahedron (away from the ligands), to the e_g ones, in the direction of the ligands. For example, while the ground-state configuration of a d^3 system is t_{2g}^3, that of the first "spin-allowed" excited state is $t_{2g}^2 e_g$. The accumulation of negative charge between the ligands and the metal center in the $^4T_{2g}$ state is expected to give rise to ligand repulsion and hence to bond labilization. In addition, the lowering of electron density along the edges toward which the emptied t_{2g} orbital is directed, may favor the nucleophilic attack of an incoming ligand. The latter situation is expected in some of the t_{2g}^3 configurations of low-energy, spin-forbidden states, such as ^2E_g. These states differ from the ground state only by spin pairing within a set of degenerate orbitals. The ligand-field contribution to the activation energy for substitution occurring from "spin-allowed" and "spin-forbidden" excited states, assuming different geometries for the possible transition state, has been semiquantitatively evaluated for Cr(III) and Co(III) complexes[18] and is discussed in Sections III.D and IV.D.

A MO scheme leads to virtually identical predictions. If only σ-bonding is considered, a spin-allowed d–d transition results in promotion of electrons from the t_{2g}, nonbonding orbitals to the e_g ones, which are σ-antibonding, determining the weakening of some metal-ligand bond. Possibly vacant t_{2g} orbitals would then be available for assisting the entry of a new ligand. The availability of an empty t_{2g} orbital for bond formation would explain the lability of a "spin-forbidden" excited state (e.g., a doublet state of Cr(III)) as well.

Some additional predictions may be made taking into account π-bonding effects. If the complex contains ligands with full (donor) π orbitals, such as halide groups, the t_{2g} orbitals become π-antibonding; the spin-allowed transition is then expected to strengthen the π-bonds. With ligands having empty (acceptor) π orbitals, such as carbonyl and cyanide, the situation is roughly reversed.[19] A detailed analysis of these aspects has been recently carried out that accounts for the photoreactivity in d^3 and d^6 systems.[20,21]

In conclusion, LF transitions involve *angular*, rather than radial, movement of electron density. The expected photoreactivity is one of heterolytic bond fission, that is, substitutional, as opposed to the homolytic bond fission associated with the radial migration of charge induced by CT excitation (see Chapters 1 and 3).

Electronic distribution has received major attention in rationalizing excited-state chemistry. However, other characteristics, such as geometry and energy, are important in determining the photochemical behavior of a given excited state. A point particularly relevant to the latter properties has been recently emphasized.[14] Ligand-field and molecular-orbital theories assume that the states between which electronic transitions occur have the same molecular symmetry and that the energy separations between these states are given by the positions of the corresponding band maxima.

The well-characterized spectroscopic behavior of Cr(III) (see Chapters 1 and 2) has provided evidence that these assumptions are unjustified, very likely for all transition-metal complexes. L-type absorption bands are generally broad and structureless, while the "spin-forbidden" ones are sharp and often well structured. The implication is that the potential energy curve (in a one-dimensional diagram) of a "spin-allowed" excited state is, in general, considerably shifted with respect to the ground-state curve. The electronic transition then produces the excited state in a high vibrational level, according to the Franck–Condon principle, (see Fig.

1-2). On the contrary, the lowest "spin-forbidden" excited states, arising from the t_{2g}^3 configuration, virtually possess the same geometry of the ground state. Luminescence data at low temperatures confirm this picture. When both fluorescence (spin-allowed) and phosphorescence (spin-forbidden) are observed, as in the classical system[22] of $Cr(urea)_6^{3+}$, the $^2E_g \rightarrow {}^4A_{2g}$ emission is again sharp, and its maximum almost coincides with that of the D absorption, while the $^4T_{2g} \rightarrow {}^4A_{2g}$ fluorescence is broad and is centered at lower energy with respect to phosphorescence.

This situation is presented in Fig. 2-6. The strong shift to lower energy for fluorescence relative to absorption indicates that vibrational relaxation of the $^4T_{2g}$ (or, better, $^4L_1$)[14] state must be accompanied by considerable distortion. Emission occurs from the thermally equilibrated excited ("thexi")[14] state to a high vibrational level of the ground state. It has been

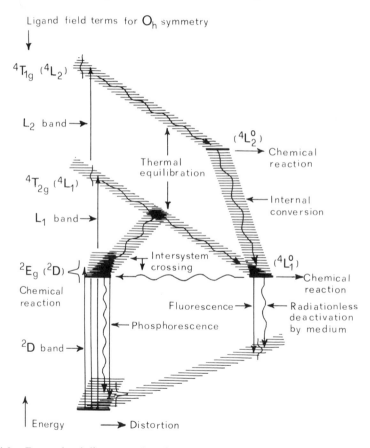

Fig. 4-3. Energy-level diagram and excited-state processes for a Cr(III) complex.[11]

pointed out[23] that such a distortion may be both radial and angular, that is, the thexi state may have both different bond lengths and different symmetry with respect to the original octahedral framework.

Figure 4-3 schematically shows the relation between distortion and relaxation processes.[11] Light absorption produces an excited state (for example, $^4L_1$) having the geometry of the ground state. In fluid solution, that is, in the conditions in which photoreactions are studied, thermal equilibration is expected to occur very rapidly and even more easily than in the solid state or in glassy matrices (where, even there, luminescence shows a large degree of distortion already present). Photoreactions are thus likely to take place from fully equilibrated excited states (such as $^4L_1^0$), having the maximum possible degree of distortion. The type of geometrical change may play an important role in determining the nature of the products.

A related point is the location of the *equilibrated* excited-state energies. Such information is obtainable when spin-forbidden transitions are involved, thanks to the coincidence of emission and absorption maxima. It is certain that the energies of the "spin-allowed" thexi states are smaller than evaluable from absorption band maxima. Only if fluorescence is observed is it possible to estimate the 0–0 transition energy from the wavelength at which the normalized absorption and emission bands cross. Unfortunately, this is possible only with a limited number of systems. The need for true energy data remains very crucial, and an empirical procedure has been proposed for evaluating them from absorption spectra, at least for Cr(III).[11,14] Another promising approach is the use of sensitizers and quenchers of well-characterized energy levels.

C. Some Experimental Aspects

The experimental investigation of photosubstitutions of coordination compounds is limited by various factors. One problem arises from the thermal reactivity of complexes. Under steady illumination a certain period of time is required for generating analytically determinable amounts of products. The final reaction mixture is, in general, the result of one or more photoreactions superimposed on a "dark" process. Therefore, only kinetically nonlabile complexes can be studied using traditional analytical methods. When necessary, the thermal contribution to the overall products is taken into account by carrying out parallel analyses on "dark" samples. This complication, which is not generally present in organic photochemistry, in principle confines the photosubstitution studies to $3d^3$ and low-spin $3d^4$, $3d^5$, and $3d^6$ octahedral complexes.[17] The systems that have been largely investigated are the most inert ones, that is, those of d^3 and d^6 configurations. Cr(III) and Co(III) ions have received more attention than any other family of coordination compounds, due to the availability of a large number of complexes of relatively easy preparation and well-characterized structure and thermal reactivity. A smaller body of literature exists on low-spin, Fe(II) compounds, essentially on cyanide and substituted cyanide complexes. Some scattered data are found for other species as, for example, the d^4, $Mn(ox)_3^{3-}$ $(ox^{2-} = C_2O_4^{2-})$ ion whose behavior, however, is not photosubstitutional.

A connected problem is that the photolysis products may recombine thermally, forming the original complex. Such systems are labeled photoreversible and, under steady irradiation, reach a photostationary state in which the photochemical (plus thermal) forward reaction rate is equal to the rate of the back thermal process. The phenomenon is known as photochromism and is discussed in Chapter 10. The photosubstitutional products of inert complexes are themselves inert. Usually, either the reaction is irreversible or the rate of recombination is negligible, especially in the early reaction stages; that is, the photostationary state is seldom reached. Conversely, possible substitution products of labile species, for example, $Ni(en)_3^{2+}$ (stable only in the presence of excess ethylenediamine), certainly regain equilibrium with the environment almost instantaneously. It is likely that the apparent insensitivity to light of many labile ions is simply due to an exceedingly fast back reaction, so that the photostationary state corresponds to virtually zero transformation. Fast analytical techniques, such as flash photolysis, should yield more information on this point. Of course, a variety of intermediate situations may occur.[4] Photoredox reactions, instead, involve irreversible steps since they produce radicals that react further, rather than recombine. They are, in fact, observed regardless of the kinetic nature of the reactant.[1,4]

Another complication, peculiar to LF photochemistry, is due to the small extinction coefficients of the $d-d$ absorption bands. Although CT irradiation can be performed in conditions of total light absorption, as is usual in organic photochemistry ($\varepsilon = 10^3 - 10^4 \, M^{-1} \, cm^{-1}$), complete absorption of incident radiation may not be always possible in the ligand-field band region ($\varepsilon = 10 - 100 \, M^{-1} \, cm^{-1}$). The fraction of light intensity instantaneously absorbed by a partially transmitting solution, at a given wavelength, is

$$\frac{dE}{dt} = I_0(1 - 10^{-\varepsilon cl}) \tag{4-1}$$

where I_0 is the incident intensity, ε the extinction coefficient of the absorbing species at that wavelength, c its instantaneous concentration, and l the cell pathlength. When reaction takes place, c decreases in time (and in space if stirring is not homogeneous), so that the total energy, E, absorbed during a finite photolysis time may be a complicated function of the system variables and difficult to calculate. The problem can be solved mathematically;[24] however, more practical procedures can be adopted. One method, which does not require the use of monochromatic radiation, consists of employing an actinometric solution having the same optical density as the sample, although the variation of absorbance in time is not, in principle, the same for the two solutions.

A more reliable determination may be achieved by differential actinometry or bolometry, that is, by monitoring both the incident and transmitted light quanta during the actual reaction time. The difficulties inherent to irradiation of D-type absorption features (see Section II.A) are even greater and have represented a considerable nuisance in photochemical studies, given the important role of spin-forbidden states (in organic photochemistry they are the principal precursors to photoreactions). Efficient indirect population of these energy levels is achieved either by intersystem crossing from higher energy, spin-allowed states (which, however, often introduces uncontrollable variables) or by energy transfer (see Chapters 1 and 2).

Furthermore, unlike redox photoreactions, photosubstitutions produce species whose absorption spectra are not usually very different from those of the reactants. Both the photoproducts and possible thermal products then subtract part of the incident radiation. The phenomenon, known as the "inner-filter effect," becomes more and more important as photolysis proceeds. For a correct evaluation of the product quantum yield, the fraction of energy absorbed exclusively by the reacting complex must be used. A mathematical treatment has been proposed that gives the quantum yields, after a finite period of irradiation, for two parallel photoreactions accompanied by a thermal process, provided the rate constant of the latter and the extinction coefficients for all the species present are known.[25]

Additionally, the partially absorbed light is expected to cause secondary photolysis of the primary photosubstitution products since the latter usually differ from the original complex by only one ligand. All these effects, in principle, can be taken into account mathematically. However, in common practice, they are minimized either by carrying out photolyses to small reaction extents or by extrapolating the results to zero irradiation time.

As a last point, regarding the determination of light intensity, it is interesting to note that all the actinometric systems commonly employed involve coordination compounds such as ferrioxalate,[26] uranyl oxalate,[27] and Reineckate ion, *trans*-$[Cr(NH_3)_2(NCS)_4]^-$.[28] The latter actinometer is particularly suitable for the 315 to 600 nm spectral range, where the LF absorption bands of $3d$ metal complexes are generally found. It is based on a photosubstitution reaction:

$$[Cr(NH_3)_2(NCS)_4]^- + H_2O \xrightarrow{h\nu} [Cr(NH_3)_2(NCS)_3(H_2O)] + NCS^- \quad (4\text{-}2)$$

Photoreleased thiocyanate ion is readily determined after complexing it with $Fe^{3+}_{(aq)}$. The introduction of the Reineckate actinometer has overcome a conspicuous limitation to photolysis studies with visible light, since the early developed systems could not be utilized satisfactorily at wavelengths higher than 500 nm.

III. LIGAND-FIELD PHOTOCHEMISTRY OF Cr(III)

The thermal chemistry of Cr(III) complexes in solution is one of slow ligand substitution. Solvation, anation, isomerization, and racemization reactions are observed.[17] The last two are probably not intramolecular, that is, they involve bond rupture and reattachment of ligands. In contrast to Co(III), a favorable redox potential makes Cr(III) species in solution very stable toward reduction.

The term diagrams of a d^3 complex ion, for different symmetries, are illustrated in Fig. 4-1, and a typical absorption spectrum of a Cr(III) species is presented in Fig. 4-2. Usually the L_1 and L_2 absorption bands (due to the $^4A_{2g} \rightarrow {}^4T_{2g}$ and $^4A_{2g} \rightarrow {}^4T_{1g}$ electronic transitions, respectively, in O_h approximation) appear in the visible or near-ultraviolet region, well separated from the intense CT absorptions.

Ligand-field band irradiation gives rise exclusively to substitution or substitution-related photoreactions, not complicated by the simultaneous occurrence of photoredox decomposition as is common for many Co(III) compounds. As in the ground-state chemistry, this behavior is attributed to the low tendency of Cr(III) to be reduced.

Photosolvation, photoisomerization, photoracemization, and photoanation reactions have been found to occur. The best investigated ones are, of course, photoaquations, due, in practice, to the great deal of available knowledge on thermal aquation. These photoreactions are generally studied in acid solution. The rates of "dark" acid hydrolysis are, in fact, smaller by some orders of magnitude compared to those of base hydrolysis; in complexes containing aquo ligands, the latter takes place at pH values as low as 4 to 5.[28,29]

In this section, the most relevant literature data is surveyed and organized according to reaction patterns. The emerging features are then used for discussing the involvement of the excited states in photoreactions and the possible mechanisms.

A.   Photoaquation

1. *Octahedral Complexes.*   Complex ions having octahedral microsymmetry undergo photoaquation of one ligand; for example,

$$Cr(NH_3)_6^{3+} + H_3O^+ \xrightarrow{h\nu} [Cr(NH_3)_5(H_2O)]^{3+} + NH_4^+ \qquad (4\text{-}3)$$

For chelate complexes, such as $Cr(en)_3^{3+}$, the primary photochemical process is the detachment of one end of the bidentate ligand

$$Cr(en)_3^{3+} + H_3O^+ \xrightarrow{h\nu} [Cr(en)_2(enH)(H_2O)]^{4+} \qquad (4\text{-}4)$$

followed by thermal cleavage of the second metal-ligand bond.[30]

$$[Cr(en)_2(enH)(H_2O)]^{4+} + H_3O^+ \xrightarrow{kT} cis\text{-}[Cr(en)_2(H_2O)_2]^{3+} + enH_2^{2+} \quad (4\text{-}5)$$

The water photoexchange of $Cr(H_2O)_6^{3+}$ [31] and the photoracemization of $Cr(ox)_3^{3-}$ [32] (Section III.B.2) may be considered related to this pattern of reactivity.

An immediate observation is that, when only one kind of coordinated ligand is present, the photoreactions have, of necessity, a stoichiometry identical to that of the thermal processes. Therefore, these systems do not provide an answer to the fundamental question as to whether or not the photochemistry of Cr(III) amounts to no more than a photoacceleration of existing thermal reaction paths.

The data, collected in Table 4-1, demonstrate two significant regularities. (1) The photoaquation quantum yields range between 0.1 and 0.5 and, except for $Cr(CN)_6^{3-}$, they are larger, the higher the position of the ligand in the spectrochemical series. The exception of $Cr(CN)_6^{3-}$ is attributed to the extensive degree of π-bonding of this complex.[33] (2) The quantum yields are virtually independent of the excitation wavelength. Where it is possible to irradiate the ^2E_g absorption band selectively, such independence is found to extend to this region.[28,34] This constancy suggests that photoreaction takes place from the lowest excited state of a given spin multiplicity, namely, either the ^2E_g or the $^4T_{2g}$ state.

It may be noticed that the determination of the quantum yield on population of the lowest doublet state by direct light absorption is critical for identifying the photoreactive state. Such experiments are subject to the difficulties outlined in Section II.A, and the results available at present indicate that the quantum yields in the doublet region do not significantly differ from those relative to irradiation of the L-type bands.

Table 4-1. Photoaquation Quantum Yields for Cr(III) Complexes of Octahedral Microsymmetry upon LF Excitation[a]

	Band Irradiated			
Complex	L_2	L_1	D	Ref.
$Cr(urea)_6^{3+}$	0.10	0.10	0.095	28
$Cr(H_2O)_6^{3+}$ [b]	0.1	0.1	—	31
$Cr(ox)_3^{3-}$ [c]	0.09	0.09	—	32
$Cr(NCS)_6^{3-}$	0.26	0.27	—	28
$Cr(NH_3)_6^{3+}$	0.3	0.26	0.29	28, 35
$Cr(en)_3^{3+}$	0.37	0.37	0.40	30, 34
$Cr(CN)_6^{3-}$	0.12	0.12	—	33, 36

[a] Temperatures are usually in the 15 to 25°C range.
[b] Water photoexchange.
[c] Photoracemization.

2. *Mixed-Ligand Complexes.* A greater variety of potentially richer information is derived from the photoreactivity of complexes with more than one kind of ligand, that is, of microsymmetry lower than octahedral. Photolysis studies have been carried out especially with acidoamine Cr(III) ions. The most significant data for C_{4v} and D_{4h} species are summarized in Tables 4-2 and 4-3.

Table 4-2. Photolysis of $Cr(NH_3)_5X^{n+}$-Type Ions[a]

X	Band Irradiated	ϕ_{NH_3}	ϕ_X	ϕ_{NH_3}/ϕ_X	Configuration of Main Product	Ref.
H_2O	L_2	0.2	—[b]	—	—	40
	L_1	0.15	—[b]	—	—	40
Cl^-	CT	0.35	0.23	—	—	42
	L_2	0.38	0.007	55	—	41, 42
	L_1	0.36	0.005	70	cis	41, 42, 43
Br^-	CT	0.20	0.26	—	—	44
	L_2	0.37	0.011	34	—	44
	L_1	0.35	0.009	39	cis	44
NCS^-	L_2	0.46	0.030	15	—	29
	L_1	0.48	0.021	23	cis	29, 45
	D	0.15	0.018	8	—	29
$OCOR^-$ [c]	L_1	0.25–0.45	<0.1	—	cis	46
N_3^-	L_1	large	—	—	—	47

[a] Temperatures 0 to 25°C, constant for a given complex.
[b] Water exchange was not investigated.
[c] $R = CF_3, CCl_3, CHCl_2, CH_2Cl, CH_3$; quantum yields unconnected with the base strength.

Table 4-3. Photolysis of *trans*-Diacidotetraaminechromium(III) Ions upon L_1 Band Irradiation

Complex	Product	ϕ	Remarks	Ref.
$[Cr(NH_3)_4Cl_2]^+$	*cis*-$[Cr(NH_3)_4(H_2O)Cl]^{2+}$	0.44	$\phi_{NH_3}=3\times10^{-3}$	38
$[Cr(NH_3)_4(H_2O)Cl]^{2+}$	*cis*-$[Cr(NH_3)_4(H_2O)Cl]^{2+}$	0.40	$\phi_{Cl^-}<4\times10^{-3}$ $\phi_{NH_3}=0.02$	38
$[Cr(NH_3)_4(H_2O)(NCS)]^{2+}$	*cis*-$[Cr(NH_3)_4(H_2O)(NCS)]^{2+}$	0.42	$\phi_{NCS}=0.02$ $\phi_{NH_3}=5\times10^{-3}$	38ᵗ
$[Cr(NH_3)_4(NCS)Cl]^+$	*cis*-$[Cr(NH_3)_4(H_2O)Cl]^{2+}$	0.27	$\phi_{NH_3}=3\times10^{-3}$	38
	cis-$[Cr(NH_3)_4(H_2O)(NCS)]^{2+}$	0.14		
$[Cr(NH_3)_4(H_2O)_2]^{3+}$	*cis*-$[Cr(NH_3)_4(H_2O)_2]^{3+}$	0.35	$\phi_{NH_3}<0.02$	38
$[Cr(en)_2Cl_2]^+$	*cis*-$[Cr(en)_2(H_2O)Cl]^{2+}$	0.32ᵃ	$\geq70\%$ *cis* prod.	48
$[Cr(en)_2(NCS)Cl]^+$	*cis*-$[Cr(en)_2(H_2O)Cl]^{2+}$	0.18	$\phi_{H^+}=6\times10^{-3}$	49
	cis-$[Cr(en)_2(H_2O)(NCS)]^{2+}$	0.04		
$[Cr(en)_2(H_2O)(OH)]^{2+}$	*cis*-$[Cr(en)_2(H_2O)(OH)]^{2+}$	0.3	No aquation	37
$[Cr(en)_2FCl]^+$	*cis*-$[Cr(en)_2(H_2O)F]^{2+}$	0.31	90% *cis* prod. $\phi_{en}=0.04$	86
$[Cr(en)_2F_2]^+$	$[Cr(en)(enH)(H_2O)F_2]^{2+}$	0.46	$\phi_F\leq0.08$	50
$[Cr(tet)Cl_2]^{+\ b}$	*cis*-$[Cr(tet)(H_2O)Cl]^{2+}$	0.06	$\phi_{H^+}<7\times10^{-3}$	51
$[Cr(cyclam)Cl_2]^{+\ c}$	*trans*-$[Cr(cyclam)(H_2O)Cl]^{2+}$	3.3×10^{-4}	$\phi_{H^+}<5\times10^{-4}$	51, 52

ᵃ Quantum yield for chloride aquation.
ᵇ tet = 1,4,8,11-tetraazaundecane.
ᶜ cyclam = 1,4,8,11-tetraazacyclotetradecane.

One outstanding feature is that photoaquations are, in general, qualitatively different from thermal aquations with respect both to the types of ligands released and the reaction stereochemistry. For such behavior, the designation "antithermal" has been proposed.[11] A specific excited-state chemistry does, therefore, exist, markedly different from the ground-state chemistry.

Generally, two, or more, photoreaction modes take place simultaneously, and the predominant photochemical process is often different from the thermally preferred one. Some typical examples of antithermal reactivity are the following. (1) All $Cr(NH_3)_5X^{2+}$-type ions studied to date undergo photoaquation of ammonia and of the acido group, X^-. NH_3 aquation is the dominant photoreaction path (see Table 4-2), while only the X^- ligand is released in the dark.

$$Cr(NH_3)_5X^{2+} + H_2O \xrightarrow{h\nu} \begin{cases} \textit{cis-}[Cr(NH_3)_4(H_2O)X]^{2+} + NH_3 & (4\text{-}6) \\ Cr(NH_3)_5(H_2O)^{3+} + X^- & (4\text{-}7) \end{cases}$$

(2) For *cis*-$[Cr(en)_2(OH)_2]^+$ the thermal reaction is essentially one of

isomerization [Eq. (4-8)], while photochemically both isomerization and aquation are observed, and the latter mode [Eq. (4-9)] predominates.[37]

$$cis\text{-}[Cr(en)_2(OH)_2]^+ \xrightarrow{h\nu} \begin{cases} trans\text{-}[Cr(en)_2(OH)_2]^+ & (4\text{-}8) \\ \\ [Cr(en)(H_2O)_2(OH)_2]^+ + en & (4\text{-}9) \end{cases}$$

(3) The *trans*-chlorothiocyanatotetraamminechromium(III) ion undergoes chloride aquation in the dark with retention of configuration

$$trans\text{-}[Cr(NH_3)_4(NCS)Cl]^+ + H_2O \xrightarrow{kT}$$

$$trans\text{-}[Cr(NH_3)_4(NCS)(H_2O)]^{2+} + Cl^- \quad (4\text{-}10)$$

Ligand-field excitation, instead, leads to both Cl⁻ and NCS⁻ aquation with comparable quantum yields. Furthermore, the reaction products have *cis* configuration.[38]

$$trans\text{-}[Cr(NH_3)_4(NCS)Cl]^+ + H_2O \xrightarrow{h\nu} \begin{cases} cis\text{-}[Cr(NH_3)_4(H_2O)(NCS)]^{2+} + Cl^- & (4\text{-}11) \\ \\ cis\text{-}[Cr(NH_3)_4(H_2O)Cl]^{2+} + NCS^- & (4\text{-}12) \end{cases}$$

(4) Irradiation of the binuclear *trans*-$[(NH_3)_5Cr(OH)Cr(NH_3)_4Cl]^{4+}$ species produces both chloride and ammonia aquation with stereochemical change in at least one photoproduct, while the thermal reaction involves only the cleavage of the hydroxo bridge with stereoretention.[39]

The quantum yields relative to different reaction modes are generally wavelength dependent. It is remarkable that also the *ratios* between yields for various paths of reactivity change with changing excitation wavelength. The data reported in Table 4-2 for NH_3 and NCS^- photoaquation of $[Cr(NH_3)_5(NCS)]^{2+}$ in the L_2, L_1, and D absorption region[29] exemplify this point.

The implication is that *different* excited-state precursors to reaction are involved. If all reactions took place from a unique excited state, the ratios of quantum yields for different reaction modes should be independent of the irradiating wavelength, although the constancy would not be required for their absolute values. In fact, in such a circumstance the reactive state may be populated to a greater or lesser extent, depending on the relative importance of the processes that competitively deactivate the state produced by absorption of radiation.

3. *The Photolysis Rules.* On the basis of the experimental data available at that time, some semiempirical rules[23] (known as Adamson's rules) were formulated in 1967, which proved able to systematize and predict the photoaquation behavior of nonoctahedral Cr(III) complexes. The principal statements are as follows:

[1] Consider the six ligands to lie in pairs at the ends of three mutually perpendicular axes. That axis having the weakest average crystal field will be the one labilized, and the total quantum yield will be about that for an O_h complex of the same average field.

[2] If the labilized axis contains two different ligands, then the ligand of greater field strength preferentially aquates. This may be a type of *trans* effect.

The predictive application of rule [1] is illustrated by the following examples.

The weaker-field axes of *trans*-[Cr(NH$_3$)$_2$(NCS)$_4$]$^-$ are the NCS–NCS ones, and the prediction is that labilization occurs along these axes. The only photoreaction taking place [Eq. (4-2)] is thiocyanate aquation.[28]

In *trans*-[Cr(NH$_3$)$_4$Cl$_2$]$^+$ the lowest ligand-field axis is the Cl–Cl one, and, as expected, the only reaction mode observed is chloride photorelease.[38]

Rule [2] applies, for example, to systems of C_{4v} symmetry, such as Cr(NH$_3$)$_5$X^{n+} and *trans*-[Cr(NH$_3$)$_4$XY]$^{n+}$ ions. All the X and Y ligands of Tables 4-2 and 4-3 have a spectrochemical position lower than NH$_3$. In all acidopentaammines (Table 4-2) the axis of weakest average field strength is thus the NH$_3$–X one. Ammonia is predicted, and found, to be the ligand preferentially aquated. *Trans*-diacidotetraamines (Table 4-3) undergo labilization along the X–Y axis. For *trans*-[Cr(NH$_3$)$_4$(NCS)Cl]$^+$ [38] and *trans*-[Cr(en)$_2$(NCS)Cl]$^+$,[49] the NCS$^-$ aquation quantum yield is consistently larger than that for Cl$^-$ release, again in agreement with the greater spectrochemical strength for the thiocyanate ligand.

The apparent exception of *trans*-[Cr(en)$_2$F$_2$]$^+$, whose main photoreaction path is ethylenediamine aquation,[50] is discussed in Section III.D.3, together with the theoretical background of the photolysis rules.

4. *Stereochemistry.* In several cases the photolysis products of D_{4h} and C_{4v} acidoamine complexes have been isomerically characterized, and

the *cis* configuration has been found to be the dominant, if not the exclusive, one. (See Tables 4-2 and 4-3). Thus, as shown in Eq. (4-6) the principal species produced in the photoaquation of $Cr(NH_3)_5X^{2+}$-type ions is *cis*-$[Cr(NH_3)_4(H_2O)X]^{2+}$.[41-46] In addition, *cis*-$[CrA_4(H_2O)X]^{2+}$ and/or *cis*-$[CrA_4(H_2O)Y]^{2+}$ are the isomers obtained photolyzing *trans*-$[CrA_4XY]^{n+}$ compounds.[37,38,48,49]

The stereochemical behavior of acidopentaammine complexes bears on the predictive limits of the above-mentioned rules. At least three interpretations, increasingly restrictive, are possible.

a. The rules imply no more than the data upon which they were constructed; that is, they predict only the *type* of ligand predominantly labilized, for example, NH_3 in $Cr(NH_3)_5X^{2+}$ ions.

b. If there are two or more ligands of the same type, the rules specify *which* actual ligand is labilized, for example, NH_3 *trans* to X^- in the above case.

c. Besides specifying the actual ligand labilized, the rules imply that the photosubstitution reaction is *stereoretentive*, for example, that the NH_3 aquation product is *trans*-$[Cr(NH_3)_4(H_2O)X]^{2+}$.

The acidopentaammines are poorly diagnostic in this respect. The *cis* configuration of the photoproducts certainly rules out interpretation c, while it is not sufficient to distinguish between views a and b. This problem has given rise to some discussion about the usefulness of the rules.[43,53]

That interpretation b is correct, is suggested by the behavior of the *trans*-diacidotetraamine species, in which the photoreleased ligands are unequivocally those on the weakest field axis. The general validity of point b has been proven by a recent investigation on the $[Cr(NH_3)_4(^{15}NH_3)Cl]^{2+}$ ion, labeled either *cis* or *trans* to chloride: 75 to 100% of NH_3 aquated upon L_1 band irradiation was originally *trans* to the Cl^- ligand.[54]

Thus it seems possible to generalize that in C_{4v} and D_{4h} ions, although LF irradiation leads to *trans* labilization, the products are always in the *cis* configuration. A uniform photoaquation mechanism, involving stereomobility, therefore, appears to apply at least to Cr(III) amines.[53] The situation is quite different in Cr(III) thermal-substitution processes, which are generally stereorigid.[55,56]

Noteworthy is the fact that stereochemical change is an important and probably necessary condition for photosubstitution to occur. This is indicated by the variation of chloride photoaquation quantum yields (see

Table 4-3) along the series

$$(4\text{-}13)$$

An increasing degree of hindrance to stereomobility, imposed by the aliphatic chains, is accompanied by a decrease in the efficiency of Cl^- substitution. The extremely low photoreactivity of compound IV coincides with the complete rigidity of the octahedron. It should be noticed that the phosphorescent intensities and lifetimes of the latter species do not significantly differ from those of the other complexes of Eq. (4-13). Even though the luminescent behavior does not provide direct information about the rates of the radiationless processes that competitively deactivate the $(^4L_1^o)$ reactive state (see Section III.D.2 and Fig. 4-3), suggests that no unusual photophysical processes are taking place.[51,52]

B. Other Substitution-Type Photoprocesses

1. *Photoisomerization.* LF excitation produces *trans* → *cis* isomerization in a number of diacidotetraamine complexes. *trans*-[Cr(NH_3)_4(H_2O)_2]^{3+},[38] *trans*-[Cr(NH_3)_4(H_2O)X]^{2+} (X = Cl, NCS),[38] and *trans*-[Cr(en)_2(H_2O)(OH)]^{2+} [37] are converted into the respective *cis* isomers, and the photoreaction is very efficient. ($\phi = 0.2$–0.4, see Table 4-3).

In all these species at least one water ligand is present on the lowest average-field axis. Isomerization is very likely to be accompanied by, water exchange with the solvent through the same mechanism governing photoaquation in the other acidoamine complexes. This suggestion can be tested experimentally by two different procedures; first, by the examination of oxygen photoexchange [investigated[31] thus far only with $Cr(H_2O)_6^{3+}$; it fits the general photochemical pattern of Cr(III)], and second, by the determination of ligand aquation and isomerization quantum yields for *trans*-[CrA_4XY]^{n+}-type ions. If the yields are equal for the two paths, the distinction from photoaquation reactions becomes only formal and is due simply to the fact that water photoexchange has not been looked for.

2. *Photoracemization.* Aqueous $Cr(ox)_3^{3-}$ photoracemizes with a virtually wavelength-independent quantum yield ($\phi = 0.09$),[32] which is in line with the behavior of other octahedral complexes (see Table 4-1).

The quantum yield for photoracemization and the rate for thermal racemization decrease by the same extent (25%) in D_2O, indicating an equal importance of hydrogen bonding in the two reactions. Also, ^{18}O photoexchange takes place, although with a quantum yield lower than that of photoracemization. Again, however, the ratio of racemization to exchange is the same in the photo- and in the dark reaction.[32] Similar solvent effects on the rates and activation energies for the photo and thermal process are observed.[32] These similarities suggest a mechanism for photoracemization related to that of thermal racemization, involving, as primary step, the cleavage of a Cr—O bond, with coordination of solvent, and possibly the formation of *ortho*-carboxylate.

(4-14)

Water exchange indicates the importance of this mechanism with respect to an intramolecular twist.

Also, $Cr(phen)_3^{3+}$ and $[Cr(phen)_2(ox)]^+$ photoracemize in aqueous solution with constant quantum yields of 0.015 and 0.21, respectively, throughout the $d-d$ absorption region.[57] The mechanism has not been investigated.

The related phenomenon of photoresolution deserves mention in this section. Racemic mixtures of $Cr(ox)_3^{3-}$,[58] and of $Cr(acac)_3$ [59] are partially resolved by irradiation with circularly polarized light of 546 nm.

Another pertinent observation is the partial resolution obtained by visible irradiation of a racemic mixture of $Cr(phen)_3^{3+}$ in the presence of an optically active anion (antimony-d-tartrate). A steady state is reached in which the mixture is enriched in one enantiomer.[60]

3. Photoanation. Only a few systems have been investigated. The photoanation of hexaaquochromium(III) by chloride or thiocyanate[61] may be taken as an example. The quantum yields are rather low ($10^{-2}-10^{-4}$), and the actual species photolyzed appears to be the ion pair formed in small amount by the complex and the entering anion. Irradiation is thought to activate an inner–outer sphere exchange of ligands.

$$Cr(H_2O)_6^{3+} + X^- \rightleftarrows [Cr(H_2O)_6X]^{2+} \xrightarrow{h\nu} Cr(H_2O)_5X^{2+} + H_2O \quad (4\text{-}15)$$

An alternative path, consisting of direct reaction between the anion and an excited complex ion, is unlikely on the grounds that the excited states possibly involved (*vide infra*) are too short-lived to survive until the reactive encounter with the anion occurs.[61]

4. Photosubstitutions in Nonaqueous Solvents. A limited number of photolysis studies in solvents different from water have been reported. However, in light of the existing data, there appear to be considerable solvent effects on both photophysical and photochemical processes.

For example, $Cr(CN)_6^{3-}$ does not phosphoresce in aqueous solution, while in DMF phosphorescence is observed up to approximately 80°C. In addition, on passing from water to DMF, the CN^- solvation quantum yield at 25°C is lowered from 0.12 to 0.08.[62]

Dimethylsulfoxide increases the solvation quantum yield of $Cr(NH_3)_6^{3+}$ by a factor of 1.7 compared to aqueous solution, while it decreases the efficiency for amine release in the $Cr(RNH_2)_5Cl^{2+}$ (R = H, CH_3, C_2H_5) series.[63]

The NCS^- photosolvation of *trans*-$[Cr(NH_3)_2(NCS)_4]^-$ and $Cr(NCS)_6^{3-}$ has been investigated in H_2O/CH_3CN mixtures.[64] A larger mole fraction of acetonitrile lowers the quantum yield of the former complex and enhances that of the latter.

Also, the photoracemization quantum yields of $Cr(ox)_3^{3-}$ in 0.2 M DMSO increase with the H^+ concentration,[65] as do the thermal racemization rate constants in water.

C.   Sensitization and Quenching of Photosubstitutions

Photoreactions of Cr(III) complexes, like phosphorescence,[66–68] can be sensitized in fluid solution by intermolecular electronic energy transfer. Both coordination compounds, such as $trans$-$[Cr(NH_3)_2(NCS)_4]^-$[68] $Ru(bipy)_3^{2+}$,[69] and organic species[66,67,70–76] can act as energy donors. The phenomenon is fairly general for transition-metal complexes and is discussed in detail in Chapter 2.

Photosensitization offers several advantages compared to direct irradiation of the spectral absorption bands. For instance, it makes possible the population of a given excited state of the acceptor complex with "purer" electronic energy, that is, with a smaller vibrational contribution than obtainable by direct light absorption. The use of photosensitizers of well-characterized donor levels affords fairly accurate identification of the thexi state energies of the acceptor species and, often, gives rise to "cleaner" photoreactions, peculiar of the selectively populated thexi states. In addition, energy transfer can give access to nonspectroscopic, or quasi-nonspectroscopic states such as the LF and, possibly, CT "spin-forbidden" ones. This technique has brought about remarkable progress in the understanding of the photochemical role of the excited states of Cr(III) ions. The most relevant results in the literature are collected in Table 4-4 and are discussed in the following sections.

Another powerful approach to the assessment of the participation of the excited states in chemical reactivity, consists of quenching a given photoreaction by suitable energy acceptors. A number of experiments in this sense have been carried out with Cr(III) complexes, and the relative data are gathered in Table 4-5. The general observation is that, under limiting conditions (infinite acceptor concentration), the photoreactions are only partially quenched, while the complex phosphorescence is totally extinguished. Again, the implications are considered in the following sections.

D.   The Photoreactive Excited States

Because of the large body of photochemical as well as luminescence data, the progress in understanding the role of the excited states in photosubstitution reactions is greater for Cr(III) than for any other family of

Table 4-4. Photosensitized Aquation of Cr(III) Complexes

Acceptor	Energy Levels		Donor, Level[b]	E(kK)	T, °C	Solvent	Ligand Aquated	ϕ_{lim}	Remarks	Ref.
	$^4L_1$(kK)[a]	^2D(kK)[a]								
$[Cr(NH_3)_5(NCS)]^{2+}$	20.5	14.6	Biacetyl, T	19.6	25°	0.1N H$_2$SO$_4$	NH$_3$	0.21	$\phi_{NH_3}/\phi_{NCS} > 100$	70,71
			Acridinium, S	21.7	25°	0.1N H$_2$SO$_4$	NH$_3$	0.21	$\phi_{NH_3}/\phi_{NCS} = 33$	71
			Acridinium, T	15.9–17.4	25°	0.1N H$_2$SO$_4$	NCS$^-$	—	no NH$_3$	71
			Riboflavin, S	—	25°	H$_2$O, pH 3.2	NH$_3$	0.24	no NCS$^-$	72
$Cr(NH_3)_5Cl^{2+}$	19.5	14.9	Naphthalene, T	21.3	18°	50% EtOH	NH$_3$	0.35	cis product	73
			Riboflavin, S	—	25°	H$_2$O, pH 3.2	NH$_3$	0.18	—	72
			Riboflavin, T	—	25°	H$_2$O, pH 3.2	Cl$^-$	—	—	72
$Cr(en)_3^{3+}$	22.1	14.9	Biacetyl, T	19.6	15°	3×10^{-3} HClO$_4$	en	0.8	$(\phi_{lim}/\phi_{dir})_A \sim 2$, $(\phi_{lim}/\phi_{dir})_P \sim 0.6$[c]	74
$Cr(NH_3)_6^{3+}$	21.6	15.2	Naphthalene, T	21.3	18°	50% EtOH	NH$_3$	0.18	—	73
$Cr(CN)_6^{3-}$	26.5	12.4	Pyrazine, T	26.2	7°	H$_2$O	CN$^-$	0.1	—	75,76
			Xantone, T	26.0	7°	50% EtOH	CN$^-$	0.003	—	75,76
			Acridine, S	25.7	7°	50% EtOH	CN$^-$	0.01	—	76
			Ru(bipy)$_3^{2+}$, T	17.2	7°	DMF	no A[c]	—	P sensitized[c]	75,76
			Erythrosin, T	16.5	7°	DMF	no A[c]	—	P sensitized[c]	76

[a] From absorption band maxima.

[b] T = lowest triplet; S = lowest singlet excited state.

[c] A = aquation; P = phosphorescence.

164

Table 4-5.   Quenching of Photoreactions of Cr(III) Complexes[a]

Complex	Reaction[b]	T, °C	Solvent	Quencher	% Reaction Quenched	Ref.
trans-[Cr(NH$_3$)$_2$(NCS)$_4$]$^-$	NCS$^-$, A	$-65°$	Mixed[c]	Cr(CN)$_6^{3-}$	50	77
Cr(CN)$_6^{3-}$	CN$^-$, A	$-60° \rightarrow 25°$	DMF	O$_2$, H$_2$O	0	62
Cr(phen)$_3^{3+}$	R	25°	H$_2$O	I$^-$	86	57
Cr(NH$_3$)$_6^{3+}$	NH$_3$, A	0°, 22°	H$_2$O	OH$^-$	30	63
Cr(en)$_3^{3+}$	en, A	15°	H$_2$O	Co^{2+}(aq), Fe^{2+}(aq)	60	78

[a] LF band irradiation; complex phosphorescence totally quenched.
[b] A = aquation; R = racemization.
[c] Methanol/water/ethylene glycol (2:1:1).

transition-metal complexes. In presenting the various theories that have been successively developed, reference is made to the patterns of reactivity illustrated in the preceding sections.

Before dealing with specific excited states, some photolysis mechanisms, which were proposed in early studies, will be mentioned. At present, they appear too simple to account for all the experimental observations.[4]

One suggestion was that "prompt bond fission"[79] followed excitation (at any wavelength). This mechanism, which was thought to encompass all the results known at that time, relative to different metal ions, is discussed in Chapter 3, Section V.1. The existence of "anti-thermal" reaction paths has demonstrated the inadequacies of this mechanism in LF photochemistry, although it probably plays a relevant role in CT photochemistry.

Another proposal was the "hot ground-state mechanism," which assumes that reaction occurs from high vibrational levels of the ground electronic state.[4,80] Also, the contribution of this mechanism is negligible. In fact, when more than one photoreaction mode takes place [e.g., in the Cr(NH$_3$)$_5$X^{2+} ions, see Eqs. (4-6) and (4-7) and Table 4-2], the *ratios* of quantum yields (for NH$_3$ and X$^-$ aquation) change with wavelength of excitation (L_2, L_1, or D bands) and with temperature. If only the absolute quantum yields, but not their ratios, were dependent on these parameters a unique state could be assumed to be committed to reaction. The variation of the ratios implies different electronic states, and, at the most, only one reaction mode may originate in the ground state. Moreover, the preferred photoaquation mode (NH$_3$ for the acidopentaammines) is not that occurring thermally (X$^-$ aquation), and energy transfer, which certainly populates excited states, favors the former mode (see Table 4-4). Photoreaction quenching experiments (see Table 4-5) also indicate that this mechanism is not relevant to Cr(III). Finally, it may be noted that the "hot" ground state is produced by internal conversion from the lowest quartet thexi state that is strongly distorted compared to the normal ground state (see Fig. 4-3). While thermal equilibration of the excited state, although including distortion, is assumed to occur very fast, the "hot ground-state mechanism" would require that the "hot" molecules linger long enough for reaction. Since equilibration brings the complex to its original geometry, there are no reasons why the ground state should equilibrate more slowly than an excited state.[14]

In the light of present knowledge, it is probable that, in some instances, the above mechanisms give a minor contribution to reactivity. The excited states are regarded as major, if not unique, precursors to reaction. Furthermore, sufficient evidence exists that *different* reaction modes take place from *different* excited states.

Even though part of the photoreactivity may be inherent to upper electronic energy levels,[23] attention is focused on the lowest LF excited states of a given spin multiplicity, that is, ^2E_g and $^4T_{2g}$ in O_h approximation, assuming that electronic relaxation to the lowest excited state within one spin manifold occurs rapidly.[81]

1. *The Doublet Hypothesis.* This early theory[18] indicates the lowest doublet excited state (^2E_g in O_h symmetry) as the one principally involved in the photoreactivity of Cr(III). When this hypothesis was formulated, studies had been reported on a limited number of systems, consisting either of octahedral complexes,[31,32,35] or of acidoammines in which only one reaction mode had been investigated.[61,79] Photoaquations thus appeared to be simple photoaccelerations of existing thermal processes. Quantum-yield values for O_h complexes (Table 4-1), and also for some other ions, such as *trans*-[Cr(NH$_3$)$_2$(NCS)$_4$]$^-$ [Eq. (4-2)], are virtually independent of whether the L_1 or L_2 absorption bands are irradiated. It was deduced that, irrespective of the excited state produced by light absorption, the complex finds itself in the lowest excited state of a given spin multiplicity, which is the reactive one.

This conclusion found support in the luminescent behavior of Cr(III), indicating that excitation to higher energy (quartet) levels is immediately followed by rapid conversion to the lowest quartet, or doublet, states, which are the only emitting ones. The reactive state was then suggested to be the doublet on the basis of several considerations.

Lifetime estimates from absorption and emission data yielded for the doublet state in solution, $\tau \leq 10^{-5}$ sec, and for the quartet state, $\tau \leq 10^{-7}$ sec. On these grounds the lowest doublet appears to have more chance to survive until either a dissociative or an associative process takes place.[18] Analogy with organic photochemistry seemed natural; the lowest ("spin-forbidden") triplet states had been recognized as the principal precursors to organic photoreactions.

As shown in Table 4-6, the ligand-field contribution to the activation energy, according to either a dissociative (square pyramidal intermediate) or to an associative (pentagonal bipyramidal intermediate) mechanism is negative for the ^2E_g state, that is, it makes the complex substitutionally labile.[17,18] On the contrary, the above energy is positive in the ground state for both types of intermediate and accounts for the inertness of Cr(III).[17]

Also, the availability of a vacant t_{2g} orbital in the ^2E_g state, produced by the spin-pairing transition within the t_{2g}^3 configuration, makes the complex labile (i.e., susceptible of nucleophilic attack by a seventh ligand) according to VB and MO theories.[17]

The increase in substitution quantum yield for O_h complexes (Table 4-1) and for NH_3 aquation in the $[Cr(NH_3)_{6-n}(H_2O)_n]^{3+}$ series[40] (H_2O exchange was not studied) with increasing ligand-field strength was taken as additional evidence for the doublet intermediate. It was postulated that the lowest doublet excited state may be depopulated by back-intersystem crossing to the lowest quartet state that, in turn, undergoes rapid radiationless deactivation to the ground state. The smaller the energy separation between the $^4T_{2g}$ and ^2E_g states, the higher the probability that this process of deactivation competes with chemical reaction.[40] It may be noticed that this correlation appears to be valid for O_h complexes, with exception of $Cr(CN)_6^{3-}$, but certainly it does not apply to mixed-ligand systems (see Table 4-2).

According to this mechanism, direct population of the lowest doublet by light absorption is expected to be more efficient than obtainable by intersystem crossing (whose efficiency is likely to be less than unity) and should increase the photoreaction yield. Efforts have been made at this kind of experiment and have been hampered by the difficulties mentioned in Section II.C. However, the results relative to the few systems accurately investigated give no indication of definite increase in efficiency either for octahedral[28,34] or for nonoctahedral[29] complexes. The constancy of quantum yields is taken as the major piece of evidence for a reaction starting from the doublet state; this however would require a 100% efficient intersystem crossing.

2. The Role of the Quartet Excited States.

Although the doublet hypothesis can account for the photoreactivity of octahedral complex ions, particularly when the thexi $^4T_{2g}$ state lies above the ^2E_g one, it appears inadequate to explain the behavior of less symmetric species. As already discussed in Section III.A.2, the wavelength dependence of ratios between yields for parallel photoreaction modes makes it necessary to

Table 4-6. Ligand-Field Contribution to the Activation Energy (in Dq units) for Cr(III) and Co(III) Complexes[a]

			Initial State		Transition States			
			Octahedron		Square Pyramid		Pentagonal Bipyramid	
d orbital energies			-6.00 4.00		-9.14 -0.86 0.86 4.57		-4.93 -2.82 5.28	
System	State	One-electron Configuration	LFSE[b]		LFSE[b]	LFAE[c]	LFSE[b]	LFAE[c]
Cr(III)	$^4A_{2g}$	t_{2g}^3	12.00		10.00	2.00	7.74	4.26
	^2E_g	t_{2g}^3	$12.00-P$		10.00	$2.00-P$	7.74	$4.26-P$
	$^4T_{2g}$	$t_{2g}^2 e_g$	2.00		10.00	-8.00	7.74	-5.74
Co(III)	$^1A_{1g}$	t_{2g}^6	$24.00-3P$		$20.00-3P$	4.00	$15.48-3P$	8.52
	$^3T_{1g}$	$t_{2g}^5 e_g$	$14.00-2P$		$20.00-3P$	$-6.00+P$	$15.48-3P$	$-1.48+P$
	$^1T_{1g}$	$t_{2g}^5 e_g$	$14.00-3P$		$20.00-3P$	-6.00	$15.48-3P$	-1.48

[a] P = Spin-pairing energy, estimated as $\sim 6.5Dq$ for Cr(III) and $\sim 7Dq$ for Co(III).
[b] LFSE = Ligand-field stabilization energy.
[c] LFAE = Ligand-field activation energy. (Data from Ref. 18.)

suppose that more than one excited state is involved. Thus, besides the lowest doublet, also the lowest quartet state must be taken into consideration. It is important to note also that the latter state has all the prerequisites of kinetic lability outlined above. The ligand-field contribution to the activation energy, according to both a dissociative and an associative mechanism, is even more negative for the $^4T_{2g}$ state than for the ^2E_g one (see Table 4-6). In addition, an empty t_{2g} orbital (available for nucleophilic attack) is present in the quartet, $t_{2g}^2e_g$ configuration, as well.

Finally, the decrease in reaction yields for O_h complexes with decreasing the $^4T_{2g}$–^2E_g energy gap may be, in principle, accounted for equally well by assuming that the lowest excited quartet is the reacting state and that radiationless deactivation via the lowest doublet is favored as the two states become closer and closer.

Evidence has been increasingly accumulated that not only are the excited quartet states involved in Cr(III) photoreactions, but that they play a major, if not exclusive, role even for complexes of octahedral microsymmetry.

The general pattern of ligand labilization for mixed-ligand complexes, rationalized by the photolysis rules (see Section III.A.3),[23] is consistent with the electronic distribution of an excited quartet state. On passing from O_h to C_{4v} or D_{4h} symmetry, the t_{2g} and e_g orbitals loose degeneracy. If the tetragonal distortion is one of weaker field strength along the z direction (as is most usual with $Cr(NH_3)_5X^{n+}$- and $trans$-$[CrA_4XY]^{n+}$-type ions) the d_{z^2}, σ-antibonding orbital is lowered in energy relative to the $d_{x^2-y^2}$ orbital. The accumulation of electron density on the z axis, produced by the $^4B_1 \rightarrow {}^4E$ transition, accounts for bond weakening along this specific direction in the lowest quartet excited state.

A second important point concerns the lifetime of the quartet thexi states and is related to the considerable degree of distortion of the latter with respect to the ground state. As already mentioned (Section II.B), the existence of distortion is revealed, at low temperatures, by the strong Stokes shift of fluorescence compared to absorption; distortion is expected to occur even to a larger extent in fluid solutions at room temperature.[14] The natural emission lifetime of a given excited state can be estimated from the integrated extinction coefficient through the oscillator strength (Chapter 1, Section V). This type of computation yields lifetimes of 10^{-6} to 10^{-7} sec for the lowest quartet and of approximately 10^{-3} sec for the lowest doublet excited states of Cr(III). However, fairly accurate results may be obtained only if the equilibrium configuration is the same in the two states involved in the electronic transition. The large distortion of the upper state, relative to the ground state, introduces factors (e.g., a not negligible difference between the matrix elements governing emission and absorption probabilities) that would allow a lifetime 10- to 100-fold longer for the quartet thexi state than that calculated from absorption spectra.[23]

Experimental support has been obtained for considerably long-lived quartet excited states. The fluorescence lifetime of solid $[Cr(urea)_6](NO_3)_3$ has been reported to be 5×10^{-5} sec.[82] Thus, as far as lifetime requirements are concerned, the lowest thexi quartet state certainly survives long enough to be deactivated by chemical reaction.

The first direct indication of the photochemical importance of the lowest quartet excited state came from an analysis of the temperature dependence of the photoaquation quantum yields and of the relative phosphoresence yields of $trans$-$[Cr(NH_3)_2(NCS)_4]^-$ in glassy solutions, over the temperature range of -195 to $25°C$.[23] A temperature rise decreases the luminescence efficiency while it increases the NCS^--aquation quantum yield. The apparent activation energies for the two processes suggest that photoreaction takes place from a state different than the emitting one. The two states appear to be competitively populated, that is, the nonradiative process reducing the phosphorescence from the ^2D level directly populates the photochemically active state, which was indicated to be thexi $^4L_1^0$.

The following picture[11,14] accounts for the observed behavior. The vibrationally excited $^4L_1$ state, generated by absorption of a photon, may undergo vibrational relaxation through two successive stages. The first one is faster and occurs still within an octahedral framework; the second is somewhat slower since it possibly involves a change in geometry and is more affected by the rigidity of the environment. Depending on the temperature, and hence on the degree of fluidity of the medium (solvent cage), two pathways may be competitively "chosen" during thermal equilibration of the $^4L_1$ state. One, preferred at low temperatures, is intersystem crossing to the emitting ^2D state and is favored because the complex still possesses essentially the ground-state, and hence the ^2D, geometry. The second, predominant at higher temperatures, is full equilibration to the reactive $^4L_1^0$ state. The situation is depicted in Fig. 4-3.[11,14] According to this picture, a fraction of the complex molecules never passes through the ^2D state.

A second piece of evidence for a quartet intermediate was obtained studying the $[Cr(NH_3)_5(NCS)]^{2+}$ ion [Eqs. (4-6), (4-7) and Table 4-2].[29] Irradiation of the L_1 band of this complex produces NH_3 and NCS^- aquation in the ratio, $23:1$, and the ratio drops to $8:1$ when the D features are directly irradiated. In particular, on passing from the L_1 to the D region, the NCS^- aquation quantum yield remains virtually the same, while that for NH_3 release decreases to approximately one-third of its original value (incidentally, this fraction is close to the contribution of the tail of the L_1 band to total doublet absorption; see Fig. 4-2). It was postulated that NH_3 and NCS^- aquation are associated with different excited states: $^4L_1$ was indicated as precursor to ammonia release and ^2D as responsible for NCS^- aquation.[29] NH_3 labilization is thus likely to occur entirely in the thexi $^4L_1^0$ level. NCS^- release upon L_1 photolysis would then be due to intersystem crossing to the ^2D state, which takes place during thermal equilibration of $^4L_1$.

Photosensitization (see Table 4-4) confirms the above conclusion. The biacetyl triplet (indicated by the quenching of the donor phosphorescence and by oxygen effects on the reaction) sensitizes exclusively ammonia aquation.[70,71] The acridinium ion singlet (fluorescence is quenched) is found instead to sensitize both NH_3 and NCS^- release in a 33:1 ratio, while its triplet promotes only NCS^- labilization (quantum yields decrease in the presence of dissolved oxygen).[71] On the basis of the donor and acceptor energies of Table 4-4 it may be inferred that biacetyl triplet populates exclusively the $^4L_1^0$ state ($\phi_{NH_3}:\phi_{NCS} > 100:1$), which is the most photoactive one. Acridinium singlet transfers energy to the $^4L_1$ level and acridinium triplet to ^2D, which is less photoactive. The conversion to the $^4L_1^0$ state is slightly preferred upon acridinium sensitization (33:1 ratio) than upon direct irradiation (23:1 ratio). The higher the donor energy, the higher the vibrational component of the nascent $^4L_1$ state, the smaller the probability of producing exclusively the $^4L_1^0$ level that undergoes its peculiar reaction. Photosensitization provides independent evidence that the thexi-state energies are considerably lower than those deduced from band maxima (see Table 4-4).

Another example of the great potentiality of energy-transfer studies is the following. Biacetyl triplet sensitizes both phosphorescence and ethylenediamine aquation of $Cr(en)_3^{3+}$ at nearly room temperature.[74] The sensitized reaction is more efficient by a factor of approximately 2 compared to direct excitation, whereas the yield for sensitized emission is lower (by a factor of ca. 0.6) than that consequent to direct light absorption. Again, these results may be taken to indicate that the distorted, $^4L_1^0$ state is the reactive one. From a comparison of the energies of Table 4-4, population of this level through energy transfer from biacetyl is more effective than through relaxation of the $^4L_1$ state, produced by direct irradiation. The path leading to the (probably exclusively) emitting ^2D state is instead less favored when biacetyl sensitization is performed.

As mentioned in Section III.C, photoreaction-quenching experiments have proven even more diagnostic in identifying the chemically active excited states. A remarkable feature, common to all the systems investigated thus far, is that complex phosphorescence is completely quenched by a sufficiently high concentration of acceptor, while photoreactions are only partially quenched (see Table 4-5).

The occurrence of doublet–doublet energy transfer between trans-$[Cr(NH_3)_2(NCS)_4]^-$ and $Cr(CN)_6^{3-}$ in fluid solution at $-65°C$ is demonstrated by phosphorescence quenching in the former complex, accompanied by the appearance of the ^2E_g emission in the latter.[68] The participation of the quartet states in the transfer process is excluded by the nature of the temperature dependence.[68] In limiting conditions, the NCS^- aquation quantum yield of the donor drops to a value of half of that observed in the absence of acceptor.[77] Thus at least one-half of the photoreaction does not originate from the doublet state of the Reineckate ion (phosphorescence is totally extinguished), but occurs from the thexi, $^4L_1^0$ state, directly reached by equilibration of the $^4L_1$ level. Furthermore, the temperature dependence of the phosphorescence lifetime suggests that about 90% of the quenchable fraction of the photoreaction proceeds from the $^4L_1^0$ state, which is repopulated by thermally activated back-intersystem crossing from the doublet (the activation energy fits the $^4L_1^0 - ^2D$ difference).[77]

Analogous data relative to aqueous $Cr(phen)_3^{3+}$,[57] $Cr(NH_3)_6^{3+}$,[63] and $Cr(en)_3^{3+}$,[78] including temperature dependences, are in agreement with this pattern. The percentages of unquenched reaction, deducible from Table 4-5, are concluded to originate directly from the $^4L_1^0$ state prior to intersystem crossing. Moreover, most, if not all, of the remaining quenchable fraction reflects a "delayed" thermal repopulation of $^4L_1^0$ by back-intersystem crossing.

A particular situation is encountered in the $Cr(CN)_6^{3-}$ ion, where the very large energy separation between the $^4T_{2g}$ and ^2E_g states (see Table 4-4) precludes establishment of thermal equilibrium between the two levels.[83] Cyanide aquation (from $^4L_1^0$) is photosensitized only by species whose donor levels lie above a certain energy threshold (lower by at least 1 kK than the L_1 absorption maximum). Lower-energy sensitizers, instead, give rise only to complex phosphorescence (from ^2D), but not to reaction; the thermal return to the upper quartet is prevented by the wide energy gap.[75,76]

In addition, dissolved oxygen and water quench the complex phosphorescence in DMF, while, under the same experimental conditions, CN^- photosolvation is completely unaffected[62] (see Table 4-5). Different temperature dependences show that emission and solvation are unrelated.[62] Again, in view of the impossibility of back-intersystem crossing, the implication is that the emitting (^2D) state is distinct from the reacting one. Thus, in $Cr(CN)_6^{3-}$, after vibrational relaxation, the excitation energy is "trapped" either in the $^4L_1^0$ state (which reacts) or in the ^2D one (which phosphoresces), and, unlike in complexes having weaker-field ligands, the two states behave independently.

In conclusion, the lowest-quartet excited state definitely plays the principal role in Cr(III) photochemistry, even with octahedral complexes. The doublet state, although very important for emission, seems to be essentially inert toward photosubstitution. There is increasing evidence that the doublet momentarily "stores" excitation energy and reacts only via the lowest-quartet thexi state, particularly near room temperature.[13]

3. *Photoreaction Quantum Yields.* Another aspect of the chemical reactivity of the quartet excited states is the identification of possible factors that determine the efficiencies of ligand labilization and the relative magnitudes of quantum yields. It must be emphasized that the experimentally accessible quantum yields represent *ratios* (or products of ratios) of rate constants of competing processes (see Chapter 2). The individual rate constants are not known, however. When photoreaction quantum yields are taken as a measure of the relative rates of ligand labilization within a group of complexes, the arbitrary assumption must be made (tacitly or unconsciously) that all the competing processes occur with the same rates (or that the lifetime of the reactive state is the same) in all the systems. In principle, any quantum yield increase could be equally well ascribed to an enhanced rate of reactive deactivation, to a slower rate of radiative or nonradiative depopulation (i.e., to a longer lifetime) of the reacting state, or to the simultaneous occurrence of both changes. Both types of processes have been invoked on occasion (see below and also Section IV.D). Unfortunately, lifetime data cannot be obtained for chromium(III) quartet excited states in solution (nor can they be reliably evaluated from absorption spectra; see the preceding section). Nevertheless, within a homogeneous series of coordination compounds, experimentally measurable photophysical processes (as the low-temperature phosphorescence from ^2E_g) are generally found to occur with comparable rates, suggesting a similarity also among the unmeasurable quantities (see, for example, Ref. 51). In these instances the above assumption, and hence the quantum yield comparisons, appear reasonable, at least from a semiquantitative point of view.

In O_h complexes the increase in photolysis yields with increasing field strength of the ligands involved, may be attributed to a progressive enhancement of the labilizing effect of the electron in the e_g, σ-antibonding orbitals as the d orbital splitting becomes larger and larger.

Also, a larger and larger field strength is expected to determine a more and more pronounced distortion of the thexi, $^4L_1^0$ state.[13] Several facts suggest that there must be a connection between degree of distortion and reactivity (see also the next section). It may be noticed that the lowest doublet states, which are scarcely photoactive, are essentially undistorted compared to the ground state.

Additionally, as already mentioned, an increase in the $^4L_1-^2D$ energy difference (the doublet energy is virtually unaffected by the ligand-field strength, see Fig. 1-6), diminishes the probability of radiationless deactivation of the quartet state via the doublet.

The behavior of $Cr(CN)_6^{3-}$ is somewhat anomalous in the series of compounds of Table 4-1. Although the cyanide ligand occupies the highest position in the spectrochemical series, the quantum yield for photoaquation is only 0.12.[33,36] Two explanations have been offered. One is that the L_1 transition probably promotes an electron from a (t_{2g}) π-bonding to a π-antibonding orbital,[19] rather than to a σ-antibonding one, and the labilizing effect is reduced.[33] The alternative interpretation is that the radiationless transitions that depopulate the quartet state and compete with reaction are favored since the high position of the latter state allows another doublet $(^2T_{2g})$ between the $^4T_{2g}$ and the ^2E_g states. More vibrational levels would thus be present between the two states, bridging the energy gap.[62]

As far as mixed-ligand complexes are concerned, a model has been recently proposed that predicts the types of labilization and the relative quantum yields.[20] The Cr(III) photolysis rules[23] are explained in terms of the σ- and π-bonding changes occurring in the lowest quartet, and doublet, excited states on the basis of the experimental orbital orderings. The possible orderings of the one-electron energy levels for various types of complexes are shown in Fig. 4-4. The application of this model to some of the systems of Tables 4-2 and 4-3 is illustrated by the following examples and is discussed with respect to the lowest quartet state. For a more comprehensive picture, the reader is referred to the original paper.[20]

The lowest-quartet excited state of C_{4v} complex ions is ^4E (arising from the splitting of the octahedral $^4T_{2g}$ state; see Fig. 4-1), and for all the $Cr(NH_3)_5X^{n+}$ species of Table 4-2 the orbital ordering is A of Fig. 4-4. The $^4B_1 \rightarrow {}^4E$ transition promotes an electron from the d_{xz} or d_{yz} (π-antibonding) orbital to the d_{z^2} (σ-antibonding) one. The increase in electron density along the $z(NH_3-X)$ axis weakens the σ-bonds in this direction, while the loss of electronic charge in the d_{xz} or d_{yz} orbitals increases the π-bonding, again, along the z axis. The overall labilization effect depends on the relative σ- and π-bonding ability of the ligands on

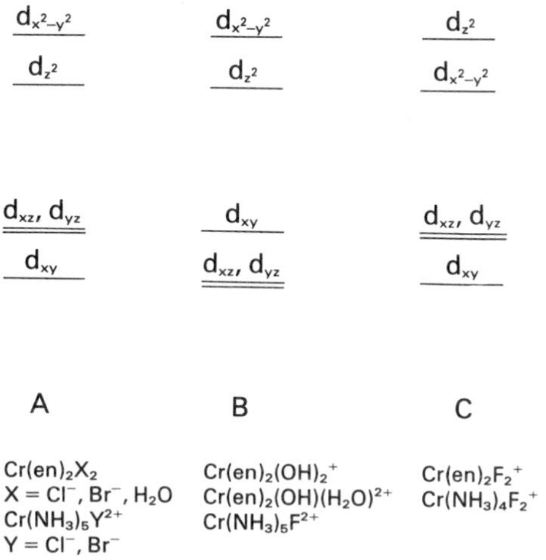

A	B	C
$d_{x^2-y^2}$	$d_{x^2-y^2}$	d_{z^2}
d_{z^2}	d_{z^2}	$d_{x^2-y^2}$
d_{xz}, d_{yz}	d_{xy}	d_{xz}, d_{yz}
d_{xy}	d_{xz}, d_{yz}	d_{xy}

Cr(en)$_2$X$_2$
X = Cl$^-$, Br$^-$, H$_2$O
Cr(NH$_3$)$_5$Y^{2+}
Y = Cl$^-$, Br$^-$

Cr(en)$_2$(OH)$_2^+$
Cr(en)$_2$(OH)(H$_2$O)$^{2+}$
Cr(NH$_3$)$_5$F^{2+}

Cr(en)$_2$F$_2^+$
Cr(NH$_3$)$_4$F$_2^+$

Fig. 4-4. Experimental energy-level orderings of the one-electron d orbitals for some Cr(III) complexes.[20]

this axis. NH$_3$ is only σ-bonded, and the metal-ligand bond is considerably weakened. All the X groups have a certain degree of π-bonding, which is strengthened by the transition; labilization of these ligands, therefore, results from two opposing effects. The substitution quantum yield for X is thus predicted, and found, to be much smaller than that for NH$_3$. Also, in *trans*-[CrA$_4$XY]$^{n+}$-type complexes (e.g., *trans*-[Cr(NH$_3$)$_4$(NCS)Cl]$^+$)[38] the relative efficiencies for X and Y release would depend on the relative π-bonding ability of the two acido groups.

This analysis accounts also for the behavior of *trans*-[Cr(en)$_2$F$_2$]$^+$,[50] which is apparently discrepant with the photolysis rules. L_1 band excitation in fact produces mainly ethylenediamine aquation, although the weak-field axis is the F–F one. Spectroscopic studies[84,85] have established that orbital-ordering C of Fig. 4-4 occurs for this ion. In particular, the lowest σ-antibonding orbital is the $d_{x^2-y^2}$ one. In the lowest-quartet excited state the xy (ethylenediamine) plane has a strong σ-antibonding character; π-bonding is strengthened along the z direction and weakened on the xy plane. Since ethylenediamine is only σ-bonded and fluoride possesses a high π donor ability, the net result of excitation is labilization on the en plane and not on the F–F axis.

Trans-[Cr(en)$_2$FCl]$^+$ has recently been reported to behave in an intermediate manner.[86] Upon irradiation of the L_1 region, approximately 90% of total bond labilization occurs along the z axis, consistent with an increased stabilization of the d_{z^2} orbital relative to $d_{x^2-y^2}$, as compared to the *trans*-difluoro complex (the orbital situation should be between A and C of Fig. 4-4). Field-strength considerations would predict F$^-$ aquation (see Section III.A.3), while the ligand photoreleased on the z axis is entirely Cl$^-$. On the other hand, π-bonding arguments seem not to conclusively account for the observed behavior.[86] It is not certain, in fact, whether the π-bonding interaction in this complex is larger for F$^-$ than for Cl$^-$, since such interaction presents opposite trends in the *trans*-[Cr(en)$_2$X$_2$]$^+$ and the Cr(NH$_3$)$_5$X^{2+} series.[85]

E.   Photosubstitution Mechanisms

Besides "electronic" theories accounting for ligand labilization, "chemical" mechanisms also should be important in determining the product yields and especially the stereochemistry of the reactions. The two aspects have certainly a large degree of interconnection. The patterns emerging at present will be briefly discussed.

It should be pointed out that labilization models implicitly assume that in the ground and in the excited states the complex geometry is the same (as is assumed by LF and MO theories). A second assumption is that the primary photoreaction step is a dissociative one,[20] as though mechanisms need not be considered. That the first assumption is unwarranted, at least for the thexi quartet states, is largely proven by luminescent and photochemical results (*vide infra*).

The general observation that in Cr(III) amine complexes release of the weak-field ligands is accompanied by stereochemical change (see Section III.A.4) suggests that a uniform, and specific, reaction mechanism, involving *trans* → *cis* isomerization is operating at least in this family of compounds. Photochemical reaction paths appear to be different from those relative to thermal (ground-state) substitutions, which are generally stereoretentive.[55,56] Thus although the stoichiometries of certain photo and thermal reactions are the same, as for *trans*-[CrA$_4$XY]$^{n+}$ ions, the stereochemistries are different [Eqs. (4-10) to (4-12)]; only with *trans*-[Cr(cyclam)Cl$_2$]$^+$ is the stereochemistry the same (*trans* product) in both cases, but now the quantum yield is anomalously low.[51,52]

In addition, rearrangement appears to be necessary for the occurrence of photosubstitution: ligands that increasingly prevent stereomobility increasingly prevent photoreaction [Eq. (4-13)].

Another, more specific, observation is that *cis*-[Cr(NH$_3$)$_4$(H$_2$O)Cl]$^{2+}$ is the common product of the L_1 photolysis of all of the following: Cr(NH$_3$)$_5$Cl^{2+},[41,43] *trans*-[Cr(NH$_3$)$_4$Cl$_2$]$^+$,[38] and *trans*-[Cr(NH$_3$)$_4$(H$_2$O)Cl]$^{2+}$,[38] with essentially the same quantum yield of 0.4 (see Tables 4-2 and 4-3). This uniformity, despite a considerable variation of the nature of the labilized ligand, NH$_3$, Cl$^-$, and (presumably) H$_2$O, suggests that the yield may be set by a virtually identical efficiency of the thermal relaxation (and distortion) of the $^4L_1^0$ state (see Fig. 4-3). The stereochemical uniformity of products then implies the same reaction path for the thexi state, including the existence of a common intermediate whose collapse yields the cis-aquochloro complex.

Regarding the nature of the intermediate, the available data suggest that the primary process is not a merely dissociative one. For example, the photoinertness of trans-$[Cr(cyclam)Cl_2]^+$ upon LF excitation, shows that loss of a chloride ligand to form a square pyramidal intermediate (without a geometry change) is precluded. Another possible dissociative mechanism, consistent with the behavior of the cyclam complex, would be that the loss of the axial ligand be concerted with collapse into another geometry, such as the trigonal bipyramid. However, applying this mechanism to the nonrigid C_{4v} and D_{4h} complexes, coordination of solvent to the intermediate is expected to yield a mixture of isomers, rather than only one product of specific (cis) configuration.

Either associative or concerted reaction pathways (involving expansion of the coordination sphere) seem more compatible with the above data and with the type of distortion expected to occur during relaxation of the $^4L_1$ state to the thexi $^4L_1^0$ one.[11,14] As already pointed out in Section II.B, such a distortion may well be both radial and angular, and may lead to a geometrical framework different from that relative to the ground state.[11,14,23] It has been proposed[11,52] that, during thermal equilibration, the original octahedron may undergo distortion towards a pentagonal bipyramid, as the coordination sphere is expanded by addition of a solvent molecule in the plane perpendicular to the weak-field axis. A rough estimate of the ligand-field stabilization energy for the lowest-quartet excited state shows that the distorted, D_{5h} geometry of this state is more stable by $5.74Dq$ compared to O_h symmetry[11] (see Table 4-6). Collapse of this structure with expulsion of the labilized ligand and rearrangement to an octahedron would yield the observed products.[11] This picture is still consistent with the inertness towards photosubstitution of trans-$[Cr(cyclam)Cl_2]^+$.

More experimental data are needed in order to establish if the configuration of the products is due to stereospecificity of the photoreaction (determined by the type of distortion of the thexi state) or simply reflects a greater thermodynamic stability of the cis, over the trans, isomer.

The solvent effects mentioned in Section III.B.4 are related to the photolysis mechanisms. A modification of the solvation sphere is thought to affect the relative shapes and positions of the potential surfaces of the excited states, which, in turn, causes considerable changes in the relative extent of the various relaxation processes.[63-65] In particular, a different degree of distortion of the thexi states may be possible in different solvents and may determine variations in reaction quantum yields (and, possibly, in the stereochemical distribution of products).

Within solvation effects, hydrogen bonding between the first coordination sphere and the entering ligand may be important, especially in connection with an associative, or concerted, mechanism. For instance, the increase of photoracemization quantum yields of $Cr(ox)_3^{3-}$ in 0.2 M DMSO with increasing H^+ concentration, is taken as indication that hydrogen bonding favors distortion of the quartet excited state, and hence heterolytic bond rupture.[65]

IV. LIGAND-FIELD PHOTOCHEMISTRY OF $3d^6$ METAL COMPLEXES

Photochemical studies of first-row d^6 metal complexes have been carried out essentially on Co(III) and, to a much lesser extent, on low-spin, Fe(II) species. Thermally, Co(III) ions are the most inert ones in the first transition row. Their solution chemistry is substitutional and is as diverse as that of Cr(III). Co(III) compounds have been widely used for investigating the reaction mechanisms of octahedral complexes.[17] The most significant difference from Cr(III) is the ease of reduction to the divalent ion. $Co(H_2O)_6^{3+}$ is reduced by water to $Co(H_2O)_6^{2+}$. Complexation usually enhances the stability of aqueous Co^{3+}; however, if easily oxidizable ligands are present, such as oxalate or iodide, thermal redox decomposition becomes possible.

The energy levels for octahedral, d^6 complex ions appear in Fig. 1-6. The ground state for low-spin systems is $^1A_{1g}$ and has the t_{2g}^6 configuration. Two spin-allowed transitions are possible to the $^1T_{1g}$ and $^1T_{2g}$ excited states, both arising from the $t_{2g}^5 e_g$ one-electron configuration. Accordingly, a typical absorption spectrum shows two LF bands in the visible or near-ultraviolet region, again indicated as L_1 and L_2, respectively, in order of increasing energy,[4] (see also Fig. 1-1a). Also, a series of spin-forbidden transitions is expected. The lowest-energy one is to the $^3T_{1g}$ state and, unlike the analogous transition of Cr(III), it involves a change in configuration with respect to the ground state. The corresponding weak spectral band is usually not detectable, since it is overlapped to a large extent by the L_1 absorption. The $^3T_{1g} \rightarrow {}^1A_{1g}$ transition is observed as phosphorescent emission only in the cases of $Co(CN)_6^{3-}$,[87,88] and of $[Co(CN)_4(SO_3)_2]^{5-}$.[89]

The strong absorption at higher energies is attributed to a CTTM transition and occurs at longer wavelengths than in the analogous Cr(III) complexes, in agreement with a higher reducibility for Co(III). The assignment is consistent with the progressive red shift of this band in complexes of $Co(NH_3)_5X^{2+}$ type, as the ionization potential of X (assumed to be the only ligand involved in the electronic transition) decreases.[90]

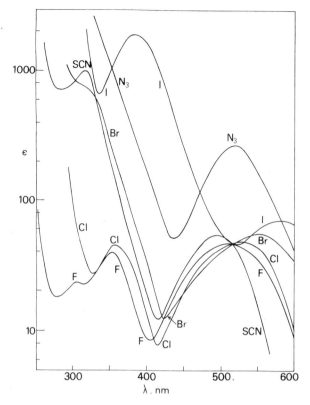

Fig. 4-5. Absorption spectra of $Co(NH_3)_5X^{2+}$ complex ions.[4]

Figure 4-5 shows that, when X = F and Cl, both LF bands are fairly well separated from the CT absorption. Movement of the latter into the visible obscures the L_2 band in the complexes with X = Br and NCS, and overlies even the L_1 band when X = I or N_3.

It is convenient to consider the behaviors of amine (acidoamine) and cyanide Co(III) species separately. The photochemistry of the former group of compounds is largely one of photoredox decomposition and is essentially related to CT excitation. The matter has been reviewed in detail[91] and is treated in Chapter 3. The d-d excited states instead show an extremely low reactivity of substitutional type. In contrast, LF-band irradiation of Co(III) cyanide complexes leads to efficient photosubstitution reactions.

Photoaquation is the only substitutional process investigated thus far with Co(III) amine complexes. Studies have been performed mostly in

Table 4-7. Photolysis of Co(III) Amine Complexes upon LF Excitation[a]

Complex	Ligand Aquated	Photoaquation Yields[b] L_2	Photoaquation Yields[b] L_1	$\phi_{Co(II)}$[b] L_2	$\phi_{Co(II)}$[b] L_1	Ref.
$Co(NH_3)_6^{3+}$	NH_3	5.4×10^{-3}	3.1×10^{-4}	$<10^{-5}$	$<3.5 \times 10^{-6}$	92, 93, 104
$Co(en)_3^{3+}$	en	1.8×10^{-4}	$\leqslant 10^{-5}$	$<10^{-5}$	$\leqslant 10^{-5}$	104
$[Co(NH_3)_5(H_2O)]^{3+}$	NH_3	—	1.8×10^{-4}	—	$<5 \times 10^{-6}$	93
$Co(NH_3)_5F^{2+}$	NH_3	—	1.9×10^{-3}	—	$<3.8 \times 10^{-5}$	92, 93
	F^{-}[c]	—	5.5×10^{-3}	—	—	92, 93
$[Co(NH_3)_5(CN)]^{2+}$	NH_3	4.6×10^{-3}	1.6×10^{-3}	$<10^{-4}$	$<10^{-5}$	94
	CN^{-}	1.7×10^{-3}	1.2×10^{-3}	$<10^{-4}$	—	94
$Co(NH_3)_5Cl^{2+}$	NH_3	—	5.0×10^{-3}	$<10^{-4}$	1.2×10^{-4}	93, 105
	Cl^{-}	0.011	1.7×10^{-3}	—	—	93, 105
$[Co(NH_3)_5(OCOCH_3)]^{2+}$	NH_3	0.011	—	1×10^{-3}	—	95
$[Co(NH_3)_5(NCS)]^{2+}$	NCS^{-}	0.015	5.4×10^{-4}	(0.03)	1.3×10^{-4}	105
$Co(NH_3)_5Br^{2+}$	Br^{-}	0.067	1.4×10^{-3}	(0.15)	$<10^{-5}$	105
$[Co(NH_3)_5(N_3)]^{2+}$	NH_3	$(<10^{-3})$	(0.2)	(0.44)	$(<10^{-3})$	96, 105
	N_3^{-}	—	$(<10^{-4})$	—	—	96, 105
$Co(NH_3)_5I^{2+}$	I^{-}	$(<10^{-3})$	$(<10^{-3})$	(0.66)	(0.10)	105
$[Co(NH_3)_5(NO_2)]^{2+}$	ONO[d]	0.11	(0.035)	(0.41)	(0.12)	101
$[Co(HEDTA)Cl]^{-}$	Cl^{-}	0.01	—	0.030	—	106, 107
$[Co(HEDTA)Br]^{-}$	Br^{-}	0.09	—	(0.035)	—	106, 107
$[Co(HEDTA)(NO_2)]^{-}$	NO_2^{-}	(4×10^{-3})	—	(4×10^{-3})	—	106, 107
$trans\text{-}[Co(NH_3)_4(H_2O)(CN)]^{2+}$	NH_3	—	1.0×10^{-3}	—	1.5×10^{-5}	94
	CN^{-}	—	0.03	—	—	94
$trans\text{-}[Co(en)_2Cl_2]^{+}$	Cl^{-}	1.1×10^{-3}	—	1.9×10^{-4}	—	93
$trans\text{-}[Co(en)_2Br_2]^{+}$	Br^{-}	(0.06)	6×10^{-4}	(7×10^{-3})	2×10^{-5}	4
$trans\text{-}[Co(en)_2(NCS)Cl]^{+}$	Cl^{-}	(6×10^{-4})	1.1×10^{-4}	(8.7×10^{-3})	$<10^{-5}$	5, 100
	NCS^{-}	(3.8×10^{-3})	1.8×10^{-4}	—	—	5, 100
$trans\text{-}[Co(cyclam)Cl_2]^{+}$[e]	Cl^{-}	4.0×10^{-4}	—	$<1.0 \times 10^{-5}$	—	93
$cis\text{-}[Co(en)_2Cl_2]^{+}$	Cl^{-}	—	2.4×10^{-3}	(3×10^{-4})	$<1 \times 10^{-6}$	32, 97
$cis\text{-}[Co(en)_2(H_2O)Cl]^{2+}$[f]	Cl^{-}	—	4.2×10^{-3}	—	$<1 \times 10^{-6}$	97

[a] Temperatures usually 20 to 25°C; aqueous acid solutions.
[b] Values in parentheses when CT and LF bands overlap.
[c] Quantum yields are pH dependent; data are for pH = 2.
[d] Linkage photoisomerization.
[e] cyclam = 1,4,8,11-tetraazacyclotetradecane.
[f] $cis \rightarrow trans$ isomerization.

acid solutions, for the same reasons of thermal kinetic stability mentioned for Cr(III).

A. Photoaquation

1. Co(III) Amines. Exclusive excitation to LF states can be achieved only with a small number of amine compounds, that is, with those having the intense CT absorption band at sufficiently short wavelengths (see Fig. 4-5). Thus, both LF absorption features may be selectively irradiated in $Co(NH_3)_6^{3+}$ and $Co(en)_3^{3+}$. The same is true for $Co(NH_3)_5X^{n+}$ complexes when X = F, Cl, H_2O. Such a particular excitation is limited to the L_1 band when X = Br or NCS and is virtually impossible when X = N_3 or I.

When "pure" LF excited states are produced, ligand photoaquation takes place, although with very low efficiency. In acidoammine complexes both ammonia and acido groups may be released, giving rise to antithermal labilization paths as is common for the analogous Cr(III) species (see Section III.A.2). For example, while thermal aquation of all $Co(NH_3)_5X^{2+}$-type ions involves replacement of the X^- ligand,[17] the complexes with X = F,[92,93] Cl,[93], CN^{94} have been found to undergo simultaneous X^- and NH_3 photoaquation. Also, the only photoreaction mode reported to occur for the X = $OCOCH_3$ [95] and X = N_3 [96] systems is ammonia aquation.

A selection of photoaquation quantum yields is collected in Table 4-7. Their values are, in general, smaller by two to three orders of magnitude than those of Cr(III) acidoamines.

Because of the apparently small chemical reactivity of the LF states, exact determination of photosubstitution yields is hindered mainly by two factors. First, using conventional light sources, long periods of irradiation are needed to obtain amounts of products determinable with the desired accuracy; relatively large quantities of thermal products, appearing in the meantime, may strongly affect, or even nullify, analytical efforts. In the second place, even a very small fraction of CT character in the LF absorption bands gives rise to efficient photoredox decomposition and aquation, both originating from a CT state (*vide infra*). It is then virtually impossible to establish the fraction of aquation products related to LF excitation. Because of these difficulties, in some cases either only one photoaquation mode (e.g., X^- aquation) could be investigated, or only upper-quantum-yield limits could be established.

Although the second complication cannot be avoided, in principle, the first type of experimental obstacle has been overcome by the use of lasers. Such intense light sources introduce a gain of as much as 10^4 in the light intensity absorbed by unit volume[92,93] (reducing the irradiation time by the same factor).

The above experimental advances disclose a definite LF photochemistry of Co(III) amines, much richer than previously thought, but at a lower efficiency level than has been observed in better-known systems. Furthermore, the antithermal behavior indicates that a specific excited-state chemistry is involved. Light is being rapidly cast on reactivity patterns, although a larger body of experimental data is still needed in order to gain an understanding comparable to that of Cr(III).

For example, the Cr(III) photolysis rules (see Section III.A.3) appear to apply to some extent also to Co(III) amines.[93] Thus NH_3 photoaquation is predicted, and found, to predominate in the acidopentaammine complexes, although not to the same extent as in the corresponding Cr(III) compounds. (cf. Tables 4-2 and 4-7). Also, photoaquation of the acido groups is the dominant mode for the *trans*-diacidotetraamine species, in agreement with expectations.

The stereochemistry of the photoaquation products has been investigated in only a few cases to date,[93,97] and no clear trend is yet apparent. However, the stereochemical course of Co(III) amine photolysis seems to be different from that relative to the analogous Cr(III) systems. While photolysis of *trans*-diacidotetraaminechromium(III) ions involves complete *trans* → *cis* isomerization (see Section III.A.4), chloride photoaquation[93] of *trans*-[Co(en)$_2$Cl$_2$]$^+$ leads to approximately 70% *trans*- and 30% *cis*-[Co(en)$_2$(H$_2$O)Cl]$^{2+}$. Virtually the same isomeric mixture is obtained in the photolysis[97] of *cis*-[Co(en)$_2$Cl$_2$]$^+$. In addition, *cis*-[Co(en)$_2$(H$_2$O)Cl]$^{2+}$ undergoes photoisomerization to *trans*, and the latter isomer is relatively photoinert.[97]

The stereochemical distribution of the thermal aquation products of diacidotetraamines appears different from that of the photolysis products. Thermally, the *trans* species yield mixtures of isomers (as is true photochemically for *trans*-[Co(en)$_2$Cl$_2$]$^+$) while the *cis* analogs generally undergo 100% stereoretentive aquation.[98]

The $4d^6$ *trans*-diacidotetraaminerhodium(III) homologs also behave differently in that they photolyze with complete stereoretention[99] (see Chapter 5).

Another significant difference from Cr(III) may be found in Table 4-7. The chloride photoaquation yields for *trans*-[Co(en)$_2$Cl$_2$]$^+$ and *trans*-[Co(cyclam)Cl$_2$]$^+$ differ by only a factor of approximately 3, while between the two Cr(III) analogs a factor of about 10^3 is found[48,51,52] (see Table 4-3). These ratios, along with the fair degree of retention of configuration, may be taken to indicate that stereomobility is not an important requirement for Co(III) photoaquation, while it is probably essential for Cr(III) (see Sections III.A.4 and III.E).

An interesting observation is the pH dependence of the quantum yields

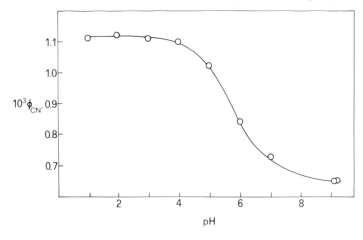

Fig. 4-6. pH dependence of the cyanide photoaquation quantum yields of $[Co(NH_3)_5(CN)]^{2+}$.[94]

for F^- aquation[93] in $Co(NH_3)_5F^{2+}$, and for CN^- release[94] in $[Co(NH_3)_5(CN)]^{2+}$. The quantum-yield values lay on a titration-like curve, such as in Fig. 4-6 for the cyanopentaammine complex, and suggest an acid–base behavior of the excited state involved. In the case of the $Co(NH_3)_5F^{2+}$ ion, the pH effect suggests that excited-state base hydrolysis may occur. Assuming a counter-base mechanism[17] a pK_a value for the excited state may be calculated.[92,93]

The behavior of $[Co(NH_3)_5(CN)]^{2+}$ may be accounted for by invoking a certain degree of protonation of the coordinated CN^- ligand, and an axial distortion (i.e., a Co—CN bond elongation) in the reactive state.[94] (The configuration of any LF excited state of Co(III) has one or more antibonding electrons.) The higher CN^- aquation yields in acid solution are explained by an enhanced weakening of the cobalt-cyanide bond, caused by the protonation, which would shift electron density along the Co—CN axis, toward the cyano group. The virtual pH independence of NH_3 photoaquation yields is then consistent with the impossibility of protonation of the ammonia ligands. An excited-state pK_a may be obtained also in this case.[94]

As already mentioned, when the CTTM band extends into the visible region, overlapping the L_2 or both L-type bands, photoaquation becomes highly efficient; however, even more efficient photoredox decomposition occurs simultaneously. It seems well established that, in this case, photosubstitution takes place through a completely different mechanism than that following LF excitation.[4,11,14] According to this mechanism, aquation and redox decomposition represent two alternative consequences of homolytic bond fission produced by CT excitation.[79] Direct evidence for two distinct aquation mechanisms has been obtained through an approach similar to that used in the case of Cr(III),[29] that is, studying the variation of the *ratios* between quantum yields for different aquation modes with changing wavelength of excitation. The ratio of Cl^- to NCS^- photoaquation yields for *trans*-$[Co(en)_2(NCS)Cl]^+$ is, in fact, 0.6 in the L_1 region and drops to 0.2 when the CT band is irradiated.[100]

It cannot be experimentally proven whether or not partial internal conversion occurs from CT to LF states. Possible products due to LF photoreactivity following CT band irradiation would, in fact, represent a negligible fraction of the total aquated species. However, the high yields for photoredox and photoaquation upon CT excitation suggest that internal conversion is likely to occur to a very limited extent (see also Chapter 3, Section IV.B).

For all the above reasons, the quantum yields of Table 4-7 appear in parentheses when light absorption generates both types of excited states. The magnitude of quantum yields parallels the degree of incursion of the CT band into the LF region (see Fig. 4-5). This is particularly evident in the cases of $Co(NH_3)_5Br^{2+}$ and $[Co(NH_3)_5(NCS)]^{2+}$.

It is significant that, at the wavelengths where the two band types overlap, the quantum yields for redox decomposition ($\phi_{Co(II)}$) are generally higher than those for aquation, whereas, upon "pure" LF irradiation, they are definitely smaller and possibly reflect a residual CT character in the transition.

Two notable exceptions to this pattern are represented by the relatively high NH_3 photoaquation yields for $[Co(NH_3)_5(OCOCH_3)]^{2+}$,[95] and especially for $[Co(NH_3)_5(N_3)]^{2+}$,[96] irradiated in the L_2 and in the L_1 regions, respectively. At the excitation wavelengths the LF bands are superimposed by the CT absorption only to a limited extent (although the L_1 absorption of the azido complex is anomalously high; see Fig. 4-5), and the Co(II) yields are found to be smaller by 1 or 2 orders of magnitude compared to the aquation efficiencies (see Table 4-7). These findings suggest that primary NH_3 photorelease in acidopentaamminecobalt(III) complexes may be more common and more efficient than previously assumed.[95,96]

The linkage photoisomerization of nitropentaamminecobalt(III)

$$[Co(NH_3)_5(NO_2)]^{2+} \xrightarrow{h\nu} [Co(NH_3)_5(ONO)]^{2+} \qquad (4\text{-}16)$$

may be discussed as a substitution-related reaction. The process is accompanied by photoredox decomposition. The quantum yields for the two reaction modes sharply increase on passing from the L_1 to the L_2, and to the CT region, but their ratio remains wavelength independent, suggesting that both processes originate from the same state, which is indicated to be CT in nature.[101] Isomerization is found to be intramolecular and due to cage recombination following homolytic bond fission, that is, a particular alternative mode to redox reaction. It should be noted that, also in this complex, the CT absorption extends into the visible. And it has been proposed that internal conversion to lower-lying, fully equilibrated CT states may be possible.[101] A recent detailed investigation[102] confirms this picture, as opposed to the conclusion that linkage isomerization takes place from a ligand-centered excited state, such as that involved in the formate isomerization of $[Co(NH_3)_5(O_2CH)]^{2+}$ [103] (the HCO_2^- and NO_2^- ligands are isoelectronic). In view of the CT nature of the reactive state, this photoreaction is considered in Chapter 3, Sections III.B and V.A.2.

2. Co(III) Cyanides.

In contrast to amine species, cyano complexes of Co(III) show a relatively high substitutional reactivity upon LF excitation. The general features of the LF spectral bands of $Co(CN)_5X^{3-}$ ions do not differ qualitatively from those of the $Co(NH_3)_5X^{2+}$ analogs. However, the maxima positions of the former are shifted by 100 to 150 nm toward the ultraviolet with respect to the corresponding absorptions of the latter, because of the very high spectrochemical position of the cyanide ligand.

$Co(CN)_6^{3-}$ undergoes efficient photoaquation of one cyanide ligand.[79,108,109]

$$Co(CN)_6^{3-} + H_2O \xrightarrow{h\nu} [Co(CN)_5(H_2O)]^{2-} + CN^- \qquad (4\text{-}17)$$

The aquopentacyano complex is the unique and terminal product,[79,108] being very stable toward further cyanide photoaquation,[110] although mechanistic investigations (see Section IV.C) indicate that water photoexchange takes place to a considerable extent.[109] Due to the apparent photoinertness of the monoaquo product, reaction (4-17) may be cleanly used for preparative purposes, with virtually quantitative yields.[111]

For aqueous $Co(CN)_5X^{n-}$ systems, with $X = Cl$,[79] Br,[79] I,[79,109] N_3,[96,112] SCN,[112] H_2O,[109] the major photoprocess involves replacement of the X ligand.

$$Co(CN)_5X^{3-} + H_2O \xrightarrow{h\nu} [Co(CN)_5(H_2O)]^{2-} + X^- \qquad (4\text{-}18)$$

The photoreaction leads to the same product obtained thermally,[113–115] and, in general, no Co(II) is produced on d–d excitation. The photolysis of hexacyanocobaltate(III) ion appears to be a particular case of reaction (4-18) for $X = CN$.

Wavelength dependence of quantum yields has been investigated to date only in a few cases. For $Co(CN)_6^{3-}$,[108] and $Co(CN)_5I^{3-}$ [109] the aquation efficiency is independent of whether the L_2 or the L_1 band is irradiated, suggesting that only the lowest excited state of a given spin multiplicity is the reactive one (cf. Section III.A.1). The L_2 absorption of the iodopentacyano complex, like a number of $Co(CN)_5X^{3-}$ ions, is strongly overlapped by a charge-transfer (possibly CTTM)[116] band, which makes interpretation of the L_2 band irradiation difficult. For these reasons, Table 4-8 shows only the quantum yields relative to L_1-band irradiation.

The $Co(CN)_6^{3-}$ photoaquation efficiencies do not depend on the pH in the 2.0 to 7.5 range,[108] or on the ionic strength of solutions,[109] while the yields for iodide photorelease of $Co(CN)_5I^{3-}$ markedly decrease with increasing ionic strength.[109] The latter observation has a strong bearing on the interpretation of the photolysis mechanism and can be more conveniently discussed in conjunction with the photoanation results (Section IV.C).

It might be anticipated that photoaquation of the $Co(CN)_5X^{3-}$ ions appears to occur through a dissociative mechanism, involving, as primary product, the $Co(CN)_5^{2-}$ species,[109] similar to that implicated in the thermal anation of $[Co(CN)_5(H_2O)]^{2-}$.[113-115] In contrast, the photolysis of $Co(CN)_6^{3-}$ is best explained in terms of a photoactivated interchange mechanism between an inner-sphere CN^- ligand and an outer-sphere water molecule.[117]

Table 4-8. Photoaquation of Co(III)-Cyanide Complexes on Irradiation of the Long-Wavelength Ligand-Field Band

Complex	Ligand Aquated	ϕ^a	pH^b	Ref.
$Co(CN)_6^{3-}$	CN^-	0.31	$2 \rightarrow 7.5$	108
$Co(CN)_5Cl^{3-}$	Cl^-	0.07	nat	79
$Co(CN)_5Br^{3-}$	Br^-	0.2	nat	79
$Co(CN)_5I^{3-}$	I^-	0.17	nat	109
$[Co(CN)_5(N_3)]^{3-}$	N_3^-	0.04	—	96
$[Co(CN)_5(NH_3)]^{2-}$	NH_3	0.35	2	94
$[Co(CN)_5(H_2O)]^{2-}$	CN^-	0.001	1.5	110
$[Co(CN)_5(OH)]^{3-}$	CN^-	0.05	13	110
trans-$[Co(CN)_4(SO_3)_2]^{5-}$	SO_3^{2-}	0.57^c	13	118
trans-$[Co(CN)_4(SO_3)(H_2O)]^{3-}$	SO_3^{2-}	0.14^c	2	118
trans-$[Co(CN)_4(SO_3)(OH)]^{4-}$	SO_3^{2-}	0.19^c	13	118

^{a} At room temperature; occasionally at 25°C.
^{b} nat = natural pH of the potassium salt, about 7.
^{c} Quantum yields for complex disappearance.

The cyanide photoaquation yield of $[Co(CN)_5(H_2O)]^{2-}$ is very low,[110] which accounts for the early report of its photoinertness.[79,108] The $[Co(CN)_5(H_2O)]^{2-}$ and $[Co(CN)_5(OH)]^{3-}$ complexes display wavelength dependence; the aquation efficiencies increase with increasing energy of excitation, and two photoactive states, having different reactivities, were postulated.[110]

The trans-$[Co(CN)_4(SO_3)X]^{n-}$ ions, where $X = SO_3^{2-}$, OH^-, and H_2O, are fairly photosensitive and undergo aquation of the sulfito group with modest wavelength dependence.[118] Increasing amounts of SO_3^{2-} ion in solution reduce the photolysis yields, suggesting a dissociative mechanism also for these species.[118]

The photochemical behavior of some mixed cyanoamminecobalt(III) complexes, other than $[Co(CN)_5(NH_3)]^{2-}$ and $[Co(NH_3)_5(CN)]^{2+}$, has been semiquantitatively explored.[94] The key observation is that, in the $[Co(CN)_n(NH_3)_{6-n}]^{3-n}$ series, the substitutional photosensitivity upon LF excitation steadily increases with increasing number of CN^- ligands.[94]

As a concluding remark, it may be mentioned that CT excitation has been recently proven to give rise to photoredox processes in Co(III)-cyano complexes, which were generally thought to be lacking in redox behavior, possibly because of an efficient conversion from CT to LF states.[119] Thus, Co(II) species are produced in the ultraviolet irradiation of $[Co(CN)_5(NO_2)]^{3-}$,[120] and of trans-$[Co(CN)_4(SO_3)(H_2O)]^{3-}$,[118] and the azide, iodide, and thiocyanate radicals are observed in the ultraviolet flash photolysis of $Co(CN)_5X^{3-}$ ions with $X = N_3$, I, NCS.[119] However, photoredox quantum yields are generally smaller, even by 1 order of magnitude, than those for the substitutional processes simultaneously taking place.

3. *Fe(II), Low-Spin Complexes.* There are relatively few nonlabile Fe(II) complexes. The strong-field ligands usually found in these species are cyanide, cyanide derivatives, and the bidentate 2,2'-bipyridine and 1,10-phenanthroline (indicated as AA).

As in Co(III) complexes, two spin-allowed LF transitions are expected: $^1A_{1g} \rightarrow {}^1T_{1g}$, and $^1A_{1g} \rightarrow {}^1T_{2g}$, in order of increasing energy (see Fig. 1-6). The corresponding absorption bands occur in the near-ultraviolet region and, especially the short-wavelength ones, often appear as shoulders on the very intense CT, or internal-ligand, absorptions. When AA-type ligands are present, the latter bands may completely cover the LF ones. Ligand-field excitation leads, in general, to photoaquation, as expected on the basis of the spectral assignment. Some selected results are summarized in Table 4-9.

The most investigated system is the $Fe(CN)_6^{4-}$ ion. In acid solution

Table 4-9. Selected Photoaquation Quantum Yields for Low-Spin, Fe(II) Complexes on LF Excitation[a]

Complex	Ligand Aquated	ϕ	pH	Ref.
$Fe(CN)_6^{4-}$	CN^-	0.4	7–10	122
$Fe(CNCH_3)_6^{2+}$	$CNCH_3$	0.03	1.7–6	124
$[Fe(CNCH_3)_4(bipy)]^{2+}$ [b]	$CNCH_3$	0.004[c]	2–6	125
$[Fe(CNCH_3)_2(bipy)_2]^{2+}$ [b]	$CNCH_3$	0.001[c]	2–6	126
$Fe(bipy)_3^{2+}$	No aquation	—	—	4

[a] Irradiation wavelength, 365 nm (L_1 band); temperature, 20°C.
[b] The internal-ligand-transition band obscures the LF band.
[c] Similar quantum yields are found for the complexes with 1,10-phenanthroline.

ferrocyanide undergoes extensive protonation. The unprotonated complex predominates at pH values greater than 6.[121] The photolysis results are considerably complicated by the various acid–base equilibria and by the instability of the products.[1,4] However, it seems well established[1,4] that the primary photolysis step is

$$Fe(CN)_6^{4-} + H_2O \xrightarrow{h\nu} [Fe(CN)_5(H_2O)]^{3-} + CN^- \qquad (4\text{-}19)$$

An interesting observation is the unusual decrease of the aquation quantum yield on passing from the LF to the CT absorption region.[122] This behavior certainly reflects the change in electron distribution following excitation deeper and deeper in the ultraviolet. Simultaneously, another reaction mode, consisting of photoelectron production[123] (i.e., a redox process, associated with CT excitation; see Chapter 3, Section III) becomes increasingly important.

The complexes of the series $[Fe(CNCH_3)_{6-2n}(AA)_n]^{2+}$, where $n = 0,$[124] $1,$[125] $2,$[126] undergo aquation of the $CNCH_3$ ligand. Although in the absorption spectrum of $Fe(CNCH_3)_6^{2+}$ the two LF bands are well resolved, in the $n = 1$ and $n = 2$ terms, the CTTL and the ligand absorptions obscure the d–d features. Photoaquation quantum yields decrease as the number of chelate ligands increases. The $n = 3$ species are photosubstitutionally inert.[4] An analogous trend is observed in the $[Fe(CN)_{6-2n}(AA)_n]$ series.[127,128] A possible explanation for this behavior is that excitation of chelate-ring electrons is essentially delocalized, and conversion from ligand-centered states to LF states is inefficient.[4]

B. Photosensitized Aquation

Electronic energy transfer from either organic or inorganic donors to Co(III) amines generally results in sensitization of photoredox decomposition of the latter, even when the donor energy level (usually a triplet) is well below that of the nearest CT band of the complex and corresponds to LF absorption wavelengths, where direct irradiation causes only inefficient photoaquation.[129-132] These findings suggest that a low-lying, nonspectroscopic CT (triplet) state is involved in the observed photoreactions.[9,11,14,129-132]

In some instances, also, photoaquation can be sensitized in amine complexes, simultaneously with redox reactions.[94,106,107] In the case of [Co(HEDTA)X]$^-$ ions (X = Cl, Br), the Ru(bipy)$_3^{2+}$ triplet is proposed to competitively populate a CTTM and a LF complex excited state, the former leading to oxidation-reduction and the latter to substitution, much more efficiently than on direct excitation[106,107] (see Tables 4-8 and 4-10).

Photosensitization of Co(III) cyanide ions, instead, gives rise exclusively to aquation. The most thoroughly investigated system is biacetyl-Co(CN)$_6^{3-}$.[133,134] The complex quenches biacetyl phosphorescence in deoxygenated solutions, showing that the donor triplet is implicated in the energy-transfer process. On the basis of the excited-state energies (Table 4-10) the $^3T_{1g}$ state of Co(CN)$_6^{3-}$ is indicated to be the acceptor level. The limiting quantum yield for sensitized cyanide aquation (0.8)[134] is much higher than that relative to direct photolysis (0.31).[108] The earlier reported, lower value (0.23)[133] is probably due to the fact that the quenching ability of the aquation product, [Co(CN)$_5$(H$_2$O)]$^{2-}$, on biacetyl phosphorescence, was not taken into account. In fact, the monoaquo complex quenches biacetyl triplets more efficiently by a factor of approximately 6×10^2 than does Co(CN)$_6^{3-}$.[134] Even very small amounts of the product, generated during the luminescence quenching measurement, greatly reduce the phosphorescent emission of the donor, introducing an error in the evaluation of the limiting yield.[134]

Mechanistic details of the energy-transfer process are treated in Chapter 2; the photochemical implications of the most significant data of Table 4-10 are discussed in Section IV.D.

No sensitization studies appear to have been reported thus far for photoaquation of diamagnetic, Fe(II) complexes.

C. Photoanation

Irradiation of Co(III) cyanide complexes in the presence of different anions gives rise to moderately efficient photoanation. Studies of this photosubstitutional behavior have been carried out with a number of systems and have produced useful information on the general photolysis mechanism. There are interesting similarities between it and the thermal anation of [Co(CN)$_5$(H$_2$O)]$^{2-}$ and the thermal aquation of the Co(CN)$_5$X^{3-} species.[113-115] Evidence has been obtained that both these reactions

Table 4-10. Photosensitized Aquation of Co(III) Complexes[a]

Acceptor	Energy Levels ^1L_t(kK)[c]	T(kK)	Donor, Level[b]	E(kK)	Ligand Aquated	ϕ_{lim}	Remarks	Ref.
$[Co(HEDTA)Cl]^-$	(17.2)	—	$Ru(bipy)_3^{2+}$, T	17.2	Cl^-	0.4	$(\phi_{Co^{2+}})_{lim} = 0.24$	106, 107
$[Co(HEDTA)Br]^-$	(16.8)	—	$Ru(bipy)_3^{3+}$, T,	17.2	Br^-	0.1	$(\phi_{Co^{2+}})_{lim} = 0.8$	106, 107
$[Co(NH_3)_5(CN)]^{2+}$	22.7	—	Acridinium, S	21.7	CN^-, NH_3	—	$(\phi_{CN^-} : \phi_{NH_3} : \phi_{Co^{2+}})_{lim}$ $= (\phi_{CN^-} : \phi_{NH_3} : \phi_{Co^{2+}})_{dir}$	94
$Co(CN)_6^{3-}$	32.3	18.5[c], 14.4[d]	Biacetyl, T	19.6	CN^-	0.8	$\phi_{lim} > \phi_{dir}$	134
		—	Acetone, T	27.6	CN^-	0.23	—	112
$[Co(CN)_5(N_3)]^{3-}$	26.1	—	Biacetyl, T	19.6	N_3^-	0.22	$\phi_{lim} = \phi_{dir}$	112
$[Co(CN)_5(SCN)]^{3-}$	26.5	—	Biacetyl, T	19.6	SCN^-	0.18	$\phi_{lim} = \phi_{dir}$	112

[a] At room temperature; occasionally at 25°C. Aqueous acid solutions.
[b] T = lowest triplet; S = lowest singlet excited state.
[c] Absorption maxima; data in parentheses are taken from graphs.
[d] Emission maximum.

proceed through a dissociative mechanism, involving the pentacoordinated $Co(CN)_5^{2-}$ intermediate. In the case of anation, the proposed reaction sequence is

$$[Co(CN)_5(H_2O)]^{2-} \rightarrow Co(CN)_5^{2-} + H_2O \qquad (4\text{-}20)$$

The intermediate is competitively scavenged by solvent and by coordinating anions.

$$Co(CN)_5^{2-} + \begin{cases} H_2O \xrightarrow{k_{H_2O}} [Co(CN)_5(H_2O)]^{2-} & (4\text{-}21) \\ X^- \xrightarrow{k_{X^-}} Co(CN)_5X^{3-} & (4\text{-}22) \end{cases}$$

The most significant photochemical results may be summarized as follows.

a. Ligand-field excitation of $Co(CN)_6^{3-}$ in the presence of fairly high X^- concentrations ($X = N_3$, NCS, I) leads to the production of the corresponding $Co(CN)_5X^{3-}$ complexes, in addition to considerable amounts of the stable $[Co(CN)_5(H_2O)]^{2-}$ species. The relative efficiencies of photoanation by the various X^- nucleophiles and of photoaquation parallel the relative scavenging rates of the $Co(CN)_5^{2-}$ intermediate produced in analogous thermal reaction systems. However, while the various ϕ_{X^-}/k_{X^-} ratios are fairly close to a constant value, the ϕ_{H_2O}/k_{H_2O} ratio is the only exception; it is larger (by a factor of ca. 4) than the value for the other ligands.[109]

b. Under similar conditions, $[Co(CN)_5(H_2O)]^{2-}$ undergoes photoanation by I^-,[109] and N_3^-.[117] In this case, the large contribution of thermal reactions prevents accurate determination of anation quantum yields.

c. In the visible (L_1 band) and ultraviolet ($L_2 + CT$ band) photolysis of $Co(CN)_5I^{3-}$, the I^- aquation quantum yield progressively diminishes in the presence of increasing amounts of NaI. Iodide photoexchange is much smaller than the reduction in aquation efficiency. In addition, almost the same drop in I^- photorelease is observed if NaI is replaced by equal concentrations of other electrolytes, such as $NaNO_3$ or $NaClO_4$. The decrease becomes even larger in the presence of di- and trivalent cation electrolytes.[109]

Results (a) and (b) may be rationalized in terms of a mechanism analogous to Eqs. (4-20) to (4-22), whereby both $Co(CN)_6^{3-}$ and $[Co(CN)_5(H_2O)]^{2-}$ undergo heterolytic bond cleavage, producing the

$Co(CN)_5^{2-}$ species that is rapidly scavenged by free anions. However, while the relative reactivities toward the X^- ions seem to be the same for the thermal and the photochemical intermediates, the latter show a larger preference for water than the former. This difference, as well as observations (c), is accounted for by a cage mechanism, still involving the $Co(CN)_5^{2-}$ intermediate.

An alternative interpretation is suggested by some analogies with the isoelectronic $Cr(CO)_6$ complex, whose LF excitation produces the $Cr(CO)_5$ fragment with unit quantum yield[135] (see Chapter 6). The intermediate is sufficiently stable to be observed in irradiated solutions[136] and reacts either with the solvent, or with other bases (B) to give products of the type, $Cr(CO)_5B$. Photolysis in a glassy hydrocarbon solution at low temperature generates the $Cr(CO)_5$ species with a square pyramidal geometry (inferred from the infrared spectrum)[137]; if the medium is softened by raising the temperature, the geometry changes to trigonal bipyramidal.[137] Photolytic bond breaking thus occurs without distortion of the original octahedron, and the nascent fragment can acquire a more stable geometry only if allowed by the fluidity of the medium.

Likewise, photolysis of Co(III) cyanides may produce a square pyramidal $Co(CN)_5^{2-}$ fragment, particularly inclined towards regaining octahedral geometry by reaction with an immediately available base, that is, by capturing either the labilized ligand itself or a water molecule of the solvation shell. Only the fraction of $Co(CN)_5^{2-}$ ions escaping cage recombination would assume a more stable geometry (presumably that characterizing the thermal reaction intermediate) and participate in ordinary scavenging competition.[109] The large availability of water in the solvation sphere would account for the high aquation yield of $Co(CN)_6^{3-}$. The reduction in the photoaquation efficiency of $Co(CN)_5I^{3-}$ in the presence of electrolytes is explained by enhancement of cage recombination of the primary photoproducts due to efficient ion pairing. The presence of a cation in the outer sphere should reduce the repulsion between the $Co(CN)_5^{2-}$ fragment and the labilized I^- ligand, reducing the probability of diffusional separation. An analysis of the experimental data shows that the ion pairs undergo 100% efficient recombination. This picture also explains the low extent of iodide photoexchange, which consistently equals the (small) difference between the photoaquation yield in the presence of $NaNO_3$ or $NaClO_4$ and that relative to solutions containing NaI. This fraction reflects the $Co(CN)_5^{2-}$ particles escaping cage recombination and undergoing competitive scavenging by I^- and H_2O.[109]

The photosubstitution paths of Co(III) cyanide complexes may be summarized by the following scheme.

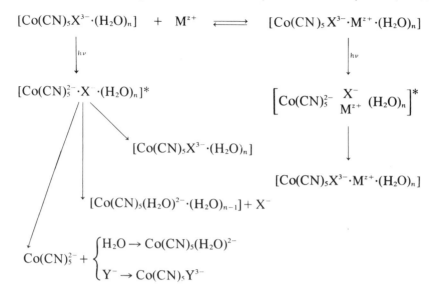

The solvent cage is indicated by brackets. The major difference from thermal reaction pathways lies in the fact that the $Co(CN)_5^{2-}$ intermediate, generated by heterolytic bond fission, must escape cage recoordination before being competitively scavenged by nucleophiles. The cage escape probability is strongly affected by the concentration and charge of cations present.

A more recent study,[117] using either flash photolysis or continuous monitoring of the species appearing in solution, confirms the above picture in the case of $Co(CN)_5X^{n-}$-type complexes, whereas it indicates that a dissociative pathway is not important for the $Co(CN)_6^{3-}$ ion. At least for moderately high concentrations of Y^-, the early photolysis stages of the latter complex give rise to $[Co(CN)_5(H_2O)]^{2-}$ but not to $Co(CN)_5Y^{3-}$. A different mechanism is proposed, involving interchange between a coordinated CN^- ligand and an outer sphere water molecule, rather than a pentacoordinated intermediate. The ligand switch is postulated to take place simultaneously with the decay of the electronic excited state involved.

D. Photolysis Mechanisms

Mechanistic information on Co(III) photosubstitutions is not as copious as that relative to Cr(III), due essentially to the scarce photoreactivity and to the general lack of luminescence in fluid solution. While some of the "chemical" aspects have been discussed in Sections IV.A and IV.C, two major questions remaining as to the excited-state behavior are the following.

a. Which LF excited state is precursor to reaction?

b. Why do LF excited states of Co(III), especially those of amine complexes, have such low photoactivities?

Photosensitization results (Section IV.B) provide a clear-cut answer to the first point, at least for cyanide complexes, indicating the "spin-forbidden," $^3T_{1g}$, state as the major state that is responsible for photochemical reactivity.[133,134] In addition, $Co(CN)_6^{3-}$ is known to luminesce from the $^3T_{1g}$ level in the solid state at low temperature. The phosphorescence lifetime, 6.8×10^{-4} sec at 77°K,[138] suggests that the triplet state may be sufficiently long-lived, even in solution at room temperature, to be able to undergo chemical reaction. The implication is that direct irradiation of the LF bands is followed by efficient intersystem crossing to the lowest triplet state. The wavelength independence of the aquation yields through the LF region indicates virtually complete $^1T_{2g} \rightarrow {}^1T_{1g}$ internal conversion.[134] The almost-unity, limiting quantum yield for photosensitization (see Table 4-10), reflects a high inherent reactivity of the $^3T_{1g}$ state, while the lower value for direct photolysis (see Table 4-8) essentially represents the efficiency of the $^1T_{1g} \rightarrow {}^3T_{1g}$ intersystem crossing.[134]

The lowest LF triplet of amine complexes, such as [Co(HEDTA)X]$^-$ (X = Cl, Br) is reported to be populated by energy transfer from the $Ru(bipy)_3^{2+}$ triplet[106,107] (although, on the grounds of the excited-state energies of Table 4-10, the inference does not seem to be as sound as for $Co(CN)_6^{3-}$). The high limiting yields for photosensitized aquation, compared to those for direct photoaquation, are taken to indicate a high intrinsic substitutional reactivity of the triplet. The apparent insensitivity to direct LF excitation is attributed to a very small degree of intersystem crossing.[106,107]

With respect to question (b), it has been stated that

> those complexes which undergo moderately efficient photosubstitution reactions in solution at room temperature also produce luminescence of the d–d type at low temperature in rigid media, while the complexes that are not photosensitive in the d–d region are also not luminescent.[14]

This observation suggests that both the absence of luminescence and the lack of photoreactivity are likely to be the consequence of a common cause, lending strong support to the conclusion that the emitting state is also the photoactive one. The general picture thus appears to be simpler than for Cr(III) (cf. Section III.D).

The inertness toward photosubstitution of Co(III) amines relative to cyanides may be, in principle, accounted for either by ineffective population, or by very efficient competitive depopulation, of the lowest LF triplet state. Various explanations have emphasized one or the other of these two possibilities, but it cannot yet be determined which is more important.

As mentioned above, energy-transfer studies show that the substitutional reactivity of the lowest triplet state is inherently high; the differences observed in direct photolysis are ascribed to a greater, or lesser, extent of intersystem crossing. Inefficient intersystem crossing in Co(III) amines may be explained by a relatively rapid rate of internal conversion from the lowest excited singlet to the ground state, which, may be understood as follows.[1] The same $t_{2g}^5 e_g$, one-electron configuration (in octahedral approximation) is associated with both $^1T_{1g}$ and $^3T_{1g}$ excited states, while the ground state ($^1A_{1g}$) has the t_{2g}^6 configuration. Nearly the same degree of metal-ligand repulsion, and hence of distortion with respect to the ground state, is expected in both excited states, whose potential energy surfaces are likely to be almost parallel. The consequence is that intersystem crossing between the $^1T_{1g}$ and $^3T_{1g}$ states is much less probable than the $^1T_{1g} \rightarrow {}^1A_{1g}$ and $^3T_{1g} \rightarrow {}^1A_{1g}$ relaxation processes. Strong-field ligands, such as CN^-, increase the energy difference between the ground and the excited states, decreasing the probability of radiationless deactivation.

The ligand-field contribution to the activation energy (see Table 4-6), estimated for the $^3T_{1g}$ state by either a dissociative or an associative mechanism,[18] is close to zero for a square pyramidal intermediate, and is definitely positive if a pentagonal bipyramid is involved. Therefore, the reactivity of the triplet state should not be very different from that of the ground state (which is kinetically inert), that is, radiationless deactivation should be comparatively fast to depopulate the triplet before it undergoes appreciable reaction. Molecules in the $^1T_{1g}$ state should be kinetically labile; but, now, a very short excited state lifetime would be unfavorable to reaction.[18]

According to either VB or MO considerations,[17] both excited states are predicted to be nonlabile since their configurations do not involve an empty t_{2g} orbital, unlike the case with Cr(III).[4]

An alternative explanation has been offered for the possibility of fast radiationless dissipation of excitation energy.[14] In the term diagram for low-spin, d^6 ions (see Fig. 1-6) for low ligand-field strengths, three spin-forbidden states are found between the $^1T_{1g}$ and the $^1A_{1g}$ energy levels, namely, two triplets and one pentuplet. Unless the ligand-field splitting energy is high enough (10 $Dq >$ ca. 25 kK) that the $^5T_{2g}$ state lies above the $^3T_{1g}$ one, a large density of vibrational levels bridges the $^1T_{1g}$-$^1A_{1g}$ energy range, favoring thermal relaxation to the ground state instead of either emission or photoreaction.

An additional reason for the relatively high photosubstitution yields of Co(III) cyanides may be found in the stability of the photoproduced $Co(CN)_5^{2-}$ intermediate, which survives long enough to be able to react either within the solvent cage, or even after escaping it.[5,14,109]

As a further point, the possibility of thexi state distortion may be briefly considered. As in the case of Cr(III), the broadness of LF absorption bands suggests an excited-state equilibrium geometry different from that of the ground state.[5] Since in the $^1T_{1g}$ state an electron occupies a σ-antibonding orbital, vibrational relaxation following generation of this state by light absorption must be accompanied by some kind of distortion. As already discussed, the thexi, $^3T_{1g}$ state should possess essentially the same geometry as the fully equilibrated $^1T_{1g}$ level since it is associated to the same one-electron configuration. This picture is consistent with the characteristics of $Co(CN)_6^{3-}$ phosphorescence, which is broad, structureless, and shifted to lower energy compared to the $^1A_{1g} \rightarrow {}^3T_{1g}$ absorption[87] (see Table 4-10). Ligand-field stabilization arguments (see Table 4-6) indicate that the framework of both excited states is likely to be still octahedral. The thexi state distortion may then be simply one of metal-ligand bond elongation within such a geometry, while for the $^4T_{2g}$ state of Cr(III) a change in symmetry appears to be favored on the same energetic grounds[5] (see Section III.E).

The loss of LFSE, subsequent to the $^4A_{2g}(t_{2g}^3) \rightarrow {}^4T_{2g}(t_{2g}^2e_g)$ transition of Cr(III) leaves only 2/12 (or ca. 17%) of the initial octahedral stabilization in the excited state, essentially allowing the molecule to switch to another geometry. The $^1A_{1g}(t_{2g}^6) \rightarrow {}^1T_{1g}(t_{2g}^5e_g)$ transition of Co(III) instead produces an excited state having still 14/24 (or ca. 58%) of the ground-state stabilization, and possible geometry changes should be less favored. In the $^3T_{1g}$ state the octahedron should be even more stabilized by spin unpairing.

In the $^1T_{1g}$ state, the pentagonal bipyramidal structure is only slightly more stable than the octahedron (by 1.48 Dq, or much less than 5.74 Dq, estimated for Cr(III) in the $^4T_{2g}$ state). Owing to the electron pairing energy, the D_{5h} symmetry for the $^3T_{1g}$ state may even become unstable with respect to the O_h one.

A virtually octahedral symmetry of the thexi states (as that of the ground state) would also account for a very rapid radiationless deexcitation, in contrast with the relatively long lifetime of the (distorted) $^4L_1^0$ state of Cr(III).[5] Experimental evidence that thexi state distortion is not as important for Co(III) as is for Cr(III), may be found in the large degree of stereoretention observed in the photoaquation of $trans$-$[Co(en)_2Cl_2]^{+}$.[93] Moreover, the small difference between the photoaquation yields of $trans$-$[Co(en)_2Cl_2]^{+}$ and $trans$-$[Co(cyclam)Cl_2]^{+}$ [93] (see Table 4-7) contrasts sharply with the factor of 10^3 found between the aquation efficiencies of the analogous Cr(III) systems[48,52] (see Table 4-3), for which stereomobility appears to be a necessary mechanistic condition.

In addition, the $Co(CN)_5^{2-}$ fragment, involved in the photolysis of $Co(CN)_5X^{3-}$ ions, is likely to be generated as a square pyramid; photolytic bond rupture probably occurs without distortion of the octahedral complex, as has been definitely ascertained in the case of the isoelectronic $Cr(CO)_5$ species.[137]

As far as the general photosubstitution mechanism is concerned, a firm conclusion can be drawn only in the case of cyanide complexes for which there is definite evidence of a dissociative primary step. Although the experimental information on Co(III) amines is still limited, dissociative reaction pathways appear to be more consistent with the observed photoproduction of isomeric mixtures. This contrasts with Cr(III) systems, where experimental evidence suggests associative photoreaction paths.

A model employing ligand-field theory arguments to account for the various modes of photolabilization of Co(III) complexes has been proposed recently.[21] In a manner similar to that used for Cr(III) complexes, predictions were made on the basis of σ- and π-bonding changes in the different excited states. Many variables, such as the nonreactive decay of excited states and excited-state distortions, influence reaction quantum yields (see Section III.D.3), and the data for the Co(III) complexes are still limited. Therefore, a critical evaluation of this model and, indeed, a reasonable reaction mechanism must await further experimental results.

REFERENCES

1. V. Balzani and V. Carassiti, *Photochemistry of Coordination Compounds*, Academic, London, 1970.
2. E. L. Wehry, *Quart. Rev. (London)* **21**, 213 (1967).
3. V. Balzani, L. Moggi, F. Scandola, and V. Carassiti, *Inorg. Chim. Acta Rev.* **1**, 7 (1967).

4. A. W. Adamson, W. L. Waltz, E. Zinato, D. W. Watts, P. D. Fleischauer, and R. D. Lindholm, *Chem. Rev.* **68**, 541 (1968).

5. A. W. Adamson, *Coord. Chem. Rev.* **3**, 169 (1968).

6. A. W. Adamson, *Rec. Chem. Progr.* **29**, 191 (1968).

7. V. Balzani, L. Moggi, and V. Carassiti, *Ber. Bunsenges. Phys. Chem.* **72**, 288 (1968).

8. D. Valentine, Jr., *Advan. Photochem.* **6**, 123 (1968).

9. A. W. Adamson, *Pure Appl. Chem.* **20**, 25 (1969).

10. D. Valentine, Jr., *Annu. Survey Photochem.* **1**, 459 (1969).

11. A. W. Adamson, *Pure Appl. Chem.* **24**, 451 (1970).

12. H. L. Schläfer, *Z. Chem.* (*Leipzig*) **10**, 9 (1970).

13. G. B. Porter, S. N. Chen, H. L. Schläfer, and H. Gaussmann, *Theor. Chim. Acta* **20**, 81 (1971).

14. P. D. Fleischauer, A. W. Adamson, and G. Sartori, *Progr. Inorg. Chem.* **17**, 1 (1972).

15. W. L. Waltz and R. G. Sutherland, *Chem. Soc. Rev.* (*London*) **1**, 241 (1972).

16. R. B. Bucat and D. W. Watts, *MTP International Reviews in Science, Inorganic Chemistry* (M. L. Tobe, Ed.), Series 1, Vol. 9, p. 159, Butterworths, London, 1972.

17. F. Basolo and R. G. Pearson, *Mechanisms of Inorganic Reactions*, 2nd ed., Wiley, New York, 1967.

18. H. L. Schläfer, *J. Phys. Chem.* **69**, 2201 (1965).

19. J. J. Alexander and H. B. Gray, *Coord. Chem. Rev.* **2**, 29 (1967).

20. J. I. Zink, *J. Amer. Chem. Soc.* **94**, 8039 (1972).

21. J. I. Zink, *Inorg. Chem.* **12**, 1018 (1973).

22. G. B. Porter and H. L. Schläfer, *Z. Phys. Chem.* (*Frankfurt*) **37**, 109 (1963).

23. A. W. Adamson, *J. Phys. Chem.* **71**, 798 (1967).

24. H. Shaw and S. Toby, *J. Chem. Educ.* **43**, 408 (1966).

25. P. Riccieri and E. Zinato, *Z. Phys. Chem.* (*Frankfurt*) **79**, 28 (1972).

26. C. G. Hatchard and C. A. Parker, *Proc. Roy. Soc. A* **235**, 518 (1956).

27. C. A. Discher, P. F. Smith, I. Lippman, and R. Turse, *J. Phys. Chem.* **67**, 2501 (1963).

28. E. E. Wegner and A. W. Adamson, *J. Amer. Chem. Soc.* **88**, 394 (1966).

29. E. Zinato, R. D. Lindholm, and A. W. Adamson, *J. Amer. Chem. Soc.* **91**, 1076 (1969).

30. W. Geis and H. L. Schläfer, *Z. Phys. Chem.* (*Frankfurt*) **65**, 107 (1969).

31. R. A. Plane and J. P. Hunt, *J. Amer. Chem. Soc.* **79**, 3343 (1957).

32. S. T. Spees and A. W. Adamson, *Inorg. Chem.* **1**, 531 (1962).

33. A. Chiang and A. W. Adamson, *J. Phys. Chem.* **72**, 3827 (1968).

34. S. N. Chen and G. B. Porter, *Abstracts Tenth Informal Conference of Photochemistry*, p. 155, Stillwater, Oklahoma, May 1972.

35. M. R. Edelson and R. A. Plane, *J. Phys. Chem.* **63**, 327 (1959).

36. H. F. Wasgestian, *Z. Phys. Chem.* (*Frankfurt*) **67**, 39 (1969).

37. A. W. Adamson, *Advan. Chem.* **49**, 237 (1965).

38. P. Riccieri and E. Zinato, *Proceedings of the XIV International Conference on Coordination Chemistry, International Union of Pure and Applied Chemistry*, p. 252, Toronto, Canada, 1972.

39. P. Riccieri and E. Zinato, *Inorg. Chim. Acta* **7**, 117 (1973).

40. M. R. Edelson and R. A. Plane, *Inorg. Chem.* **3**, 231 (1964).
41. H. F. Wasgestian and H. L. Schläfer, *Z. Phys. Chem. (Frankfurt)* **57**, 282 (1968).
42. H. F. Wasgestian and H. L. Schläfer, *Z. Phys. Chem. (Frankfurt)* **62**, 127 (1968).
43. M. F. Manfrin, L. Moggi, and V. Balzani, *Inorg. Chem.* **10**, 207 (1971).
44. P. Riccieri and H. L. Schläfer, *Inorg. Chem.* **9**, 727 (1970).
45. E. Zinato and P. Riccieri, *Inorg. Chem.* **12**, 1451 (1973).
46. E. Zinato, C. Furlani, G. Lanna, and P. Riccieri, *Inorg. Chem.* **11**, 1746 (1972).
47. A. Vogler, *J. Amer. Chem. Soc.* **93**, 5912 (1971).
48. A. D. Kirk, K. C. Moss, and J. G. Valentin, *Can. J. Chem.* **48**, 1524 (1971).
49. M. T. Gandolfi, M. F. Manfrin, L. Moggi, and V. Balzani, *J. Amer. Chem. Soc.* **94**, 7152 (1972); *Inorg. Chem.* **13**, 1342 (1974).
50. S. C. Pyke and R. G. Linck, *J. Amer. Chem. Soc.* **93**, 5281 (1971).
51. C. Kutal and A. W. Adamson, *Inorg. Chem.* **12**, 1990 (1973).
52. C. Kutal and A. W. Adamson, *J. Amer. Chem. Soc.* **93**, 5581 (1971).
53. A. D. Kirk, *J. Amer. Chem. Soc.* **93**, 283 (1971).
54. E. Zinato, P. Riccieri, and A. W. Adamson, *J. Amer. Chem. Soc.* **96**, 375 (1974).
55. R. D. Archer, *Coord. Chem. Rev.* **4**, 243 (1969).
56. C. S. Garner and D. A. House, *Transition Metal Chem.* **6**, 59 (1970).
57. N. A. P. Kane-Maguire and C. H. Langford, *J. Amer. Chem. Soc.* **94**, 2125 (1972).
58. K. L. Stevenson and J. F. Verdieck, *J. Amer. Chem. Soc.* **90**, 2974 (1968).
59. K. L. Stevenson, *J. Amer. Chem. Soc.* **94**, 6652 (1972).
60. N. A. P. Kane-Maguire, B. Dunlop, and C. H. Langford, *J. Amer. Chem. Soc.* **93**, 6293 (1971).
61. A. W. Adamson, *J. Inorg. Nucl. Chem.* **13**, 275 (1960).
62. H. F. Wasgestian, *J. Phys. Chem.* **76**, 1947 (1972).
63. C. H. Langford and L. Tipping, *Can. J. Chem.* **50**, 887 (1972).
64. V. S. Sastri, R. W. Henwood, S. Behrendt, and C. H. Langford, *J. Amer. Chem. Soc.* **94**, 753 (1972).
65. S. Sastri and C. H. Langford, *J. Phys. Chem.* **74**, 3945 (1970).
66. D. J. Binet, E. L. Goldberg, and L. S. Forster, *J. Phys. Chem.* **72**, 3017 (1968).
67. T. Ohno and S. Kato, *Bull. Chem. Soc. Japan* **42**, 3385 (1969).
68. S. N. Chen and G. B. Porter, *J. Amer. Chem. Soc.* **92**, 3196 (1970).
69. J. N. Demas and A. W. Adamson, *J. Amer. Chem. Soc.* **93**, 1800 (1971).
70. A. W. Adamson, J. E. Martin, and F. D. Camassei, *J. Amer. Chem. Soc.* **91**, 7530 (1969).
71. J. E. Martin and A. W. Adamson, *Theor. Chim. Acta* **20**, 119 (1971).
72. E. L. Wehry, *J. Amer. Chem. Soc.* **95**, 2137 (1973).
73. E. Zinato, P. Tulli, and P. Riccieri, *J. Phys. Chem.* **75**, 3504 (1971).
74. V. Balzani, R. Ballardini, M. T. Gandolfi, and L. Moggi, *J. Amer. Chem. Soc.* **93**, 339 (1971).
75. N. Sabbatini and V. Balzani, *J. Amer. Chem. Soc.* **94**, 7587 (1972).
76. N. Sabbatini, M. A. Scandola, and V. Carassiti, *J. Phys. Chem.* **77**, 1307 (1973).

77. S. N. Chen and G. B. Porter, *Chem. Phys. Letters* **6**, 41 (1970).
78. R. Ballardini, G. Varani, H. F. Wasgestian, L. Moggi, and V. Balzani, *J. Phys. Chem.* **77**, 2947 (1973).
79. A. W. Adamson and A. H. Sporer, *J. Amer. Chem. Soc.* **80**, 3865 (1958); *J. Inorg. Nucl. Chem.* **8**, 209 (1958).
80. V. Balzani, V. Carassiti, and F. Scandola, *Gazz. Chim. Ital.* **96**, 1213 (1966).
81. M. Kasha, *Radiat. Res. Suppl.* **2**, 243 (1960).
82. R. Dingle, *J. Chem. Phys.* **50**, 1952 (1969).
83. F. D. Camassei and L. S. Forster, *J. Chem. Phys.* **50**, 2603 (1969).
84. L. Dubicki, M. A. Hitchman, and P. Day, *Inorg. Chem.* **9**, 188 (1970).
85. D. A. Rowley, *Inorg. Chem.* **10**, 397 (1971).
86. G. Wirth and R. G. Linck, *J. Amer. Chem. Soc.* **95**, 5913 (1973).
87. M. Mingardi and G. B. Porter, *J. Chem. Phys.* **44**, 4354 (1966).
88. G. A. Crosby, *J. Chim. Phys.* **64**, 160 (1967).
89. F. Zuloaga and M. Kasha, *Photochem. Photobiol.* **7**, 549 (1968).
90. C. K. Jørgensen, *Advan. Chem. Phys.* **5**, 33 (1963).
91. J. F. Endicott, *Israel J. Chem.* **8**, 209 (1970).
92. A. W. Adamson, *Proceedings of the XIV International Conference on Coordination Chemistry, International Union of Pure and Applied Chemistry*, p. 240, Toronto, Canada, 1972.
93. R. A. Pribush, C. K. Poon, C. M. Bruce, and A. W. Adamson, *J. Amer. Chem. Soc.* **96**, 3027 (1974).
94. R. Sriram, Ph.D. Dissertation, University of Southern California, 1972.
95. E. R. Kantrowitz, M. Z. Hoffman, and J. F. Endicott, *J. Phys. Chem.* **75**, 1914 (1971).
96. G. Ferraudi and J. F. Endicott, *J. Chem. Soc., Chem. Commun.*, 674 (1973).
97. P. Sheridan and A. W. Adamson, *J. Amer. Chem. Soc.* **96**, 3032 (1974).
98. C. K. Poon, *Inorg. Chim. Acta Rev.* **4**, 123 (1970).
99. T. L. Kelly and J. F. Endicott, *J. Phys. Chem.* **76**, 1937 (1972).
100. A. Vogler and A. W. Adamson, *J. Phys. Chem.* **74**, 67 (1970).
101. V. Balzani, R. Ballardini, N. Sabbatini, and L. Moggi, *Inorg. Chem.* **7**, 1398 (1968).
102. F. Scandola, C. Bartocci, and M. A. Scandola, *J. Amer. Chem. Soc.* **95**, 7898 (1973).
103. E. R. Kantrowitz, M. Z. Hoffman, and K. M. Schilling, *J. Phys. Chem.* **76**, 2492 (1972).
104. M. F. Manfrin, G. Varani, L. Moggi, and V. Balzani, *Mol. Photochem.* **1**, 387 (1969).
105. A. W. Adamson, *Discuss. Faraday Soc.* **29**, 163 (1960).
106. P. Natarajan and J. F. Endicott, *J. Amer. Chem. Soc.* **94**, 3635 (1972).
107. P. Natarajan and J. F. Endicott, *J. Amer. Chem. Soc.* **95**, 2470 (1973).
108. L. Moggi, F. Bolletta, V. Balzani, and F. Scandola, *J. Inorg. Nucl. Chem.* **28**, 2589 (1966).
109. A. W. Adamson, A. Chiang, and E. Zinato, *J. Amer. Chem. Soc.* **91**, 5467 (1969).
110. M. Wrighton and D. Bredesen, *Inorg. Chem.* **12**, 1707 (1973).
111. J. H. Bayston, R. N. Beale, N. K. King, and M. E. Winfield, *Aust. J. Chem.* **16**, 954 (1963).

112. M. Wrighton, D. Bredesen, G. S. Hammond, and H. B. Gray, *J. Chem. Soc., Chem. Commun.*, 1018 (1972).

113. A. Haim and W. K. Wilmarth, *Inorg. Chem.* **1,** 573, 583 (1962).

114. A. Haim, R. J. Grassi, and W. K. Wilmarth, *Advan. Chem.* **49,** 31 (1965).

115. R. J. Grassi, A. Haim, and W. K. Wilmarth, *Inorg. Chem.* **6,** 237 (1967). R. Barca, J. Ellis, M. Tsao, and W. K. Wilmarth, *Inorg. Chem.* **6,** 243 (1967).

116. D. F. Gutterman and H. B. Gray, *J. Amer. Chem. Soc.* **93,** 3364 (1971).

117. M. Wrighton, G. S. Hammond, and H. B. Gray, *J. Amer. Chem. Soc.* **93,** 5254 (1971).

118. M. Wrighton, H. B. Gray, G. S. Hammond, and V. Miskowski, *Inorg. Chem.* **12,** 740 (1973).

119. G. Ferraudi and J. F. Endicott, *J. Amer. Chem. Soc.* **95,** 2371 (1973).

120. F. Scandola, private communication.

121. J. Jordan and G. J. Ewing, *Inorg. Chem.* **1,** 587 (1962).

122. G. Emschwiller and J. Legros, *Compt. Rend.* **261,** 1535 (1965).

123. G. Stein, *Advan. Chem.* **50,** 238 (1965).

124. V. Carassiti, G. Condorelli, and L. L. Condorelli-Costanzo, *Ann. Chim. (Rome)* **55,** 329 (1965).

125. G. Condorelli and L. L. Condorelli-Costanzo, *Ann. Chim. (Rome)* **56,** 1140 (1966).

126. G. Condorelli and L. L. Condorelli-Costanzo, *Ann. Chim. (Rome)* **56,** 1159 (1966).

127. V. Balzani, V. Carassiti, and L. Moggi, *Ann. Chim. (Rome)* **54,** 251 (1964).

128. V. Balzani, V. Carassiti, and L. Moggi, *Inorg. Chem.* **3,** 1252 (1964).

129. A. Vogler and A. W. Adamson, *J. Amer. Chem. Soc.* **90,** 5943 (1968).

130. M. A. Scandola and F. Scandola, *J. Amer. Chem. Soc.* **92,** 7278 (1970).

131. H. D. Gafney and A. W. Adamson, *J. Phys. Chem.* **76,** 1105 (1972).

132. J. N. Demas and A. W. Adamson, *J. Amer. Chem. Soc.* **95,** 5159 (1973).

133. G. B. Porter, *J. Amer. Chem. Soc.* **91,** 3980 (1969).

134. M. A. Scandola and F. Scandola, *J. Amer. Chem. Soc.* **94,** 1805 (1972).

135. W. Strohmeier and D. von Hobe, *Z. Phys. Chem. (Frankfurt)* **34,** 393 (1962).

136. M. A. El-Sayed, *J. Phys. Chem.* **68,** 433 (1964).

137. I. W. Stoltz, G. R. Dobson, and R. K. Sheline, *J. Amer. Chem. Soc.* **84,** 3589 (1962); *ibid.* **85,** 1013 (1963).

138. M. Mingardi and G. B. Porter, *Spectrosc. Letters* **1,** 293 (1968).

5

PHOTOCHEMISTRY OF THE HEAVIER ELEMENTS

Peter C. Ford, Ray E. Hintze, and John D. Petersen

Department of Chemistry
University of California
Santa Barbara, California

I. INTRODUCTION

In this chapter we examine the photochemical properties that have been found for compounds of the heavier d-block elements and of the f-block elements. In doing this, we also examine certain of the closely related photophysical properties of these compounds and attempt to delineate differences between the photochemistries of the heavier and lighter elements when apparent. Although a wide range of systems are discussed, we primarily focus on quantitative mechanism studies, the majority of which involve Werner-type complexes of the group VIII metals. For this reason, our attention is directed almost exclusively toward those studies carried out in homogeneous solution. The photochemistries of carbonyl and other organometallic complexes of certain heavier transition elements are discussed in the Chapter 6 and photochemistries of solid-state systems in Chapter 9.

In recent years, certain heavier transition-metal complexes have been the recipient of very rapidly increasing attention. In particular, complexes of the "rarer platinum elements," ruthenium, rhodium, and iridium, are the subject of a substantial part of this chapter, in contrast to the small number of references to those systems that were available when the comprehensive review by Adamson and co-workers[1] and the monograph by Balzani and Carassiti[2] were written in 1968 and 1969, respectively. The surge of interest in these complexes can be attributed to more than just the normally expanding forays of new and/or established photochemists into different systems. More importantly, this interest can be attributed to the developing understanding of the thermal reactions and properties[3] (including synthesis procedures) of these compounds and to the availability of new, high-quality photophysical information from luminescence spectroscopists.[4,5] In this context, one might safely predict that the photochemistries of both mononuclear and polynuclear complexes of other heavier elements will be the subject of increasing quantitative attention as their thermal and spectral properties become better understood.

The chapter consists of two major parts. The first and larger section is concerned with coordination compounds of the second and third transition-metal rows. The second section deals with some photochemistry of the f-block elements.

II. COMPOUNDS OF THE HEAVIER d-BLOCK ELEMENTS

A. Comparisons Among the Transition Series

Although there are many differences between the characteristics of analogous elements of the second and third transition rows, the properties of these elements and their compounds differ far less from each other than from the first-row analogs. Compare radii: the appearance of the rare-earth elements immediately prior to the $5d$ elements leads to the "lanthanide contraction." As a result, ionic and covalent radii of the $5d$ elements nearly equal those of the $4d$ elements in analogous oxidation states and compounds.[6] In contrast, significant increases in atomic radii are observed between the first and second transition series. Another distinction between the heavier and lighter transition elements is the absence of high-spin electronic configurations for the ground-state complexes of the second and third transition series. We can, in part, attribute this behavior to the greater splitting of the $4d$ and $5d$ orbital energies by a given ligand field. For example, the octahedral splitting energies of the hexaammine complexes, $M(NH_3)_6^{3+}$, of the d^6 homologs, Co(III), Rh(III), and Ir(III), are 22.9, 34.0, and 41.2 kK, respectively, increasing about 50% in going from the $3d$ to the $4d$ complex and another 20% in going from the $4d$ to the $5d$ analog.[7] In addition, the larger sizes of the $4d$ and $5d$ orbitals lead to smaller electron pairing energies than in $3d$ orbitals; thus Racah B parameters, measures of interelectronic repulsion, decrease in going from Co(III) (615 cm^{-1}) to Rh(III) (430 cm^{-1}) and Ir(III) (470 cm^{-1}) for the hexaammine complexes.[7]

Another important feature of the ground electronic states is evidenced by the magnetic properties of analogous paramagnetic compounds. The magnetic moments of first-row compounds are usually close to the predicted "spin-only" values, implying quenching of orbital angular momentum and little spin-orbit coupling. In contrast, spin-orbit coupling constants are much higher for the heavier elements, especially in the third transition series, and spin-orbit coupling is sufficiently influential to the magnetic properties of the heavier transition series that magnetic moments measured at room temperature often fall well below the spin-only predicted moments. Therefore magnetic moments alone are of little diagnostic value in determining the complexes' electronic configurations.[8]

The heavier transition elements also have a more extensive chemistry for their higher oxidation states, and it generally has been found that few lower oxidation states are stable in simple Werner-type complexes. Consequently, the aquo ions of Cr(II) and Cr(III) are stable, characterizable species; yet the comparable complexes of tungsten are unknown, while the aquo ion of Mo(III) was isolated only recently. Nonetheless, there is a great deal of known chemistry for the heavier transition elements of low oxidation states *when* these complexes include polarizable, π-acceptor-type ligands as carbonyl, olefins, and phosphines. In addition, Werner-type complexes of lower oxidation state group VIII metals are reasonably well known. Stereochemistry and coordination number provide another comparison between the transition series. For the heavier transition elements, octahedral coordination remains the most predominantly observed configuration; ground-state tetrahedral coordination is largely confined to oxide species of high oxidation states, for example, OsO_4, RuO_4, and NbO_4^{3-}. For lower oxidation states, four coordination is primarily square planar in d^8 species such as Pt(II), Pd(II), Rh(I), Ir(I), and Au(III), owing to the tendency toward spin-paired complexes even with weak-field ligands as Cl^-. The d^8 Ni(II) analogs are square planar only with very strong field ligands, as in $Ni(CN)_4^{2-}$. Coordination numbers greater than six are also common for the heavier transition elements. The octacyano complexes of Mo(IV) and W(IV), whose photochemical properties are discussed in this chapter, are eight-coordinate with dodecahedral or square antiprismatic geometries. At the other end of the scale, some d^{10} complexes such as the Pt(0) phosphines and those of Ag(I) and Au(I) have coordination numbers from two to four.

B. Excited States and Absorption and Luminescence Spectra

The absorption spectra of the $4d$ and $5d$ metal complexes display the types of bands described in earlier chapters, namely, ligand field (d–d) transitions (LF), charge transfer to ligand (CTTL), charge transfer to metal (CTTM), and internal ligand transitions, normally, $\pi_L \rightarrow \pi_L^*$. Observance of any or all of these depends on the nature of the metal, including its oxidation state, and on the nature of the coordinated ligands—however, clear identification of the electronic transitions is commonly very difficult. This difficulty is not only due to the experimental problem of resolving overlapping absorption bands but also to the frequent lack of clear distinction of the excited states' character. Classification of electronic states according to metal localized d-orbitals derives from electrostatic crystal-field theory and is not fully valid in the case of extensive metal-ligand covalency.[9,10] Numerous studies have demonstrated such covalency, and examples even involve significant delocalization of "d orbital" electrons into ligands having extensive π-conjugation.[11]

With regard to excited-state processes, the magnitude of spin-orbit coupling may be the most important distinction between the different transition rows. Spin-orbit coupling constants are a function of the fourth power of an atom's effective nuclear charge,[8] and free-ion values reported for various group VIII metals are about two to three times greater for the $4d$ elements and about eight times greater for the $5d$ elements than for the $3d$ elements.[3] Coupling between the spin and orbital angular momenta makes accurate factoring of the total wave function into spin and orbital components impossible (see Chapter 1). Since the factoring can only be approximate, the spin-selection rule for electronic transitions cannot be strict. Consequently, intensities of "spin-forbidden" bands relative to "spin-allowed" bands of various complexes increase as spin-orbit coupling constants increase. In essence, the spin-forbidden transition becomes partially allowed due to a mixing of terms in the ground and excited states having the same spin-orbit representations. For this reason, relatively intense spin-forbidden bands are observed in the absorption spectra of the $5d$ Os(II) complex, Os(bipy)$_3^{2+}$, but comparable bands in the spectrum of the Ru(II) analog are poorly resolved and much less intense.[12]

In discussing photochemical and spectral processes (including those of the heavier elements), it is common practice to refer to excited states in terms of spin multiplicities. Indeed, in subsequent sections of this chapter, we also conform to this practice. However, one should be reminded of repeated warnings[13,14] that, while such labeling is convenient and indeed is probably valid for the $3d$ and $4d$ metals, it may prove especially deceiving for the $5d$ metals where spin-orbit coupling is much more important.

Luminescence spectral studies of numerous heavier transition-metal complexes have been reported; however, nearly all these studies are concerned with d^6 group VIII metal complexes. In discussing the photochemistry of specific systems we also consider any relevant information derived from luminescence studies. Unfortunately, most photochemistry of heavy-metal complexes has been studied in fluid solution and at near-ambient temperature, while luminescence experiments have largely been carried out in rigid media at low temperatures approximating that of liquid nitrogen (77°K). Clearly, there is a prominent need for a better correspondence of these conditions so that photophysical and photochemical data may be compared. Such has been done for several Cr(III) systems, and there is much potential for this type of correlation with heavy-metal complexes. In this context, it is important to note that relatively few complexes have been observed to luminesce in ambient-temperature fluid solutions. Thus it may be necessary in many cases to choose conditions intermediate between those normally used for photochemistry and for luminescence studies in order to obtain both types of data under similar conditions.

Various information can be deduced from the luminescence data for metal complexes. Band shapes, lifetimes, and supplementary data such as medium effects can be used to identify the emitting state. For the $4d^6$ and $5d^6$ complexes, observed emissions are nearly always triplet-to-singlet phosphorescence from the lowest lying thermally equilibrated electronic excited state of the complex.[12,14-18] Depending on the identity of the central metal and its ligand field, phosphorescent emissions from *d–d* states, from ligand $\pi-\pi^*$ states, and from metal-to-ligand charge-transfer states have all been identified. In fact, Crosby and co-workers[14] have obtained *d–d*, CTTL, and $\pi-\pi^*$ emissions from different Ir(III) complexes containing a 1,10-phenanthroline in the coordination sphere by variations of the balance of the ligand field (Fig. 5-1). Such modifications

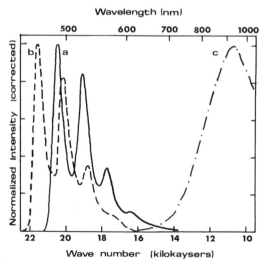

Fig. 5-1. Luminescence spectra of three Ir(III) phenanthroline complexes: (a) $\pi_L-\pi_L^*$ emission from [IrCl$_2$(5,6-dimethylphen)$_2$]Cl; (b) CTTL emission from [IrCl$_2$(phen)$_2$]Cl; (c) *d–d* emission from K[IrCl$_4$(phen)]. All spectra obtained at 77°K (reproduced with permission from [14]. Copyright 1970 by the American Association for the Advancement of Science).

in the emission properties by chemical modifications of the ligand field have been referred to by these workers as "tuning" and variations in emission energies by ligand substituents as "fine tuning."[16] It is clear that such "tuning" has great potential for examining the chemical and electronic properties of transition-metal excited states.

C. The Photochemistry of d^6 Hexacoordinate Complexes

As we have seen in previous chapters, the photochemistries of d^6 metal-ion complexes, especially those of Co(III), have long intrigued chemists interested in the photoreactions of transition-metal coordination compounds. Here we are concerned with the photochemistry that has been reported for complexes of the heavier d^6 ions, Ru(II), Rh(III), Ir(III), and Pt(IV). Luminescence data has also been accumulated for complexes of each of these metals, including many whose photoreactions have been quantitatively studied. Of other d^6 group VIII metal ions, luminescence, but not quantitative photochemistry, has been reported for Os(II) complexes (no doubt the high cost of osmium and uncertainty regarding thermal chemistry has served to inhibit the photochemists in this regard). Pd(IV) complexes are much less stable than the Pt(IV) analogs.

All of the d^6 ions of the group VIII metals form strong-field, spin-paired complexes, but, as expected, show fundamental differences in the character of excited states observable in absorption and emission spectra. Rh(III) is neither easily oxidized to the +4 oxidation state nor easily reduced to the +2 state. Consequently, charge-transfer states, when observed, are normally high energy, and ligand-field states are lower. Thus in luminescence spectroscopy of various Rh(III) complexes the emitting (and lowest energy) state has proved to be a LF triplet.[10,15,17,18,21,22] [In several cases, however, such as the tris o-phenanthroline and tris 2,2'-bipyridyl complexes of Rh(III), the lowest states have been identified as triplet $\pi-\pi^*$, ligand-localized states.[23]] In contrast, Ir(III) complexes of unsaturated ligands such as pyridine and bipyridine often display prominent, relatively low energy CTTL bands in their absorption spectrum, and, in some cases, CTTL "triplets" are the lowest states.[12,14,16] The difference between the homologous Rh(III) and Ir(III) complexes no doubt lies in the greater ease by which Ir(III) is oxidized to Ir(IV).[24] Complexes of Ru(II) and Os(II) with unsaturated ligands have prominent low-energy CTTL bands in their absorption spectra, and the lowest excited states of phenanthroline and bipyridine complexes have been identified by luminescence spectroscopy as CTTL triplets.[15] Again, both Ru(II) and Os(II) can be oxidized with relative ease to a higher oxidation state. Ru(II) complexes of saturated ammines [e.g., Ru(NH$_3$)$_6^{2+}$] display

Table 5-1. Spectral Features of Some Representative $4d^6$ and $5d^6$ Complexes[a]

	Absorption[b]			Emission[c]		
	$\lambda_{max}(\log \varepsilon)^d$	Assignment	Ref.	λ_{max}^d	Emitting State	Ref.
Ru(II)						
Ru(NH$_3$)$_6^{2+}$	385 (1.59)	LF	25	—	—	—
	275 (2.83)	LF	25	—	—	—
Ru(NH$_3$)$_5$py^{2+}	407 (3.89)	CTTL	28	—	—	—
	244 (3.66)	$\pi_L \to \pi_L^*$	28	—	—	—
Ru(bipy)$_3^{2+}$	~450 (4.17)	CTTL	29, 12	590	CTTL	29
	286 (4.87)	$\pi_L \to \pi_L^*$	29, 12	645	CTTL	29
	244 (4.40)	CTTL	29, 12	—	—	—
Rh(III)						
Rh(NH$_3$)$_6^{3+}$	306 (2.13)	LF	30	612	LF	18
	256 (2.00)	LF	30	—	—	—
cis-Rh(bipy)$_2$Cl$_2^+$	382 (2.03)	LF	31	710	LF	31
	313 (4.46)	$\pi_L \to \pi_L^*$	31	—	—	—
trans-Rh(py)$_4$Cl$_2^+$	~465 (~0.5)	LF[e]	31	650	LF	31
	403 (1.91)	LF	31	—	—	—
	257 (4.13)	$\pi_L \to \pi_L^*$	31	—	—	—
Ir(III)						
Ir(en)$_3^{3+}$	~302 (1.45)	LF[e]	30	595	LF	31
	249 (2.23)	LF	30	—	—	—
	218 (2.18)	LF	31	—	—	—
Ir(bipy)$_2$Cl$_2^+$	457 (2.67)	—	31	470	CTTL	31
	389 (3.42)	CTTL	31	502	CTTL	31
	287 (4.34)	$\pi_L \to \pi_L^*$	31	—	—	—
Pt(IV)						
PtCl$_6^{2-}$	465 (1.76)[f]	LF[e]	26	690	LF	27
	358 (2.67)[f]	LF	26	—	—	—
	265 (4.44)[f]	CTTM	26	—	—	—

[a] Absorption spectra in aqueous solution, luminescence spectra in frozen alcohol/water glasses.
[b] Multiplicity-allowed transitions (singlet → singlet) unless noted.
[c] Phosphorescence (triplet to singlet) in all cases.
[d] λ in nanometers.
[e] Multiplicity forbidden (singlet → triplet).
[f] $(n\text{-}C_4H_8)_4N^+$ salt in CH_3THF/CH_3OH solvent.

ligand-field bands in their absorption spectra.[25] In the other direction, complexes of the more highly charged Pt(IV) display prominent CTTM bands in their spectra.[26] However, the luminescence spectra reported to date [PtCl$_6^{2-}$, PtBr$_6^{2-}$, PtCl$_4$py$_2$, and PtCl$_4$(phen)] show a LF triplet as the luminescing lowest state.[27]

Table 5-1 summarizes the spectral features of some representative d^6 complexes of the heavier group VIII metals. Also included is information regarding the character of the luminescent state for complexes where this determination has been made. These data serve to illustrate the comments regarding the spectral patterns observed for different metals discussed above.

For low-spin d^6 complexes of octahedral-type coordination, the three t_{2g} orbitals are fully occupied. Consequently all ligand-field excited states (e.g., those with the configuration, $t_{2g}^5 e_g^1$) and charge-transfer-to-metal states must involve electronic population of an e_g orbital having antibonding character with respect to metal-ligand σ bonds. Whether accompanied by reduction of the central metal ion or not, labilization of coordinated ligands is an expected reaction characteristic of either of these types of excited states. In contrast, there is no obvious suggestion that a charge-transfer-to-ligand state, which results from promotion of a t_{2g} electron into low-lying, empty ligand π-orbitals, should be particularly substitution labile, since analogous d^5 complexes, for example, Ru(III) and Ir(IV), are (with some exceptions) relatively substitution inert. If redox processes occur, *oxidation* of the central metal ion would be an expected reaction pathway for the CTTL state.

1. **Ammine Complexes of Rh(III).** The +2 oxidation state is virtually unknown for iridium or rhodium. This characteristic contrasts sharply from that of the first-row congener, cobalt, for which a labile +2 oxidation state has extensive chemistry. Thus, while ultraviolet irradiation of Co(III) complexes (CTTM excitation) normally leads to relatively efficient formation of Co(II)

$$Co(NH_3)_5Cl^{2+} \xrightarrow[\phi=0.25]{h\nu(254\,nm)} Co^{2+} + \text{other products} \qquad (5\text{-}1)$$

analogous photolysis of Rh(III) complexes leads only to Rh(III) products

$$Rh(NH_3)_5Cl^{2+} + H_2O \xrightarrow[\phi=0.11]{h\nu(254\,nm)} Rh(NH_3)_5H_2O^{3+} + Cl^- \qquad (5\text{-}2)$$

Another feature distinguishing Rh(III) photochemistry is the relative reactivity of Rh(III) ligand-field states compared to the apparent inertness of LF states for most Co(III) complexes (Chapter 4). However, it has been suggested[20] that the differences between the apparent reactivities of Rh(III) LF states and those of Co(III) may lie not in the intrinsic reactivities of the LF states, but in the efficiency of intersystem crossing to the reactive LF triplets from singlet states produced by initial excitation. Intersystem crossing from LF singlets to triplets occurs with nearly unitary efficiency with Rh(III) complexes, but with markedly lower efficiency for Co(III) complexes. These differences may be attributable to the much larger spin-orbit coupling for the heavier rhodium atom.

Several workers[19,20,32] have studied the photochemistry of the halopentaamminerhodium(III) ions, $Rh(NH_3)_5X^{2+}$ (X = Cl^-, Br^-, or I^-), in aqueous solution. Rh(III) products result whether CTTM or LF bands are irradiated; however, quantum yields and the course of the reactions were

Table 5-2. Quantum Yields from Irradiation of Pentaammine Rh(III) Complexes
in Aqueous Solution at Ambient Temperature

Complex	Type of Absorption Band Irradiated	λ, nm	Φ_X	Φ_{NH_3}	Ref.
$Rh(NH_3)_5Cl^{2+}$	LF	254	0.11	$\leq 10^{-3}$	20
	LF	280	0.12	$\leq 10^{-3}$	20
	LF	350	0.16	$\leq 10^{-3}$	20
	LF	380	0.14	$< 10^{-3}$	20
$Rh(NH_3)_5Br^{2+}$	LF	360	0.019	0.18	20
	LF	420	0.019	0.17	20
$Rh(NH_3)_5I^{2+}$	CTTM	214	—	0.41	19, 20
	CTTM	254	—	0.43	19, 20
	CTTM	280	—	0.52	19, 20
	CTTM+LF	350	—	0.58	19, 20
	LF	385	—	0.82	19, 20
	LF	420	—	0.87	19, 20
	LF	470	—	0.85	19, 20
$Rh(NH_3)_6^{3+}$	LF	313	0.075 ± 0.007	—	33
		254	0.07	—	33
$Rh(NH_3)_5py^{3+}$	LF	313	0.12	<0.02	33
$Rh(NH_3)_5(CH_3CN)^{3+}$	LF	313	0.47	$<10^{-2}$	33
	LF	254	0.45	—	33
$Rh(NH_3)_5(C_6H_5CN)^{3+}$	LF	313	0.35	$\leq 10^{-3}$	33

found to depend both upon the identity of X and on the wavelength of
irradiation (Table 5-2). Ligand-field excitation of $Rh(NH_3)_5Cl^{2+}$ leads to
chloride aquation [Eq. (5-2)], but LF excitation of the iodo complex

$$Rh(NH_3)_5I^{2+} \xrightarrow{h\nu} trans\text{-}[Rh(NH_3)_4(H_2O)I]^{2+} + NH_3 \qquad (5\text{-}3)$$

leads to aquation of *trans*-NH$_3$ [Eq. (5-3)], while both pathways are
observed with the bromo complex

$$Rh(NH_3)_5Br^{2+} \xrightarrow{h\nu} \begin{cases} Rh(NH_3)_5H_2O^{3+} + Br^- \\ trans\text{-}[Rh(NH_3)_4(H_2O)Br]^{2+} + NH_3 \end{cases} \qquad (5\text{-}4)$$

Sensitization experiments at room temperature[20] demonstrated that each
of these complexes quenches the phosphorescence of biacetyl but does
not affect the fluorescence. The same products and quantum yields were
obtained from the sensitized photoreactions as from the direct ligand-field
photolyses. These data clearly suggest that the products arise from triplet,
ligand-field excited states and that the efficiency of intersystem crossing

from singlet LF states populated by direct excitation to the reactive triplet states is unity ($\Phi_{isc} = 1.0$).

The mechanism proposed[20] to account for the product distributions from LF photolyses of $Rh(NH_3)_5X^{2+}$ involves two reactive, triplet, LF excited states (Fig. 5-2). The state whose reaction leads to *trans*-NH₃ aquation is designated as ^3X, the one leading to halide aquation as ^3Y. The geometry of these excited states was suggested to be strongly distorted along the labile bond axis (nearly dissociative), and the energy separation to be a function of the halogen. The product distributions are explained if the energy gap is large for $Rh(NH_3)_5Cl^{2+}$, so that ^3Y is the only reactive state with significant population, small for $Rh(NH_3)_5Br^{2+}$, and very small (or perhaps the order of the states reversed) for $Rh(NH_3)_5I^{2+}$. The orbital character of the proposed states, ^3X and ^3Y, is not obvious. However, the idea of two reactive states is consistent with the suggestion,[18] based on spectral data from pentaammine Rh(III) complexes, that for strong tetragonal distortion by the sixth ligand a B_2 state of T_{2g} parentage (from the

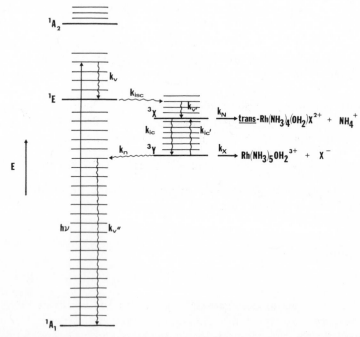

Fig. 5-2. Qualitative energy-level scheme for the photochemistry of halopentaammine Rh(III) complexes. First-order rate constants are indicated for vibrational relaxation (k_v, k_v', k_v''), internal conversion (k_{ic}, k_{ic}'), intersystem crossing (k_{isc}, k_n) and product formation k_N and k_X. (Reproduced with permission from [20]. Copyright by the American Chemical Society.)

octahedral field) could have an energy comparable to the E state of T_{1g} parentage normally expected to be lowest in energy.

Table 5-2 also lists quantum yields obtained for ligand-field excitation of pentaammine Rh(III) complexes of the uncharged ligands, NH_3, pyridine, acetonitrile, and benzonitrile. Like the chloro complex, these undergo predominantly photoaquation of the sixth ligand[33]

$$Rh(NH_3)_5L^{3+} \xrightarrow{h\nu} Rh(NH_3)_5H_2O^{3+} + L \qquad (5-5)$$

A simple spectrochemical series[34,35] for these Rh(III) complexes based upon the energy of the first LF band ($^1T_1 \leftarrow {^1A_1}$ in octahedral symmetry) gives the following order for the various halide and uncharged ligands: $I^- < Br^- < Cl^- < NH_3 \simeq py \simeq CH_3CN \simeq C_6H_5CN$. This series suggests that, unlike the iodo and bromo complexes, the electronic distortion of the various $Rh(NH_3)_5L^{3+}$ complexes away from the octahedral microsymmetry of the hexaammine is small. Despite this, only the sixth ligand, L (or Cl^- in $Rh(NH_3)_5Cl^{2+}$), is subject to photoaquation.[7] One explanation may lie in examining ligand base strength. The photoaquation quantum yields follow the order, $NH_3 < py < C_6H_5CN \sim CH_3CN$, the approximate inverse of the Brönsted basicities of the free ligands, $NH_3 > py \gg CH_3CN \sim C_6H_5CN$. The relatively high ligand-field strength of the nitriles must in large part be due to their polarizability (softness),[35] while the relative reactivity of the lowest excited state appears to be dominated by the σ-donor ability of the unique ligand.

Stereochemical consequences of ligand-field excitation of Rh(III) amines were examined by Kutal and Adamson, who studied the photochemistry of aqueous dichlorotetraamine Rh(III) complexes, $RhA_4Cl_2^+$, where $A = NH_3$, ethylenediamine/2, or cyclam/4 (cyclam is 1,4,8,11-tetraazacyclotetradecane).[36] In each case, chloride aquation was the only observed pathway, and the reaction proceeded with retention of configuration

$$trans\text{-}[RhA_4Cl_2]^+ \xrightarrow{h\nu} trans\text{-}[RhA_4(H_2O)Cl]^{2+} + Cl^- \qquad (5-6)$$

Quantum yields (Table 5-3) were dependent on the identity of the equatorial amines; however, the overall decrease in Φ_{Cl} in going from $trans\text{-}[Rh(NH_3)_4Cl_2]^+$ to $trans\text{-}[Rh(cyclam)Cl_2]^+$ was a relatively small factor of about 10. These results contrast with the ligand-field photochemistry of the analogous Cr(III) complexes,[37,38] where $trans\text{-}[Cr(NH_3)_4Cl_2]^{2+}$ and $trans\text{-}[Cr(en)_2Cl_2]^{2+}$ photoaquate Cl^- with high quantum yields to give the isomerized products, $cis\text{-}[CrA_4(H_2O)Cl]^{2+}$, but the stereorigid cyclam complex is at least a factor of 400 less photoactive. Reactions of the excited state consistent with the Cr(III) results would be either a dissociative pathway involving collapse to a trigonal bipyramidal

Table 5-3. Quantum Yields for Dihalo Rh(III) Amine Complexes

Complex	Type of Absorption Band Irradiated	λ, nm	Φ_X	$\Phi_{H^+}{}^a$	Ref.
trans-Rh(NH$_3$)$_4$I$_2^+$	CTTM	254	0.20	—	19
	CTTM	280	0.32	—	19
	CTTM	340	0.33	—	19
	LF	470	0.48	—	19
trans-Rh(NH$_3$)$_4$Cl$_2^+$	LF	358	0.17	—	36
	LF	407	0.13	$<2 \times 10^{-3}$	36
	?	254	0.031	—	40
	LF	350	0.04	—	40
trans-Rh(en)$_2$Cl$_2^+$	LF	407	0.047	$<3 \times 10^{-3}$	36
	?	254	0.03	—	40
	LF	350	0.1	—	40
trans-Rh(en)$_2$Br$_2^+$	?	254	0.054	—	40
	LF?	350	0.1	—	40
trans-Rh(en)$_2$I$_2^+$	CTTM	254	0.23	—	40
	CTTM	350	0.3	—	40
cis-Rh(en)$_2$Cl$_2^+$	?	254	0.1	—	40
	LF	350	0.056	—	40
trans-Rh(cyclam)Cl$_2^+$	LF	407	0.011	$<10^{-4}$	36

^{a} Φ_{H^+} is a measure of amine group aquation.

intermediate concerted with loss of Cl$^-$ or an associative pathway involving attack of solvent H$_2$O to form a pentagonalbipyramidal intermediate. Neither mechanism is consistent with the Rh(III) quantum yields. The stereochemistry and relative insensitivity to the nature of the equatorial ligands in Rh(III) ligand-field photochemistry can be interpreted as indicating a dissociative pathway via a pentacoordinate square pyramidal intermediate. Evidence for the ability of such an intermediate to maintain its stereochemical integrity has been reported for thermal substitution reactions of isoelectronic Ru(II) ammine complexes.[39] Photolyses of other trans dihalo complexes (trans-[Rh(en)$_2$Br$_2$]$^+$ and trans-[Rh(en)$_2$I$_2$]$^+$) have also been shown[40] to result in halide photoaquation with retention of configuration in reactions analogous to Eq. (5-6). However, photolyses of cis-[Rh(en)$_2$Cl$_2$]$^+$ results in substantial cis to trans isomerization with quantum yields exceeding photosubstitution

$$cis\text{-}[Rh(en)_2Cl_2]^+ \xrightarrow{h\nu} trans\text{-}[Rh(en)_2Cl_2]^+ + \text{aquation products} \quad (5\text{-}7)$$

Thus stereorigidity is not a general photochemical characteristic of Rh(III) amine complexes, and the failure of trans complexes to undergo photoisomerization may merely reflect the intrinsic stability of the trans products.

A clear understanding of the rates and mechanisms of all processes leading to deactivation of excited states is of fundamental importance to understanding the photochemical behavior of a system. Theories of deactivation processes leading to the ground state are discussed in Chapters 1 and 2; however, recent experiments with Rh(III) amine complexes provide illustration of the importance of these processes under photochemically significant conditions.[33] We can illustrate this point with the energy-level diagram for the octahedral ion, $Rh(NH_3)_6^{3+}$ (Fig. 5-3). If one presumes that, as with the halopentaammine complexes, intersystem crossing to a $^3T_{1g}$ reactive state has unitary efficiency for this ion, then the quantum yield of photoreaction is equal to

$$\Phi_p = \frac{k_p}{k_p + k_n + k_r} \tag{5-8}$$

Since attempts to observe luminescence from $Rh(NH_3)_6^{3+}$ at room temperature have proved unsuccessful[18,41] under these conditions, the rate of

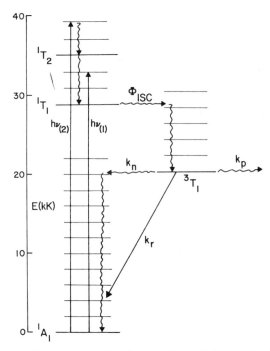

Fig. 5-3. Energy-level diagram for the photoreactions of $Rh(NH_3)_6^{3+}$. $h\nu(1)$ and $h\nu(2)$ represent excitation into the $^1T_1$ and $^1T_2$ states, respectively, k_p is aquation of excited state, $^3T_1$, to give $[Rh(NH_3)_5H_2O]^{3+}$, and k_n and k_r represent nonradiative and radiative deactivation to the ground state.

radiative deactivation (k_r) must be substantially less than the respective rates of nonradiative deactivation to ground state (k_n) or aquation to photoproducts (k_p). Thus we can rewrite Eq. (5-8) as

$$\Phi_p = \frac{k_p}{k_p + k_n} \qquad (5\text{-}9)$$

and rearrange this to

$$\frac{k_n}{k_p} = \frac{1 - \Phi_p}{\Phi_p} \qquad (5\text{-}10)$$

The quantum yield for photoaquation of $Rh(NH_3)_6^{3+}$ at ambient temperature (Table 5-2) is 0.075, and use of Eq. (5-10) translates this into a k_n/k_p ratio of 12.3. Thus it is clear that interpretative discussion of photoreaction quantum yields must consider other deactivation pathways as well to be complete.

This question has been considered[17] for the luminescence spectroscopy of Rh(III) ammine complexes in methanol/water glasses at 110°K. These complexes all phosphoresce from the lowest triplet LF state, which is found at a large Stokes shift relative to the lowest energy-absorption band. Radiative lifetimes depend on the ligand field and are short for complexes of heavier atoms, a phenomenon attributed to greater spin-orbit coupling. Despite the large Stokes shift, the radiationless deactivation process has been assigned to a *weak coupling mechanism*, where deactivation of the electronic excited state occurs via vibrational activation of the N—H bonds (the highest frequency vibrational modes of the complex) (Chapter 1). This conclusion is primarily based on the observation that excited states of the perdeuterated complexes, $Rh(ND_3)_5X^{2+}$, are significantly longer lived (a factor of 62 larger for $Rh(ND_3)_6^{3+}$) than those of the corresponding perprotio complexes, $Rh(NH_3)_5X^{2+}$. Since the longer lifetimes are accompanied by larger emission quantum yields, they must result from decreases in the nonradiative deactivation rate, k_n.

On the basis of photochemical observations, it has been argued[20] that in aqueous solution at 25°C radiationless deactivation occurs predominantly via a *strong-coupling mechanism* (Chapters 1 and 2), one that involves large horizontal displacement of the potential well of the excited state relative to the ground state due to major distortion along a metal-ligand bond axis. This conclusion is based on the observation of high quantum yields for photoaquation (presumably by a dissociative mechanism) and the apparent insensitivity of photoaquation quantum yields to perdeuteration of the ammines. Since photochemistry is not observed under the luminescence conditions, it is argued that the temperature and medium change from 110 to 300°K results in an apparent change of the dominant deactivation mechanism. This is a trivial conclusion for the iodo complex,

$Rh(NH_3)_5I^{2+}$, which undergoes photoaquation with a quantum yield near unity, so that deactivation occurs almost exclusively by photoreaction. On the other hand, perdeuteration of coordinated ammines has been shown to enhance the photoaquation quantum yields of $Rh(NH_3)_5Cl^{2+}$ by about 40%[20] and of $Rh(NH_3)_6^{3+}$ by about 100%[33] in ambient aqueous solution. Since the deuteration effect is apparently due to decreases in the radiationless deactivation rates, a weak-coupling deactivation mechanism in these cases must remain competitive with dissociation and other deactivation pathways.

Although irradiation of the CTTM bands of $Rh(NH_3)_5I^{2+}$ and *trans*-$[Rh(NH_3)_4I_2]^+$ gives the same net photoaquation products as irradiation of LF bands, quantum yields are lower for charge-transfer excitation. This can only mean that the efficiency of intersystem crossing/internal conversion from CTTM states to the reactive LF triplet is less than unity, and that direct radiationless deactivation from the CTTM states is competitive with intersystem crossing. Flash photolysis of $Rh(NH_3)_5I^{2+}$ in the CTTM region when traces of I^- are present produced a transient identified as I_2^-. Since the I_2^- was produced only when the CTTM bands were irradiated, it can be concluded that a redox pathway must be responsible for a major fraction of the NH_3 photoaquation under these conditions[19]

$$Rh(NH_3)_5I^{2+} \xrightarrow{h\nu(CTTM)} [Rh(NH_3)_4]^{2+} + NH_3 + \cdot I \qquad (5\text{-}11)$$

$$\cdot I + I^- \longrightarrow I_2^- \qquad (5\text{-}12)$$

$$[Rh(NH_3)_4]^{2+} + I_2^- + H_2O \longrightarrow trans\text{-}[Rh(NH_3)_4(H_2O)I]^{2+} + I^- \qquad (5\text{-}13)$$

or

$$[Rh(NH_3)_4]^{2+} + \cdot I + H_2O \longrightarrow trans\text{-}[Rh(NH_3)_4(H_2O)I]^{2+}$$

Since only *trans*-tetraamminerhodium(III) products were observed, the proposed Rh(II) intermediate appears to maintain its integrity over the course of its lifetime, presumably as a tetragonally distorted six-coordinate or square planar, low-spin d^7 species.

2. Ir(III) and Other Rh(III) Complexes.

Amine complexes of Ir(III) have been shown to be photoreactive, but so far they have been the subject of relatively little quantitative study. Broad-band irradiation of the complexes, *trans*-$[Ir(en)_2X_2]^+$ ($X = Cl, Br, I$), in aqueous solution leads to stereospecific aquation[42]

$$trans\text{-}[Ir(en)_2X_2]^+ + H_2O \xrightarrow{h\nu} trans\text{-}[Ir(en)_2(H_2O)X]^{2+} + X^- \qquad (5\text{-}14)$$

Since the haloaquo product reacts stereospecifically with excess ligand, Y^-, this photoreaction provides a useful route for preparation of mixed diacidobis(ethylenediamine) complexes, $trans$-$[Ir(en)_2XY]^+$. The Rh(III) analogs can be synthesized in a like manner.[42,43]

Irradiation of the azido complexes, $[Rh(NH_3)_5N_3]^{2+}$ and $[Ir(NH_3)_5N_3]^{2+}$, in aqueous hydrochloric acid leads not only to substitution reactions, but also to cleavage of the coordinated azide.[44] The products isolated from the photolysis of the latter ion are $[Ir(NH_3)_5(NH_2Cl)]^{3+}$ plus molecular nitrogen. The chloramine product was attributed to the intermediate formation of coordinated nitrenes

$$(NH_3)_5IrN_3^{2+} \xrightarrow{h\nu} [(NH_3)_5IrN:]^{2+} + N_2$$

$$[(NH_3)_5IrN:]^{2+} \xrightarrow{H^+, HCl} [(NH_3)_5Ir(NH_2Cl)]^{3+}$$

(5-15)

Flash-photolysis studies of the Rh(III) complex have been reported[45] to show formation of a transient species in aqueous solution believed to be the nitrene intermediate.

Given the recent contributions in the luminescence spectroscopy of other Rh(III) and Ir(III) complexes[10,12,14,16,21,23] and the known photosensitivity of some,[46-52] the photochemistry of these systems will no doubt receive increasing attention in the near future. Most observed photoactivity involves simple ligand-substitution reactions, both for Rh(III) and for Ir(III), and some photoreactions have proved useful for the synthesis of new compounds. Quantitative studies have been reported[40,51] for complexes of the type, cis-$[M(bipy)_2X_2]^+$, cis-$[M(phen)_2X_2]^+$, and $trans$-$[M(py)_4X_2]^+$ (X = Cl, Br, I; M = Rh, Ir) in aqueous solution. Each of these undergoes halide aquation when irradiated at 254 nm (internal ligand, $\pi-\pi^*$ absorption region) or 350 nm ($d-d$ for Rh(III),[15] but CTTL for Ir(III) complexes[16]) with a general reactivity order: $I^- > Br^- > Cl^-$. None of these complexes undergoes photoisomerization. The similarity in photoreactivity between analogous Rh(III) and Ir(III) complexes is puzzling given that the luminescent (therefore lowest) states are of different nature, $[Ir(phen)_2X_2]^+$ and $[Ir(bipy)_2X_2]^+$ emitting from CTTL states, while $[Rh(bipy)_2X_2]^+$ and $[Rh(phen)_2X_2]^+$ emit from $d-d$ states. The ligand-field excited states would be expected to be substitution labile; however, given that Ir(IV) complexes are not labile, it is unclear that CTTL states of Ir(III) should be (see discussion on Ru(II) complexes below).

3. *Ammine Complexes of Ru(II).* Here we turn our attention to a series of Ru(II) complexes whose absorption spectra display considerable charge-transfer character, but, unlike Co(III) complexes, the charge transfer in these cases is metal-to-ligand. Pentaammine Ru(II) complexes of π-unsaturated ligands, $Ru(NH_3)_5L^{2+}$, when L is an aromatic heterocycle,[28] an organonitrile such as benzonitrile or acetonitrile,[53] or even

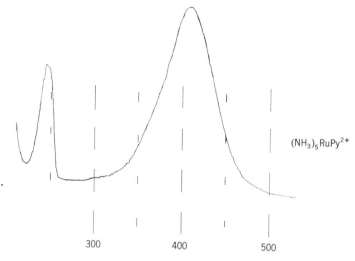

Fig. 5-4. Electronic spectrum for $[Ru(NH_3)_5py]^{2+}$ in aqueous solution.

dinitrogen (N_2),[54] have prominent CTTL bands in their absorption spectra. As an example, the electronic spectrum of $Ru(NH_3)_5py^{2+}$ is shown in Fig. 5-4. Two absorption bands are clearly defined: a CTTL transition with a λ_{max} at 408 nm $(\varepsilon = 7.78 \times 10^3 \, M^{-1} \, cm^{-1})$ and a $\pi_L - \pi_L^*$ band at 244 nm $(\varepsilon = 4.58 \times 10^3 \, M^{-1} \, cm^{-1})$. The internal ligand band is slightly higher energy than analogous ones in pyridinium ion (a multiplet centered at 255 nm), an observation suggesting $d\pi - p\pi$ orbital interaction.[28] No LF bands are apparent in this spectrum. The spectrum of the benzonitrile complex, $Ru(NH_3)_5bz^{2+}$, follows a similar pattern. If the sixth ligand is not unsaturated, such as with $Ru(NH_3)_6^{3+}$, the spectrum is characterized by less intense absorption assigned to the spin-allowed LF transitions[25] $(^1A_{1g} \rightarrow {}^1T_{1g}$ and $^1A_{1g} \rightarrow {}^1T_{2g}$ predicted for octahedral symmetry).

Initial photochemical studies of Ru(II) ammine complexes showed that broad-band ultraviolet- and visible-range irradiation (high-pressure mercury sources) of $Ru(NH_3)_5L^{2+}$ complexes results in several reactions including oxidation of Ru(II) to Ru(III) and aquation of one or more ligands. For example, irradiation of the dinitrogen complexes, $Ru(NH_3)_5N_2^{2+}$ or $[Ru(NH_3)_5]_2N_2^{2+}$ in deaerated aqueous solution was reported [55] to give a redox pathway

$$Ru(NH_3)_5N_2^{2+} + X^- \xrightarrow{h\nu} Ru(NH_3)_5X^{2+} + ? \qquad (5\text{-}16)$$

where $X^- = Cl^-$ in dilute HCl or OH^- in neutral water without chloride. The appearance of the product, $Ru(NH_3)_5X^{2+}$, does not necessarily imply involvement of X^- in the primary photochemical step, since if

$Ru(NH_3)_5H_2O^{3+}$ was produced initially, it would react rapidly to form $Ru(NH_3)_5X^{2+}$ under the experimental conditions.[25] The other product(s) of the redox process were not identified, but N_2 reduction was speculated. Similar photooxidation of Ru(II) to Ru(III) is the result [56] of broad-band ultraviolet irradiation of deaerated aqueous solutions of $Ru(NH_3)_5py^{2+}$, $Ru(NH_3)_5CH_3CN^{2+}$, $Ru(NH_3)_5H_2O^{2+}$, and $Ru(NH_3)_6^{2+}$. Mass spectral analysis was carried out on the gas above the product solution of $Ru(NH_3)_5py^{2+}$ photolysis, and hydrogen (H_2) is a product of this reaction.[56] Analysis of product solutions showed that simple aquation is also a major photoreaction mode under these conditions.

Photoaquation is the principal reaction mode resulting from irradiation[56,57] of scrupulously deaerated aqueous solutions of $Ru(NH_3)_5py^{2+}$ and of $[Ru(NH_3)_5bz]^{2+}$ at wavelengths (366 and 405 nm) corresponding to excitations of their CTTL bands. *No photooxidation of Ru(II) to Ru(III) was observed as a result of irradiation of the CTTL band.* For $[Ru(NH_3)_5bz]^{2+}$, pH-independent benzonitrile aquation is the principal pathway, but for $Ru(NH_3)_5py^{2+}$, competitive aquation of the ammonias leading to both *cis* and *trans* aquotetraamminepyridine products is also observed.

$$Ru(NH_3)_5py^{2+} \xrightarrow{h\nu, 405\text{ nm}} \begin{cases} \overset{(a)}{\longrightarrow} Ru(NH_3)_5H_2O^{2+} + py \\ \overset{(b)}{\longrightarrow} cis\text{-}[Ru(NH_3)_4(H_2O)py]^{2+} + NH_3 \quad (5\text{-}17) \\ \overset{(c)}{\longrightarrow} trans\text{-}[Ru(NH_3)_4(H_2O)py]^{2+} + NH_3 \end{cases}$$

A particularly interesting feature is that the quantum yield of pyridine aquation, Φ_{py}, has pH-dependent (increasing at low pHs) and pH-independent contributions (Fig. 5-5), while that for ammonia aquation, Φ_{NH_3}, is independent of pH.

A metal-to-ligand charge-transfer transition can be conceptualized as leading to an excited state (CTTL*) having an oxidized metal ion and a radical-ion ligand in a coordination complex

$$\left[(NH_3)_5Ru(II)N\hexagon \right]^{2+} \xrightarrow[\text{CTTL}]{h\nu} \left[(NH_3)_5Ru(III)N\hexagon^{-} \right]^{2+}$$
CTTL*

The chemical reaction properties of such an excited state are not obvious. One potential consequence suggested by the formulation, CTTL*, is photooxidation of the Ru(II), but no photooxidation results from monochromatic irradiation of the $Ru(NH_3)_5py^{2+}$ CTTL band in deaerated aqueous solution. Another possibility is reaction of the coordinated ligand. It has been calculated that the CTTL excited state of the pentaamminepyrazine Ru(II) ion is about 5 orders of magnitude more

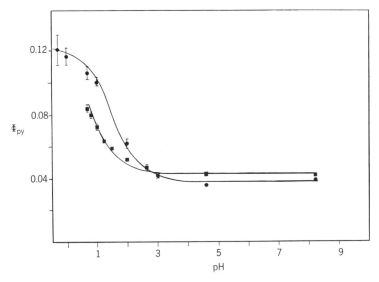

Fig. 5-5. Plot of Φ_{py} versus pH for the 405 nm photolyses of $[Ru(NH_3)_5py]^{2+}$ in aqueous chloride solution (25°C). Circles: Φ_{py} (spectroscopic) for $\mu = 2.00$ M. Squares: Φ_{py} (Spectroscopic) for $\mu = 0.20$ M.[57a]

basic than is the ground state[28]

$$\left[(NH_3)_5RuN\bigcirc N^{2+}\right]^* + H^+ \rightleftharpoons \left[(NH_3)RuH\bigcirc NH^{3+}\right]^* \qquad (5\text{-}18)$$

This suggests that the aromatic ring in the CTTL excited state should be more susceptible than the ground state to electrophilic reactions. In acidic D_2O solutions, electrophilic H/D exchange of pyridine protons was observed as a photochemical pathway[57]; however, the quantum yield for deuteration is small (~ 0.001), suggesting that a carbon-protonated species is not an important intermediate in the acid-dependent pyridine photoaquation pathways.

It appears unlikely that **CTTL*** should be aquation labile. The electron promoted from the low-spin d^6 Ru(II) has π symmetry with respect to the metal-ligand bond, and the σ metal-ammonia bonds and the σ component of the Ru-py bond should be unaffected except for a possible enhancement owing to the more positive nature of the ruthenium in the excited state. It is improbable that a ligand in a charge-transfer state such as **CTTL*** would be noticeably more substitution labile than in the corresponding Ru(III) compound, which is substitution inert under the photolysis conditions. Thus it appears that if a single excited state is responsible for the competitive aquation of py and NH_3 from $Ru(NH_3)_5py^{2+}$, this state is likely to be a ligand-field excited state generated by interconversion from the manifold of CTTL states. Estimates of the energies of the lowest triplet CTTL and LF states[57a] suggest that the LF triplet may have the lower energy (Fig. 5-6). Analogous complexes of *para*-substituted pyridines, $Ru(NH_3)_5(py-X)^{2+}$, where X is a strongly electron-withdrawing group, display CTTL absorption bands at much lower energy than $Ru(NH_3)_5py^{2+}$ (e.g., λ_{max} at 540 nm for X = *p*-CHO). These complexes are at least several orders of magnitude less photoreactive than the pyridine complex when photolyzed in pH 3 aqueous solution,[58] indicating that perhaps in these cases a relatively unreactive CTTL state is the lowest state.

Irradiation of $Ru(NH_3)_5py^{2+}$ in pH 3 aqueous solution at 334 nm,

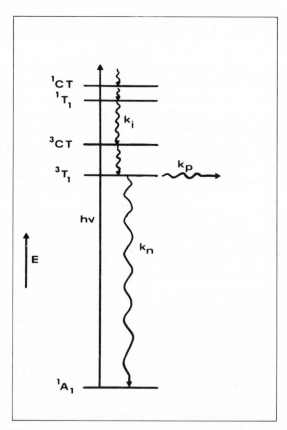

Fig. 5-6. Excited-state diagram proposed for the photoaquation of $[Ru(NH_3)_5py]^{2+}$; k_r represents reactions leading to photoproducts.

313 nm, 254 nm, and 248 nm gives quantum yields[59] for pyridine aquation comparable to those observed for irradiation of the CTTL band. This result suggests that there is relatively efficient interconversion from the higher states (254 and 248 nm correspond to the π_L–π_L^* transition) to a lowest energy-reactive state. However, although photooxidation of Ru(II) to Ru(III) is not observed for a wavelength of irradiation 334 nm or greater, photooxidation is found with a small quantum yield (~0.01) at 313 nm and with increasing quantum yields for lower wavelength.[59] This demonstrates that the photoredox processes mentioned previously must involve an excited state such as a charge transfer to solvent rather than the CTTL transitions dominating the absorption spectrum.

For the CTTL photolysis-induced aquations of $Ru(NH_3)_5py^{2+}$, the pH dependence of Φ_{py} and the pH independence of Φ_{NH_3} require a mechanism where the role of H^+ in pyridine photoaquation occurs after the NH_3 aquation and the acid-dependent pyridine aquation pathways have branched. Two mechanistic schemes for pyridine aquation consistent with the experimental data have been proposed. In the first, photolysis generates a relatively long-lived intermediate capable of competitive acid-dependent and acid-independent aquation as well as acid-independent return to substrate

$$[Ru(NH_3)_5(py)]^{2+} \underset{k_3}{\overset{h\nu,\Phi_A}{\rightleftharpoons}} \{(NH_3)_5Ru(II) \cdots py\} \qquad (5\text{-}19)$$

$$\mathbf{A}$$

$$H_2O + A \xrightarrow{k_1} [Ru(NH_3)_5(H_2O)]^{2+} + py \qquad (5\text{-}20)$$

$$H^+ + A \underset{}{\overset{K}{\rightleftharpoons}} AH^+ \qquad (5\text{-}21)$$

$$AH^+ \xrightarrow{k_2} [Ru(NH_3)_5(H_2O)]^{2+} + pyH^+ \qquad (5\text{-}22)$$

$$\Phi_{py} = \Phi_A\left(\frac{k_1 + k_2K[H^+]}{k_1 + k_3 + k_2K[H^+]}\right) \qquad (5\text{-}23)$$

An implicit assumption of this scheme is that **A**, a Ru(II) intermediate, is the precursor to both acid-independent and acid-dependent pyridine aquation pathways. Since an analogous species, $[(NH_3)_4py\text{-}Ru(II) \cdots NH_3]^{2+}$, can be envisioned as an intermediate for the ammonia aquation pathways, absence of an acid dependence for Φ_{NH_3} could be due to a lifetime insufficient for reaction with H^+. Pyridine, however, has a possible bonding mode not available to NH_3; the pyridine ring could turn 90° with respect to the normal Ru(II)—N bond axis and enter into a weak π complex with the $Ru(NH_3)_5^{2+}$. Presumably a ligand-field excited state would be the precursor for both ammonia aquation and formation of **A**. In another proposed scheme,[57a] two excited states would be responsible for the aquations: a ligand-field state for NH_3 and acid-independent py aquation, and a lower energy CT state capable of reversible protonation for the acid-dependent py aquation. An additional requirement imposed by the pH independence of Φ_{NH_3} is that the higher LF state cannot be significantly populated from the low CT state (Fig. 5-7).

Flash-photolysis experiments[60] with $Ru(NH_3)_5py^{2+}$ have shown transient bleaching of the CTTL band followed by relatively slow decay to substrate and products at a rate, in part, inversely proportional to $[H^+]$. This observation was interpreted in terms of an intermediate having a free radical coordinated to a Ru(III) center with the pyridine nitrogen assuming an "insulating" tetrahedral configuration capable of reversible protonation

$$(NH_3)_5Ru(III)—\overset{..}{N} \quad + H^+ \; \rightleftharpoons \; (NH_3)_5Ru(III)—\overset{\overset{\displaystyle H}{|}}{N} \quad \quad (5\text{-}24)$$

This suggestion is more or less compatible with the second mechanistic scheme proposed above (except that the lifetime requires the transient be considered an intermediate), but alone it is insufficient to describe the

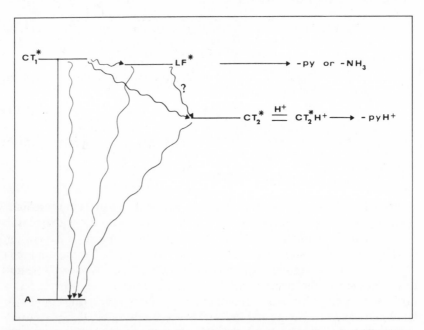

Fig. 5-7. Another proposed scheme for acid-catalyzed photoaquation of pyridine from a charge-transfer excited state of $[Ru(NH_3)_5py]^{2+}$.[57a]

other photochemistry of the complex. The flash data is also reconcilable with a scheme similar to Eqs. (5-19) to (5-22), where the intermediate, A, is reversibly protonated and is thus stabilized relative to substrate or products but capable of independent decomposition to products. Transient bleaching was also observed[60] in the flash photolysis of $Ru(NH_3)_5bz^{2+}$, but the lifetime is pH independent. The pH independence of the benzonitrile complex photoaquation and transient lifetime is consistent with the suggestion that the nitrogen atom is the protonation site in transients produced by the photolysis of the pyridine complex.

4. *Other Ru(II) Complexes Having CTTL Bands.* The Ru(II) complexes of chelating nitrogen aromatic heterocyclic ligands also display CTTL bands in their absorption spectra, but their photochemistries contrast sharply to the ammine complexes. For example, irradiation of *cis*-bis(bipyridine)-bis(4-stilbazole)Ru(II)

$$cis\text{-}\left[(bipy)_2Ru\left(N\bigcirc\text{-}CH{=}CH\text{-}\bigcirc\right)_2\right]^{2+}$$

T (for *trans*-stilbazole)
C (for *cis*-stilbazole)

in butyronitrile solution leads to *cis/trans* isomerization about the 4-stilbazole carbon–carbon double bond as the only observed photoreaction.[61] No photosubstitution was observed. Photolysis of these complexes at different wavelengths gave the following results: Irradiation at 313 nm, a wavelength corresponding closely to the $\pi-\pi^*$ transition of the free stilbazole ligand, causes efficient *trans* to *cis* isomerization of the olefinic linkage (i.e., **T** → **C**). In contrast, irradiation in the visible region (405, 436, and 546 nm), corresponding to excitation of the CTTL bands, causes nearly complete conversion of **C** to **T**, and it is clear from the photostationary ratios that the $\pi_L-\pi_L^*$ state has a reactivity distinctly different from that of the CTTL state with regard to isomerization of the ligand double bond.

The efficient photoisomerization, **T** → **C**, at 313 nm (*trans* to *cis* for the double bond) is comparable to the photochemical behavior of the free ligand under $\pi-\pi^*$ excitation.[61] The *cis* to *trans* isomerization from CTTL excitation parallels the behavior observed for thermal isomerization. The latter is consistent with the behavior expected for a charge-transfer state where an electron has been transferred to the lowest antibonding orbital of the 4-stilbazole ligand.

$Ru(bipy)_3^{2+}$, does not undergo permanent changes in either coordination sphere or oxidation state when irradiated, but its photochemistry is relatively rich. It has been established that $Ru(bipy)_3^{2+}$ can be used as a triplet sensitizer for the photoreactions of $PtCl_4^{2-}$ [62] and other transition-metal complexes.[63-67] Another observation has been that excited $Ru(bipy)_3^{2+}$ can be quenched by molecular oxygen with the concomitant formation of singlet oxygen.[68]

$$[Ru(bipy)_3^{2+}]^* + O_2(^3\Sigma) \rightarrow Ru(bipy)_3^{2+} + O_2(^1\Delta) \tag{5-25}$$

However, complications with the sensitization process have turned up. For example, it was proposed[69] that $Ru(bipy)_3^{2+}$ when photolyzed is capable of acting as a reductant for certain pentaamminecobalt(III) complexes.

$$Co(NH_3)_5X^{2+} + [Ru(bipy)_3^{2+}]^* \rightarrow Co(II) + Ru(bipy)_3^{3+} \tag{5-26}$$

This proposal was based on the observation of the $Ru(bipy)_3^{3+}$ product when the reaction was carried out in strongly acidic solution (1 N aqueous H_2SO_4). In less acidic solution, the Ru(III) complex is a sufficiently strong oxidant to be reduced by solvent back to $Ru(bipy)_3^{2+}$.

An alternative explanation for formation of the Ru(III) is that it results from reaction between ground state, Ru(II), and a transient radical produced by the sensitized photoredox decomposition of the Co(III) complex.[69,70] Flash photolysis and quencher studies[70] appear to support the latter mechanism for formation of $Ru(bipy)_3^{3+}$ in this case. For example, flash photolysis[70a] of $Co(NH_3)_5Br^{2+}$ and $Ru(bipy)_3^{2+}$ together in 1 N H_2SO_4 containing dilute bromide produces Br_2^-, which oxidizes $Ru(bipy)_3^{2+}$ at near diffusion controlled rates. The results were rationalized on the basis of the following equations:

$$Ru(bipy)_3^{2+} + h\nu \rightarrow [Ru(bipy)_3^{2+}]^* \tag{5-27}$$

$$[Ru(bipy)_3^{2+}]^* + Co(NH_3)_5Br^{2+} \rightarrow Ru(bipy)_3^{2+} + Co^{2+} + \cdot Br \tag{5-28}$$

$$\cdot Br + Br^- \rightleftharpoons Br_2^- \tag{5-29}$$

$$Br_2^- + Ru(bipy)_3^{2+} \rightleftharpoons Ru(bipy)_3^{3+} + 2Br^- \tag{5-30}$$

A redox mechanism has also been proposed[67] for the $Ru(bipy)_3^{2+}$-sensitized photolysis of $Co(C_2O_4)_3^{3-}$. Here again, $Ru(bipy)_3^{3+}$ is produced along with the Co^{2+} that would be the product of either reduction of the Co(III) substrate by Ru(II) or of sensitized decomposition of $Co(C_2O_4)_3^{3-}$. Although the latter process would also be expected to result in the formation of free radicals resulting from oxidation of the ligands, it was argued[67] that these radicals (probably $C_2O_4^-$ or CO_2^-) would be reducing, not oxidizing, in character. In support of this argument, it was reported that direct photolysis of the $Co(C_2O_4)_3^{3-}$ CTTM bands with 254 nm light (conditions leading to efficient redox decomposition of $Co(C_2O_4)_3^{3-}$) in the presence of Ru(bipy)$_3^{2+}$ gave very little Ru(III) formation, a result inter-preted to indicate the inability of the ligand radicals to oxidize Ru(bipy)$_3^{2+}$. As a general conclusion, it appears that neither mechanism for formation of Ru(III) in these "sensitization" experiments has an exclusive fran-chise, and that the specifics of each system studied need examination to establish which one may be operable under a particular set of conditions.

Another photochemical phenomenon[71] involving Ru(bipy)$_3^{2+}$ is the chemiluminescence observed when the Ru(III) analog, Ru(bipy)$_3^{3+}$, is reduced with certain reagents including hydrazine in aqueous acid solu-tion and hydroxide ion. The light emitted in this reaction is identical to that of normal phosphorescence under comparable conditions. Similar chemiluminescence can be generated electrochemically.[72]

5. The Hexahalo Complexes of Pt(IV), PtX$_6^{2-}$. In contrast to the Ru(II) complexes, charge transfer to metal states play important roles in the photochemistry of Pt(IV) complexes. Spectra of the complexes, PtCl$_6^{2-}$ and PtBr$_6^{2-}$, are shown in Fig. 5-8. For PtCl$_6^{2-}$ in 2-CH$_3$THF, bands at 465 nm (21.5 kK) and 358 nm (27.9 kK) have been assigned to singlet-triplet and singlet-singlet d-d transitions, respectively, while the very broad and intense band at 265 nm (37.7 kK) is a CTTM absorption[26] (Table 5-1). Similar assignments can be made for analogous absorptions in the PtBr$_6^{2-}$ spectrum.[26] The photosensitivity of such complexes was noted as early as 1832,[2] but our concern here is with results and conclusions of relatively recent quantitative investigations.

Aqueous hexachloroplatinate(IV), PtCl$_6^{2-}$, undergoes both photoaqua-tion and photocatalyzed exchange with chloride ion in the solution

$$\text{PtCl}_6^{2-} + \text{H}_2\text{O} \xrightarrow{h\nu} \text{PtCl}_5(\text{H}_2\text{O})^- + \text{Cl}^- \qquad (5\text{-}31)$$

$$\text{PtCl}_6^{2-} + {}^*\text{Cl}^- \xrightarrow{h\nu} [\text{PtCl}_5{}^*\text{Cl}]^{2-} + \text{Cl}^- \qquad (5\text{-}32)$$

In studies of the latter reaction (which has quantum yields up to 1300),[74] it was found[73] that exchange is greatly accelerated even by diffuse room light, and that both thermal and photochemical exchange are inhibited by oxidizing agents such as Fe(CN)$_6^{3-}$, IrCl$_6^{2-}$, and Cl$_2$. Thus, it was proposed, that a chain mechanism involving a catalytic intermediate such as PtCl$_5^{2-}$ is responsible

$$\text{PtCl}_6^{2-} \xrightarrow{h\nu} \text{PtCl}_5^{2-} + \cdot\text{Cl} \qquad (5\text{-}33)$$

$$\text{PtCl}_5^{2-} + {}^*\text{Cl}^- \rightleftharpoons [\text{PtCl}_4{}^*\text{Cl}]^{2-} + \text{Cl}^- \qquad (5\text{-}34)$$

$$[\text{PtCl}_4{}^*\text{Cl}]^{2-} + \text{PtCl}_6^{2-} \rightleftharpoons [\text{PtCl}_5{}^*\text{Cl}]^{2-} + \text{PtCl}_5^{2-} \qquad (5\text{-}35)$$

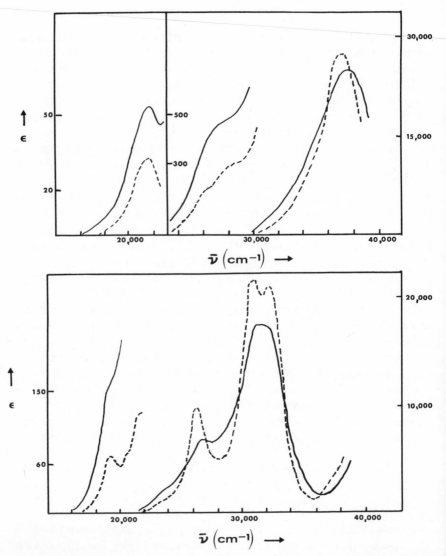

Fig. 5-8. Upper: Electronic spectrum of $[(n-C_4H_9)_4N]_2[PtCl_6]$ in 2–CH$_3$THF–CH$_3$OH: (———) 300°K; (– – – –) 77°K. Lower: Electronic spectrum of $[(n-C_4H_9)_4N]_2[PtBr_6]$ in 2–CH$_3$THF–CH$_3$OH: (———) 300°K; (– – – –) 77°K. (Reproduced with permission from [26]. Copyright by the American Chemical Society.)

The Pt(III) intermediate apparently can also be generated by adding $PtCl_4^{2-}$ to a dark solution of $PtCl_6^{2-}$

$$PtCl_4^{2-} + PtCl_6^{2-} \rightarrow 2PtCl_5^{2-} \qquad (5-36)$$

Inhibition by oxidizing agents purportedly is due to the oxidation of $PtCl_5^{2-}$ to Pt(IV).

A study[75] of the photoaquation of $PtCl_6^{2-}$ [Eq. (5-31)] in 1 M perchloric acid has shown that irradiation at 270 or 365 nm results in overall quantum yields for this reaction substantially higher than unity, but that the presence of dissolved Cl_2 suppresses the photoaquation almost completely. Again it was concluded that a chain mechanism involving a labile Pt(III) species would account for this behavior. If $PtCl_5^{2-}$ is produced, the following reactions could serve as the chain propagation steps for aquation:

$$PtCl_5^{2-} + H_2O \rightleftharpoons PtCl_4(H_2O)^- + Cl^- \qquad (5-37)$$

$$PtCl_4(H_2O)^- + PtCl_6^{2-} \rightleftharpoons PtCl_5(H_2O)^- + PtCl_5^{2-} \qquad (5-38)$$

Conceivably, redox reaction (5-38) [and (5-35) as well] could proceed via a chloride-bridged electron-transfer step.[76] Inhibition by Cl_2 would again be explained by oxidation of Pt(III) catalytic intermediates.

Photolysis at 270 nm and at 365 nm corresponds to direct excitation of a CTTM transition and of a LF transition on the shoulder of a CTTM band, respectively, and reaction (5-33) is consistent with the chemistry expected for a CTTM state (Chapter 3). In addition, trapping experiments[75] indicated the generation of free chlorine atoms in the photolysis experiments. Excitation of the $^1A_1 \rightarrow {}^3T_1$ band with 450-nm light also gives photoaquation, but at much smaller quantum yields. This process is also largely suppressed by the presence of Cl_2, again indicating a redox-type mechanism. It was suggested that the photoreactivity even at this wavelength is the consequence of the charge-transfer band tailing out to long wavelengths.

A flash-photolysis study[77] of $PtCl_6^{2-}$ has confirmed the presence of a redox reaction. A continuum flash of aqueous $PtCl_6^{2-}$, with light absorption predominantly by the CTTM band at 262 nm, gives a transient absorption band in the wavelength range, 350 to 450 nm ($\lambda_{max} \sim 410$ nm). The spectrum of this transient is virtually identical to that of the Pt(III) species obtained by pulse radiolysis[78] of $PtCl_4^{2-}$ and of $PtCl_6^{2-}$. The Pt(III) transient of the flash experiment decays according to a second-order process

$$2Pt(III) \xrightarrow{k_1} Pt(IV) + Pt(II) \qquad (5-39)$$

The relatively slow rate of disproportionation ($k_1 = 4.6 \times 10^6 \, M^{-1} \, sec^{-1}$ at

20°C) makes it virtually certain that Pt(III) is the chain carrier responsible for the large quantum yields of photoexchange with Cl^- and of photoaquation. In the presence of excess Cl^-, the flash photolysis also produced Cl_2^- radicals, presumably from reaction of $\cdot Cl$ [Eq. (5-33)] with Cl^-.

Hexabromoplatinate(IV), $PtBr_6^{2-}$, also undergoes both photoaquation and photoexchange. However, photoaquation of aqueous $PtBr_6^{2-}$ apparently occurs via direct heterolysis of the Pt(IV)—Br bond and does not involve a chain-carrying intermediate.[79,80]

$$PtBr_6^{2-} + H_2O \xrightarrow{h\nu} PtBr_5H_2O^- + Br^- \qquad (5\text{-}40)$$

In contrast, the exchange of free Br^- in aqueous solution with coordinated bromide occurs with variable quantum yields greatly exceeding unity, a clear indication of chain reaction.[81] A mechanism analogous to Eqs. (5-33) to (5-35) was originally proposed to account for this photoexchange, although subsequent flash-photolysis results[82] were interpreted to be inconsistent with this pathway. Since neither free bromine atoms nor Br_2^- were detected in the flash experiments, an alternate pathway was proposed involving generation of a reactive, chain-carrying tetrabromo-platinum(II) intermediate by concerted departure of two *cis*-bromine atoms

$$PtBr_6^{2-} \rightarrow [PtBr_4^{2-}] + Br_2 \qquad (5\text{-}41)$$

The importance of $[PtBr_4^{2-}]$ as a chain carrier is questionable,[83] since this species could be expected to react with Pt(IV) to produce Pt(III) intermediates. In this context, it is particularly interesting that flash photolysis[82] of $PtBr_6^{2-}$ did produce a transient with characteristics remarkably similar to those of the Pt(III) transient detected[77] in the flash photolysis of $PtCl_6^{2-}$. On these bases, it appears that, regardless of the method of production, Pt(III) species are probably present in the photolysis solutions of $PtBr_6^{2-}$ under conditions leading to bromide exchange.

Photoaquation of $PtBr_6^{2-}$ occurs with reproducible quantum yields less than unity and is not quenched[79] by the oxidizing agents H_2O_2 or $IrCl_6^{2-}$. These observations led to the conclusion[79,80] that a redox reaction was not involved in the photoaquation mechanism, but instead the reaction proceeded by direct heterolysis of the Pt(IV)—bromide bond [Eq. (5-40)]. In addition, quantum yields for this process measured in aqueous acid solutions for irradiations at 313, 365, 433, and 530 nm (corresponding to CTTM and LF bands) proved to be independent of wavelength ($\phi = 0.4$). The conclusion drawn was that photoaquation does not take place in the various excited states reached directly by irradiation since the different characters of these states should be reflected in different reactivities.

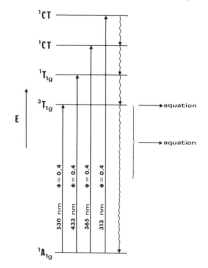

Instead, radiationless processes of unitary efficiency must lead from the
higher states to the lowest, the ligand-field "triplet" state (Fig. 5-9).
Aquation would occur from this state, having a $t_{2g}^5 e_g^1$ configuration, or
from a vibrationally "hot" ground electronic state produced by radiationless deactivation of the LF triplet.[79,80]

A comparison[2] of the photoaquation and photoexchange processes for
PtBr$_6^{2-}$ clearly shows that the two reactions display different behaviors and
that their respective mechanisms cannot be coupled. The failure to detect
photochemical products not attributable to the photoaquation pathway
(noting the Br$^-$ exchange cannot be detected except by isotope methods)
indicates that the principal direct photoreaction is heterolytic aquation
and that other primary photolysis pathways including those responsible
for Br$^-$ exchange are minor. This conclusion is consistent with the
proposed redox chain mechanism for photoexchange since only very
small primary yields of a chain-carrying intermediate may be required to
give very large overall quantum yields.

Somewhat more puzzling is the contrast between the photoaquation
mechanisms for PtBr$_6^{2-}$ and for PtCl$_6^{2-}$, particularly since bromide would
be expected to be more easily oxidized than chloride in a reaction such as
Eq. (5-33). In this context, the following points should be considered:
First, since PtBr$_6^{2-}$ undergoes photoexchange by a chain mechanism, it is
probable that, under the conditions reported for the photoaquation
study,[79] some chain-carrying intermediate, presumably Pt(III), is also
produced. Second, interpretation[77] of the flash photolysis of PtCl$_6^{2-}$
suggested that heterolytic photoaquation may be the dominant primary

photolysis pathway and that generation of the Pt(III) intermediate may be a relatively minor primary pathway. Consequently, the differences in the photoaquation pathways may not be due either to the unique generation of Pt(III) intermediates in PtCl$_6^{2-}$ photolysis or to the unique predominance of heterolytic dissociation in PtBr$_6^{2-}$ photolysis, but may be due to differences in the ability of the Pt(III) intermediates to act efficiently as chain carriers in a redox-pathway photoaquation. The key might lie in the proposed unsymmetrical redox step (5-38), which for the Cl$^-$ and Br$^-$ systems would be

$$[PtCl_4(H_2O)]^- + PtCl_6^{2-} \rightleftharpoons PtCl_5(H_2O)^- + [PtCl_5]^{2-}$$

or

$$[PtBr_4(H_2O)]^- + PtBr_6^{2-} \rightleftharpoons PtBr_5(H_2O)^- + [PtBr_5]^{2-}$$

The affinity of Br$^-$ (vs. H$_2$O) for Pt(IV) relative to that for Pt(III) might be sufficient to make this reaction much less favorable for the bromide system. Similarly, for the Br$^-$ analog of Eq. (5-37)

$$[PtBr_5]^{2-} + H_2O \rightleftharpoons [PtBr_4(H_2O)]^- + Br^-$$

the relative affinities of the halide ion for Pt(III) in comparison to H$_2$O might be enough greater for Br$^-$ than for Cl$^-$ to make this process less favorable for the Br$^-$ system. Clearly, if these chain steps are slow relative to the chain-termination processes, the reduced Pt(III) intermediate cannot contribute significantly to the bromide photoaquation yields.

D. The Photochemistry of Other Hexacoordinate Complexes

1. d^5 *Complexes: Ru(III) and Ir(IV).* In this section, we turn our attention to octahedral complexes of Ru(III) and Ir(IV), which have the electronic configuration, t_{2g}^5. Metal ions with this configuration have a vacancy in a π-symmetry orbital that can serve as the acceptor orbital in a CTTM transition. As a consequence, we can predict that their spectra are likely to display $\sigma_L \rightarrow \pi_M^*$ or $\pi_L \rightarrow \pi_M^*$ CTTM bands, as well as the ligand-field ($\pi_M^* \rightarrow \sigma_M^*$) and $\pi_L \rightarrow \sigma_M^*$ CTTM bands expected for a d^6 ion such as Pt(IV). For example, the spectrum of IrCl$_6^{2-}$ (Fig. 5-10) has absorption maxima at 588, 488, 434, and 410 nm assigned[84] to $\pi_L \rightarrow \pi_M^*$ CTTM transitions, at 360 and 306 nm assigned to LF transitions, and a very intense band at 232 nm assigned to a $\pi_L \rightarrow \sigma_M^*$ CTTM transition. Relatively intense ($\varepsilon > 10^3$) $\pi_L \rightarrow \pi_M^*$ CTTM bands are also the lowest energy absorptions in the spectra of the halopentaammine Ru(III) complexes, Ru(NH$_3$)$_5$X^{2+} (X$^-$ = Cl$^-$, Br$^-$, or I$^-$).[85]

The chemical reactivities of the σ_L or $\pi_L \rightarrow \sigma_M^*$ CTTM excited states or of the LF excited states ($\pi_M^* \rightarrow \sigma_M^*$) would not, in general, be expected to

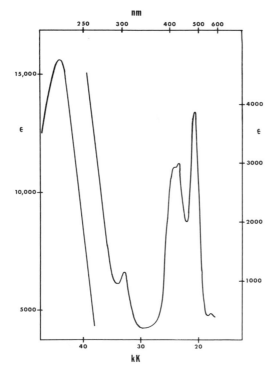

Fig. 5-10. Electronic spectrum of aqueous $IrCl_6^{2-}$. (Reproduced with permission from [2], p. 312. Copyright by Academic Press.)

differ significantly from those of similar states in analogous $4d^6$ or $5d^6$ complexes. However, the σ_L or $\pi_L \rightarrow \pi_M^*$ CTTM excited states should have a pattern of reactivity distinct from that of a CTTM state where a σ_M^* orbital is populated. For a Ru(III) or Ir(IV) complex, the former CTTM state can be described simplistically as a Ru(II) or Ir(III) complex coordinated to a free-radical ligand. The lability of Ru(II) is not particularly great (for example, aqueous $Ru(NH_3)_5H_2O^{2+}$ has been predicted to exchange coordinated H_2O with solvent H_2O with a rate constant of 1 to $10 \ sec^{-1}$ at $25°C)^{39b}$ and that of Ir(III) is less. Thus, the excited state might be expected to undergo internal redox to regenerate the original complex faster than dissociation to the reduced metal complex plus a free radical in solution. This expectation is borne out by the relative insensitivities of Ir(IV) and Ru(III) complexes to photolysis of $\pi_L \rightarrow \pi_M^*$ CTTM absorption bands.[86-88]

The photochemistry of aqueous $IrCl_6^{2-}$ is summarized in the following reactions:[86]

$$IrCl_6^{2-} + H_2O \xrightarrow{\ h\nu\ }_{a} IrCl_5(H_2O)^- + Cl^-$$

$$h\nu \Big| c \qquad \searrow^{h\nu}_{b}$$

$$\qquad\qquad\qquad IrCl_5(H_2O)^{2-} + \cdot Cl \qquad (5\text{-}42)$$

$$IrCl_6^{3-} + H^+ + \cdot OH$$

The three do not occur simultaneously. Irradiation of aqueous $IrCl_6^{2-}$ at 254 nm ($\pi_L \rightarrow \sigma_M^*$ CTTM transition) gives $IrCl_5(H_2O)^-$ and $IrCl_5(H_2O)^{2-}$ in parallel zero-order reactions [Eqs. (5-42a) and (5-42b)]. Changes in chloride concentration do not affect the total quantum yield for the two products ($\Phi_{sum} = 0.029$), but the ratio of the two pathways changes. Formation of the redox product, $IrCl_5(H_2O)^{2-}$, is favored at higher [Cl$^-$]. The dependence of the redox/aquation ratio and the independence of Φ_{sum} has been interpreted as indicating a common intermediate for these two reactions. The following mechanism was proposed[86]:

$$IrCl_6^{2-} \underset{k_d}{\overset{h\nu(CTTM)}{\rightleftharpoons}} [Ir^{3+}(Cl^-)_5(Cl^0)]^{2-}$$

$$\swarrow k_r$$

$$\qquad\qquad\qquad\qquad\qquad\qquad \xrightarrow{k_{red}[Cl^-]} IrCl_5(H_2O)^{2-} + \cdot Cl \quad (5\text{-}43)$$

$$[Ir^{3+}(Cl^-)_5(H_2O)\cdot(Cl^0)]^{2-} \longrightarrow$$

$$\qquad\qquad\qquad\qquad\qquad\qquad \xrightarrow{k_{aq}} IrCl_5(H_2O)^- + Cl^-$$

$$\textbf{(I)}$$

where $[Ir^{3+}(Cl^-)_5(Cl^0)]^{2-}$ represents the $\pi_L \rightarrow \sigma_M^*$ CTTM excited state and **I** is an intermediate composed of an Ir(III) complex with a chlorine atom solvent-caged in the second coordination sphere. The intermediate would be formed by an irreversible step (k_r) in competition with a deactivation process (k_d). An internal redox reaction of **I** followed by diffusion of Cl$^-$ from the solvent cage would result in aquation (k_{aq}), while diffusion of $\cdot$Cl out of the solvent cage, in some manner assisted by free chloride (k_{red}[Cl$^-$]), would lead to a net redox reaction. It was suggested[86] that a possible role of chloride in the redox reaction could be due to an electron transfer from an outer Cl$^-$ to the solvent-caged chlorine atom; however, this process appears energetically unfavorable. More likely would be the reaction of **I** with Cl$^-$ to give the transient Cl_2^- plus the Ir(III) product.

Excitation of the LF transitions by irradiation at 313 and 365 nm in the absence of added chloride led only to $IrCl_5(H_2O)^-$ with quantum yields of 0.01 and 0.001 at these respective wavelengths. However, in 1.2 M Cl$^-$ solution, irradiation at the same wavelengths gave the redox product, $IrCl_6^{3-}$ [Eq. (5-42c)], with quantum yields of 0.02 at 313 nm and 0.005 at 365 nm. Finally, irradiation at 433 nm and 495 nm ($\pi_L \rightarrow \pi_M^*$ CTTM) caused no detectable changes in the solution for the longer wavelength and only traces ($\Phi \sim 1 \times 10^{-4}$ in 1.2 M Cl$^-$) of an unidentified redox reaction for 433 nm photolysis.

The wavelength dependence of the photoreaction modes and their quantum yields indicates that IrCl$_6^{2-}$ photoreactions are characteristics of states populated by initial excitation and are competitive with decay processes. The lowest spectroscopic state, apparently a $\pi_L \rightarrow \pi_M^*$ CTTM, is essentially unreactive, as predicted by the rather naive considerations mentioned previously. In this context, it is interesting to compare the photochemical properties of the $5d^5$ IrCl$_6^{2-}$ to those of the $5d^6$ Pt(IV) analogs, PtCl$_6^{2-}$ and PtBr$_6^{2-}$. Both Pt(IV) complexes display very high quantum yield, chain mechanisms (Cl$^-$ exchange and aquation for PtCl$_6^{2-}$, and Br$^-$ exchange for PtBr$_6^{2-}$), apparently as the result of charge-transfer generation of chain-carrying, reduced platinum intermediates. Unfortunately, chloride photoexchange with IrCl$_6^{2-}$ has not been examined; however, the quantum yields for aquation are small and reproducible, and there is no indication of a chain process. This result is certainly expected on the basis that Ir(III), the redox product of CTTM excitation, has a $5d^6$ configuration and is not substitution labile, thus cannot be a good chain-carrying intermediate. Despite this, comparison of IrCl$_6^{2-}$ photochemistry to that of PtBr$_6^{2-}$ raises some interesting points. First, both display some form of redox behavior when $\pi_L \rightarrow \sigma_M^*$ states are irradiated, Ir(IV) being reduced in small quantum yield to Ir(III), PtBr$_6^{2-}$ undergoing chain mechanism exchange with solution Br$^-$. However, photoaquation of IrCl$_6^{2-}$ has wavelength-dependent quantum yields that at their largest (<0.03 for excitation to the $\pi_L \rightarrow \sigma_M^*$ CTTM state) are rather small, while the photoaquation of PtBr$_6^{2-}$ (not a chain reaction) occurs with a relatively large, wavelength-independent quantum yield (0.4). The latter observation was interpreted as indicating highly efficient interconversion of higher states to the lowest excited state (presumably a ligand-field triplet) responsible for this reaction. Since less than 3% of the IrCl$_6^{2-}$ higher states decay by photoreaction, there is no evidence to deny the possibility that the remainder undergo interconversion to the lowest excited state, very likely the unreactive $\pi_L \rightarrow \pi_M^*$ CTTM. Thus, the mechanisms of these respective photoreactions may have the unifying feature that decay to the lowest state occurs with high efficiency in both cases, but the difference in the reactivities is due primarily to the differences in the reactivities of the lowest states.

Among $4d^5$ systems, the photochemistries of several Ru(III) ammine complexes, Ru(NH$_3$)$_6^{3+}$, Ru(NH$_3$)$_5$Cl^{2+}, and cis-[Ru(NH$_3$)$_4$Br$_2$]$^+$ have been studied.[87,88] Aqueous Ru(NH$_3$)$_5$Cl^{2+} displays a moderately intense $\pi_{Cl} - \pi_{Ru}^*$ CTTM band (λ_{max} 328 nm, $\varepsilon \sim 2 \times 10^3$ M^{-1} cm^{-1}) as its most prominent spectral feature. Irradiation of Ru(NH$_3$)$_5$Cl^{2+} in aqueous solution showed it to have a low photoreactivity.[87] Aquation of ammonia (primarily *cis*) is the principal photoreaction, and photolyses at 230, 254, and 321 nm gave quantum yields (0.02 to 0.032) essentially wavelength independent, given the experimental uncertainty.[87] Aquation of Cl^- was also observed

$$Ru(NH_3)_5Cl^{2+} \xrightarrow{h\nu} Ru(NH_3)_5H_2O^{3+} + Cl^- \qquad (5\text{-}44)$$

however, the quantum yields were small and erratic, at times falling below 10^{-2}. Since the Ru(II) complex ion, $Ru(NH_3)_5H_2O^{2+}$, catalyzes Eq. (5-44)[25] the erratic nature of Φ_{Cl^-} suggests that traces of the redox product, Ru(II), may result from the CTTM excitation. If so, the quantum yields for the redox reaction must be extremely low since no Ru(II) was detected. In addition, flash photolysis[87] ($\lambda \geq 300$ nm) of $Ru(NH_3)_5Cl^{2+}$ provided no evidence for the generation of a chlorine radical.

The wavelength independence of the ammonia aquation quantum yields for $Ru(NH_3)_5Cl^{2+}$ was attributed to efficient interconversion of higher states to the lowest one, presumably a $\pi_{Cl} \rightarrow \pi_{Ru}^*$ CTTM state.[87] The low quantum yields would then be explained as the result of the relatively substitution-inert character of an excited state that formally can be considered to be a Ru(II) complex (e.g., $[Ru(II)(NH_3)_5(Cl^0)]^{2+}$). However, this excited state should be relatively short-lived and pentaammineruthenium(II) complexes are not labile with respect to NH_3 aquation.[25] Given the essentially photoinert nature of the $\pi_L \rightarrow \pi_M^*$ CTTM state of $IrCl_6^{2-}$, it is not inconceivable that the analogous state for $Ru(NH_3)_5Cl^{2+}$ may be similarly unreactive. If so, the photosubstitution behavior may be the consequence of a spectrally undetected ligand-field excited state that, with the electronic configuration, $t_{2g}^4 e_g^1$, should be labile.

2. A $4d^3$ Complex: $TcCl_6^{2-}$.

In contrast to the extensive photochemistry of $3d^3$ complexes, principally those of Cr(III), virtually nothing is known of the photochemical properties of $4d^3$ and $5d^3$ complexes. The principal reason for this state of affairs no doubt lies in the scarcity of published reports of the synthesis, characterization, and thermal reactions of the latter systems,[24] especially of mononuclear complexes amenable to study in homogeneous solution. As a result, there have been no extensive photochemical studies of such complexes for either Mo(III) or W(III), the second- and third-row congeners of Cr(III). An exception is the $4d^3$ Tc(IV) ion, $TcCl_6^{2-}$. Irradiation of the thermally stable $TcCl_6^{2-}$ in aqueous perchloric acid at 253 nm and at 340 nm led to aquation as the only primary photoreaction detected[89]

$$TcCl_6^{2-} + H_2O \xrightarrow{h\nu} TcCl_5(H_2O)^- + Cl^- \qquad (5\text{-}45)$$

The two irradiation wavelengths, which correspond to a $\pi_{Cl} \rightarrow \sigma^*_{Tc}$ CTTM transition (253 nm) and to a $\pi_{Cl} \rightarrow \pi^*_{Tc}$ CTTM transition (340 nm), gave the respective quantum yields, 0.25 and 0.064, thus indicating that the higher state does not undergo efficient internal conversion to the lower state. (This characteristic of a $\pi_L \rightarrow \sigma^*_M$ state has been mentioned previously for Rh(III) ammine complexes.) Notably, the higher quantum yield for photoaquation is the result of populating an e_g orbital.

E. The Photochemistry of Square Planar Complexes

The square planar configuration, though rare among the lighter transition elements, is the commonly observed geometry for d^8 complexes of the heavier transition elements. Most complexes of Rh(I), Ir(I), Pd(II), Pt(II), and Au(III) are square planar; however, only Pt(II) has been the subject of extensive quantitative photochemical investigation. Electronic transitions in the absorption spectra of Pt(II) complexes include ligand-field and charge-transfer bands and perhaps $5d$–$6p$ transitions.[90] For example, the absorption spectrum[13] of tetrachloroplatinate(II), $PtCl_4^{2-}$, displays well-separated LF and CT transitions with a maximum at 480 nm ($\varepsilon = 15 \ M^{-1} \ cm^{-1}$ aqueous solution) assigned to a singlet-to-"triplet" LF transition, two maxima at 394 nm ($\varepsilon = 57 \ M^{-1} \ cm^{-1}$) and 337 nm ($\varepsilon = 62 \ M^{-1} \ cm^{-1}$) to singlet-to-singlet LF transitions, and very intense bands at 200 to 250 nm ($\varepsilon > 10^3 \ M^{-1} \ cm^{-1}$) to CTTM transitions.[13,90] The luminescence spectrum of $K_2[PtCl_4]$ is observable at 77°K for the solid and in frozen aqueous solution but not at room temperature.[91] The emitting state was assigned to the lowest LF triplet, and it was suggested that the molecular geometry of this state may be D_{2d} (distorted tetrahedron) rather than D_{4h} (square planar). A similar suggestion was made[90b] in interpretation of the absorption spectrum.

The suggestion of significant geometric distortion in Pt(II) excited states has particular relevance to the photochemistry of these complexes and is a recurring theme in our ensuing discussions. Consequences observed as the result of irradiating Pt(II) complexes include ligand-substitution reactions, *cis/trans* isomerization of the complex, and oxidation-reduction processes. These photolysis modes are discussed separately in the following sections.

1. *Photosubstitution of Coordinated Ligands.*

Pt(II) complexes are relatively substitution inert, but undergo slow thermal ligand replacements that proceed with stereochemical retention. Extensive investigations have led to the conclusion that most of these reactions proceed with an associative mechanism via a pentacoordinate intermediate.[92]

Scandola, Traverso, and Carassiti[93] have examined the quantitative photochemistry of $PtCl_4^{2-}$ in aqueous solution and have found aquation as the only photoreaction

$$PtCl_4^{2-} + H_2O \xrightarrow{h\nu} PtCl_3(H_2O)^- + Cl^- \qquad (5\text{-}46)$$

Irradiation of the ligand-field region at 313, 404, and 474 nm gave essentially wavelength-independent photoaquation quantum yields of 0.20, 0.19, and 0.17, respectively. In contrast, irradiation at 254 nm, in the charge-transfer region, gives a much larger quantum yield of 0.9. These authors have argued that excitation in the LF region is followed by a radiationless cascade to a common state, presumably the lowest lying triplet, which is reactive itself or undergoes intersystem crossing to a vibrationally "hot" reactive ground state (Fig. 5-11). The proposed role of the triplet state is substantiated by the observation that aquation of $PtCl_4^{2-}$ in aqueous solution can be photosensitized by the triplet donors biacetyl[94] and tris(bipyridyl)ruthenium(II).[62]

The larger quantum yield resulting from photolysis of $PtCl_4^{2-}$ at 254 nm indicates that irradiation of the CTTM bands leads to population of a

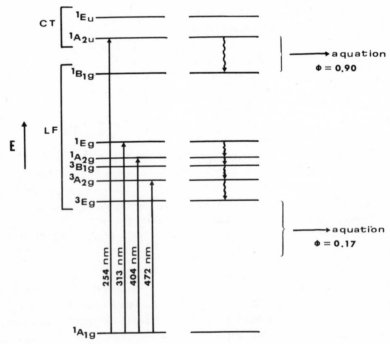

Fig. 5-11. Energy-level diagram and photoaquation mechanism for $PtCl_4^{2-}$.[93]

reactive excited state not accessible by irradiation at 313 nm. This state must lie in the energy range between 254 nm and 313 nm and does not deactivate efficiently to the lower energy LF states. Scandola et al.[93] have proposed the scheme illustrated in Fig. 5-11 with state assignments based on previous spectral and theoretical studies. In this scheme, the $^1B_{1g}$ LF state and the $^1A_{2u}$ CT state both lie in the appropriate region to be the reactive higher energy state, and there is no evidence to distinguish these possibilities. However, in studies of octahedral Rh(III) complexes it was shown[19] that, while internal conversion/intersystem crossing between upper and lower LF states is very efficient, internal conversion between CTTM and LF excited states is not. Although the Pt(II) and Rh(III) systems are hardly analogous, this consideration suggests that the different photoreactivity of the higher states might be the property of the CTTM states manifold. Nonetheless, the B_{1g} state, which in the square planar complexes arises from a $\sigma_{Pt}^n \to \sigma_{Pt}^*$ ($d_{z^2} \to d_{x^2-y^2}$) transition, has no analogy in an octahedral system and has a very different character from other LF states of the square planar system. Consequently, it would not be unreasonable for this state to have reactivity and photophysical properties different from the lower energy LF states.

The electronic character of the excited states of $PtCl_4^{2-}$, $\sigma_{Pt}^n \to \sigma_{Pt}^*$ and $\pi_{Pt}^* \to \sigma_{Pt}^*$ for the LF states and $\pi_{Cl} \to \sigma_{Pt}^*$ for the CTTM states, is suggestive of ligand lability owing to weakening of metal-ligand σ-bonds. However, a direct correlation between bond weakening and ligand lability is implicit only for a dissociative mechanism. If a dissociative mechanism is operable, it appears that both the $^1B_{1g}$ and $^1A_{2u}$ excited states should be relatively substitution labile compared to the lowest LF state, and this would explain the higher quantum yields observed for 254 nm irradiation. However, the normal pathway followed in thermal substitutions of square planar complexes is associative,[92] and while there need not be close similarity between the thermal and photochemical substitution reactions, this possibility as well as the possible distortions of excited states away from the square planar geometry altogether must be scrutinized. For example, it was suggested[93] that the $^1B_{1g}$ excited state ($d_{z^2} \to d_{x^2-y^2}$) might readily undergo association with a water molecule along the principal axis to form pentacoordinate $PtCl_4(H_2O)^{2-}$, a square pyramidal intermediate. Subsequent loss of Cl^- would give the aquated product, $PtCl_3(H_2O)^-$.

The possibility of an associative mechanism in square planar photosubstitutions is reinforced by some observations[95] in the photochemistry of the Ir(I) complex, $Ir(CH_3NC)_4^+$. In the dark, solutions of this species in various solvents (including methanol, dimethylsulfoxide, acetonitrile, dimethylformamide, and acetone) were stable for several days. However, irradiation with visible light led to the formation of a trigonalbipyramidal, pentacoordinate adduct, $[Ir(CH_3NC)_4L]^+$ (L = solvent or added ligand such as CH_3NC or CO).

$$Ir(CH_3NC)_4^+ + L \xrightarrow{h\nu} [Ir(CH_3NC)_4L]^+ \qquad (5\text{-}47)$$

In several cases, this pentacoordinate adduct was unstable with respect to the initial complex and underwent slow thermal dissociation back to $Ir(CH_3NC)_4^+$ plus L. Therefore, the system is photochromic. The essential point, however, is that the excited state (of undetermined nature) is capable of an associative reaction.

The photochemistry of Zeise's anion, $PtCl_3(C_2H_4)^-$, has also been examined in acidic aqueous solution.[96] Aquation was the principal photo-reaction mode, and both cis-chloride and ethylene aquation were observed

$$PtCl_3(C_2H_4)^- + H_2O \xrightarrow{h\nu} cis\text{-}[PtCl_2(H_2O)(C_2H_4)] + Cl^- \qquad (5\text{-}48)$$

$$PtCl_3(C_2H_4)^- + H_2O \xrightarrow{h\nu} PtCl_3(H_2O)^- + C_2H_4 \qquad (5\text{-}49)$$

Aquation of the $trans$ chloride is the expected thermal reaction owing to ethylene's strong "$trans$ effect"[92]

$$PtCl_3(C_2H_4)^- + H_2O \rightleftharpoons trans\text{-}[PtCl_2(H_2O)(C_2H_4)] + Cl^- \qquad (5\text{-}50)$$

However, the equilibrium represented in Eq. (5-50) is established so rapidly that the photochemical labilization of the $trans$-Cl^- could not be determined, and only the $antithermal$ reactions [Eqs. (5-48) and (5-49)] were studied.

The absorption spectrum of $PtCl_3(C_2H_4)^-$ shows a series of broad overlapping bands beginning at ~500 nm which can be resolved into a number of Gaussian curves.[96] The lowest energy transitions have been assigned to singlet-to-triplet LF bands and were unreactive when subjected to direct photolysis. Irradiation at lower wavelength (below 360 nm) in the region of spin-allowed, ligand-field transitions, gave aquation of cis-chloride and of ethylene. The quantum-yield ratio of these two processes, $\Phi_{Cl}/\Phi_{ethylene}$, proved to be wavelength dependent, while sensitization experiments show that photoaquation of ethylene is sensitized by acetone and acetophenone, but aquation of cis-chloride [Eq. (5-48)] is not. These results clearly indicate that the two aquations do not occur from a common reactive state, and it was suggested[96] that ethylene aquation occurs from the lowest singlet LF state and chloride aquation from a vibrationally "hot" ground state. The failure of the "triplet" LF states to be reactive is puzzling.

The lability of ethylene in a LF excited state, but not the ground state of $PtCl_3(C_2H_4)^-$, can be explained on the basis of excited-state geometry.

Interpretation both of absorption spectra and of the large Stokes shift observed for the emission spectrum[96] leads to the conclusion that the LF excited states are considerably distorted toward a tetrahedral geometry. In such a configuration, π-backbonding from Pt(II) to ethylene should be reduced compared to that in the stable square planar geometry, a perturbation leading to greater ligand lability.

2. Photoisomerization Reactions. Cis-trans photoisomerization of square planar Pt(II), in some cases, can occur reversibly in either direction

$$\begin{array}{ccc} \text{X} \diagdown \quad \diagup \text{Y} & & \text{X} \diagdown \quad \diagup \text{Y} \\ \quad \text{Pt} & \underset{h\nu}{\overset{h\nu}{\rightleftarrows}} & \quad \text{Pt} \\ \text{X} \diagup \quad \diagdown \text{Y} & & \text{Y} \diagup \quad \diagdown \text{X} \end{array} \qquad (5\text{-}51)$$

but often occurs in only one direction. Systems undergoing photoisomerization in both directions, thus eventually achieving a photostationary state (pss), are the dihalobis(tertiary phosphine)platinum(II) complexes, PtX_2L_2.[97] Cis- and trans-dichlorobis(triethylphosphine)platinum(II) can both be isomerized with quantum yields of about 10^{-2} in a variety of solvents when the ligand-field bands ($\lambda > 304$ nm) are irradiated.[97a]

$$\begin{array}{ccc} \text{Cl} \diagdown \quad \diagup \text{PEt}_3 & & \text{Cl} \diagdown \quad \diagup \text{PEt}_3 \\ \quad \text{Pt} & \underset{h\nu}{\overset{h\nu}{\rightleftarrows}} & \quad \text{Pt} \\ \text{Cl} \diagup \quad \diagdown \text{PEt}_3 & & \text{Et}_3\text{P} \diagup \quad \diagdown \text{Cl} \end{array} \qquad (5\text{-}52)$$

The relative concentrations of cis and trans isomers at the photostationary state are a function of the solvent identity, but for a particular solvent the steady-state cis/trans ratio is the same whether approached from a pure cis or from a pure trans starting isomer. The photostationary state is not equivalent to the thermodynamic equilibrium. For example, at equilibrium in benzene Eq. (5-52) would consist of 92.5% trans, while the photostationary state observed[97] was about 66.2% trans. The effect of solvent on the cis/trans pss ratio reflects solvent polarity, with high-dielectric solvents favoring the dipolar cis isomer (dipole moment 10.7 D) over the nonpolar trans isomer, while the opposite is true for low-dielectric-constant solvents. Although the appropriate data has not been reported, the spectral differences between the cis and trans isomers should make the pss cis/trans ratios dependent on irradiation wavelength as well.[61]

Mechanisms that may be proposed for the isomerization of square planar complexes are either intramolecular or intermolecular processes. The simplest appearing intramolecular process is a "twist" mechanism where the initial square planar complex rearranges to its isomer via an intermediate or transition state having a tetrahedral configuration

$$
\begin{array}{c}
X \diagdown \quad \diagup Y \\
\quad Pt \\
X \diagup \quad \diagdown Y
\end{array}
\rightleftharpoons
\begin{array}{c}
X \diagdown \quad \diagup Y \\
\quad Pt \\
X \diagup \quad \diagdown Y
\end{array}
\rightleftharpoons
\begin{array}{c}
X \diagdown \quad \diagup Y \\
\quad Pt \\
Y \diagup \quad \diagdown X
\end{array}
\qquad (5\text{-}53)
$$

Thermal isomerizations of Pt(II) complexes are not common and occur in most cases by intermolecular mechanisms (see below), yet there is evidence that in rare cases[98] a twist mechanism may be operable. Certainly, this mechanism is very attractive for photoisomerization given the spectral evidence[90b,91] suggesting tetrahedral-type configurations for Pt(II) ligand-field excited states.

The more common mechanism for thermal isomerization requires the presence of excess ligand as catalyst.[92] This intermolecular mechanism apparently involves successive substitutions

$$
\begin{array}{c}
X \\ | \\
{}^{-}X{-}Pt{-}Y + Y \\ | \\ Y
\end{array}
\rightleftharpoons
\begin{array}{c}
X \\ | \\
Y{-}Pt{-}Y + X \\ | \\ Y
\end{array}
\rightleftharpoons
\begin{array}{c}
X \\ | \\
Y{-}Pt{-}Y + Y \\ | \\ X
\end{array}
\qquad (5\text{-}54)
$$

Several steps are required for the transformation owing to the high stereospecificity of Pt(II) thermal substitutions. An associative isomerization mechanism not involving cleavage of original metal-ligand bonds could conceivably be attack of a nucleophile on the square planar complex to give a trigonal bipyramidal intermediate that undergoes pseudorotation. Loss of the nucleophile from the pseudorotated intermediate would give the isomerized product

$$
\begin{array}{c}
Y \diagdown \quad \diagup X \\
\quad Pt \\
Y \diagup \quad \diagdown X
\end{array}
\underset{-N}{\overset{+N}{\rightleftharpoons}}
\begin{array}{c}
X \\ | \\
X{-}Pt \\ | \diagdown \\ Y \quad N
\end{array}
\rightleftharpoons
\begin{array}{c}
X \diagdown X \\
Pt{-}Y \\
Y \diagup N
\end{array}
\rightleftharpoons
$$

$$
\begin{array}{c}
X \diagdown Y \\
Pt{-}N \\
X \diagup Y
\end{array}
\underset{+N}{\overset{-N}{\rightleftharpoons}}
\begin{array}{c}
Y \diagdown \quad \diagup X \\
\quad Pt \\
X \diagup \quad \diagdown Y
\end{array}
\qquad (5\text{-}55)
$$

For thermal reactions, this mechanism is unlikely owing to the very high stereospecificity of Pt(II) substitution reactions; however, it is not inconceivable that a Pt(II) excited state might be more susceptible to this pathway. If N is a solvent molecule, the kinetics of Eq. (5-55) would be indistinguishable from an intramolecular process.

Another possible intermolecular mechanism for isomerization is a dissociative pathway giving a trigonal planar, three-coordinate intermediate capable of reactions with the dissociated ligand to give the original or a new isomer

$$\begin{array}{c}Y\diagdown\quad\diagup X\\ \quad Pt\\ Y'\diagup\quad\diagdown X\end{array} \underset{+X}{\overset{-X}{\rightleftharpoons}} \begin{array}{c}Y\diagdown\\ \quad Pt{-}X\\ Y'\diagup\end{array} \overset{+X}{\longrightarrow} \begin{array}{c}X\diagdown\quad\diagup Y\\ \quad Pt\\ Y'\diagup\quad\diagdown X\end{array} \qquad (5\text{-}56)$$

For thermal isomerization, the dissociative pathway is improbable given the bulk of evidence substantiating associative mechanisms; however, both CTTM and LF excited states of Pt(II) complexes should be considerably more susceptible to dissociative pathways. If the ligand dissociated is one end of a chelating ligand, the reaction would be considered an intramolecular mechanism.

The photoisomerizations of cis- and $trans$-dichlorobis(pyridine)-platinum(II) have been examined,[97,99] and quantitative studies of the cis complex in chloroform solution have provided evidence for the simultaneous operations of several isomerization mechanisms.[99] Irradiation with 313 nm light of the cis complex in chloroform (20°C) results in moderately efficient isomerization under the same conditions in which the $trans$ isomer undergoes photoisomerization with relatively low quantum yields ($\sim 10^{-3}$).

$$cis\text{-}[Pt(py)_2Cl_2] \underset{h\nu}{\overset{h\nu}{\rightleftharpoons}} trans\text{-}[Pt(py)_2Cl_2] \qquad (5\text{-}57)$$

Both complexes are thermally inert under the photolysis conditions.

Spectral analyses of changes in cis-$[Pt(py)_2Cl_2]$ solutions induced by 313-nm irradiation show immediate formation of $trans$ isomer but are complicated by the absence of an isosbestic point and by the fact that, after the photolysis ends, the solutions continue to change for a period of hours. At the conclusion of this slow, secondary thermal reaction, the product solutions analyze as a mixture of cis- and $trans$-$[Pt(py)_2Cl_2]$ only. Consequently, the secondary reactions involve the formation of these isomers from some intermediate(s) produced by the photolysis. Addition of pyridine after irradiation greatly increases the rate of the secondary reaction, but does not change the eventual product distribution. Photolysis in the presence of excess pyridine gives only isomerization with a clean isosbestic point. The quantum yield for the overall cis-to-$trans$ isomerization of these experiments was 0.046 ± 0.005. These results were interpreted as indicating that the $trans$ isomer and the photosubstitution products, cis- and $trans$-$[Pt(py)Cl_2S]$ (S = solvent), are formed as primary photolysis products (Fig. 5-12). The slow secondary reactions are the

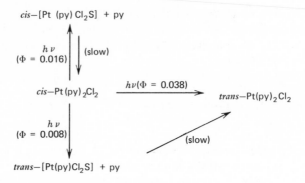

Fig. 5-12. Observed pathways for the photoisomerization of *cis*-[Pt(py)$_2$Cl$_2$] in chloroform (S = CHCl$_3$).

stereospecific thermal substitutions of the dissociated pyridine for the coordinated solvent molecule. The formation of both *cis*- and *trans*-[Pt(py)Cl$_2$S] was suggested to be the result of a dissociative substitution mechanism where the tricoordinate intermediate [see Eq. (5-56)] can react with solvent to give either isomer.[99] The direct photoisomerization could be the result either of a twist mechanism [Eq. (5-53)] or of the dissociative formation of an ion pair, Pt(py)$_2$Cl$^+$—Cl$^-$, that would recombine to give isomerized product or starting material in the poor ion-solvating chloroform solvent.[99]

The photochemical behavior of the bis(glycinato) complexes, *trans*-[Pd(gly)$_2$], *trans*-[Pt(gly)$_2$], and *cis*-[Pt(gly)$_2$] have been studied.[80,100–103] In aqueous solution, the *trans* complexes are not photosensitive to LF excitation at 313 nm, but undergo decomposition to black precipitate when irradiated at 254 nm. In contrast, excitation in either the ligand-field (313 nm) or the charge-transfer (254 nm) region leads to isomerization of *cis*-[Pt(gly)$_2$] with the respective quantum yields, 0.13 ± 0.01 and 0.12 ± 0.01, in neutral aqueous solution

$$(5\text{-}58)$$

Cis-[Pt(gly)$_2$] is known to undergo thermal isomerization to the *trans* complex also, but this occurs only with prolonged heating and only in the presence of excess glycine.[101] Quantum yields of photoisomerization, however, are independent of the presence or absence of free glycine. In addition, when the thermal and photochemical isomerizations were carried out in the presence of radioactively labeled free glycine, the thermal *trans* product was found to contain labeled glycine, while the photochemical product contained none.[101] Therefore, the two reactions must proceed by distinctly different mechanisms, the thermal reaction by an intermolecular pathway and the photochemical isomerization by an intramolecular pathway. It was further concluded that the photochemical reactions involve a twist mechanism, since three-coordinate intermediates of dissociative processes would be expected to be subject to reaction with free glycine.

The facts that *cis*-to-*trans* photoisomerization of *cis*-[Pt(gly)$_2$] occurs with moderate efficiency ($\Phi = 0.13$) but the corresponding *trans*-to-*cis* process does not occur has been interpreted by Balzani and co-workers to mean that the thermally equilibrated excited state responsible for isomerization is "transoid." That is, although the lowest energy configuration of the excited state has a tetrahedral-like configuration, it is distorted toward the *trans* square planar arrangement. Nonradiative deactivation of the transoid excited state would give a transoid vibrationally excited ground state that relaxes to the *trans* isomer and that has insufficient energy to isomerize to the *cis* isomer. Thus, population of this lowest excited state by excitation of either *cis*- or *trans*-Pt(gly)$_2$ should lead only to *trans*-Pt(gly)$_2$, and the quantum yield of isomerization from the *cis* configuration would reflect the efficiency of the interconversion of higher states to the transoid excited state.

The geometrical nature of the excited state responsible for photoisomerization of *cis*-[Pt(gly)$_2$] has been further evaluated utilizing an extended Hückel semiempirical molecular orbital model.[104] These calculations offered somewhat similar conclusions as the Balzani model, but predicted the lowest electronic excited state to have not one but two minima. This is illustrated in Fig. 5-13, where the *cis* isomer is represented by a twist angle θ of 0° and the *trans* by $\theta = 180°$. The shallower of the excited state minima is cisoid and was calculated to have a twist angle between 80° and 90°, while the deeper minimum is transoid with a twist angle between 90° and 100°. The calculations also indicated that the barrier height in going from the cisoid minimum to the transoid minimum is rather low, thus making the *cis*-to-*trans* isomerization favorable. The reverse is not true, and the photoisomerization is essentially irreversible, therefore confirming the general aspects of the Balzani mechanism. It was suggested, however, that radiationless deactivation to ground state could occur from the cisoid minimum in competition to the interconversion to

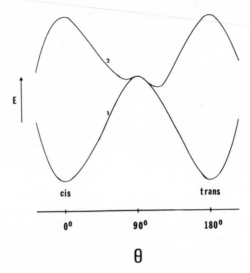

Fig. 5-13. Pictoral representation of the calculated electronic energy for the ground state (1) and lowest excited state (2) of Pt(gly)$_2$.

the transoid excited state, thus accounting for the quantum yield less than unity.

In a recent study, the photoisomerization of cis-[Pt(gly)$_2$] was shown to be subject both to sensitization by organic triplet donors and to quenching by Ni^{2+}(aq) ions.[102] The lowest energy donor for which sensitization was noted is xanthone (26 kK), and the limiting quantum yield for this sensitizer (~0.1) is roughly the same as that observed for direct photolysis. Sensitization was not observed with donors in the energy range, 19.2 to 22.9 kK. Although quenching was observed with Ni^{2+}(aq) (8.5 kK), it was not observed for Mn^{2+}(aq) (18.9 kK). These data were taken to indicate that the excited state responsible for cis-to-trans isomerization has a geometry significantly distorted from the ground-state, square planar configuration. Therefore, the vertical deactivation energy (between ~9 and ~19 kK) is significantly shifted from the vertical excitation energy (between ~23 and 26 kK). The failure for donors in the range of 19.2 to 22.9 kK to act as sensitizers for reaction (5-58) can be attributed in part to low rate constants for donor-acceptor energy transfer owing to Franck–Condon factors.[102]

3. *Photoredox Reactions.* In several cases photolysis of square planar complexes in homogeneous solution leads to redox reactions, but no quantitative mechanism studies of these systems have been reported. For

example, irradiating Zeise's anion, $PtCl_3(C_2H_4)^-$ (discussed above), with 254-nm light causes some oxidation of the metal to $Pt(IV)$ ($\Phi_{ox} \sim 10^{-3}$).[96] The chemical nature of the reduced species was not determined. The 254-nm irradiation corresponds to transitions in the $PtCl_3(C_2H_4)^-$ spectrum identified as charge-transfer bands. Since the central metal ion is oxidized, it is likely that the relevant transition is charge transfer to ligand.

Most other photoredox reactions reported for square planar complexes apparently involve irradiation of CTTM transitions and reduction of the metal with associated oxidation of ligands. For example, photoredox reactions have been observed[105] for the Pt(II) and Pd(II) oxalate complexes, $ML_2(C_2O_4)$ ($L_2 = [P(C_6H_5)_3]_2$, $[As(C_6H_5)_3]_2$, or $(C_6H_5)_2PCH_2\text{-}CH_2P(C_6H_5)_2$). Irradiations of these complexes with a Pyrex-filtered, medium-pressure mercury lamp in deaerated ethanol caused the evolution of nearly 2 moles of carbon dioxide per mole of complex. No CO_2 was produced under the same conditions in the dark. Photolytic production of CO_2 was accompanied by formation of metallic palladium for $Pd[P(C_6H_5)_3]_2(C_2O_4)$ and colored precipitates of Pt(0) compounds when Pt(II) complexes were photolyzed. The photochemical reactions of $Pt[P(C_6H_5)_3]_2(C_2O_4)$ were studied in further detail,[105] and it was shown that products could be attributed to the *in situ* formation of the Pt(0) intermediate $Pt[P(C_6H_5)_3]_2$

$$Pt[P(C_6H_5)_3]_2(C_2O_4) \xrightarrow{h\nu} 2CO_2 + Pt[P(C_6H_5)_3]_2 \qquad (5\text{-}59)$$

In the absence of other ligands, the actual product isolated is the dinuclear complex, $Pt_2[P(C_6H_5)_3]_4$, while in the presence of excess triphenylphosphine, $Pt[P(C_6H_5)_3]_4$ was isolated. Both products can be attributed to the intermediacy of $Pt[P(C_6H_5)_3]_2$. The same intermediate is formed by both photolytic and thermal redox decomposition of the Pt(II) carbonate complex, $Pt[P(C_6H_5)_3]_2CO_3$, in deaerated ethanol solution.[106] In these reactions, however, the solvent is oxidized

$$Pt[P(C_6H_5)_3]_2CO_3 + C_2H_5OH \rightarrow$$
$$[Pt[P(C_6H_5)_3]_2]_2 + CO_2 + H_2O + CH_3CHO \qquad (5\text{-}60)$$

Photoreactions of these types have much promise as practical routes to the synthesis of low-valent platinum organometallic complexes or catalysts.

F. Photochemical Reactions of Octacyano Molybdenum and Tungsten Complexes

Eight-coordinate complexes have not been widely studied; however, among these the octacyano molybdenum and tungsten complexes,

$M(CN)_8^{3-}$ or $M(CN)_8^{4-}$ [$M = Mo(V)$, $W(V)$, $Mo(IV)$, or $W(IV)$], are best known. In general, the photochemical properties of the molybdenum complexes are nearly duplicated by those of the corresponding tungsten species. The photochemistry of some other molybdenum and tungsten complexes have also been reported, but mostly these are complexes of carbonyls or π-bonding organic ligands and are discussed in Chapter 6.

The complex, $Mo(CN)_8^{4-}$, is thermally stable in aqueous solution, but undergoes wavelength-dependent photoaquation and photooxidation when irradiated. The photooxidation process is associated with low wavelength photolysis[107–109] (≤ 313 nm), and apparently the principal reaction involves photoelectron formation

$$Mo(CN)_8^{4-} \xrightarrow[\lambda \leq 313\,nm]{h\nu} Mo(CN)_8^{3-} + e^-(aq) \qquad (5\text{-}61)$$

The photoelectron formation was confirmed both by direct observation of the hydrated electron in flash photolysis experiments[108] and by the more quantitative method of N_2O scavenging.[107] An analysis of the $Mo(CN)_8^{4-}$ electronic spectrum has concluded that bands at 431, 368, 308, and 267 nm are spin-allowed ligand-field transitions, while the very strong absorption at 240 nm is a charge-transfer band.[110] A more recent study has identified[111] a CTTS transition prominent at 270 nm. The photoelectron formation can thus be correlated with the presence of this CTTS excited state reached either by direct excitation or by internal conversion from the higher-energy charge-transfer state.

Irradiation of $Mo(CN)_8^{4-}$ or $W(CN)_8^{4-}$ at wavelengths greater than 300 nm (excitation of LF transitions) leads to photosubstitution processes. For example, photolysis of aqueous $Mo(CN)_8^{4-}$ gives first a red intermediate, then a blue product identified[112] as $Mo(CN)_4O(OH)^{3-}$. Adamson and co-workers[81,113] and Carassiti, Balzani, and co-workers[114–116] have suggested that LF excitation initially results in the photoaquation of one cyanide

$$Mo(CN)_8^{4-} + H_2O \xrightarrow{h\nu} Mo(CN)_7(H_2O)^{3-} + CN^- \qquad (5\text{-}62)$$

The heptacyano complex was considered to be the red intermediate, and subsequent thermal reactions of this species (rapid at high pH) were thought to give the blue product

$$Mo(CN)_7(H_2O)^{3-} + 3OH^- \rightarrow 2H_2O + Mo(CN)_4O(OH)^{3-} + 3CN^- \qquad (5\text{-}63)$$

Several observations are consistent with this mechanism. For example, there is an immediate increase in pH due to the hydrolysis of cyanide released as the irradiation of aqueous $Mo(CN)_8^{4-}$ is initiated,[115] and this increase parallels formation of the red product. Even more convincing is the isolation and characterization of this red intermediate. Mitra, Sharma, and Mohan[117] precipitated this species as the silver salt, $Ag_3[Mo-(CN)_7(H_2O)]$, after irradiating an acidic solution of $Mo(CN)_8^{4-}$ with wavelengths (340–430 nm) corresponding to LF absorptions. The isolation of $Mo(CN)_7H_2O^{3-}$ was possible because of this anion's relative stability in acidic solution. The same authors have also reported[118] quantum-yield studies for the formation both of CN^- and of $Mo(CN)_7(H_2O)^{3-}$ as products of the $Mo(CN)_8^{4-}$ photolysis at 365 nm. The reaction was carried out in aqueous medium at 0°C, where the secondary thermal reactions of $Mo(CN)_7(H_2O)^{3-}$ are minimal. Spectra taken during photolysis showed the disappearance of $Mo(CN)_8^{4-}$ and the appearance of $Mo(CN)_7(H_2O)^{3-}$ with isosbestic points, indicating occurrence of Eq. (5-62) only. Cyanide analysis showed CN^- to be released simultaneously with the formation of $Mo(CN)_7(H_2O)^{3-}$ and with the same quantum yield. Values of this quantum yield $(0.85 \pm 0.05$ at 0°C) were independent of complex concentration $(5 \times 10^{-5}$ to 5×10^{-2} M) and solution pH (pH 1.3–3.8) and were essentially identical to those obtained (0.8–0.9) at 25°C and pH 6.75 using a different technique.[115]

The photoaquation of $Mo(CN)_8^{4-}$ has been sensitized by the triplet donors naphthalene $(E_T = 21.3 \text{ kK})$, phenanthrolene $(E_T = 21.7 \text{ kK})$, anthraquinone $(E_T = 21.7 \text{ kK})$, and benzophenone $(E_T = 24.2 \text{ kK})$.[119] The phosphorescence of biacetyl $(E_T = 19.2 \text{ kK})$ was found to be quenched by $Mo(CN)_8^{4-}$, but no photoaquation occurred. These results have been taken as an indication that the lowest LF triplet state of $Mo(CN)_8^{4-}$ is not reactive and that photoaquation reflects the reactivity of a higher LF triplet state.[119]

As suggested above, the photochemistry of $W(CN)_8^{4-}$ is closely analogous. Irradiation at 254 nm leads to photoelectron production $(\Phi = 0.34)$,[107,108] and photolysis at 365 nm[118] releases CN^- $(\Phi = 0.78)$ to form a red intermediate that decays to the violet product, $W(CN)_4O(OH)^{3-}$.

The Mo(V) complex, $Mo(CN)_8^{3-}$, has long been known to be susceptible to photochemical reduction.[1,2] However, there have been several conflicting reports on the photoreaction mode of this ion in aqueous solution. One report[120] has indicated that cyanogen, C_2N_2, is a reaction product, thus suggesting the photoredox pathway

$$Mo(CN)_8^{3-} \xrightarrow{h\nu} Mo(CN)_7^{3-} + CN \cdot \qquad (5\text{-}64)$$

The cyanide radical could dimerize to give cyanogen, and the molybdenum products would result from subsequent reactions of the heptacyano intermediate. In another study,[121] $Mo(CN)_8^{4-}$ and H^+ were found to be the products of 366-nm irradiation in aqueous solution. Quantum yields increased from 1.9 to 4.4 over the pH range, 4.0 to 7.0, and photooxidation of water was proposed

$$Mo(CN)_6^{3-} + H_2O \xrightarrow{h\nu} Mo(CN)_8^{4-} + H^+ + OH\cdot$$

$$OH\cdot + Mo(CN)_8^{3-} \longrightarrow Mo(CN)_8^{4-} + H^+ + \tfrac{1}{2}O_2$$

(5-65)

These steps would predict a maximum quantum yield of 2, and some involved chain processes must be proposed to account for the higher quantum yields.

A recent study of $Mo(CN)_8^{3-}$ photoreduction by Gray and Spence[122] has shown that the primary photoproduct is $Mo(CN)_7(H_2O)^{3-}$ in acidic or neutral aqueous solution and $Mo(CN)_7(OH)^{4-}$ in basic solution. Qualitative experiments show that, when aqueous $Mo(CN)_8^{3-}$ is irradiated at 365 or 436 nm under acidic or neutral conditions, red $Mo(CN)_7(H_2O)^{3-}$ is first formed, then over a longer period blue $Mo(CN)_4O(OH)^{3-}$ results. As mentioned for the photoaquations of $Mo(CN)_8^{4-}$ described above, secondary thermal reactions to the blue final product are considerably more rapid in basic solution. Both cyanogen, C_2N_2, and another nitrogen-containing oxidation product were identified as reaction products, but all attempts to detect O_2 or H_2O_2 as photolysis products proved negative. These observations discriminate against the mechanism described by Eq. (5-65); however, the product quantum yield was found to be larger than unity under each condition studied, varying from about 1.5 at pH 4.0 to 4.3 at pH 9.0.

If a process such as Eq. (5-64) were the only photoredox mechanism, a quantum yield of 1.0 would be predicted, and other cyanide oxidation products should not be found

$$Mo(CN)_8^{3-} + H_2O \xrightarrow{h\nu} Mo(CN)_7(H_2O)^{3-} + \tfrac{1}{2}C_2N_2$$

(5-66)

However, a crucial observation[122] is that addition of CN^- in the dark to aqueous $Mo(CN)_8^{3-}$ at pH 9.00 results in very rapid reduction of an equivalent amount of $Mo(CN)_8^{3-}$

$$Mo(CN)_8^{3-} + CN^- \xrightarrow{fast} Mo(CN)_8^{4-} + \tfrac{1}{2}C_2N_2$$

(5-67)

This reaction occurs to a lesser extent in neutral solution and does not occur in acid owing to the low equilibrium concentration of cyanide relative to HCN. In basic solution, the initial photoreduction product, $Mo(CN)_7(OH)^{4-}$, decomposes, rapidly releasing cyanide ion[122]

$$Mo(CN)_7(OH)^{4-} + 2OH^- \rightarrow Mo(CN)_4O(OH)^{3-} + 3CN^- + H_2O$$

(5-68)

This additional cyanide can reduce more $Mo(CN)_8^{3-}$, thus accounting for the high photoreduction quantum yields in base.

A most interesting feature of the photoreduction of $Mo(CN)_8^{3-}$ is that the region of excitation (365 and 438 nm) corresponds to LF absorption bands.[110] Consequently, it appears that for this moderately strong oxidant (standard reduction potential = 0.73 V)[123] ligand-field excitation results in photoredox reactions.

III. COMPOUNDS OF THE f-BLOCK ELEMENTS— THE LANTHANIDES AND ACTINIDES

A. Some General Properties

The lanthanides are the fourteen elements following lanthanum, whose electronic configurations are derived from successively adding $4f$ electrons to lanthanum's $[Xe]5d^1 7s^2$. The chemistries of these elements closely resemble lanthanum, the most common oxidation state being +3, which has the electronic configuration, $[Xe](4f)^n$, with n ranging from zero for La^{3+} to 14 for Lu^{3+}. Compounds of the divalent ions, M^{2+}, and tetravalent ions, M^{4+}, are known for some lanthanides,[24] but these oxidation states are most stable when the ion's electronic configuration has an empty $(4f)^0$, half-filled $(4f)^7$, or filled $(4f)^{14}$ subshell. The properties of the lanthanide ions are strongly influenced by the fact that the $4f$ orbitals are effectively shielded from external (e.g., ligand-field) forces by the larger, filled, $5s$ and $5p$ subshells.[124] For this reason, the M^{3+} ions have properties resembling La^{3+} regardless of the $4f$ subshell population. Compounds of the lanthanides are essentially ionic in character, and π-bonding is unknown for these compounds. The aqueous M^{3+} ions in solution have coordination numbers of 8 or 9, depending on ionic radii, and generally are very substitution labile.[125]

Spin-orbit coupling constants for the lanthanides are very large; consequently, electronic states are defined by values of the total angular momentum, J.[124] As a result of $4f$ subshell shielding, electronic states of lanthanide ions are little affected by changes in the surrounding ligand field. Ligand fields cause such small perturbations of the various J states arising from different f^n configurations that $f–f$ transitions are extremely sharp, much like the free ions. In addition to $f–f$ transitions, both charge-transfer and $4f \rightarrow 5d$ transitions have been identified in the spectra of some lanthanide ions.[126] Shielding by the filled $5s$ and $5p$ subshells also insulates $f \rightarrow f$ excited states and decreases the rates of radiationless deactivation. Thus lanthanide ions having from 2 to 12 f-electrons show line emission in the solid state, and at least five show line emission in aqueous solution.[127] Energy transfers occurring intramolecularly from chelate-ligand excited states or intermolecularly from excited donor molecules in solution have also been observed to lead to luminescence from lanthanide f level excited states.[127] The luminescence spectroscopy of the lanthanide compounds has received much attention, in part due to industrial applications in color television and lasers.

The actinides are comprised of actinium and the 14 elements whose electronic configurations derive from successively adding $5f$ electrons to actinium's $[Rn]6d^17s^2$. These display significant differences in chemical and physical properties from the lanthanides. Like the lanthanides, $+3$ is the oxidation state common to all the actinides, but unlike them, $+2$ and $+4$ are not the only other oxidation states seen: certain actinides form compounds with oxidation states $+5$, $+6$, and even $+7$.[24] In addition, the $5f$ orbitals (especially in the earlier actinides) are not as effectively shielded (owing to relatively greater spatial extension) by the filled $6s$ and $6p$ orbitals as are the lanthanide $4f$ orbitals. Consequently, the actinide $5f$ orbitals interact with ligand fields more in the manner that d orbitals interact, including contributions of covalent overlap with ligand orbitals and even some π-bonding. Thus, the actinides are much more inclined toward forming complexes with various ligands than are the more ionic lanthanides.

Spectra of the actinide complexes generally have not been well characterized.[124,128] Charge-transfer and f–f transitions have both been identified. For the M^{3+} ions, transitions within the $5f$ shell are sharp, "lanthanide-like" lines for the heavier actinides, but considerably broadened and ligand-field influenced for earlier actinides where $5f$ orbital contribution to ligand-metal bonding is more significant.[129]

Quantitative studies of the photochemical properties of lanthanide and actinide compounds are limited to few systems.[1,2] By far, the most studied are the photoreactions of the uranyl ion, UO_2^{2+}, and its complexes and the compounds of Ce(III) and Ce(IV). With a few exceptions photosubstitution reactions have not been examined, in part certainly, because of the thermal lability of the lanthanide ions. The majority of the reactions examined have involved photoredox transformations of the metal or photocatalysis or sensitization of ligand reactions.

B. The Photochemistry of Ce(IV) and Ce(III)

The electronic configurations of Ce(III) and Ce(IV) are $[Xe]4f^1$ and $[Xe]$, respectively. In aqueous solution, Ce(III) is found as the hydrated ion, $Ce(H_2O)_n^{3+}$, with a relatively small tendency to hydrolyze, but it associates to some degree with anions in solution, particularly chelating species.[130] The character of Ce(IV) in aqueous solution is less certain owing to its tendency to hydrolyze and to form dimeric, trimeric, and polymeric species and to associate with anions.[131–133] It is probable that the hydrated ion, $Ce(H_2O)_n^{4+}$, exists only in strong aqueous perchloric acid. Ce(IV) is a powerful one-electron oxidizing agent; the Ce(IV)/Ce(III) couple has standard reduction potential, 1.7 V in 1.0 M $HClO_4$.[132]

Ceric ion solutions are yellow with intense, broad, and unstructured absorption in the ultraviolet, tailing into the visible. Given the inert-gas electronic configuration of Ce^{4+}, this absorption is undoubtedly CTTM in character. Thus, it is no surprise that redox reactions of this ion are markedly enhanced by photolysis.[134] For example, aqueous perchloric acid solutions of Ce(IV), although thermodynamically unstable, can be stored in the dark for months with little change in Ce(IV) concentration. However, absorption of ultraviolet light causes the oxidation of water[135]

$$4Ce(IV) + 2H_2O \xrightarrow{h\nu} 4Ce(III) + 4H^+ + O_2 \qquad (5\text{-}69)$$

Heidt and Smith[135b] measured quantum yields for 254-nm photolysis of $HClO_4$ solutions containing varying concentrations of Ce(IV) and Ce(III). Quantum yields were observed to level at a maximum of about 0.14 at higher Ce(IV) concentrations and were independent of light intensity. They concluded that the photochemically active species were Ce(IV) dimers and that Ce(III) served to deactivate reactive states of these dimers. The dimers were suggested to serve as two-electron acceptors in the redox reaction. These workers also noted that cerous ion is photooxidized by 254-nm photolysis in aqueous $HClO_4$ and proposed the accompanying formation of hydrogen

$$2Ce(III) + 2H^+ \xrightarrow{h\nu} 2Ce(IV) + H_2 \qquad (5\text{-}70)$$

They suggested that when the relationship, $\varepsilon_3 \Phi_3 [Ce(III)] = \varepsilon_4 \Phi_4 [Ce(IV)]$ (ε_3 and ε_4 are extinction coefficients and Φ_3 and Φ_4 are quantum yields for Ce(III) and Ce(IV), respectively, at the irradiation wavelength), is satisfied, [Ce(III)] and [Ce(IV)] would remain unchanged, while solvent water would be decomposed to H_2 and O_2.

$$2H_2O \xrightarrow[h\nu]{Ce(IV)/Ce(III)} O_2 + 2H_2 \qquad (5\text{-}71)$$

Heidt and McMillan[136,137] later confirmed photoreactions (5-70) and (5-71).

The conclusion that ceric dimers are responsible for the photoactivity of Ce(IV) in aqueous perchloric acid has been challenged by Evans and Uri,[138] who proposed an electron-transfer process producing hydroxyl radicals

$$CeOH^{3+} \xrightarrow{h\nu} Ce^{3+} + \cdot OH \qquad (5\text{-}72)$$

The concentration effects of Ce(IV) and Ce(III) were suggested to be the result of competitive reactions of the hydroxyl radical

$$CeOH^{3+} + \cdot OH \longrightarrow Ce^{3+} + H_2O_2 \qquad (5\text{-}73)$$

$$Ce(III) + \cdot OH \longrightarrow Ce(IV) + OH^- \qquad (5\text{-}74)$$

$$H_2O_2 + 2Ce(IV) \xrightarrow{fast} 2Ce(III) + H_2O + \tfrac{1}{2}O_2 \qquad (5\text{-}75)$$

This mechanism proved consistent with Heidt and Smith's data,[135b] although the argument of Evans and Uri that no dimer is present under the experimental conditions can be shown by equilibrium data[131a] to be incorrect. However, the same equilibrium data shows that the Ce(IV) is but a few percent dimerized under conditions where quantum yields were noted to maximize. In general, the pattern of free-radical production as the result of Ce(IV) photolysis (see below) clearly points to the operation of processes such as Eq. (5-72). Furthermore, electron-spin-resonance studies[139] of irradiated frozen solutions of ceric perchlorate in aqueous $HClO_4$ show the formation of trapped H_2O^+ radicals, a species that would dissociate to $\cdot OH$ plus H^+ in solution and that must be formed by electron transfer from the hydration sphere to the ceric ion.

The photoreduction of aqueous Ce(IV) [Eq. (5-69)] has also been studied in 0.4 M sulfuric acid solutions[140] in which Ce(IV) is monomeric and exists as a mixture of sulfate complexes. The quantum yields for formation of Ce(III) (resulting from 254-nm irradiation) were enhanced by the presence of Br^-, Cl^-, formic acid, and Tl(I). In each case, this effect was interpreted as indicating the trapping of the hydroxyl radical, thus preventing its reaction with Ce(III). Flash photolysis of 6 M nitric acid solution of $[NH_4]_2[Ce(NO_3)_6]$ with ultraviolet light results in the formation of NO_3 radicals.[141] However, examination of decay plots for the transient produced when aqueous nitrate solutions of Ce(IV) were flashed[142] led to the interpretation that the NO_3 radical results from secondary oxidation of NO_3^- by $\cdot OH$ produced in primary photolysis [Eq. (5-72)]. The question of $\cdot NO_3$ being a primary photoproduct remains controversial, and recent results of electron-spin-resonance studies of photolyzed frozen solutions of $[NH_4]_2[Ce(NO_3)_6]$ in aqueous $HNO_3/HClO_4$ mixtures (88°K) appear to indicate that $\cdot NO_3$ is produced without a radical precursor such as $\cdot OH$.[143]

Ce(IV) is a well-known oxidizing agent for organic substrates,[144] and it has been established that the oxidation of numerous organic substrates is accelerated when solutions of Ce(IV) and these compounds are exposed to light.[134,145] For example, benzoic acid is apparently stable to refluxing aqueous Ce(IV) solutions, but when exposed to bright sunlight (at room temperature) oxidation to fumaric acid occurs.[134b] Sheldon and Kochi have examined the photo and thermal reactions of Ce(IV) in various carboxylic acids (RCO_2H) including glacial acetic acid.[134a] Photolysis of these deaerated solutions at 350 or 254 nm caused reduction of Ce(IV) and accompanying reactions of the carboxylate ligand or solvent. For example, 350-nm irradiation of ceric acetate in glacial acetic acid produced Ce(III), methane, and carbon dioxide as the major products

$$Ce(CH_3CO_2)_4 \xrightarrow{h\nu} Ce(CH_3CO_2)_3 + CH_4 + CO_2 \qquad (5\text{-}76)$$

Small amounts of methylacetate, ethane, acetoxyacetic acid, and succinic acid were also found. Analogous decarboxylations of the appropriate carboxylate were the principal pathways observed in n-butyric acid and isobutyric acid. Examination of products in the absence and presence of radical scavengers led to the proposal that the photodecomposition of the Ce(IV) carboxylates occurred by the following steps:

$$Ce(IV)(RCO_2^-) \xrightarrow{h\nu} Ce(III) + \cdot RCO_2 \qquad (5\text{-}77)$$

$$\cdot RCO_2 \longrightarrow \cdot R + CO_2 \qquad (5\text{-}78)$$

Subsequent reactions of the alkyl radicals, $\cdot R$ [oxidation by Ce(IV), capture by radical scavenger, dimerization, etc.], determine the eventual stoichiometry.

The photoreaction of Ce(IV) in aqueous solution with carboxylic acids and in the neat organic liquids have also been studied by electron-spin-resonance spectroscopy of the frozen solutions (77°K).[146] For example, broad-band photolysis (xenon source, $\lambda > 300$ nm) of a 0.1 M ceric perchlorate and 0.1 M carboxylic acid (RCO_2H) substrate solution in 4 M aqueous $HClO_4$ led to detection of the alkyl radical, $\cdot R$, but gave no evidence of the acyloxy radical, $\cdot RCO_2$. Thus, it was concluded that the primary photochemical act is not Eq. (5-77), but a concerted oxidative decarboxylation

$$R-\overset{\overset{\displaystyle O}{\|}}{\underset{\underset{\displaystyle H}{|}}{C}}-O + Ce(IV) \xrightarrow[\lambda > 300 \text{ nm}]{h\nu} \cdot R + CO_2 + H^+ + Ce(III) \qquad (5\text{-}79)$$

C. Other Lanthanides

Since, in certain cases, excitation of lanthanide states serves to photosensitize ligand luminescence,[127] one might expect photosensitization of ligand reactions to also be possible. Such an observation has been reported for the polymerization of vinyl compounds with a variety of +3 lanthanides[147] and Eu(II).[148] In addition, photoredox reactions for lanthanide ions in solution may be anticipated for those elements having more than one stable oxidation state.

Photooxidation of Eu(II) in acidic aqueous solution has been reported[149] as the result of 366-nm irradiation ($\Phi \sim 0.2$)

$$2Eu^{2+} + 2H^+ \xrightarrow{h\nu} 2Eu^{3+} + H_2 \qquad (5\text{-}80)$$

The reverse reaction results from ultraviolet photolysis[150] (254 nm) of aqueous Eu^{3+}. This excitation wavelength corresponds to a CTTM transition[126b] for Eu^{3+}, and the proposed primary process is

$$Eu^{3+} + H_2O \xrightarrow{h\nu} Eu^{2+} + \cdot OH + H^+ \tag{5-81}$$

However, since Eu^{2+} has an extinction coefficient several orders of magnitude greater than Eu^{3+} at the irradiation wavelength,[150] as the photolysis proceeds the back reaction becomes important. In the presence of isopropanol, hydrogen formation results (after an induction period), a process which can be attributed to the reactions

$$Eu^{2+} + H_3O^+ \xrightarrow{h\nu} Eu^{3+} + \cdot H + H_2O \tag{5-82}$$

$$\cdot H + (CH_3)_2CHOH \longrightarrow H_2 + (CH_3)_2\dot{C}OH \tag{5-83}$$

The organic radical produced in Eq. (5–83) apparently reacts with Eu^{3+} in solution, giving acetone and regenerating Eu^{2+}. Therefore, the Eu^{2+} concentration builds to a point where it is absorbing essentially all the excitation light, and a steady-state Eu^{2+} concentration is achieved where the net photoreaction is the Eu^{2+}-catalyzed decomposition of isopropanol to H_2 and acetone. Thus, for the photolysis of aqueous Eu^{3+} in the presence of H-atom scavengers such as isopropanol, production of H_2 is diagnostic of the formation of Eu^{2+} despite the fact that the primary photoreaction [Eq. (5-81)] oxidizes the solvent. A similar approach has been used to examine briefly the photoreactions[150] of the aqueous lanthanide ions, Sm^{3+}, Yb^{3+}, Tb^{3+}, and Pr^{3+}. Photolysis with 185-nm light in the presence of isopropanol led to enhancement of H_2 formation (over background) for solutions of Sm^{3+} and Yb^{3+}, but no reaction for Tb^{3+} and Pr^{3+}. Since Sm^{3+} and Yb^{3+} can be reduced to the divalent state in aqueous solution but Tb^{3+} and Pr^{3+} cannot,[24] enhancement of H_2 formation in the presence of the former ions was concluded to be evidence for initial primary photoreactions in analogy to Eq. (5-81).

D. The Photochemistry of Uranyl Complexes

In aqueous solution U(VI) is found primarily as the uranyl ion, UO_2^{2+}, which is sufficiently stable to persist through a variety of chemical perturbations.[24] The O—U—O group is linear with short, very strong U—O bonds. In aqueous solution, the U—O bonds are inert to exchange with $H_2^{18}O$ in the absence of reduced forms of uranium. Numerous complexes between UO_2^{2+} and various ligands have been observed, and crystallographic studies have shown that five- and six-coordination in the UO_2^{2+} equatorial plane is the most common configuration.[151] In aqueous solutions, the equatorial ligands are labile, and mixed complexes are common. This lability and the rather small spectral changes accompanying perturbations of the equatorial ligand field present serious problems in identifying uranyl species in solution, and failure to characterize chemically relevant UO_2^{2+} complexes has been offered as a major criticism discounting the value of many early studies of uranyl photochemistry.

The photochemistry of uranyl species was extensively reviewed[152] in 1964 and reviewed more recently in Refs. 1 and 2. Photoreduction of U(VI) accompanied by ligand oxidation and/or photosensitized ligand reactions are the general processes observed. A case of uranyl photosensitization of a transition-metal-complex reaction has also been reported.[153] Particularly careful attention has been paid to the uranyl oxalate system, as this has proved to be a reliable chemical actinometer under appropriate conditions.

The spectrum of UO_2^{2+} in aqueous perchloric acid shows a number of weak absorption bands in the visible region plus some broader more intense bands in the ultraviolet region.[152,154] The weak visible and near-ultraviolet bands are transitions from bonding orbitals to unfilled $5f$ or $6d$ uranium orbitals.[151b] These absorption bands are not greatly affected by changing the equatorial ligands. Complexes of UO_2^{2+} show charge-transfer bands (CTTM) with energies dependent on the relevant ligands, primarily in the ultraviolet region.[155] Uranyl ion and some of its complexes also luminesce when irradiated in aqueous solutions at room temperature.[127] The emission consists of a series of bands spaced in a manner similar to the visible absorption bands but appearing at lower energy. The emission bands are apparently transitions within the UO_2^{2+} moiety, as the equatorial ligands have relatively small effects on their general pattern.

1. *The Uranyl Oxalate System, a Chemical Actinometer.* The uranyl oxalate actinometer as first characterized[156] consists of aqueous solutions of uranyl sulfate (0.01 M) and oxalic acid (0.05 M) in a medium that is acidic (pH 1.4) owing to oxalic acid dissociation. Irradiation of this solution results in decomposition of oxalate; however, the stoichiometry of the reactions involved is not simple. Major products include CO_2, CO, HCO_2H, and U^{4+}, depending on reaction conditions.[157,158] In the original method, quantum yields for oxalate loss are determined by titration with an oxidizing agent such as aqueous permanganate. In this manner, absolute quantum yields were determined over the range 208 to 436 nm. The quantum yield for oxalate loss is rather insensitive to pH over the range of 1 to 5,[157] temperature,[156c] concentration of reactants, and light intensity. However, the titrametric method of analysis makes the uranyl oxalate actinometer much less sensitive than the ferric oxalate actinometer. Other, more sensitive, analytical techniques have been developed for the uranyl oxalate actinometer; one involves differential absorption spectrophotometry[159] to increase sensitivity by at least an order of

magnitude, while the other utilizes gas chromatographic analysis with flame ionization detection to determine the CO produced in the photolysis.[158,160]

The overall photochemistry occurring in the uranyl oxalate actinometer solution apparently can be summarized by three reactions[157,158]

$$H_2C_2O_4 \xrightarrow[UO_2^{2+}]{h\nu} CO + CO_2 + H_2O \tag{5-84}$$

$$H_2C_2O_4 \xrightarrow[UO_2^{2+}]{h\nu} HCO_2H + CO_2 \tag{5-85}$$

$$H_2C_2O_4 + UO_2^{2+} \xrightarrow{h\nu} UO_2H^+ + H^+ + 2CO_2 \tag{5-86}$$

The contributions from each pathway have been found to be independent of intensity, photolysis wavelength, and temperature, but dependent on reactant concentration and solution pH.

Heidt and co-workers[157] have concluded that the absorption of light by uranyl oxalate complexes, even at wavelengths as long as 436 nm, results in generation of reactive ligand (oxalate) to metal (uranyl) charge-transfer excited states. In this context, the observation[161] of reversible absorption changes in flash photolysis studies of aqueous uranyl oxalate was interpreted as consistent with a mechanism in which U(VI) is reduced to U(V) in the primary photochemical reaction. A secondary reaction is required for the reoxidation of U(V) to U(VI)

$$(UO_2^{2+})(C_2O_4^{2-}) \xrightarrow{h\nu} (UO_2^+)(CO_2^-) + CO_2 \tag{5-87}$$

$$(UO_2^+)(CO_2^-) + H_2O \longrightarrow (UO_2^{2+})(HCO_2^-) + OH^- \tag{5-88}$$

$$(UO_2^+)(CO_2^-) + H^+ \longrightarrow (UO_2^{2+})(OH^-) + CO \tag{5-89}$$

The primary reaction [Eq. (5-87)] produces the U(V) intermediate plus CO_2, and the competitive secondary reactions regenerate U(VI) and produce formic acid and CO, respectively. The pH dependence of CO would be explained by the acid-dependent character of Eq. (5-89). The pH dependence of CO formation has also been considered by Volman and Seed,[158] who suggested that CO is a photoproduct of the protonated oxalate complex, $UO_2(HC_2O_4)^+$, while formate is a photoproduct of $UO_2(C_2O_4)$

$$UO_2(HC_2O_4)^+ + H^+ \xrightarrow{h\nu} CO_2 + CO + H_2O + UO_2^{2+} \tag{5-90}$$

$$UO_2(C_2O_4) + H^+ \xrightarrow{h\nu} UO_2(HCO_2)^+ + CO_2 \tag{5-91}$$

Current information does not appear to distinguish clearly these suggested mechanisms.

2. *Uranyl Photochemistry with Other Substrates.* The photochemistries of a number of other uranyl carboxylic acid systems have been examined.[152,162,163] For example, aqueous solutions of UO_2^{2+} and formic acid undergo photoredox reactions when irradiated

$$UO_2^{2+} + HCO_2H + 2H^+ \xrightarrow{h\nu} U^{4+} + CO_2 + 2H_2O \qquad (5\text{-}92)$$

It has been suggested that the primary photochemical reaction involves a one-electron transfer (presumably via a CTTM state) of a uranyl formate complex[2]

$$(UO_2^{2+})(HCO_2^-) \xrightarrow{h\nu} UO_2^+ + \cdot HCO_2 \qquad (5\text{-}93)$$

Formation of U^{4+} and CO_2 presumably would come from disproportionation of the U(V) species to U^{4+} plus UO_2^{2+} and of the $\cdot HCO_2$ radicals

$$2 \cdot HCO_2 \rightarrow CO_2 + HCO_2H \qquad (5\text{-}94)$$

Conceivably, the $\cdot HCO_2$ radical could also react with UO_2^{2+} to give U(V), CO_2, and H^+ in a manner analogous to steps suggested[157] for the redox reaction observed in the uranyl oxalate system. A quantum yield approximating unity has been observed for Eq. (5-92) under optimum conditions.

The photochemistry of aqueous solutions of acetic acid and UO_2^{2+} also displays a photoredox pathway[152]

$$UO_2^{2+} + 2H^+ + 2CH_3CO_2H \xrightarrow{h\nu} C_2H_6 + 2CO_2 + 2H_2O + U^{4+} \qquad (5\text{-}95)$$

but, in addition, it displays a UO_2^{2+}-sensitized carboxylation pathway

$$CH_3CO_2H \xrightarrow[UO_2^{2+}]{h\nu} CH_4 + CO_2 \qquad (5\text{-}96)$$

Similar photoredox and photodecarboxylation reactions have been observed with other carboxylic acids in solution with UO_2^{2+}. The photoredox products can be rationalized by a primary photolytic reaction analogous to Eq. (5-93), giving $\cdot RCO_2$ radicals that can subsequently lose CO_2 to give the $\cdot R$ radical that may dimerize.

The photoreactions between uranyl and a variety of organic compounds, including aldehydes, alcohols, and carbohydrates, are well known.[152] For example, irradiating deaerated aqueous solutions of uranyl ion and aliphatic alcohols generally leads to photooxidation of the alcohol to a corresponding aldehyde or ketone and reduction of U(VI) to U(IV), for example[164]

$$CH_3CH_2OH + UO_2^{2+} \xrightarrow{h\nu} CH_3C\overset{\displaystyle O}{\underset{\displaystyle H}{\diagup\diagdown}} + U(IV) \text{ species} \qquad (5\text{-}97)$$

Measurements of uranyl luminescence quenching by alcohols follow a Stern–Volmer relationship with respect to alcohol concentration, the quenching constants, K_q, for various alcohols correlate closely with the photoreactivity, Φ, of that alcohol, and neither seems to be related to the possible stabilities of uranyl alcohol complexes. These results were interpreted to indicate that prior complex formation between UO_2^{2+} and alcohols was not involved in the quenching process, but that energy transfer from excited uranyl to alcohol is collisional. Structure-reactivity relationships, deuterium-isotope effects, and correlation of K_q and Φ all point to a chemical-quenching mechanism involving abstraction of an α-hydrogen of the alcohol

$$(UO_2^{2+})^* + R_1R_2CHOH \rightarrow UO_2H^{2+} + R_1R_2\dot{C}OH \qquad (5\text{-}98)$$

This suggestion is supported by electron-spin-resonance identification[165] of the $R_1R_2\dot{C}OH$ species when the irradiation is carried out in matrices at $77°K$. Flash photolysis[166] of uranyl perchlorate using a ruby-doubled laser (λ 347.1 nm, flash duration 5 nsec) allowed direct observation of the absorption spectrum of the uranyl excited state, $(UO_2^{2+})^*$. The lifetime of this species is shortened by added substrates in a manner consistent with the operation of Eq. (5-98).

IV. SOME CONCLUDING REMARKS

In discussing the photochemistries of the heavier elements, we have by necessity focused on the relatively few, well-studied systems for which data in homogeneous solution is available. From these, it is clear that these systems provide a rich and expanding area of interest to photochemists. The types of photoreactions that may be studied are dependent on several features, among them the instrumental techniques available and a knowledge of the ground-state properties of the systems of interest. The continuing improvements of experimental sophistication make problems previously insurmountable easily solvable. For example, the advent of high-energy, pulse lasers have made submicrosecond flash photolysis a reality, and many short-lived intermediates and/or excited states before inaccessible are now potentially within reach for direct study.

In general, photoreactions of the lanthanide and actinide elements involve redox processes, either oxidation of the metal ion or decomposition of ligands owing to cyclic, photolysis-induced redox reactions. Thus, the elements for which photoreactions have been characterized are largely those of cerium, europium, and uranium, whose aqueous chemistries each include several stable oxidation states. Substitutional behavior has not been studied due to the high lability of these ions in their ground states. Complexes of the heavier d-block elements, however, run a full gamut of photoreactivities, including reactions of coordinated ligands, redox reactions, and substitutional processes. Energy transfer has been observed with both types of systems. It is clear that the range of photoreactions for these systems is as wide or even wider than observed for the lighter transition elements.

In comparing the excited-state chemistries of the heavier and lighter elements, spin-orbit coupling emerges as an issue of potentially very great importance that photochemists in this area have largely ignored. Indeed, the authors of this chapter have adopted the use of the multiplicity terms "singlet," "triplet," and so forth, used also by other photochemists to describe states of many systems. However, given the magnitude of the spin-orbit coupling constants, it is likely that these simple terms may be virtually meaningless for the third-row transition elements and questionable even for complexes of the second row. If so, then such terms as "intersystem crossing" have a more complicated meaning (or perhaps none at all) than indicating a change in spin multiplicity. Advances in photochemical theory, by necessity, will have to treat this feature of the heavier elements more quantitatively. Nonetheless, terms such as "ligand-field excited state," "charge-transfer state," and "internal-ligand $\pi-\pi^*$ state" all appear to retain qualitative validity in describing excited-state reactions and properties of many heavier metal complexes. Although mixing of state characters may occur, a perhaps surprising number of the photoreactions considered in this chapter are analyzable in terms of the expected character of excited states corresponding to one or another of these descriptions.

Lastly, a fail-safe prediction is that the photochemistry discussed in this chapter only scratches the surface of the heavier metal systems which shall be the subjects of mechanistic and/or synthetic photochemistry in the near future. For example, most quantitative work with square planar complexes revolves around Pt(II) complexes; yet, in recent years numerous other square planar complexes [such as those of Ir(I)] have been synthesized and their thermal reactions studied. Expanded photochemical studies of these systems will allow a critical examination of the generalities of mechanistic thought regarding square planar complex photochemistry. In this context, advances in the mechanistic photochemistry of the heavier elements shall depe..d in large part on the synthesis and characterization of new systems chosen to provide insight into and criticisms of generalizations that have been drawn to date.

REFERENCES

1. A. W. Adamson, W. L. Waltz, E. Zinato, D. W. Watts, P. D. Fleischauer, and R. D. Lindholm, *Chem. Rev.* **68,** 541 (1968).
2. V. Balzani and V. Carassiti, *Photochemistry of Coordination Compounds*, Academic, London, 1970.
3. W. P. Griffith, *The Chemistry of the Rarer Platinum Metals*, p. 34, Wiley-Interscience, New York, 1967.
4. P. D. Fleischauer and P. Fleischauer, *Chem. Rev.* **70,** 199 (1970).
5. P. D. Fleischauer, A. W. Adamson, and G. Satori, *Progr. Inorg. Chem.* **17,** 1 (1972).
6. J. Kleinberg, W. J. Argersinger, and E. Griswold, *Inorganic Chemistry*, pp. 487–490, Heath, Boston, 1960.
7. A. B. P. Lever, *Inorganic Electronic Spectroscopy*, p. 305, Elsevier, Amsterdam, 1968.
8. B. N. Figgis, *Introduction to Ligand Fields*, pp. 248–292, Wiley-Interscience, New York, 1967.
9. C. K. Jørgensen, *Progr. Inorg. Chem.* **4,** 73 (1962).
10. M. K. DeArmond and J. E. Hillis, *J. Chem. Phys.* **54,** 2247 (1971).
11. R. D. Foust and P. C. Ford, *J. Amer. Chem. Soc.* **94,** 5686 (1972).
12. J. N. Demas and G. A. Crosby, *J. Amer. Chem. Soc.* **93,** 2841 (1971).
13. D. L. Webb and L. A. Rossiello, *Inorg. Chem.* **10,** 2213 (1971).
14. G. A. Crosby, R. J. Watts, and D. W. Carstens, *Science* **170,** 1195 (1970).
15. J. N. Demas and G. A. Crosby, *J. Amer. Chem. Soc.* **92,** 7262 (1970).
16. R. J. Watts and G. A. Crosby, *J. Amer. Chem. Soc.* **94,** 2606 (1972).
17. T. R. Thomas, R. J. Watts, and G. A. Crosby, *J. Chem. Phys.* **59,** 2123 (1973).
18. T. R. Thomas and G. A. Crosby, *J. Mol. Spectrosc.* **38,** 118 (1971).
19. T. L. Kelly and J. F. Endicott, *J. Amer. Chem. Soc.* **94,** 1797 (1972).
20. T. L. Kelly and J. F. Endicott, *J. Phys. Chem.* **76,** 1937 (1972).
21. J. E. Hillis and M. K. DeArmond, *Chem. Phys. Letters,* **10,** 325 (1971).
22. J. E. Hillis and M. K. DeArmond, *J. Luminescence* **4,** 273 (1971).
23. D. H. W. Carstens and G. A. Crosby, *J. Mol. Spectrosc.* **34,** 113 (1970).
24. F. A. Cotton and G. Wilkinson, *Advanced Inorganic Chemistry*, 3rd ed., Wiley-Interscience, New York, 1972.
25. P. C. Ford, *Coord. Chem. Rev.* **5,** 75 (1970).
26. D. L. Swihart and W. R. Mason, *Inorg. Chem.* **9,** 1749 (1970).
27. (a) F. Zuloaga and M. Kasha, *Photochem. Photobiol.* **7,** 549 (1968); (b) I. N. Douglas, J. V. Nicholas, and B. G. Wybourne, *J. Chem. Phys.* **48,** 1415 (1968); (c) L. A. Rossiello, *J. Chem. Phys.* **51,** 5191 (1969).
28. P. Ford, D. Rudd, R. Gaunder, and H. Taube, *J. Amer. Chem. Soc.* **90,** 1187 (1968).
29. F. E. Lytle and D. M. Hercules, *J. Amer. Chem. Soc.* **91,** 253 (1969).
30. C. K. Jørgensen, *Acta Chem. Scand.* **10,** 500 (1956).
31. M. K. DeArmond and J. E. Hillis, *J. Chem. Phys.* **54,** 2247 (1971).

32. L. Moggi, *Gazz. Chim. Ital.* **97**, 1089 (1967).

33. J. D. Petersen and P. C. Ford, reported at the 165th National Meeting of the American Chemical Society, p. INORG. 39, Dallas, Texas, April 1973.

34. J. A. Osborn, R. D. Gillard, and G. Wilkinson, *J. Chem. Soc.* 3168 (1964).

35. R. D. Foust and P. C. Ford, *Inorg. Chem.* **11**, 899 (1972).

36. C. Kutal and A. W. Adamson, *Inorg. Chem.* **12**, 1454 (1973).

37. C. Kutal and A. W. Adamson, *J. Amer. Chem. Soc.* **93**, 5581 (1971).

38. A. D. Kirk, *J. Amer. Chem. Soc.* **93**, 283 (1971).

39. (a) P. C. Ford and C. Sutton, *Inorg. Chem.* **8**, 1544 (1969); (b) R. J. Allen and P. C. Ford, *Inorg. Chem.* **11**, 679 (1972).

40. M. M. Muir and W. L. Huang, *Inorg. Chem.* **12**, 1831 (1973).

41. J. D. Petersen and P. C. Ford, unpublished observations.

42. (a) R. A. Bauer and F. Basolo, *J. Amer. Chem. Soc.* **90**, 2437 (1968); (b) *Inorg. Chem.* **8**, 2231 (1967).

43. G. S. Kovalenko, I. B. Baranovskii, and A. V. Babaeva, *Russ. J. Inorg. Chem.* **16**, 148 (1971).

44. J. L. Reed, F. Wang, and F. Basolo, *J. Amer. Chem. Soc.* **94**, 7173 (1972).

45. G. Ferraudi and J. F. Endicott, *J. Amer. Chem. Soc.* **95**, 2371 (1973).

46. J. A. Broomhead and W. Grumley, *Inorg. Chem.* **10**, 2002 (1971).

47. J. A. Broomhead and W. Grumley, *Chem. Commun.*, 1211 (1968).

48. L. H. Berka and G. E. Philippon, *J. Inorg. Nuc. Chem.* **32**, 3355 (1970).

49. G. L. Blackmer, J. L. Sudmeier, R. N. Thibedeau, and R. M. Wing, *Inorg. Chem.* **11**, 189 (1972).

50. (a) P. R. Brookes and B. L. Shaw, *Chem. Commun.*, 919 (1968); (b) P. R. Brookes, C. Masters, and B. L. Shaw, *J. Chem. Soc. A*, 3756 (1971).

51. M. M. Muir and W. L. Huang, *Inorg. Chem.* **12**, 1930 (1973).

52. M. Lamache and F. Larèze, *Compt. Rend.* **275**, 115 (1972).

53. R. E. Clarke and P. C. Ford, *Inorg. Chem.* **9**, 227 (1970).

54. I. M. Treitel, M. T. Flood, R. E. Marsh, and H. B. Gray, *J. Amer. Chem. Soc.* **91**, 6512 (1969).

55. C. Sigwart and J. Spence, *J. Amer. Chem. Soc.* **91**, 3991 (1969).

56. P. C. Ford, D. H. Stuermer, and D. P. McDonald, *J. Amer. Chem. Soc.* **91**, 6209 (1969).

57. (a) D. A. Chaisson, R. E. Hintze, D. H. Stuermer, J. D. Petersen, D. P. McDonald, and P. C. Ford, *J. Amer. Chem. Soc.* **94**, 6665 (1972); (b) P. C. Ford, D. A. Chaisson, and D. H. Stuermer, *Chem. Commun.*, 530 (1971).

58. G. Malouf and P. C. Ford, *J. Amer. Chem. Soc.* **96**, 601 (1974).

59. R. E. Hintze and P. C. Ford, reported at the Symposium on Inorganic Photochemistry, the 164th National Meeting of the American Chemical Society, New York, August 1972.

60. P. Natarajan and J. F. Endicott, *J. Amer. Chem. Soc.* **94**, 5909 (1972).

61. P. P. Zarnegar, C. R. Bock, and D. G. Whitten, *J. Amer. Chem. Soc.* **95**, 4367 (1973).

62. J. N. Demas and A. W. Adamson, *J. Amer. Chem. Soc.* **93**, 1800 (1971).

63. J. N. Demas, Proceedings of the XIVth International Conference on Coordination Chemistry, Toronto, p. 166, 1972.

64. P. Natarajan and J. Endicott, *J. Amer. Chem. Soc.* **95**, 2470 (1973).

65. N. Sabbatini and V. Balzani, *J. Amer. Chem. Soc.* **94**, 7587 (1972).

66. F. Bolletta, M. Maestri, and L. Moggi, *J. Phys. Chem.* **77**, 861 (1973).

67. J. N. Demas and A. W. Adamson, *J. Amer. Chem. Soc.* **95**, 5159 (1973).

68. J. N. Demas, D. Diemente, E. W. Harris, *J. Amer. Chem. Soc.* **95**, 6864 (1973).

69. H. Gafney and A. W. Adamson, *J. Amer. Chem. Soc.* **94**, 8238 (1972).

70. (a) P. Natarajan and J. Endicott, *J. Phys. Chem.* **77**, 971 (1973); (b) *ibid.*, 1823 (1973).

71. F. E. Lytle and D. M. Hercules, *Photochem. Photobiol.* **13**, 123 (1971).

72. N. E. Tokel-Takvoryan, R. E. Hemingway, and A. J. Bard, *J. Amer. Chem. Soc.* **95**, 6582 (1973).

73. R. L. Rich and H. Taube, *J. Amer. Chem. Soc.* **76**, 2608 (1954).

74. R. Dreyer, K. König, and H. Schmidt, *Z. Phys. Chem.* (*Leipzig*) **227**, 257 (1964).

75. L. E. Cox, D. G. Peters, and E. L. Wehry, *J. Inorg. Nucl. Chem.* **34**, 297 (1972).

76. F. Basolo and R. G. Pearson, *Mechanisms of Inorganic Reactions*, 2nd ed., Chap. 6, Wiley, New York 1967.

77. R. C. Wright and G. S. Laurence, *J.C.S. Chem. Commun.*, 132 (1972).

78. G. E. Adams, R. B. Broszkiewicz, and B. D. Michael, *Trans. Faraday Soc.* **64**, 1256 (1968).

79. V. Balzani, F. Manfrin, and L. Moggi, *Inorg. Chem.* **6**, 354 (1967).

80. V. Balzani and V. Carassiti, *J. Phys. Chem.* **72**, 383 (1968).

81. A. W. Adamson and A. H. Sporer, *J. Amer. Chem. Soc.* **80**, 3865 (1958).

82. (a) S. A. Penkett and A. W. Adamson, *J. Amer. Chem. Soc.* **87**, 2514 (1965); (b) P. D. Fleischauer, Ph.D. Thesis, University of Southern California, 1968, as quoted in Ref. 1.

83. D. Valentine, *Advan. Photochem.* **6**, 123 (1968).

84. C. K. Jørgensen, *Absorption Spectra and Chemical Bonding in Complexes*, p. 298, Addison-Wesley, Reading, Mass., 1962.

85. H. Hartman and C. Bushbeck, *Z. Phys. Chem.* (*Frankfurt*) **11**, 120 (1957).

86. L. Moggi, G. Varani, M. F. Manfrin, and V. Balzani, *Inorg. Chim. Acta* **4**, 335 (1970).

87. W. L. Wells and J. F. Endicott, *J. Phys. Chem.* **75**, 3075 (1971).

88. A. Ohyoshi, N. Takebayashi, and Y. Hiroshima, *Chem. Letters*, 675 (1973).

89. (a) M. Koyama, Y. Kanchiku, and T. Fujinaga, *Coord. Chem. Rev.* **3**, 285 (1968); (b) T. Fujinaga, M. Koyama, and Y. Kanchiku, *Bull. Chem. Soc. Jap.* **40**, 2970 (1967); (c) Y. Kanchiku, *Bull. Chem. Soc. Jap.* **42**, 2831 (1969).

90. (a) H. Basch and H. B. Gray, *Inorg. Chem.* **6**, 365 (1967); (b) D. S. Martin, Jr., M. A. Tucker, and A. J. Kassman, *Inorg. Chem.* **5**, 1298 (1966).

91. D. L. Webb and L. A. Rossiello, *Inorg. Chem.* **9**, 2622 (1970).

92. F. Basolo and R. G. Pearson, *Mechanisms of Inorganic Reactions*, 2nd ed., pp. 351–453, Wiley, New York, 1967.

93. F. Scandola, O. Traverso, and V. Carassiti, *Mol. Photochem.* **1**, 11 (1969).

94. V. S. Sastri and C. H. Langford, *J. Amer. Chem. Soc.* **91**, 7533 (1969).

95. W. M. Bedford and G. Roushias, *Chem. Commun.*, 1224 (1972).

96. P. Natarajan and A. W. Adamson, *J. Amer. Chem. Soc.* **93**, 5599 (1971).

97. (a) P. Haake and T. A. Hylton, *J. Amer. Chem. Soc.* **84**, 3774 (1962); (b) S. H. Mastin and P. Haake, *Chem. Commun.*, 202 (1970); (c) P. Haake, S. H. Mastin, P. Lui, and C. C. Deatherage, Abstracts of Papers, 164th ACS National Meeting, p. INORG. 71, New York, August 1972.

98. R. Ellis, T. A. Weil, and M. Orchin, *J. Amer. Chem. Soc.* **92**, 1078 (1970).

99. L. Moggi, G. Varani, N. Sabbatini, and V. Balzani, *Mol. Photochem.* **3**, 141 (1971).

100. V. Balzani, V. Carassiti, L. Moggi, and F. Scandola, *Inorg. Chem.* **4**, 1243 (1965).

101. F. Scandola, O. Traverso, V. Balzani, G. L. Zucchini, and V. Carassiti, *Inorg. Chim. Acta* **1**, 76 (1967).

102. F. Bolletta, M. Gleria, and V. Balzani, *J. Phys. Chem.* **76**, 3934 (1972).

103. F. Bolletta, M. Gleria, and V. Balzani, *Mol. Photochem.* **4**, 205 (1972).

104. F. S. Richardson, D. D. Schillady, and A. Waldrop, *Inorg. Chim. Acta* **5**, 279 (1971).

105. D. M. Blake and C. J. Nyman, *J. Amer. Chem. Soc.* **92**, 5359 (1970); *ibid.*, *Chem. Commun.*, 483 (1969).

106. D. M. Blake and R. Mersecchi, *Chem. Commun.*, 1045 (1971).

107. W. L. Waltz and A. W. Adamson, *J. Phys. Chem.* **73**, 4250 (1969).

108. W. L. Waltz, A. W. Adamson, and P. D. Fleischauer, *J. Amer. Chem. Soc.* **89**, 3923 (1967).

109. M. Shirom and Y. Siderer, *J. Chem. Phys.* **57**, 1053 (1972).

110. J. R. Perumareddi, A. D. Liehr, and A. W. Adamson, *J. Amer. Chem. Soc.* **85**, 249 (1963).

111. A. Bettelheim and M. Shirom, *Chem. Phys. Letters* **9**, 166 (1971).

112. S. J. Lippard and B. J. Russ, *Inorg. Chem.* **6**, 1943 (1967).

113. A. W. Adamson and J. R. Perumareddi, *Inorg. Chem.* **4**, 247 (1965).

114. Arguments summarized in V. Balzani and V. Carassiti, *Photochemistry of Coordination Compounds*, pp. 123–132, Academic, London, 1970.

115. V. Balzani, M. F. Manfrin, and L. Moggi, *Inorg. Chem.* **8**, 47 (1969).

116. L. Moggi, F. Bolletta, V. Balzani, and F. Scandola, *J. Inorg. Nucl. Chem.* **28**, 2589 (1966).

117. R. P. Mitra, B. K. Sharma, and H. Mohan, *Can. J. Chem.* **47**, 2317 (1969).

118. R. P. Mitra, B. K. Sharma, and H. Mohan, *Aust. J. Chem.* **25**, 499 (1972).

119. R. D. Wilson, V. S. Sastri, and C. H. Langford, *Can. J. Chem.* **49**, 679 (1971).

120. J. R. Perumareddi, Ph.D. dissertation, University of Southern California, 1962, as reported in Ref. 1.

121. (a) V. Carassiti and V. Balzani, *Ann. Chim. (Rome)* **51**, 518 (1961), as reported in Ref. 2; (b) V. Balzani and V. Carassiti, *ibid.*, 533 (1961), as reported in Ref. 2.

122. G. W. Gray and J. T. Spence, *Inorg. Chem.* **10**, 2751 (1971).

123. W. M. Latimer, *Oxidation Potentials*, 2nd ed., p. 253, Prentice-Hall, Englewood Cliffs, N.J., 1952.

124. B. N. Figgis, *Introduction to Ligand Fields*, pp. 324–333, Wiley-Interscience, New York, 1967.

125. K. Kustin and J. Swinehart, *Progr. Inorg. Chem.* **13**, 107 (1970).

126. (a) J. C. Barnes, *J. Chem. Soc.*, 3880 (1964); (b) C. K. Jørgensen and J. S. Brinen, *Mol. Phys.* **6**, 629 (1963); (c) C. K. Jørgensen, *Mol. Phys.* **5**, 271 (1962 .

127. C. A. Parker, *Photoluminescence of Solutions*, pp. 470–481, Elsevier, Amsterdam 1968.

128. M. Fred, *Lanthanide/Actinide Chemistry*, pp. 180–202, Advances in Chemistry Series, No. 71, American Chemical Society, Washington, D.C., 1967.

129. W. T. Carnall and P. R. Fields, *Lanthanide/Actinide Chemistry*, pp. 86–101, Advances in Chemistry Series, No. 71, American Chemical Society, Washington, D.C., 1967.

130. J. Kleinberg, W. J. Argersinger, and E. Griswold, *Inorganic Chemistry*, pp. 632–634, Heath, Boston, 1960.

131. (a) T. J. Hardwick and E. Robertson, *Can J. Chem.* **29**, 818, 828 (1951); (b) E. L. King and M. L. Pandow, *J. Amer. Chem. Soc.* **74**, 1966 (1952); (c) K. B. Wiberg and P. C. Ford, *Inorg. Chem.* **7**, 369 (1968).

132. E. Wadsworth, F. R. Duke, and C. A. Goetz, *Anal. Chem.* **29**, 1824 (1957).

133. B. D. Blaustein and J. W. Gryder, *J. Amer. Chem. Soc.* **79**, 540 (1957).

134. (a) R. A. Sheldon and J. K. Kochi, *J. Amer. Chem. Soc.* **90**, 6688 (1968); (b) S. P. Rao, T. R. Lodka, and J. N. Gaur, *Naturwissenschaften* **48**, 404 (1961).

135. (a) E. Baur, *Z. Phys. Chem.* **63**, 683 (1908); (b) L. J. Heidt and M. E. Smith, *J. Amer. Chem. Soc.* **70**, 2476 (1948).

136. L. J. Heidt and A. F. McMillan, *J. Amer. Chem. Soc.* **76**, 2135 (1954).

137. L. J. Heidt and A. F. McMillan, *Science* **117**, 75 (1953).

138. M. G. Evans and N. Uri, *Nature* (*London*) **166**, 602 (1950).

139. P. N. Moorthy and J. J. Weiss, *J. Chem. Phys.* **42**, 3127 (1965).

140. (a) T. J. Sworski, *J. Amer. Chem. Soc.* **77**, 1074 (1955); (b) *ibid.*, 3655 (1957).

141. T. W. Martin, R. E. Rummell, and R. C. Gross, *J. Amer. Chem. Soc.* **86**, 2595 (1964).

142. L. Dogliotti and E. Hayon, *J. Phys. Chem.* **71**, 3802 (1967).

143. T. W. Martin, L. L. Swift, and J. H. Venable, *J. Chem. Phys.* **52**, 2138 (1970).

144. For a review, see W. H. Richardson, in *Oxidation in Organic Chemistry*, (K. B. Wiberg, Ed.), pp. 243–276, Academic, New York, 1965.

145. T. W. Martin, J. M. Burk, and A. Henshall, *J. Amer. Chem. Soc.* **88**, 1097 (1966).

146. (a) D. Greatorex and T. J. Kemp, *Chem. Commun.*, 383 (1969); (b) D. Greatorex and T. J. Kemp, *Trans. Faraday Soc.* **67**, 56 (1971); (c) *ibid.*, **67**, 1576 (1971); (d) *ibid.*, **68**, 121 (1972).

147. M. Hrekovov, D. Feldman, and C. Sinionescu, *Rev. Roum. Chim.* **10**, 77 (1965); *ibid.*, *Stud. Cercet. Chim.* **14**, 77 (1965), as reported in Ref. 2, p. 318.

148. F. S. Dainton and D. G. L. James, *Trans. Faraday Soc.* **54**, 649 (1958).

149. D. L. Douglas and D. M. Yost, *J. Chem. Phys.* **17**, 1345 (1949); *ibid.*, **18**, 1687 (1950).

150. (a) Y. Haas, G. Stein and M. Tomkiewicz, *J. Phys. Chem.* **74**, 2558 (1970); (b) Y. Haas, G. Stein and R. Tenne, *Israel J. Chem.* **10**, 529 (1972).

151. (a) N. K. Dalley, M. H. Mueller, and S. H. Simonsen, *Inorg. Chem.* **10**, 323 (1971); (b) I. I. Chernyeav, *Complex Compounds of Uranium* (translated from Russian by L. Mandel), Davey, New York, 1966.

152. E. Rabinowitch and R. L. Belford, *Spectroscopy and Photochemistry of Uranyl Compounds*, Macmillan, New York, 1964.

153. R. Matsushima, *Chem. Letters*, 115 (1973).

154. J. T. Bell and R. E. Biggers, *J. Mol. Spectrosc.* **18**, 247 (1965); *ibid.*, **22**, 262 (1967); *ibid.*, **25**, 312 (1968).

155. J. C. Barnes and P. Day, *J. Chem. Soc.*, 3886 (1964).

156. (a) W. G. Leighton and G. S. Forbes, *J. Amer. Chem. Soc.* **52,** 3139 (1930); (b) F. P. Brackett, Jr., and G. S. Forbes, *ibid.*, **55,** 4459 (1933); (c) B. M. Norton, *ibid.*, **56,** 2294 (1934); (d) G. S. Forbes and L. J. Heidt, *ibid.*, **56,** 2363 (1934); (e) C. A. Discher, P. F. Smith, I. Lippman, and R. Turse, *J. Phys. Chem.* **67,** 2501 (1963).

157. L. J. Heidt, G. W. Tregay, and F. A. Middleton, Jr., *J. Phys. Chem.* **74,** 1876 (1970).

158. D. H. Volman and J. R. Seed, *J. Amer. Chem. Soc.* **86,** 5095 (1964).

159. J. N. Pitts, Jr., J. D. Margerum, R. P. Taylor, and W. Brim, *J. Amer. Chem. Soc.* **77,** 5499 (1955).

160. K. Porter and D. H. Volman, *J. Amer. Chem. Soc.* **84,** 2011 (1962).

161. C. A. Parker and C. G. Hatchard, *J. Phys. Chem.* **63,** 22 (1959).

162. S. Sakuraba and R. Matsushima, *Bull. Chem. Soc. Jap.* **43,** 1950 (1970).

163. S. Sakuraba and R. Matsushima, *Bull. Chem. Soc. Jap.* **44,** 1278 (1971).

164. (a) S. Sakuraba and R. Matsushima, *Bull. Chem. Soc. Jap.* **43,** 2359 (1970); (b) *ibid.*, **44,** 2915 (1971); (c) R. Matsushima and S. Sakuraba, *J. Amer. Chem. Soc.* **94,** 2622 (1972); (d) *ibid.*, **93,** 5421 (1971); (e) R. Matsushima, *J. Amer. Chem. Soc.* **94,** 6010 (1972).

165. D. Greatorex, R. J. Hill, T. J. Kemp, and T. J. Stone, *J.C.S. Faraday I,* **68,** 2059 (1972).

166. D. M. Allen, H. D. Barrows, A. Cox, R. J. Hill, T. J. Kemp, and T. J. Stone, *J.C.S. Chem. Commun.*, 59 (1973).

6

PHOTOCHEMISTRY OF
CARBONYL COMPLEXES

Arnd Vogler

Fachbereich Chemie
Universität Regensburg
Germany

I. INTRODUCTION

The photosensitivity of metal carbonyls has been known almost as long as the class of coordination compounds itself. Among no other group of inorganic compounds may one find so many light-sensitive materials. Hence photochemical reactions of metal carbonyls have found wide applications for synthetic purposes. However, whereas much research has been done to understand the thermal reactions,[1-3] the mechanism leading to photochemical reactions of metal carbonyls is not yet well investigated. The papers that have appeared on the photochemistry of metal carbonyls are largely restricted to preparative aspects, although some preliminary attempts have been made to correlate the photochemical reactions with the electronic structures of these compounds. Two excellent reviews of the photochemistry of metal carbonyls are of special importance.[4,5] In 1964 the first review was published by Strohmeier,[6] who did pioneering work in this field, and in 1969 von Gustorf and Grevels presented a comprehensive survey on this subject.[7] During the last few years considerable progress has been made, particularly in the investigation of primary photochemical steps. However, the lack of sufficient information on the complicated electronic structures of metal carbonyls remains an obstacle to the understanding of their photochemical reactions.

Since the metal in most carbonyl complexes, such as $Cr(CO)_6$, has the formal oxidation state, zero, a simple electrostatic picture can not account for the stability of these compounds. Only MO theory provides a model that is in agreement with the chemical and physical properties of metal

269

carbonyls. The necessary condition to form a stable complex with a metal in a low oxidation state is that the ligands be able to form π acceptor bonds in addition to σ donor bonds. The ligands must provide empty π orbitals that are low enough to interact with lower lying filled $d\pi$ orbitals of the metal. The occupied, nonbonding $d\pi$ orbitals are thus lowered in energy to give π-bonding MOs that contribute very much to the stability of the complex. The metal character of this π-bonding MO may still predominate, but some electron density of the metal is shifted toward the ligands. The extension of the metal d orbitals toward the ligands thus introduces an appreciable covalent character into the metal-ligand bond. This delocalization of d-electron density into the ligands induces an electrostatic attraction between ligands and metal. It follows that the formation of π acceptor bonds ("back donation" of charge), in turn, facilitates increased σ-bonding, which again contributes to the stability of the complex. Both effects, π- and σ-bonding, are responsible for the very large d orbital splittings that occur in metal carbonyls.

It is generally accepted that this picture describes the bonding situation of metal carbonyls qualitatively. However, the quantitative description is still controversial among investigators in this field. Although different semiempirical MO calculations seem to agree with the experimental data obtained for a certain metal carbonyl, the detailed interpretation of the nature of the bonding varies with the calculational procedure used. For a critical survey of the different methods and their limitations the reader is referred to a paper by Fenske.[8]

Photoelectron spectroscopy provides a good tool to determine the origin and the energies of the occupied MOs; ground electronic states are fairly well described. Much less reliable information is available on the empty MOs. The large d orbital splittings and the introduction of low-lying antibonding $\pi^*(CO)$ states of different symmetries lead to an accumulation of many empty MOs in a narrow energy range compared to complexes with σ-bonding ligands only. It follows that the absorption spectra are ill resolved because many electronic transitions give rise to absorption bands in the same wavelength region. Weak bands may not be identified at all. None of the absorption spectra of metal carbonyls show well-separated ligand-field bands. In addition, the intensity criterion for the distinction between LF and CT bands may not work very well because Laporte forbidden transitions may become more allowed in metal carbonyls due to very effective vibronic coupling with allowed transitions as a consequence of the strong covalent interaction between ligand and metal. In any case, many reasonable assignments of absorption bands of metal carbonyls were made by the application of temperature- and solvent-dependent absorption spectroscopy and spectroscopy with polarized light on the basis of selection rules and similar restrictions. In particular, the work of H. B. Gray has contributed much to the understanding of the electronic spectra of metal carbonyls.

Very little information is available on the photophysical processes following light absorption. Emission spectra have been obtained only for a few substituted metal carbonyls. It does seem that higher excited states of metal carbonyls are deactivated to the lowest excited state as the precursor of a photochemical reaction since, in some important cases, the quantum yield has been shown to be wavelength independent. The deactivation of higher excited states may be facilitated by the strong interaction of states of different origin and symmetry, Hence our discussion of excited states is largely confined to the lowest excited states of metal carbonyls. This restriction is also justified by the fact that most of the photochemical work is limited to the irradiation of the long-wavelength bands of metal carbonyls. In the following sections, an attempt is made to explain the photochemical reactions of metal carbonyls on the basis of the nature of those excited states that may initiate the observed photoreactions. Hence many interesting photochemical reactions are not mentioned here if the available information is not sufficient to allow a meaningful conclusion or at least a reasonable speculation about the mechanism of a photochemical reaction. The reader is referred to the reviews cited above if he is interested only in preparative aspects of the photochemistry of metal carbonyls.

II. HEXACARBONYLS OF CHROMIUM, MOLYBDENUM, AND TUNGSTEN

Upon irradiation of $M(CO)_6$ (M = Cr, Mo, W) dissolved in organic solvents, the release of one CO ligand is observed. The mechanism of this reaction may be closely related to that of the photochemical substitution of one CN^- group upon irradiation of the isoelectronic $Co(CN)_6^{3-}$ in aqueous solutions.[9] The photoactive excited state of $Co(CN)_6^{3-}$ was shown to be of the ligand-field type since the photochemical reaction was produced upon irradiation of the LF bands that are well separated from and lower in energy than the lowest CT bands. But the photoactive LF state is probably not the singlet that is reached directly by a spin-allowed transition from the ground state. The lowest LF triplet state that may be populated by intersystem crossing seems to be the precursor of the photoaquation; this state was shown to be the active one for the sensitized photolysis of $Co(CN)_6^{3-}$.[10]

A. Electronic Structure

$Co(CN)_6^{3-}$ and $M(CO)_6$ are both octahedral complexes with six d electrons at the central metal. Carbon monoxide and CN^- are both ligands that are able to form π acceptor bonds. But the extent of π-bonding and hence of electron delocalization is much larger for $M(CO)_6$ due to the low oxidation state of M and the better π acceptor properties of CO. It is thus not surprising that the electronic structures of $Co(CN)_6^{3-}$ and $M(CO)_6$ differ in some important details.[11] The influence of π-bonding in $M(CO)_6$ is shown in a simplified MO scheme (Fig. 6-1).

Photoelectron spectra show that the highest occupied orbitals of $M(CO)_6$ are π-bonding t_{2g} orbitals.[13] These t_{2g} orbitals are assumed to be composed of 75% d_{xy}, d_{yz}, d_{xz} and 25% $\pi^*(CO)$ orbitals in the case of $Cr(CO)_6$.[14] The absorption spectra of $M(CO)_6$ species[15] exhibit two intense bands at around 35,000 cm^{-1} and 44,000 cm^{-1}. Most importantly, the first band is unsymmetric and shows some structure on the low-energy tail. The close spacing and small resolution of these shoulders have made assignments quite difficult.

A detailed interpretation of the $M(CO)_6$ spectra was given by Beach and Gray.[15] The long-wavelength shoulder of $W(CO)_6$ and $Mo(CO)_6$ was assigned to the lowest spin-forbidden ($^1A_{1g} \rightarrow {}^3T_{1g}$) LF transition. A similar assignment of a weak long-wavelength absorption of $Cr(CO)_6$ was made in an older study.[16] The next two shoulders were attributed as vibrational structure of the lowest spin-allowed ($^1A_{1g} \rightarrow {}^1T_{1g}$) LF transition ($t_{2g} \rightarrow e_g$ in Fig. 6-1). Both intense bands, around 35,000 cm^{-1} and 44,000 cm^{-1}, were assigned to symmetry- and spin-allowed CT transitions of the metal-to-ligand $t_{2g} \rightarrow \pi^*(CO)$ type. However, the position of the band maximum near 35,000 cm^{-1} was almost solvent independent, whereas the band near 44,000 cm^{-1} showed a solvent-dependent shift. In addition, the solvent-dependent band was about ten times as intense as the other one. These facts were taken as evidence that the band near

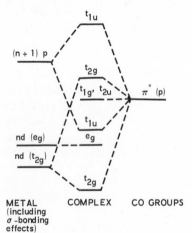

METAL (including σ-bonding effects)   COMPLEX   CO GROUPS

Fig. 6-1. Simplified molecular orbital diagram for typical $M(CO)_6$ complex.

35,000 cm^{-1} belongs to the $t_{2g} \rightarrow t_{1u}$ CT transition; this transition is not a complete electron transfer since the $t_{1u}\pi^*$(CO) orbital is not a pure ligand orbital due to the admixture of metal character. The more intense band around 44,000 cm^{-1} was assigned to the $t_{2g} \rightarrow t_{2u}$ CT transition. This transition is associated with an almost complete electron transfer to the ligand since the t_{2u} orbital is a π^*(CO) orbital that does not interact with metal orbitals. The other possible low-energy $t_{2g} \rightarrow \pi^*$(CO) CT transitions are symmetry forbidden. Their absorption bands are difficult to identify. It should be mentioned here that different assignments were made by Schreiner and Brown.[17] These authors calculated the t_{2u} orbital as the lowest empty MO. According to this assumption, the longest wavelength absorption of Cr(CO)$_6$ was assigned to the $t_{2g} \rightarrow t_{2u}$ CT transition.

B.   Photochemical Behavior

The quantum yield for the photodissociation of M(CO)$_6$ to M(CO)$_5$ and CO was found to be unity and independent of the solvent.[6,18] It seems likely that the excitation of higher CT states leads to a complete internal conversion to the lowest excited state as the precursor of the photochemical reaction. Since the lowest excited state is assumed to be a LF state, the release of a CO ligand may be caused by the labilization of the M—CO bond that occurs if an electron is removed from bonding $d\pi$ orbitals and added to antibonding $d\sigma$ orbitals. This assumption is supported by another observation. Evidence was obtained that the reduction of M(CO)$_6$ by sodium, which should lead to an occupation of the antibonding e_g orbitals, causes a labilization of M(CO)$_6$ and the subsequent formation of [M(CO)$_5$]$^-$.[19] The internal conversion from higher CT states to the lowest LF state, a process that is not common for the metal ammines,[4,5] may be facilitated by the strong mixing of LF and CT states. In analogy to Co(CN)$_6^{3-}$, the photoactive excited state of M(CO)$_6$ may be expected to be the lowest excited LF triplet. In a sensitized photoreaction in benzene it was shown that the energy of the lowest excited triplet of benzophenone can be transferred to the lowest triplet of Cr(CO)$_6$,[20] which then undergoes the well-known photodissociation with a limiting quantum yield of one. It was assumed that the direct or unsensitized photolysis also originates from this triplet state, which may be reached via intersystem crossing from excited singlet states.

The absence of any photoredox reaction upon CT excitation of M(CO)$_6$ may be taken as evidence for the complete internal conversion to the lowest excited LF state as precursor for the photochemical release of CO. However, in the case of Ni(CO)$_4$ (see below), it can be shown that the release of CO must be initiated by a metal-to-CO CT excited state. Hence the possibility that CT excitation of $M(CO)_6$ leads directly to the observed photochemical reaction cannot be excluded at present. In the lowest $d\pi$ to $\pi^*(CO)$ CT transition ($t_{2g} \rightarrow t_{1u}$) no extensive charge separation is achieved due to the large degree of electron delocalization. The photodissociation may simply occur because an electron is transferred from a π-bonding to a π-antibonding MO of $M(CO)_6$. Irradiation of the very intense and solvent-dependent band near 44,000 cm^{-1}, which corresponds to a CT transition from a $d\pi$ to a noninteracting $\pi^*(CO)$ orbital, also leads to the release of CO.[21] The behavior may be due to the deactivation of this excited state to lower ones which initiate the photodissociation, since the transfer of an electron to an noninteracting $\pi^*(CO)$ orbital should not labilize M—CO bonds. Photoelectron production that might be expected to occur upon irradiation of the solvent-dependent band is not likely to be observed, according to rules of Waltz and Adamson.[22]

The nature of the $M(CO)_5$ fragment produced in the primary step of the photolysis of $M(CO)_6$ has been investigated extensively. Strohmeier, in the early 1960s, studied the photodissociation of $M(CO)_6$ to give $M(CO)_5$ and CO.[23] A thermal back-reaction may regenerate $M(CO)_6$. However, in the presence of a potential ligand, L, the $M(CO)_5$ fragment can react with L to yield stable $M(CO)_5L$. Irradiation of $M(CO)_6$ in isopentane methylcyclohexane/glasses at 77°K gave a new species that was assumed to be $M(CO)_5$.[24] Infrared data were consistent with a C_{4v} square pyramidal structure of this fragment. Slow warming led to a structure change of C_{4v} $M(CO)_5$ at the softening point of the glass. According to infrared measurements, a D_{3h} trigonal bipyramidal $M(CO)_5$ was formed. On further warming the solution became fluid, and $M(CO)_5$ recombined with the released CO to regenerate $M(CO)_6$. The photolysis of $M(CO)_6$ is accompanied by a reversible change from colorless to yellow, reversed in the thermal back-reaction. This photochromic behavior was demonstrated with $Cr(CO)_6$ dissolved in methyl methacrylate polymers[25] and investigated by flash photolysis of $Cr(CO)_6$ in fluid solutions and polymers.[26]

In contrast to the above results it was shown only recently that even at low temperatures in rigid media an isolated C_{4v} $M(CO)_5$ fragment probably does not exist.[17,21,28] One of the molecules in the medium is assumed to fill the coordination gap. Since any molecule present in the solvent or matrix cage of $M(CO)_6$ may photosubstitute one CO ligand, it was suggested that the released CO escapes from the solvent cage. The absorption spectrum and the stability of $M(CO)_5A$, where A denotes a medium molecule, varies with the ligand properties of A. If $M(CO)_6$ was photolyzed in methyltetrahydrofurane (MTHF) glass at 90°K, $M(CO)_5MTHF$ was formed.[28] On melting the glass no immediate change occurred since MTHF is a moderately good ligand. On further standing a slow regeneration of $M(CO)_6$ took place.

In a series of very sophisticated experiments, J. J. Turner and co-workers studied the photolysis products of $M(CO)_6$ in gas matrices at low temperatures.[21,27,29] In an argon matrix a $M(CO)_5$ species was formed. Its infrared spectrum was consistent with a square pyramidal structure. In the case of chromium the $M(CO)_5$ fragment exhibited a long-wavelength absorption at 542 nm. The interaction of $M(CO)_5$ with an argon atom that may fill the coordination sphere to give $M(CO)_5Ar$ must be very small due to the negligible coordinating ability of an inert gas. With increasing interaction in other media the long-wavelength absorption of $M(CO)_5$ was shifted to higher energies. Interestingly, the absorption spectra of photolyzed $Cr(CO)_6$ in an alkane glass ($\lambda_{max} = 485$ nm) and in a methane matrix ($\lambda_{max} = 492$) were very similar. The comparison of the spectra of $Cr(CO)_5$ in argon and methane matrices indicates a considerable interaction between methane and $Cr(CO)_5$. The existence of $Cr(CO)_5CH_4$, which, however, is stable only at low temperatures, may not be too surprising since stable transition-metal complexes with BH_4^- as ligand are well known[30]; BH_4^- is isoelectronic with CH_4. Evidence for the formation of $M(CO)_5N_2$ was found if $M(CO)_6$ was photolyzed in a nitrogen matrix at low temperatures.[21] If $W(CO)_6$ was photolyzed at 20°K in an argon matrix doped with $C^{18}O$, two products were formed: a $W(CO)_5$ fragment identical with that obtained in pure argon and $W(CO)_5(C^{18}O)$. All these experiments provide sufficient evidence that any molecule present in the solvent cage of $M(CO)_6$ may fill the coordination gap of $M(CO)_5$. It follows that the released CO should have enough kinetic energy to escape from the solvent cage even in rigid media. Otherwise, a recombination may be expected.

The transformation of the square pyramidal $M(CO)_5$ or, better, $[M(CO)_5(solvent)]$ to the trigonal bipyramidal isomer at the softening point of alkane glasses was rejected recently.[27,28] Although proof is still missing, it was suggested that in fluid solutions polynuclear complexes are formed.[28] Thus $(OC)_5W(CO)W(CO)_5$ was proposed to be the long-lived intermediate in the photolysis of $W(CO)_6$ in inert solvents at room temperature, thereby accounting for the slow regeneration of $W(CO)_6$ even in the presence of an excess of CO.[21] A weakly interacting solvent molecule as ligand in $M(CO)_5L$ should be replaced easily by CO. The binuclear complex could be formed by the reaction of $W(CO)_5$ with $W(CO)_6$, which may coordinate via oxygen of one of its CO ligands to the $W(CO)_5$ fragment.

In addition to the low-temperature photolysis of $M(CO)_6$ in rigid media, attempts have been made to characterize the primary photoproduct of

M(CO)$_6$ by flash photolysis. Nasielski and co-workers observed two transients.[31] The first one ($\lambda_{max} = 483$ nm) did not react with CO. It was converted with a half-life of 6 msec to the second transient ($\lambda_{max} = 445$ nm), which decayed only slowly by second-order kinetics. Added CO accelerated the disappearance of the second transient and the regeneration of Cr(CO)$_6$. These intermediates were assumed to be the C_{4v} and D_{3h} isomers of Cr(CO)$_5$ since their absorption spectra were similar to those obtained in the early study of the photolysis of Cr(CO)$_6$ in low-temperature glasses of hydrocarbons.[24] However, the existence of a D_{3h} isomer of Cr(CO)$_5$ is doubtful (see p. 274). In addition, it is hard to explain why this isomer, but not the C_{4v} form, should react with CO to regenerate Cr(CO)$_6$. Finally, the suggestion that the second transient may be a binuclear chromium carbonyl,[21,27] which must be formed in a bimolecular reaction, is not consistent with the kinetic data which indicate a unimolecular conversion of the first to the second transient.

The results of a very recent flash-photolysis study of Cr(CO)$_6$, which were obtained with a higher time resolution, indicated that neither transient is the proposed C_{4v} or D_{3h} isomer of Cr(CO)$_5$; instead, they are products formed by the reaction of another very short-lived intermediate with impurities of the solvent.[32] The photolysis of Cr(CO)$_6$ in cyclohexane produced the initial transient with a lifetime >200 μsec. This transient reacts either with CO or with impurities of the solvent. Although this newly detected intermediate is assumed to be the true Cr(CO)$_5$ species produced in the primary step of the photolysis of Cr(CO)$_6$ in solution, its real identity is also not known since its absorption maximum at 503 nm does not correspond to those obtained by low-temperature photolysis of Cr(CO)$_6$ in alkane glasses or rare-gas matrices.

III. SUBSTITUTED HEXACARBONYLS OF CHROMIUM, MOLYBDENUM, AND TUNGSTEN

A. Absorption Spectra

The longest wavelength absorption of monosubstituted hexacarbonyls M(CO)$_5$L is shifted to lower energies compared to M(CO)$_6$.[33-36] Whereas this red shift is small if L is a phosphine or another ligand with good π-accepting properties, the long-wavelength absorption appears at much lower energies if L is a simple σ-donating ligand as an amine, ether, or ketone ($\sim$425 nm for Cr(CO)$_5$L [33]). A further substantial shift to lower energy was observed for the first absorption band of the M(CO)$_5$ fragment obtained as photoproduct of M(CO)$_6$ in rare-gas matrices[21,27,29]; this fragment may be regarded as M(CO)$_5$L, where L is a rare-gas atom with negligible interaction with the metal. The longest wavelength maximum of $Cr(CO)_5$ was found at 542 nm, but was later corrected to be at 623 nm.[33]

These long-wavelength absorptions of $M(CO)_5L$ have been assigned either to d–d[33,36–38] or to d–π*(CO) CT transitions.[35,35] Although the fairly high intensity of these bands ($\varepsilon > 1000$ M^{-1} cm^{-1}) may favor a CT assignment, the wavelength dependence on the nature of L is much better explained by a d–d assignment. The degeneracy of the t_{2g} and e_g orbitals of $M(CO)_6$ is removed in $M(CO)_5L$ of C_{4v} symmetry. The t_{2g} states are split into e ($d_{xz,yz}$) and b_2 (d_{xy}), while the e_g states give a_1 (d_{z^2}) and b_1 ($d_{x^2-y^2}$).

The observation that all $W(CO)_5L$ complexes exhibit their first absorption band at nearly the same wavelength, if L is a σ-donating ligand, was taken as evidence that the axial ligand field is weak and dominated by the CO trans to L.[36–38] The ordering, b_2, e, a_1, b_1, means that the longest wavelength absorption should belong to the lowest energy transition $(e^4 b_2^2) \rightarrow (e^3 b_2^2 a_1)$. The assumption of a weak axial ligand field dominated by the CO *trans* to L requires that the $W(CO)_5$ fragment of C_{4v} symmetry and other $W(CO)_5L$ complexes (L = σ-donating ligand) all show their long-wavelength absorption at the same energy. While this absorption is only moderately shifted for $W(CO)_5$ (440 nm)[21] compared to $W(CO)_5L$ ($\sim$400 nm)[36] the shift is much more pronounced for the analogous chromium complexes (see above).

There is no obvious reason why the electronic structures of $Cr(CO)_5L$ should be completely different from those of analogous $W(CO)_5L$. Therefore, another model has been proposed that seems to be in better agreement with the experimental data.[33] The interaction of the four equatorial CO ligands with the metal is assumed to remain essentially unaffected by the axial ligand, L. It follows that the $b_2(d_{xy})$ and $b_1(d_{x^2-y^2})$ orbitals should have the same energies as the t_{2g} and e_g orbitals of $M(CO)_6$. The $e(d_{xz,yz})$ orbitals are lifted above the b_2 orbital according to the poorer π-accepting property of L. The $a_1(d_{z^2})$ orbital is lowered below the b_1 orbital if L is a rare-gas atom not interacting with the a_1 orbital, or is raised above the b_1 orbital if L is a good σ-donating ligand such as an amine.

This interpretation is consistent with the position of the longest-wavelength absorption being dependent on the nature of L. This band corresponds to the $e \rightarrow b_1$ transition, which is shifted to lower energies with decreasing π-accepting capacity of L and reaches a nearly constant value for pure σ-donating ligands. The long-wavelength band is shifted further to the red and assigned to the $e \rightarrow a_1$ transition if L is a rare-gas atom. However, it must be emphasized again that a distinction between LF and CT transitions may become meaningless if the metal orbitals involved in these transitions are extensively mixed with ligand orbitals as indicated by the high extinction coefficients of the corresponding absorption bands.

Complexes of the type, $M(CO)_4L_2$, where L_2 denotes a bidentate aromatic diamine such as 2,2'-bipyridyl (bipy) or 1,10-phenanthroline (phen), exhibit a very broad and intense long-wavelength absorption near 450 nm. The very low energy of the π^* states of L_2 justifies the $d-\pi^*(L_2)$ CT assignment of this band.[39,40] The impressive solvochromic behavior of these complexes with solvent-dependent shifts of the long-wavelength band up to 4000 cm^{-1} gives additional support for the CT assignment.[40,41]

B. Emission Spectra

Metal carbonyls and their derivatives generally do not exhibit any luminescence. Recently, however, some substituted hexacarbonyls have been shown to luminesce at low temperatures. The red emission of $M(CO)_4(bipy)$ (M = Cr, Mo, W) was attributed to the spin-forbidden triplet singlet CT transition from the π^* state of bipy to the $d\pi$ orbitals of the metal.[42] Also, some monosubstituted tungsten carbonyls, $W(CO)_5L$, where L is a simple σ donor (amines, ethers, ketones), luminesce at approximately 530 nm in glasses of methylcyclohexane at 77°K.[36,37] Even $W(CO)_5$ produced by the photolysis of $W(CO)_6$ in methylcyclohexane glasses was found to emit, again at 533 nm. However, the identity of this latter emitting species is not yet clear. First, "true" noninteracting $W(CO)_5$ would be expected to emit at longer wavelength than the other $W(CO)_5L$ complexes because of the shift of the first absorption maximum (see above). Second, $W(CO)_5$ was found to emit only if the CO released in the photolysis of $W(CO)_6$ was removed. To do this the low-temperature glass was probably melted, but it is not yet known to what species $W(CO)_5$ is converted in fluid solution.[21,28] It was shown recently that impurities (possibly ketones), which are always present in very small concentrations in hydrocarbon solvents, can effectively quench $M(CO)_5$.[32] It is possible, therefore, that some $W(CO)_5L$ species caused the emission that was attributed to $W(CO)_5$. The emission of the $W(CO)_5L$ complexes was assumed to be associated with the lowest triplet-to-singlet LF transition, $a_1 \rightarrow e$. However, since the lowest LF transition may be actually $b_1 \rightarrow e$,[33] the emission could belong to the corresponding spin-forbidden transition. It was suggested that at low temperatures all excited species decay by way of the lowest triplet irrespective of excitation wavelength.[37] The observed variation of relative quantum yields and lifetimes of emission of the various $W(CO)_5L$ complexes was attributed to the influence of L on the rate of radiationless deactivation processes. In solution at room temperature no luminescence was observed. It was suggested that this quenching is associated with fast photochemical reactions that successfully compete with emission. The emitting triplet state was observed not only in emission but also in absorption as a weak shoulder at the long-wavelength tail of the lowest spin-allowed band. This spin-forbidden transition of the corresponding chromium and molybdenum complexes was seen neither in absorption nor in emission. Both effects are probably associated with the smaller spin-orbit coupling of the lighter metals. The absence of emission may be due to the lack of efficient intersystem crossing from higher excited states to the emitting triplet. As another reasonable explanation, one can assume that the triplet is populated by intersystem crossing from higher excited states (this is a quite common behavior even if the population from the ground state is strongly forbidden[43]), but its radiative deactivation may be too slow to compete with other nonradiation processes.

C.   Photochemical Reactions

Excited LF states of $M(CO)_5L$ may lead to a destabilization of M—CO and M—L bonds. The transition, $d_{xz,yz} \rightarrow d_{x^2-y^2}$, is expected to labilize preferentially equatorial M—CO bonds if L is a σ-donating ligand that does not interact with the orbitals involved in this transition. With growing π-accepting capacity of L, the $d_{xz,yz}$ orbitals become increasingly bonding. Hence $d_{xz,yz}$ to $d_{x^2-y^2}$ transitions may also destabilize axial M—CO and M—L bonds. Also, these bonds should be weakened by $d_{xz,yz}$ to d_{z^2} transitions. CT transitions from the metal to $\pi^*(CO)$ states, which should occur at higher energies, could also lead to a labilization of M—CO and M—L bonds if the $\pi^*(CO)$ states are extensively mixed with metal orbitals. Even a CT transition from a d to a noninteracting $\pi^*(CO)$ orbital may initiate a substitution reaction by facilitating a nucleophilic attack of a substituting ligand at the metal.

Both photochemical reactions of $M(CO)_5L$, that is, substitution of L and of CO, are observed. However, the mechanism of these photoreactions and the interdependence of L versus CO substitution is not at all clear. Strohmeier postulated that $M(CO)_5L$ photodissociates into $M(CO)_4L$ and $M(CO)_5$ simultaneously with an overall quantum yield of one.[6,44,45] This assumption was based on two observations. Upon irradiation of $M(CO)_5$(pyridine) in the presence of an excess of pyridine, $M(CO)_4$(pyridine)$_2$ was produced with quantum yields between 0.05 and 0.3. While the first intermediate, $M(CO)_4$(pyridine), yields the product, $[M(CO)_4$(pyridine)$_2]$, the second intermediate, $M(CO)_5$, should react with additional pyridine to give back the starting complex, which does not contribute to the net reaction. If $Cr(CO)_5$(pyridine) was irradiated in the presence of added CO, $Cr(CO)_6$ was regenerated. This observation was taken as evidence for the formation of the second intermediate, $Cr(CO)_5$. The quantum yield of the production of $M(CO)_4(pyridine)_2$ was found to be solvent dependent, and it was considerably lower for excitation at 436 nm than at 366 nm. The postulation that both intermediates are formed with an overall quantum yield of unity leads to the conclusion that the formation of $M(CO)_5$ should be favored at longer wavelengths. This assumption implicates the participation of two different excited states as precursors for both reaction modes.

Basically the same conclusions were drawn from a recent study by Wrighton, Hammond, and Gray.[38] $W(CO)_5(pyridine)$ dissolved in isooctane was found to give cis-$[W(CO)_4(pyridine)_2]$ upon irradiation in the presence of added pyridine. The quantum yield increased with decreasing wavelength of irradiation, from $\phi = 0$ at 436 to $\phi = 0.3$ at 254 nm. In the presence of 1-pentene instead of pyridine, $W(CO)_5(pentene)$ was produced; the quantum yields increased with increasing irradiating wavelength from $\phi = 0.3$ at 254 nm to $\phi = 0.63$ at 436 nm. The substitution of coordinated pyridine by pentene was explained by the assumption that the irradiation (436 nm) of the longest-wavelength band of $W(CO)_5(pyridine)$ is associated with the lowest energy transition ($d_{xz,yz}$ to d_{z^2}) that should labilize the ligands on the z axis. However, it should not be the pyridine but the axial CO ligand that is expected to be dissociated since the axial ligand field was assumed to be dominated by the CO *trans* to pyridine.[36,37] Shorter-wavelength irradiation was suggested to lead to the population of the next higher excited state ($d_{xz,yz}$ to $d_{x^2-y^2}$), which initiates the release of an equatorial CO ligand, yielding cis-$[W(CO)_4(pyridine)_2]$ in the presence of an excess of pyridine. But the internal conversion to the lowest excited state ($d_{xz,yz}$ to d_{z^2}) must be very fast to explain the small quantum yield of CO and the large efficiency of pyridine substitution upon shorter-wavelength irradiation. However, some critical remarks concerning these results still must be made.

The quantum yields of the formation of $W(CO)_4(pyridine)_2$ in isooctane reported by Wrighton, Hammond, and Gray[38] are much lower than those found by Strohmeier and von Hobe[45] for the same reaction in THF or benzene. Since the concentrations of pyridine used in the latter experiments were four times as high as those in the former ones, there is some evidence that the quantum yield is not only solvent dependent but varies also with the concentration of the substituting ligand. The concentration of pentene used for determinations of the quantum yield of formation of $W(CO)_5(pentene)$ was more than fifteen times as high as that of pyridine in the reaction with $[W(CO)_5(pyridine)]$.[38] Hence the significance of these data seems to be doubtful. No explanation was given for the absence of any $[W(CO)_4(pyridine)(pentene)]$, which, according to the mechanism suggested above, should have been formed in addition to [W(CO)$_5$(pentene)] upon shorter-wavelength irradiation of [W(CO)$_5$(pyridine)] in the presence of pentene. The assumption that the production of cis-[W(CO)$_4$(pyridine)$_2$] is directly related to the photodissociation of an equatorial CO ligand of [W(CO)$_5$(pyridine)] seems to be supported by the observation that the photoexchange of [M(CO)$_5$(piperidine)] (M = Mo, W) with labeled CO takes place preferentially at an equatorial coordination position.[46] In the case of M(CO)$_5$[P(C$_6$H$_5$)$_3$] (M = Mo, W), an axial CO was exchanged. It is possible that this difference is associated with the σ-donor character of piperidine and the π-acceptor strength of P(C$_6$H$_5$)$_3$. However, no such relationship is apparent if the substituting ligand is a phosphine or an amine instead of CO.

Other properties, such as steric ones, may be responsible for the stereospecific photoproduction of disubstituted hexacarbonyls. In contrast to the observation that W(CO)$_5$(pyridine) undergoes a photosubstitution of pyridine with high quantum yields, the study by Darensbourg et al.[46] led to the conclusion that the light sensitivity of M(CO)$_5$L complexes, including W(CO)$_5$L, is restricted to M—CO bonds, while the M—L bonds are photochemically very stable. Much of this confusion can be avoided by simultaneous quantum-yield determinations for the conversion of M(CO)$_5$L to M(CO)$_4$LL' and M(CO)$_5$L' under variation of the solvent and determining the nature and the concentration of the substituting ligand. It is clear that L must be different from L' to obtain unambiguous results.

An interpretation of the photochemistry of M(CO)$_5$L complexes can not yet be given because of the various conflicting observations. However, there is also some indication that the photochemical substitution may not proceed by a dissociative release of coordinated ligands but, instead, by an associative reaction of the substituting ligand with the complex in its excited state. A heptacoordinated intermediate may rearrange to yield M(CO)$_4$LL' or M(CO)$_5$L'. In contrast to the pure hexacarbonyls, their monosubstituted derivatives have not been observed to undergo any photodissociation in inert solvents. In addition, while M(CO)$_5$L complexes were clearly demonstrated to undergo a photosubstitution of CO in the presence of strong σ-donating or π-accepting ligands, L', to yield M(CO)$_4$LL' [46] (L = L'),[6,7] CO substitution is apparently not achieved in the presence of weak nucleophiles such as pentene.[38] A dissociative reaction should not depend that much on the nature of the substituting ligand. Additional support for an associative mechanism comes from the variation of the quantum yield with the solvent and, probably, the concentration of the substituting ligand.

Upon irradiation of the long-wavelength absorption of the M(CO)$_5$ fragment produced by the photolysis of M(CO)$_6$ in low-temperature glasses of inert hydrocarbons or rare-gas matrices, the regeneration of $M(CO)_6$ was observed.[21,29,36] Since $M(CO)_5$ may be regarded as a monosubstituted hexacarbonyl with a noninteracting ligand in the sixth coordination position, the observed photoreversal can be considered as photosubstitution of a $M(CO)_5L$ complex that takes place even at low temperatures. However, this photoreversal seems to be a quite general behavior of metal carbonyls photolyzed in low-temperature media and independent of the geometry and electronic structure of the metal carbonyl fragment.[47,48] Therefore, it was suggested that the photoregeneration of $M(CO)_6$ does not involve excited-state chemistry.[21,28] According to this assumption, the photoreversal is facilitated by a local softening of the low-temperature medium that occurs on conversion of the absorbed light to heat by radiationless deactivation of excited $M(CO)_5$ to its ground state. The CO molecule, which is assumed to be ejected from the matrix cage when $M(CO)_6$ is photolyzed, may then recombine with the $M(CO)_5$ fragment by diffusion, as it does at higher temperatures without irradiation.

The photosubstitution of hexacarbonyls does not end when two CO ligands are replaced. Under favorable conditions, continued irradiation of dissolved hexacarbonyls leads to the substitution of five or even all six CO groups if a replacing ligand, such as $P(OCH_3)_3$ or $P(OCH_3)_2F$, has good π-acceptor properties and can stabilize the zerovalent metal.[49]

The arene-$M(CO)_3$ complexes constitute a category of substituted hexacarbonyls deserving special attention. These compounds may be regarded as pseudooctahedral complexes where three ligand positions are occupied by an arene ring. The metal-arene bonding is achieved by the interaction of the π orbitals of the aromatic ring with metal orbitals. The absorption spectrum of $C_6H_6Cr(CO)_3$ shows an intense long wavelength band at 320 nm with a shoulder at 376 nm. These bands were assigned to a CT transition from Cr to the benzene ring and, to a lesser extent, to the carbonyl groups.[50] The next band at 260 nm was assigned to a CT transition from Cr to the carbonyls. Upon irradiation of solutions of $C_6H_6Cr(CO)_3$ at 366 nm into the long-wavelength absorption of the complex, coordinated CO was exchanged with labeled CO,[51] or replaced by another ligand present in the solution.[6,7] In addition, a photoexchange of coordinated benzene with labeled benzene was observed.[52] It becomes increasingly difficult to achieve a photoexchange of CO or arene for the corresponding complexes of the heavier metals, Mo and W. However, the relative rate of exchange of toluene in $CH_3C_6H_5W(CO)_3$ was increased by shorter-wavelength irradiation at 254 nm. These results indicate either a hot-molecule mechanism or the participation of higher excited states in the photoexchange of coordinated arenes. The intermediate formation of a $M(CO)_3$ fragment was postulated to occur in this photoreaction. But it seems more likely that the photoexchange of arenes takes place by a bimolecular reaction.

IV. CARBONYLS OF MANGANESE AND RHENIUM

Diamagnetic $Mn_2(CO)_{10}$ is composed of two square pyramidal $Mn(CO)_5$ units linked by a metal-metal bond. The whole molecule is staggered and, therefore, of D_{4d} symmetry. The highest occupied orbitals are the six d_{xz}, d_{yz}, d_{xy} π orbitals (e_1, e_2, e_3), and the $d\sigma$ (M—M) orbital (a_1), which is the bonding combination of the d_{z^2} orbitals of both manganese atoms and represents the metal-metal bond of $Mn_2(CO)_{10}$. Photoelectron spectroscopy provides evidence that the $d\sigma$(M—M) lies slightly above the $d\pi$ orbitals.[53] Since the lowest empty MO is anticipated to be the antibonding $d\sigma^*$(M—M) orbital (b_2), the intense long-wavelength band in the absorption spectrum of $Mn_2(CO)_{10}$ at 340 nm ($\varepsilon \sim 2 \times 10^4 \, M^{-1} \, cm^{-1}$), which is polarized along the metal-metal axis, is assigned to the allowed transition from $d\sigma$(M—M) to $d\sigma^*$(M—M).[54] Due to the small energy difference between $d\sigma$(M—M) and $d\pi$ orbitals, $d\pi \rightarrow d\sigma^*$(M—M) transitions may contribute to the intensity of this first absorption. A shoulder on the long-wavelength tail of this band has been assigned to a $d\pi \rightarrow d\sigma^*$(M—M) transition, but such an assignment would require that the $d\pi$ levels lie above the $d\sigma$(M—M) orbital. The intense near-ultraviolet absorption of $Re_2(CO)_{10}$ at 310 nm ($\varepsilon \sim 17,000$) is also assigned to the $d\sigma$(M—M) $\rightarrow d\sigma^*$(M—M) transition.

The formation of $Mn(CO)_5$ or $Re(CO)_5$ radicals, or at least a destabilization of the metal-metal bond, may be expected if the irradiation of $Mn_2(CO)_{10}$ or $Re_2(CO)_{10}$ is associated with the $d\sigma$(M—M) $\rightarrow d\sigma^*$(M—M) transition. In addition, a labilization of M—CO bonds is possible if the irradiation leads to some extent to the excitation of a π-bonding d electron in the $d\pi \rightarrow d\sigma^*$(M—M) transition.

Although no photochemical reaction of $Mn_2(CO)_{10}$ was observed in inert solvents, evidence for the formation of manganesecarbonyl radicals was obtained if $Mn_2(CO)_{10}$ was photolyzed with long-wavelength irradiation ($\lambda = 436$ nm) in halogen-containing solvents.[55-57] The photolysis of $Mn_2(CO)_{10}$ in the presence of HBr in cyclohexane led to the almost quantitative formation of $Mn(CO)_5Br$.[58] Irradiation of $Mn_2(CO)_{10}$ in the halocarbon solvent, CCl_4, produced $Mn(CO)_5Cl$ and CCl_3 radicals, which are used to initiate polymerization reactions.

It may be expected that the primary step in these photolyses is the homolytic cleavage into two $Mn(CO)_5$ radicals initiated by the $d\sigma(M—M) \to d\sigma^*(M—M)$ transition. An alternative possibility for the halogen abstraction is the interaction of the halogen-containing compound with $Mn_2(CO)_{10}$ in its $d\sigma \to d\sigma^*$ excited state.[59] However, the analysis of the kinetic data and the quantum-yield determinations of this photoreaction led to the conclusion that the photoprimary step is an unsymmetric cleavage of the metal-metal bond producing $Mn(CO)_4$ and $Mn(CO)_6$ radicals.[55-57] Their subsequent reactions are complicated processes. In contrast to these results, evidence was obtained recently that the photolysis of $Re_2(CO)_{10}$ in CCl_4 at 313 or 366 nm, corresponding to the $d\sigma \to d\sigma^*$ transition, leads to a symmetric metal-metal bond cleavage in the photoprimary step.[60] With 313-nm radiation $Re_2(CO)_{10}$ disappeared with a quantum yield of 0.6, while $Re(CO)_5Cl$, which is assumed to be formed by the reaction of $Re(CO)_5$ with CCl_4, was produced with a quantum yield of 1.2. Other observations also support a light-induced homolytic cleavage of M—M bonds. Upon irradiation of a mixture of $Mn_2(CO)_{10}$ and $Re_2(CO)_{10}$ in hexane some $MnRe(CO)_{10}$ was formed.[61] The photolysis of $(OC)_5Mn—Mn(CO)_3L_2$ (L_2 = phen or bipy) in ether or benzene yielded $Mn_2(CO)_{10}$ and $[Mn(CO)_3L_2]_2$.[62,63] The irradiation of $(OC)_5Mn—Re(CO)_3(phen)$ in THF produced $Mn_2(CO)_{10}$ and $[Re(CO)_3(phen)]_2$.[64]

A CTTS excited state in CCl_4, which initiates the photochemical oxidation of ferrocene in CCl_4,[7] can be excluded as precursor for the photooxidation of $Mn_2(CO)_{10}$ or $Re_2(CO)_{10}$ to $Mn(CO)_5Cl$ or $Re(CO)_5Cl$ because the absorption spectra of $M_2(CO)_{10}$ dissolved in inert solvents such as benzene does not show a new CT-to-solvent band on addition of CCl_4.[56]

Although $Mn_2(CO)_{10}$ does not undergo any photochemical reaction in inert solvents, a photosubstitution of coordinated CO does take place in the presence of potential ligands. This behavior may reflect the contribution of $d\pi$–$d\sigma^*(M—M)$ transitions to the long-wavelength absorption of $Mn_2(CO)_{10}$. Up to three CO groups may be replaced, depending on the substituting ligand. In the presence of simple nitrogen bases only one equatorial CO was substituted.[65] Ligands with good π-acceptor properties (phosphines, arsines) can replace one[65] or two[66] axial CO groups. PF_3 can photosubstitute up to three CO ligands in axial and equatorial positions, while the thermal reaction led only to the substitution of axial CO ligands.[67] It is also feasible that the photosubstitutions do not take place at the intact $Mn_2(CO)_{10}$ molecule.[67] If manganesecarbonyl radicals, $Mn(CO)_5$, are formed in the photoprimary step, these radicals may undergo a substitution of CO. The recombination of the substituted radicals should then yield the most stable isomer. A similar mechanism was also suggested for thermal reactions of manganese carbonyls.[68,69]

V. SUBSTITUTED MANGANESECARBONYLS

The manganese atom in the pseudooctahedral complexes, $Mn(CO)_5X$ ($X = Cl$, Br, I), can be regarded as a formally unipositive ion with six d electrons. Photoelectron spectroscopy shows that the highest filled orbitals of these complexes are π orbitals of primarily halogen character.[53,70] Slightly below these orbitals are the $p\sigma(X)$ and $d\pi$ orbitals of manganese. The lowest empty antibonding states are expected to be $d\sigma$ below $\pi^*(CO)$ orbitals.[71] From this ordering, low-energy X-to-Mn, X-to-CO, and Mn-to-CO CT transitions and LF transitions can be anticipated. The longest-wavelength band in the absorption spectrum of $Mn(CO)_5X$ was assigned to a $\pi(X)-\pi^*(CO)$ CT transition.[71] Other low-energy bands were assumed to be less intense and hidden under the X-to-CO CT band, which showed a solvent-dependent shift. This shift was attributed to the interaction of the π electrons of the halogen with the solvent.

Although no detailed investigation of the photochemistry of the manganese carbonyl halide complexes has yet been published, it was shown that irradiation of dissolved $Mn(CO)_5X$ may lead to the release of CO.[72] $Mn(CO)_5Br$ was observed to undergo a photoexchange with labeled CO.[73] In contrast to the thermal exchange, the photoexchange did not proceed stereospecifically. The thermal dimerization of $Mn(CO)_5Br$ to $[Mn(CO)_4Br]_2$, which occurs with release of CO and the formation of two Br bridges linking the metal centers, was shown to take place faster if the long-wavelength band of $Mn(CO)_5Br$ was irradiated.[72] Also, the irradiation of $Re(CO)_5Cl$ produced $[Re(CO)_4Cl]_2$.[60] Interestingly, $[Mn(CO)_4Br]_2$ underwent a photoredox reaction according to the equation[72]

$$2[Mn(CO)_4Br]_2 \rightarrow Mn_2(CO)_{10} + 2MnBr_2 + 6CO \qquad (6\text{-}1)$$

At present, no explanation for this photoredox behavior can be given. The long-wavelength band that was irradiated in the photolysis of $[Mn(CO)_4Br]_2$ was assigned to a $d\pi-\pi^*(CO)$ CT transition.[71]

The ligand, X, in the complexes, $Mn(CO)_5X$ ($X = H$, CH_3, CF_3), does not have any π-bonding properties. From photoelectron spectroscopy,[53] the highest filled orbitals were assigned to be either $d\pi$-bonding[53,74] or σ-bonding orbitals of X.[75] The $d\pi$ and $\sigma(X)$ levels are certainly close in energy. The lowest empty antibonding orbitals are again assumed to be $d\sigma$ below $\pi^*(CO)$ states.[71] The excitation of a bonding $\sigma(X)$ electron may lead to a homolytic splitting of the Mn—X bond, whereas $d\pi-d\sigma$ or $d\pi-\pi^*(CO)$ transitions could lead to a labilization of Mn—CO bonds. The long-wavelength band in the absorption spectra of $Mn(CO)_5X$ was assigned to a $d\pi-\pi^*(CO)$ transition.[71] This band may obscure other bands of lower intensity.

The photolysis of $HMn(CO)_5$ does not seem to lead to a homolytic splitting of the H—Mn bond. The loss of an equatorial CO group with formation of trigonal bipyramidal $HMn(CO)_4$ was observed when $HMn(CO)_5$ was irradiated at 229 nm in an argon matrix at 15°K.[47] A regeneration of $HMn(CO)_5$ occurred for long-wavelength irradiation at $\lambda > 285$ nm. The photolysis of $HMn(CO)_5$ in the presence of PF_3 led to the substitution of CO by PF_3.[76] In contrast to the photochemical stability of the H—Mn bond of $HMn(CO)_5$, the irradiation of $HMn(CO)_3(PR_3)_2$ in CCl_4 produced $ClMn(CO)_3(PR_3)_2$ and, most likely, $CHCl_3$.[77] In analogy to the photolysis of $Re_2(CO)_{10}$ in CCl_4,[60] the irradiation of $HMn(CO)_3(PR_3)_2$ may lead to an excitation of a H—Mn σ-bonding electron and the subsequent homolysis of the H—Mn bond. The radicals, H and $Mn(CO)_3(PR_3)_2$, may then react with CCl_4 to yield $ClMn(CO)_3(PR_3)_2$ and $CHCl_3$.

When $CH_3Mn(CO)_5$ was irradiated (230 nm $< \lambda <$ 280 nm) in an argon matrix at 15°K, the loss of one CO ligand was observed.[78,79] This photolysis may be reversed with long-wavelength irradiation at $\lambda >$ 280 nm.[79] The release of CO was also shown to occur if $CF_3Mn(CO)_5$ was photolyzed under similar conditions.[78] In the presence of PF_3 at room temperature the photolysis of $HC_2F_4Mn(CO)_5$ led to a substitution of CO by PF_3.[76] Although the low-temperature photolysis of $CR_3Mn(CO)_5$ complexes provides definite evidence only for the release of CO, other observations suggest that the irradiation of $CH_3Mn(CO)_5$ may also lead to the cleavage of the CH_3—Mn bond, possibly caused by the excitation of a σ-bonding electron. The photolysis of $CH_3Mn(CO)_5$ in the presence of $CClF=CF_2$ led to an insertion of the olefin into the carbon-manganese bond with formation of $CH_3CF_2CFClMn(CO)_5$.[80] The photochemical insertion was also achieved with hexafluoro-1,3-butadiene.[81]

$C_5H_5Mn(CO)_3$ may be regarded as a pseudooctahedral complex. The uninegative aromatic cyclopentadienyl ring occupies three ligand positions. The central manganese is formally unipositive with six d electrons. The electronic spectrum of $C_5H_5Mn(CO)_3$ is known,[82] but no band assignments have been made. A characteristic feature of the photochemistry of $C_5H_5Mn(CO)_3$ is its striking similarity to that of the hexacarbonyls, $M(CO)_6$. Upon irradiation in organic solvents, $C_5H_5Mn(CO)_3$ releases one CO ligand with a quantum yield of unity that is independent of the irradiation wavelength.[45] The formation of a $C_5H_5Mn(CO)_2$ fragment was observed upon irradiation of $C_5H_5Mn(CO)_3$ in glasses of inert hydrocarbons at low temperatures.[83] Softening of the glass at higher temperatures led to a regeneration of $C_5H_5Mn(CO)_3$. The photolysis of $C_5H_5Mn(CO)_3$ in the presence of potential ligands was used to synthesize numerous complexes of the general formula, $C_5H_5Mn(CO)_2L$.[6,7] Further photolysis may lead to the substitution of the second or even the third CO ligand. For example, $C_5H_5Mn(PF_3)_3$ was prepared by the irradiation of $C_5H_5Mn(CO)_3$ in THF solution in the presence of PF_3.[84]

$Mn(CO)_4NO$ is isoelectronic and isostructural with $Fe(CO)_5$. However, the ordering of the electronic energy levels of $Mn(CO)_4NO$ is not known and may be quite different from that of $Fe(CO)_5$ because of the influence of the NO group. Nevertheless, the main features of the photochemistries of the two complexes seem to be quite similar. The irradiation of $Mn(CO)_4NO$ in low-temperature glasses leads to the release of one CO ligand and the formation of a $Mn(CO)_3NO$ fragment.[85] If the photolysis of $Mn(CO)_4NO$ is carried out at room temperature, this fragment apparently adds to an intact $Mn(CO)_4NO$ molecule to yield $Mn_2(CO)_7(NO)_2$.[86]

Photochemical substitution reactions of $Mn(CO)_4NO$ were investigated in a detailed study by Keeton and Basolo.[87] Upon irradiation of dissolved $Mn(CO)_4NO$ in the presence of a substituting ligand, L, $[Mn(CO)_3(NO)L]$ was produced with a quantum yield that was dependent upon the nature and the concentration of the nucleophile, L. If L was the relatively weak nucleophile, $As(C_6H_5)_3$, the quantum yield at 427-nm irradiating wavelength was 0.15 and independent of the concentration of $As(C_6H_5)_3$. In the case of the stronger nucleophile, $P(C_6H_5)_3$, the quantum yield increased with increasing concentration of $P(C_6H_5)_3$. The quantum yield at low concentrations of $P(C_6H_5)_3$ was found to have a lower limit. This limiting value was nearly equal to the concentration-independent quantum yield for the formation of $Mn(CO)_3(NO)As(C_6H_5)_3$.

Since two absorption bands in the spectrum of $Mn(CO)_4NO$ are overlapping at the irradiating wavelength, it was suggested that two different excited states may be involved in the photochemical substitution. The higher-energy band, which was estimated to account for 20% of the absorption at the irradiating wavelength, was assumed to belong to a LF transition. This LF excited state should then lead to a spontaneous release of CO and formation of a $Mn(CO)_3NO$ intermediate, which reacts with the nucleophile according to a SN1 mechanism. The quantum yield is then expected to be independent of the concentration and nucleophilicity of the substituting ligand. The second excited state, which was assumed to be associated with the lower-energy band, was tentatively assigned to a metal-to-NO CT transition. The CT excited complex should be accessible to the attack of a strong nucleophile such as $P(C_6H_5)_3$. Some evidence was obtained for the intermediate formation of $[Mn(CO)_4(NO)][P(C_6H_5)_3]$, which finally releases CO to yield $[Mn(CO)_3(NO)][P(C_6H_5)_3]$. This SN2 path for the formation of the photoproduct would explain the concentration-dependent quantum yield at higher $P(C_6H_5)_3$ concentrations. The possible formation of $Mn_2(CO)_7(NO)_2$ as an intermediate (see above) that may interfere with the discussed mechanism was not examined.

$Fe(CO)_5$ has a trigonal bipyramidal structure. The formally zerovalent iron atom has a d^8 electron configuration. Semiempirical calculations have shown that the highest filled MOs are derived from the metal $d_{x^2-y^2}$ and d_{xy} orbitals which interact with σ orbitals of suitable CO ligands.[88] Since π interaction is negligible, the resultant two MOs (e' in D_{3h}), which contain about 30% of ligand σ-orbital character, are slightly antibonding. The next lower filled orbitals (e'') are derived from d_{xz} and d_{yz} metal orbitals. Since these two metal orbitals interact only with π(CO) orbitals, the resultant MOs should be bonding. This ordering of the highest filled orbitals of $Fe(CO)_5$ was confirmed by photoelectron spectroscopy.[13] The lowest unfilled MO was assumed to be derived from the d_{z^2} metal orbital.[88] The d_{z^2} orbital of iron interacts strongly with σ orbitals of both axial CO ligands to form an antibonding MO (a_1'), which, however, has almost 60% ligand character. The next MOs, of mainly π^*(CO) character, are placed at higher energies.

The electronic spectrum[88] of $Fe(CO)_5$ exhibits a long-wavelength absorption of moderately high intensity ($\varepsilon = 3800$ M^{-1} cm^{-1}) at 35,500 cm^{-1}. This band was assigned to lowest-energy transition, $e'(d_{x^2-y^2}, d_{xy}) \rightarrow a_1'(d_{z^2})$, which is spin- and symmetry-allowed. The same transition was also assigned[13] to a weak shoulder at 28,200 cm^{-1}, which was observed in an older study,[89] but may be due to impurities of $Fe_2(CO)_9$ (solid $Fe_2(CO)_9$ has an absorption band at 25,300 cm^{-1} [90]). The next absorption bands of higher intensities at shorter wavelength should belong to Fe-to-CO CT transitions.

Upon irradiation, $Fe(CO)_5$ loses one CO ligand in the primary photochemical step. It seems likely that this photoreaction is associated with the lowest energy transition $e'-a_1'$.[88] It is probably not the removal of an electron from a nearly nonbonding MO but the addition to the strongly antibonding d_{z^2} orbital that causes the release of a CO ligand that should be an axial one.

The mechanism of the photochemical release of CO by $Fe(CO)_5$ and the nature of the ironcarbonyl fragment, which is formed as an intermediate in the primary photochemical step, have been investigated in several studies. In the absence of substituting ligands, the photolysis of $Fe(CO)_5$ yields $Fe_2(CO)_9$. Labeled CO was incorporated into $Fe_2(CO)_9$ only to a small extent when the gas phase photolysis of $Fe(CO)_5$ was carried out in the presence of ^{14}CO.[91] It was suggested that electronically excited $Fe(CO)_5$ does not lose CO; instead, it reacts with a ground-state molecule of $Fe(CO)_5$ to yield $Fe_2(CO)_9$ with loss of CO. Later studies, however, provided compelling evidence that the primary photochemical step of excited $Fe(CO)_5$ is indeed the loss of one CO ligand. The small incorporation of labeled CO into $Fe_2(CO)_9$ may be explained by the assumption that the reaction of the $Fe(CO)_4$ intermediate with $Fe(CO)_5$ is much faster than its reaction with CO.[7] In solutions of inert hydrocarbons the formation of $Fe_2(CO)_9$ was slow enough to demonstrate that an efficient photoexchange between $Fe(CO)_5$ and $C^{18}O$, as well as a photochemical scrambling of coordinated CO between $Fe(CO)_5$ and $Fe(C^{18}O)_5$, takes place.[92] These observations are consistent with the assumption that $Fe(CO)_5$ photodissociates into $Fe(CO)_4$ and CO. The thermal regeneration of $Fe(CO)_5$ then accounts for the exchange of CO.

Additional evidence for the loss of CO in the first step was provided by the results of the photolysis of $Fe(CO)_5$ in low-temperature glasses and matrices. The infrared spectra of irradiated $Fe(CO)_5$ in hydrocarbon glasses at 77°K indicated the formation of a $Fe(CO)_x$ species that could not be identified.[24] On melting the rigid glass, a new infrared band developed at $1834\ cm^{-1}$, which was probably due to the formation of $Fe_2(CO)_9$.[7] In an argon matrix containing 20% CO no indication for a photochemical change was obtained.[90] The infrared spectra of $Fe(CO)_5$ in a pure argon matrix at 15 or 20°K showed that $Fe(CO)_5$ could be photolyzed only to a small extent.[93] The photoproduct was not identified. However, $C^{18}O$ was photochemically incorporated in $Fe(CO)_5$ when the argon matrix was doped with labeled CO.[90] Apparently, a fast reversal of the photolysis occurred even in rigid media at low temperatures. Only recently, Poliakoff and Turner showed that this regeneration was promoted by the light of the Nernst glower that was used in the analyzing infrared spectrometer.[48] Upon irradiation with a medium-pressure mercury arc, an efficient photolysis of $Fe(CO)_5$ in a neon matrix at 4°K was observed using infrared spectroscopy provided that the ultraviolet and visible light was removed from the infrared spectrometer beam by suitable filters. The analysis of the infrared spectra led to the conclusion that the photoproduct was a $Fe(CO)_4$ species that has a distorted tetrahedral structure. This distortion is probably due to the Jahn–Teller effect, which is expected to occur in tetrahedral d^8 complexes. The distortion was not caused by the matrix since similar results were obtained for a variety of matrices at 20°K. A reversal of the photolysis took place by warming the matrix or upon irradiation of the $Fe(CO)_4$ fragment, which has an absorption maximum at 320 nm. The photoreversal was assumed to be a thermal process initiated by the conversion of light to heat. Upon irradiation of $Fe(CO)_5$ in argon matrices doped with ethylene, the formation of $Fe(CO)_4C_2H_4$ was observed.[94]

The detection of a $Fe(CO)_4$ species by flash photolysis of $Fe(CO)_5$ in the gas phase[95] or in solution has not yet been successful.[96] A long-lived transient with a half-life of 0.3 sec that was observed in the flash photolysis of $Fe(CO)_5$ in benzene[97] was probably a complex with benzene weakly coordinated to $Fe(CO)_4$.[98]

Whereas the photolysis of $Fe(CO)_5$ in solution yields $Fe_2(CO)_9$ as final product, complexes of the type, $Fe(CO)_4L$, are formed when the irradiation is carried out in the presence of substituting ligands. Numerous substituted ironcarbonyls with a great variety of ligands have been prepared in this way.[6,7] According to the results presented above, a $Fe(CO)_4$ intermediate reacts with suitable ligands to give $Fe(CO)_4L$.

Another interesting aspect of these photosubstitutions is their stereochemical course since, as already mentioned, absorption of light into the long-wavelength band of $Fe(CO)_5$ should lead to the release of an axial CO ligand. In what may be a contradiction to this prediction the light-induced CO exchange between labeled CO and $Fe(CO)_5$ or $Fe(CO)_4[P(C_6H_5)_3]$ in solutions of hydrocarbons was shown to lead to a statistical distribution of labeled CO in axial and equatorial positions.[92] However, in the presence of other ligands the formation of $Fe(CO)_4L$, induced thermally[99,100,101] or photochemically,[7,102,103] proceeds stereospecifically. Ligands with σ-donor properties were found to occupy an axial position, while some π-bonding ligands such as olefins were shown to be coordinated in an equatorial position of the trigonal bipyramidal structure of $Fe(CO)_4L$.[104] Since $Fe(CO)_5$ seems to be stereochemically nonrigid in solution at room temperature,[92,105] it is apparently unclear whether these stereospecific photosubstitutions are dependent on the position of the released CO ligand prior to photolysis. In addition, such a dependence can not be expected if the $Fe(CO)_4$ intermediate adopts a nearly tetrahedral structure, as was suggested by the low-temperature studies.[48] It is probably the most stable isomer formed in the photoreaction. Frequently, a second CO ligand may be photosubstituted also to yield complexes of the formula, $Fe(CO)_3L_2$.[6,7]

With one exception no quantum yields have been reported for photoreactions of $Fe(CO)_5$. As early as 1929, quantum yields for the photochemical decomposition of liquid $Fe(CO)_5$ were obtained.[106] The quantum yield for CO evolution increased from 0.83 at 436 nm to 0.99 at 254-nm irradiating wavelength. Since the photolysis of $Fe(CO)_5$ in the liquid state may be a complicated process, it is difficult to draw conclusions from these results.

Finally, it should be mentioned that an $Fe(CO)_4$ species was also postulated as an intermediate in the photolysis[107] and in the thermal reactions[108] of $Fe_2(CO)_9$. However, the results of a low-temperature study in inert gas matrices did not give any evidence for the formation of $Fe(CO)_4$. The infrared spectra of photolyzed $Fe_2(CO)_9$ indicated the loss of CO under formation of $Fe_2(CO)_8$.[90]

VII. NICKELTETRACARBONYL

$Ni(CO)_4$ has a tetrahedral structure. The nickel atom has a completely filled d shell and thus contributes ten valence electrons to the bonding of $Ni(CO)_4$. In T_d symmetry the d orbitals are split into t_2 and e orbitals.[13] The highest filled orbitals of $Ni(CO)_4$ are three degenerate t_2 levels that are formed by the antibonding interaction of the d_{xz}, d_{yz}, and d_{xy} orbitals of nickel with σ orbitals of CO. These antibonding t_2 levels may also interact weakly with $\pi(CO)$ orbitals. In spite of this interaction, the d character should prevail in the t_2 states.

The two degenerate e orbitals that are placed below the t_2 orbitals are probably nearly pure nonbonding $d_{x^2-y^2}$ and d_{z^2} orbitals that interact only slightly with $\pi(CO)$ orbitals. Different calculations have been carried out to determine the energies and populations of the occupied MOs and the charge distribution within $Ni(CO)_4$.[14,17,109] Although the actual values obtained varied with the calculational procedure used, the splitting between the t_2 and e orbitals certainly is not large. By photoelectron spectroscopy the energy separation was found to be 0.8 eV (~ 6400 cm^{-1}).[13]

Since the e and t_2 orbitals are completely filled with ten d electrons provided by the zero-valent nickel, no d–d transitions are possible in $Ni(CO)_4$. The lowest empty orbitals are derived from π^* states of the CO ligands. Assignments have been made for the absorption bands in the electronic spectrum of $Ni(CO)_4$.[17] Although other assignments could have been obtained using different calculational methods, it seems clear that all absorption bands in the experimentally accessible region of the spectrum belong to $d \rightarrow \pi^*(CO)$ CT transitions. For this reason the photochemical behavior of $Ni(CO)_4$ is a valuable test to examine whether it is necessary to invoke LF excited states as precursors for the photodissociation of metal carbonyls. Since $Ni(CO)_4$ is known to release CO upon irradiation, $d \rightarrow \pi^*(CO)$ CT excited states are clearly responsible for this photodissociation. Although this result is not proof that other metal carbonyls can photodissociate under $d \rightarrow \pi^*(CO)$ excitation, it does demonstrate that the long-standing postulation that only LF excited states can initiate the release of CO is not generally valid.

Upon irradiation of $Ni(CO)_4$ in an inert gas matrix at 15°K the formation of a nonplanar (C_{3v}) $Ni(CO)_3$ fragment was observed.[110] A partial recombination of the released CO ligand with $Ni(CO)_3$ was achieved when the matrix was warmed to 30°K. The photolysis of $Ni(CO)_4$ in a nitrogen matrix at 20°K led to the formation of $Ni(CO)_3N_2$.[111] Investigations of the decomposition of $Ni(CO)_4$ by gas-phase flash photolysis gave results that were consistent with the formation of $Ni(CO)_3$ and $Ni(CO)_2$ intermediates.[112] The decomposition was inhibited by added CO. This inhibition is probably due to the reversal of the photolysis. Photochemical substitution reactions of $Ni(CO)_4$ have not yet been investigated. This deficiency is probably associated with the fact that thermal substitutions of $Ni(CO)_4$ are generally fast.[1-3] The thermal lability of Ni—CO bonds may be related to the occupation of the antibonding t_2 orbitals in the ground state of $Ni(CO)_4$.

VIII. PHOTOISOMERIZATION OF COORDINATED OLEFINS

The photochemical *trans-cis* isomerization of coordinated olefins has been observed for a variety of carbonyl complexes. In all cases, it is likely that orbitals of the olefin are involved in the electronic transition initiating the transformation of the olefinic ligand. Generally, two different situations may lead to the observed isomerizations, which require a free or less hindered rotation around the olefinic double bond. First, the double bond may be weakened by removal of an electron from a π-bonding orbital of the olefin. This may be achieved by π-to-π^* intraolefin, π(olefin)-to-metal CT and π(olefin)-to-π^*(CO) CT transitions. Second, the double bond may be weakened by placing an electron into the antibonding π^* orbital of the olefin. This occurs in π-to-π^* intraolefin, metal-to-π^*(olefin) CT, and π- or σ(CO)-to-π^*(olefin) CT transitions.

Although reliable assignments of electronic absorption bands of these olefin complexes have not yet been reported, rough estimates can be made since the ionization potentials and the energies of the lowest $\pi \to \pi^*$ transitions of many olefins are known.[113,114] It seems likely that low-energy transitions of olefin-metal-carbonyl complexes involving π orbitals of the olefin are possible. In addition, it is also feasible that the isomerization does not take place at the coordinated olefin. The electronically excited complex may dissociate into a metal-carbonyl fragment in its ground state and the olefin in an excited state. The lowest excited triplet of many olefins is known to lead to *cis-trans* isomerizations.[115] Such a mechanism would require that the lowest triplet of the olefin is lower than the initially excited state of the complex. The isomerized olefins may then recombine with the complex fragment.

Photoisomerization of coordinated *trans*- and *cis*-stilbene was observed for $Mo(CO)_5$stilbene and $W(CO)_5$stilbene; the $M(CO)_5L$ compounds were produced by photolysis of the hexacarbonyls in the presence of the olefins.[116] Starting with either *trans*- or *cis*-stilbene, a photostationary state was reached. The photocatalysis by $W(CO)_6$ led to an equilibrium of about 40% *trans*- and 60% *cis*-stilbene. It was shown that the isomerization probably takes place at the coordinated olefin. Absorption of light was limited to the long-wavelength band ($\lambda_{max} \sim 415$ nm) of the complex. Neither the hexacarbonyls nor the free stilbenes absorb in this region. The overall quantum yields were of the order of 0.01.

Although no assignment was made for the long-wavelength band of these olefin complexes, low-energy $d-\pi^*$(stilbene) CT transitions should be possible. Such transitions may weaken the olefinic double bond and facilitate the isomerization. It was proposed that the excited state of the complex produces an intermediate characterized by a metal-carbon σ-bond with free rotation about the olefin bond. Intermediates of this type were suggested to result from metal-to-olefin CT excitation that leads to a positively charged metal carbonyl fragment and an olefin anion.[7,114] This nucleophilic radical anion may then attack the metal to form a metal-carbon σ-bond in the transition state.[114] Hence the regeneration of the starting complex proceeds under isomerization of the olefin. $Cr(CO)_6$ was found to be inactive in the photocatalysis of stilbene isomerization.[116] This result was explained as being due to the large difference in the ability of the three metals to bind the olefin. In addition, the photoactive excited state was assumed to be not a singlet, but a triplet that may not be populated in the chromium complex because of the smaller probability of spin-forbidden transitions. However, this latter explanation should be accepted with reservation since there is some evidence that intersystem crossing from the lowest singlet excited state to the lowest triplet of $Cr(CO)_6$ occurs with unit efficiency.[20] Alternatively, the inability of $Cr(CO)_6$ to photocatalyze the isomerization of stilbene may be due to smaller $d-d$ splittings in chromium compared to molybdenum or tungsten complexes.[11] If the $d\sigma$ levels are lower in energy than the π^*(stilbene) states, the irradiation of the complex could lead to LF excited states that may not initiate the isomerization.

Irradiation of $W(CO)_6$ in the presence of 1,3-pentadiene or 2,4-hexadiene led to *cis-trans* isomerizations of the conjugated dienes with quantum yields being smaller than 0.1 at 313 nm irradiating wavelength.[117] The lowest triplet excited state of $W(CO)_5$diene was assumed to be the precursor of the photoisomerization. Although *cis*-to-*trans* was preferred over *trans*-to-*cis* isomerization, as is also the case for the free diene in its lowest triplet excited state, the intermediate formation of an excited free diene was ruled out for the isomerization photocatalyzed by $W(CO)_6$.

Trans-to-*cis* photoisomerization of coordinated olefins was also observed for $W(CO)_5$(*trans*-4-styrylpyridine) and $W(CO)_5$(*trans*-2-styrylpyridine).[38] In contrast to the complexes discussed above, the styrylpyridines, NC_5H_4—CH=CH—C_6H_5, are assumed to be coordinated via nitrogen instead of the olefinic double bond. The longest wavelength absorption of these complexes was assigned to the lowest LF transition in analogy to other N-coordinated $W(CO)_5$ complexes. The quantum yield of isomerization of the olefin increased with increasing wavelength and was highest upon irradiation of the longest wavelength band of the complex [$\phi = 0.49$ at 436 nm for $W(CO)_5$(*trans*-4-styrylpyridine)]. In addition, a photosubstitution of the olefinic nitrogen base by 1-pentene was observed with quantum yields that also increased with increasing wavelength of irradiation [$\phi = 0.16$ at 436 nm for $W(CO)_5$(*trans*-4-styrylpyridine)]. It was assumed that higher excited states of the complex undergo internal conversion to the lowest LF excited state that initiates the photosubstitution. Since, however, the lowest triplets of the free styrylpyridines known to lead to *trans-cis* isomerizations have presumably lower energies (~ 50 kcal) than the lowest LF triplets of the complexes (58 kcal), it was suggested that the photoisomerization of the coordinated ligand takes place by transfer of a large portion of the excitation energy from the complex to the olefin. Additional evidence for an efficient energy transfer to the olefin was obtained by the observation that these complexes did not show any emission from their lowest LF triplet states. However, the low energies of the $\pi-\pi^*$ transitions of the styrylpyridines and the perturbation of the π-electron system upon coordination, which is indicated by a significant shift of the $\pi-\pi^*$ absorption of free *trans*-4-styrylpyridine (255 nm) to lower energies in the coordinated state (316 nm), do not exclude that $d-\pi^*$(olefin) CT contributes to the lowest energy transition of these complexes. Then it would not be necessary to invoke energy transfer to an intraolefin excited state as explanation for the observed photoisomerization.

Cis-trans photoisomerizations of coordinated olefins are not limited to carbonyls of Mo and W. Irradiation of dimethyl maleate iron tetracarbonyl yielded dimethyl fumarate iron tetracarbonyl.[7,114,118] This *cis*-to-*trans* isomerization was assumed to be initiated by a metal-to-π^*(olefin) CT transition. However, the efficiency of this isomerization seems to be small, whereas the dimethyl maleate ligand was photoexchanged with labeled dimethyl maleate with quantum yields between 0.27 and 0.47. It was suggested that the initially excited CT state may undergo an efficient internal conversion to a lower LF excited state, which causes the release of the equatorial olefinic ligand without isomerization. The first-order kinetics of this photoexchange is consistent with the intermediate formation of a $Fe(CO)_4$ fragment. The same type of photoexchange was observed for dimethyl fumarate and methyl acrylate iron tetracarbonyl.

Irradiation of either (cis-CHBr=CHBr)$Fe(CO)_4$ or ($trans$-CHBr=CHBr)$Fe(CO)_4$ led to an intramolecular insertion of iron into one of the carbon-bromine bonds of the coordinated olefin to yield

$(CO)_4BrFe$—CH=CHBr as primary photoproduct.[114] Since the iron was shown to be *trans* to the bromine of the σ-bonding —CH=CHBr ligand, a rotation about the olefinic double bond must have occurred in the case of $(cis$-CHBr=CHBr)Fe$(CO)_4$ as starting complex. It was concluded that these insertion reactions originate from d–π^*(olefin) CT transitions; they were assigned to the lowest energy absorptions in the electronic spectra of (cis- and $trans$-CHBr=CHBr)Fe$(CO)_4$. Since the π^* level of CHBr=CHBr is antibonding for both the C—C and C—Br bonds as well, the CT excited state was assumed to facilitate free rotation about the olefinic bond and insertion of the positively charged iron into a C—Br bond by a nucleophilic attack of the olefinic radical anion.

REFERENCES

1. F. Basolo and R. G. Pearson, *Mechanisms of Inorganic Reactions*, Wiley, New York, 1967.

2. D. A. Brown, *Inorg. Chim. Acta Rev.* **1**, 35 (1967).

3. R. J. Angelici, *Organometal. Chem. Rev. A* **3**, 173 (1968).

4. A. W. Adamson, W. L. Waltz, E. Zinato, D. W. Watts, P. D. Fleischauer, and R. D. Lindholm, *Chem. Rev.* **68**, 541 (1968).

5. V. Balzani and V. Carassiti, *Photochemistry of Coordination Compounds*, Academic, New York, 1970.

6. W. Strohmeier, *Angew. Chem.* **76**, 873 (1964).

7. E. Koerner von Gustorf and F.-W. Grevels, *Fortschr. Chem. Forsch.* **13**, 366 (1969).

8. R. F. Fenske, *Pure Appl. Chem.* **27**, 61 (1971).

9. A. W. Adamson, A. Chiang, and E. Zinato, *J. Amer. Chem. Soc.* **91**, 5468 (1969).

10. G. B. Porter, *J. Amer. Chem. Soc.* **91**, 3980 (1969).

11. H. B. Gray and N. A. Beach, *J. Amer. Chem. Soc.* **85**, 2922 (1963).

12. E. W. Abel, R. A. N. McLean, S. P. Tyfield, P. S. Braterman, A. P. Walker, and P. J. Hendra, *J. Mol. Spectrosc.* **30**, 29 (1969).

13. D. R. Lloyd and E. W. Schlag, *Inorg. Chem.* **8**, 2544 (1969).

14. I. H. Hillier and V. R. Saunders, *Mol. Phys.* **22**, 1025 (1971).

15. N. A. Beach and H. B. Gray, *J. Amer. Chem. Soc.* **90**, 5713 (1968).

16. S. Yamada, H. Yamazaki, H. Nishikawa, and R. Tsuchida, *Bull. Chem. Soc. Jap.* **33**, 481 (1960).

17. A. F. Schreiner and T. L. Brown, *J. Amer. Chem. Soc.* **90**, 3366 (1968).

18. W. Strohmeier and D. von Hobe, *Chem. Ber.* **94**, 2031 (1961).

19. P. A. Breeze and J. J. Turner, *J. Organometal. Chem.* **44**, C7 (1972).

20. A. Vogler, *Z. Naturforsch. B* **25**, 1069 (1970).

21. M. A. Graham, M. Poliakoff, and J. J. Turner, *J. Chem. Soc. A*, 2939 (1971).

22. W. L. Waltz and A. W. Adamson, *J. Phys. Chem.* **73**, 4250 (1969).

23. W. Strohmeier and K. Gerlach, *Chem. Ber.* **94**, 398 (1961).

24. I. W. Stolz, G. R. Dobson, and R. K. Sheline, *J. Amer. Chem. Soc.* **84**, 3589 (1962); *ibid.* **85**, 1013 (1963).

25. A. G. Massey and L. E. Orgel, *Nature (London)* **191**, 1387 (1961).

26. J. A. McIntyre, *J. Phys. Chem.* **74**, 2403 (1970).

27. M. A. Graham, R. N. Perutz, M. Poliakoff, and J. J. Turner, *J. Organometal. Chem.* **34**, C34 (1972).

28. M. J. Boylan, P. S. Braterman, and A. Fullarton, *J. Organometal. Chem.* **31**, C29 (1971).

29. M. A. Graham, A. J. Rest and J. J. Turner, *J. Organometal. Chem.* **24**, C54 (1970).

30. B. D. James and M. G. H. Wallbridge, *Progr. Inorg. Chem.* **11**, 99 (1970).

31. J. Nasielski, P. Kirsch, and L. Wilputte-Steinert, *J. Organometal. Chem.* **29**, 269 (1971).

32. J. M. Kelly, H. Herman, and E. Koerner von Gustorf, *J.C.S. Chem. Commun.*, 105 (1973).

33. F. A. Cotton, W. T. Edwards, F. C. Rauch, M. A. Graham, R. N. Perutz, and J. J. Turner, *J. Coord. Chem.* **2**, 247 (1973).

34. P. S. Braterman and A. P. Walker, *Discuss. Faraday Soc.* **47**, 121 (1969).

35. D. J. Darensbourg and T. L. Brown, *Inorg. Chem.* **7**, 959 (1968).

36. M. Wrighton, G. S. Hammond, and H. B. Gray, *J. Amer. Chem. Soc.* **93**, 4336 (1971).

37. M. Wrighton, G. S. Hammond, and H. B. Gray, *Inorg. Chem.* **11**, 3122 (1972).

38. M. Wrighton, G. S. Hammond, and H. B. Gray, *Mol. Photochem.* **5**, 179 (1973).

39. H. Saito, J. Fujita, and K. Saito, *Bull. Chem. Soc. Jap.* **41**, 359 (1968).

40. J. Burgess and S. F. N. Morton, *J.C.S. Dalton*, 1712 (1972).

41. H. Saito, J. Fujita, and K. Saito, *Bull. Chem. Soc. Jap.* **41**, 863 (1968).

42. Y. Kaizu, I. Fujita, and H. Kobayashi, *Z. Phys. Chem. (Frankfurt)* **79**, 298 (1972).

43. P. D. Fleischauer and P. Fleischauer, *Chem. Rev.* **70**, 199 (1970).

44. W. Strohmeier and D. von Hobe, *Chem. Ber.* **94**, 2031 (1961).

45. W. Strohmeier and D. von Hobe, *Z. Phys. Chem. (Frankfurt)* **34**, 393 (1962).

46. G. Schwenzer, M. Y. Darensbourg, and D. J. Darensbourg, *Inorg. Chem.* **11**, 1967 (1972).

47. A. J. Rest and J. J. Turner, *Chem. Commun.*, 375 (1969).

48. M. Poliakoff and J. J. Turner, *J.C.S. Dalton*, 1351 (1973).

49. R. Mathieu and R. Poilblanc, *Inorg. Chem.* **11**, 1858 (1972).

50. D. G. Carrol and S. P. McGlynn, *Inorg. Chem.* **7**, 1285 (1968).

51. W. Strohmeier and D. von Hobe, *Z. Naturforsch. B* **18**, 770 (1963).

52. W. Strohmeier and D. von Hobe, *Z. Naturforsch. B* **18**, 981 (1963).

53. S. Evans, J. C. Green, M. L. H. Green, A. F. Orchard, and D. W. Turner, *Discuss. Faraday Soc.* **47**, 112 (1969).

54. R. A. Levenson, H. B. Gray, and G. P. Ceasar, *J. Amer. Chem. Soc.* **92**, 3653 (1970).

55. C. H. Bamford, P. A. Crowe, and R. P. Wayne, *Proc. Roy. Soc. A* **284**, 455 (1965).

56. C. H. Bamford, P. A. Crowe, J. Hobbs, and R. P. Wayne, *Proc. Roy. Soc. A* **292**, 153 (1966).
57. C. H. Bamford and J. Paprotny, *Chem. Commun.*, 140 (1971).
58. J. C. Kwok, Thesis, University of Liverpool, 1971.
59. C. H. Bamford and S. U. Mullik, *J.C.S. Faraday Trans. I*, 1127 (1973).
60. M. Wrighton and D. Bredesen, *J. Organometal. Chem.* **50**, C35 (1973).
61. G. O. Evans and R. K. Sheline, *J. Inorg. Nucl. Chem.* **30**, 2862 (1968).
62. W. Hieber and W. Schropp, *Z. Naturforsch. B* **15**, 271 (1960).
63. W. Hieber, W. Beck, and G. Zeitler, *Angew. Chem.* **73**, 364 (1961).
64. T. Kruck, M. Höfler, and M. Noack, *Chem. Ber.* **99**, 1153 (1966).
65. M. L. Ziegler, H. Haas, and R. K. Sheline, *Chem. Ber.* **98**, 2454 (1965).
66. A. G. Osborne and M. H. B. Stiddard, *J. Chem. Soc.*, 634 (1964).
67. R. J. Clark, J. P. Hargaden, H. Haas, and R. K. Sheline, *Inorg. Chem.* **7**, 673 (1968).
68. D. Hopgood and A. J. Poë, *Chem. Commun.*, 831 (1966).
69. J. P. Fawcett, A. J. Poë, and M. V. Twigg, *J.C.S. Chem. Commun.*, 267 (1973).
70. D. L. Lichtenberger, A. C. Sarapu, and R. F. Fenske, *Inorg. Chem.* **12**, 702 (1973).
71. G. B. Blakney and W. F. Allen, *Inorg. Chem.* **10**, 2763 (1971).
72. C. H. Bamford, J. W. Burley, and M. Coldbeck, *J.C.S. Dalton*, 1846 (1972).
73. A. Berry and T. L. Brown, *Inorg. Chem.* **11**, 1165 (1972).
74. M. B. Hall and R. F. Fenske, *Inorg. Chem.* **11**, 768 (1972).
75. M. B. Hall, M. F. Guest, and I. H. Hillier, *Chem. Phys. Letters* **15**, 592 (1972).
76. W. J. Miles and R. J. Clark, *Inorg. Chem.* **7**, 1801 (1968).
77. W. Hieber, M. Höfler, and J. Muschi, *Chem. Ber.* **98**, 311 (1965).
78. J. F. Ogilvie, *Chem. Commun.*, 323 (1970).
79. A. J. Rest, *J. Organometal. Chem.* **25**, C30 (1970).
80. J. B. Wilford, P. M. Treichel, and F. G. A. Stone, *J. Organometal. Chem.* **2**, 119 (1964).
81. P. J. Craig, M. Green, A. J. Rest, and F. G. A. Stone, *J. Organometal. Chem.* **12**, 548 (1968).
82. R. T. Lundquist and M. Cais, *J. Org. Chem.* **27**, 1167 (1962).
83. P. S. Braterman and J. D. Black, *J. Organometal. Chem.* **39**, C3 (1972).
84. T. Kruck and V. Krause, *Z. Naturforsch. B* **27**, 302 (1972).
85. A. J. Rest, *Chem. Commun.*, 345 (1970).
86. P. M. Treichel, E. Pitcher, R. B. King, and F. G. A. Stone, *J. Amer. Chem. Soc.* **83**, 2593 (1961).
87. D. P. Keeton and F. Basolo, *Inorg. Chim. Acta* **6**, 33 (1972).
88. M. Dartiguenave, Y. Dartiguenave, and H. B. Gray, *Bull. Soc. Chim. France*, 4223 (1969).
89. W. Hieber and D. von Pigenot, *Chem. Ber.* **89**, 193 (1956).
90. M. Poliakoff and J. J. Turner, *J. Chem. Soc. A*, 2403 (19.1).
91. D. F. Keeley and R. E. Johnson, *J. Inorg. Nucl. Chem.* **11**, 33 (1959).
92. K. Noack and M. Ruch, *J. Organometal. Chem.* **17**, 309 (1969).
93. A. J. Rest and J. J. Turner, *Proceedings of the Fourth International Conference on Organometallic Chemistry*, Bristol, Crane, Russak and Co., New York 1969.

94. M. J. Newlands and J. F. Ogilvie, *Can. J. Chem.* **49**, 343 (1971).
95. A. B. Callear and R. J. Oldman, *Trans. Faraday Soc.* **63**, 2888 (1967).
96. E. Koerner von Gustorf, as cited in M. Poliakoff and J. J. Turner, *J. Chem. Soc. A*, 2403 (1971).
97. E. Koerner von Gustorf and E. M. Langmuir, as cited in E. Koerner von Gustorf and F.-W. Grevels, *Forstschr. Chem. Forsch.* **13**, 366 (1969).
98. E. Koerner von Gustorf and F.-W. Grevels, as cited in M. Poliakoff and J. J. Turner, *J.C.S. Dalton*, 1351 (1973).
99. F. A. Cotton and R. V. Parish, *J. Chem. Soc.*, 1440 (1960).
100. E. Weiß, K. Stark, J. E. Lancaster, and H. D. Murdoch, *Helv. Chim. Acta* **46**, 288 (1963).
101. A. Reckziegel and M. Bigorgne, *J. Organometal. Chem.* **3**, 341 (1965).
102. E. H. Schubert and R. K. Sheline, *Inorg. Chem.* **5**, 1071 (1966).
103. E. Koerner von Gustorf, M. C. Henry, and C. Di Pietro, *Z. Naturforsch. B* **21**, 41 (1966).
104. H. Haas and R. K. Sheline, *J. Chem. Phys.* **47**, 2996 (1967).
105. R. Bramley, B. N. Figgis, and R. S. Nyholm, *Trans. Faraday Soc.* **58**, 1893 (1962).
106. O. Warburg and E. Negelein, *Biochem. Z.* **204**, 495 (1929).
107. I. Wender and P. Pino, *Organic Syntheses via Metal Carbonyls*, Vol. 1, Wiley-Interscience, New York, 1968.
108. P. S. Braterman and W. J. Wallace, *J. Organometal. Chem.* **30**, C17 (1971).
109. I. H. Hillier and V. R. Saunders, *Chem. Commun.*, 642 (1971).
110. A. J. Rest and J. J. Turner, *Chem. Commun.*, 1026 (1969).
111. A. J. Rest, *J. Organometal. Chem.* **40**, C76 (1972).
112. A. B. Callear and R. J. Oldman, *Proc. Roy. Soc. A* **265**, 71 (1962).
113. E. Koerner von Gustorf, M. C. Henry, and D. J. McAdoo, *Liebigs Ann. Chem.* **707**, 190 (1967).
114. F.-W. Grevels, Dissertation, University of Bochum, 1970.
115. J. G. Calvert and J. N. Pitts, *Photochemistry*, Wiley, New York, 1967.
116. M. Wrighton, G. S. Hammond, and H. B. Gray, *J. Amer. Chem. Soc.* **93**, 3285 (1971).
117. M. Wrighton, G. S. Hammond, and H. B. Gray, *J. Amer. Chem. Soc.* **92**, 6068 (1970).
118. G. O. Schenck, E. Koerner von Gustorf, and M.-J. Jun, *Tetrahedron Letters*, 1059 (1962).

7

PHOTOCHEMISTRY OF 1,3-DIKETONATE CHELATES

Richard L. Lintvedt

Department of Chemistry
Wayne State University
Detroit, Michigan

I. INTRODUCTION

The spectral complexities of 1,3-diketonate metal chelates make photo-chemical studies difficult, to say the least. In general, both the ligands and the coordinated metal are rich in spectral bands. In addition, for many systems charge-transfer bands occur in a region of overlapping in-traligand bonds and "*d–d*" bands. Despite these and other spectral difficulties, a great deal may be learned from photochemical studies of 1,3-diketonate chelates.

The 1,3-diketone ligands exist in two forms, keto and enol. The latter contains a strong intramolecular hydrogen bond. The value of the

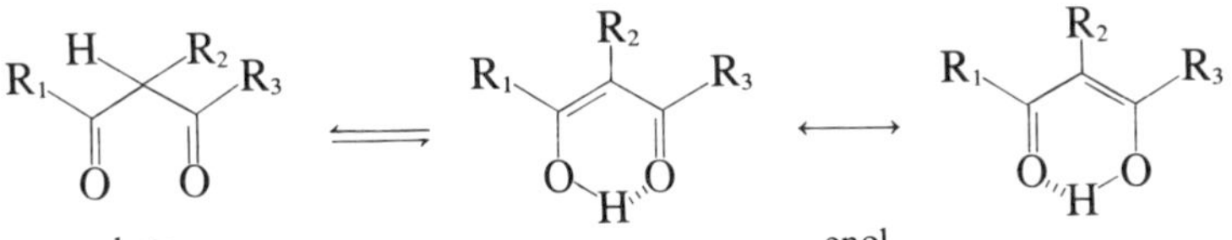

keto ⇌ enol

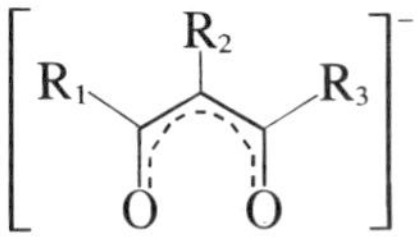

equilibrium constant for keto ⇌ enol depends strongly on the polarity of the solvent. The acidity of the enolic proton is greatly influenced by the nature of the substituent groups, R_1, R_2, and R_3.[1,2] In the presence of a base the enolic proton may be removed, yielding an anion whose solution structure is most likely planar and delocalized. The electronegativity of the oxygens assures that most of the negative charge is distributed near them. Formation of an anion enhances coordination to metal ions; however, the only requirement is that there be a proton acceptor present. The tremendous versatility of the 1,3-diketones is attested to by the fact that 1,3-pentanedione has been coordinated to every naturally occurring metal and metalloid.

The physical and chemical properties of the 1,3-diketonate chelates have been extensively studied.[3,4] Results of structural studies are consistent with the generalized formula shown below.[5] Structural,[5] spectral,[6-9]

and chemical[10] evidence is consistent with the delocalized and essentially planar nature of the six-membered ring. The number of chelated ligands, x, may vary from one to four, depending on the metal present. The substituent groups, R_1, R_2, and R_3, may be practically any group imaginable.

Some of the features of this class of chelates that make photochemical studies of potential interest are discussed below:

1. The ligand substituent groups, R_1, R_2, and R_3, are readily changed, resulting in systematic and predictable spectral shifts. The spectral shifts are dependent on the electronic effect of the substituent groups[9] and take place without altering the coordination geometry about the metal. In this sense, one can "design" complexes and study the effects of spectral shifts.

2. The 1,3-diketones offer great flexibility with respect to the number of compounds containing different metal ions in similar coordination environments that can be investigated. Such flexibility makes possible close comparative studies in which only the metal ion differs from compound to compound.

3. Mixed-ligand chelates can be prepared in which different 1,3-diketones are bonded to one metal. Since the spectra of the ligands differ, it is potentially possible to excite one part of the molecule and observe the effect of localized excitation.

4. Rich spectra in the 40,000 to 10,000 cm^{-1} region for transition-metal chelates allow for varied reaction schemes depending on irradiating wavelength.

5. The 1,3-diketonate chelates exhibit a wide range of physical properties that may be useful in photochemical studies of coordination

compounds. Compounds are available that are soluble in a wide range of solvents including alcohols, ethers, and alkanes. Inasmuch as most inorganic photochemical studies to date have been carried out in aqueous media, no significant information exists about the importance of solvent polarity, for example. The 1,3-diketonate chelates offer the opportunity of critically assessing the affect of solvent variation in photochemical reactions of coordination compounds.

In addition to solubility, certain 1,3-diketonate chelates exhibit quite high vapor pressure and may be readily sublimed at about 5 mm Hg in the temperature region, 100 to 150°C. Such volatility is rare in compounds containing normal oxidation-state metal ions. As a result of this volatility, gas-phase photochemistry of normal oxidation state metal complexes is feasible using 1,3-diketonate chelates.

There are perhaps additional reasons for considering the 1,3-diketonates for photochemical studies. We mention those above only as a means of indicating that there are valid reasons for investigating systems as complicated as these.

II. GENERAL SPECTRAL CONSIDERATIONS

Hückel[11-14] and self-consistent field[14] calculations of the acetylacetonate anion (acac$^-$) have led to the approximate placement of the five π levels. There is general agreement that the lowest spin-allowed transition, $\pi \rightarrow \pi^*$, lies in the 30,000 to 37,000 cm^{-1} region. A second $\pi \rightarrow \pi^*$ transition is calculated to occur at 50,000 cm^{-1}. This is above normal excitation energies used in photochemical studies and will not be considered further. The energies of the $\pi \rightarrow \pi^*$ transition in the anionic form are slightly lower than in the protonated ligand and nearly identical to the energies observed in the transition-metal chelates.[15] This is considered adequate justification for treating the transitions as being localized on the ligands.

Similar to the $\pi \rightarrow \pi^*$ transition is an $n \rightarrow \pi^*$ transition resulting from the promotion of an oxygen nonbonding electron to a π^* orbital within the ligand. Kasha has discussed methods of assigning these transitions based on the solvent dependence and the shape of the observed absorption band.[16] Unfortunately, little information is available about the excited π^* state of the 1,3-diketones. By comparing these excited states with the π^* states of simple organic molecules, some qualitative conclusions can be drawn.[17] The promotion of an electron to a π^* orbital reduces the resistance to distortion by decreasing the π-bonding order by one. Excitation of a nonbonding oxygen electron, $n \rightarrow \pi^*$, decreases the π-bond order by only one-half of a π-bond. Thus the resistance of this configuration to distortion would be expected to be larger.

Also observed within these chelates are charge-transfer transitions. These result from the transfer of an electron from an orbital primarily centered on the ligand to one centered on the metal (designated L → M), or from an orbital primarily centered on the metal to one centered on the ligand (designated M → L). Orgel[18] and Jorgensen[19] have suggested that the energy of the transition, ΔE, is given by $\Delta E = I - E + \Delta$, where E is the electron affinity of the acceptor and I is the ionization potential of the donor. The energy, Δ, describes changes in the nuclear coordinates, solvation sphere, and electron spin of the complex. These transitions, as well as those previously discussed, involve excitation of a single electron.

Charge-transfer transitions, observed at energies higher than those of ligand-field transitions, may occur at energies higher or lower than the ligand $\pi \rightarrow \pi^*$. Jorgensen[20] has suggested that a charge-transfer transition can occur with a change in the spin angular momentum, that is, a spin-forbidden transition. Hammond et al. have shown that 1,3-diketonate

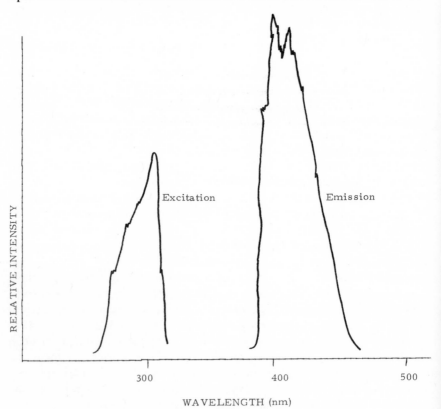

Fig. 7-1. Luminescence spectrum of Na(acac) in ethanol.

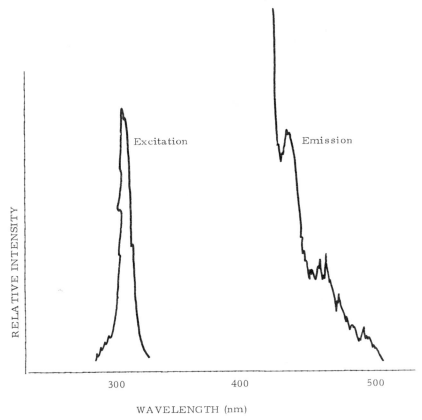

WAVELENGTH (nm)

Fig. 7-2. Luminescence spectrum of Na(tfac) in ethanol.

complexes can quench triplet-state emission.[21] Adamson[22] has shown that organic compounds with relatively stable triplet states can sensitize the reduction of Co(III) amine complexes. Although these data seem to indicate a low-lying triplet charge-transfer level, at present, no certain spin-forbidden charge-transfer transitions have been observed.

A spin-forbidden transition in the 1,3-diketonate complexes does occur within the ligand. The transition, designated $^1\pi \rightarrow {}^3\pi^*$, occurs between the singlet ground state and the low-lying triplet π^* level. This weak transition, $\varepsilon < 1.0$, is usually masked in the chelates by the stronger $\pi \rightarrow \pi^*$ and charge-transfer transitions, but is observed in the luminescence spectra of ethanol solutions of the free ligand and its sodium salt.[23,24] The excitation and emission spectra of three sodium 1,3-diketonates at 77°K in absolute ethanol are shown in Figs. 7-1 to 7-3. The

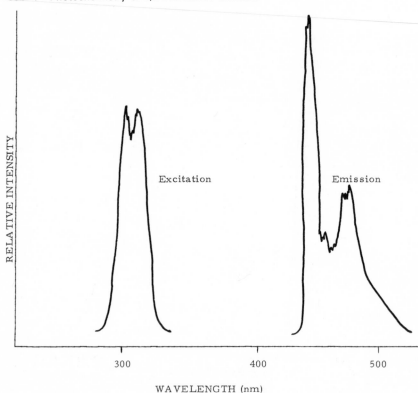

Fig. 7-3. Luminescence spectrum of Na(hfac) in ethanol.

luminescence observed in Na(tfac)† is weak and of questionable validity, but is not due to the solvent. The excitation spectrum of Na(acac) is not identical to the absorption spectrum, which may mean the emission is sensitized by a solvent impurity. Observation of phosphorescence allows one to assign a low-lying triplet state (ca. 22,500 cm^{-1}) associated with the ligand. Since the metal ion does not appreciably change the energy of the intraligand $\pi \rightarrow \pi^*$ transition, it is reasonable to assume that the $^1\pi \rightarrow ^3\pi^*$ band is not shifted either.

The excited states of the 1,3-diketonate chelates result from the absorption of a photon by a particular chromophore of the molecule. Luminescence studies of the 1,3-diketonate chelates of chromium[25] and the rare earths[26,27] show that the absorption of light in a ligand $\pi \rightarrow \pi^*$ transition can lead to metal-ion phosphorescence. This process of energy migration is referred to as intramolecular energy transfer. Crosby et al.[28]

† See end of chapter for list of abbreviations.

have shown that for the rare-earth chelates transfer occurs through a ligand triplet level. With the exception of the Cr(III) chelates, neither energy transfer nor luminescence has been observed in the chelates of the first transition series. Thus, radiative processes are not important in the dissipation of electronic energy for these molecules. However, electronic energy, resulting from the absorption of a photon, can be dissipated via a photochemical reaction.

The first-row transition series chelates of 1,3-diketonates are rich in electronic transitions in the visible region as well, that is, $d-d$ transitions. However, since the photochemistry observed is almost exclusively redox, these transitions are of less interest than the intraligand and charge-transfer transitions that are thought to give rise to the photoredox.

III. PHOTOCHEMICAL REACTIONS OF 1,3-DIKETONATES

A. d^1: VO(acac)$_2$ [23]

Bis(acetylacetonato)oxo V(IV) undergoes a fairly rapid thermal decomposition in alcoholic solvents. The decomposition produces V(V) as evidenced by the disappearance of absorption bands in the visible region characteristic of a d^1 system.[29] Photochemical experiments,[23] however, gave no indication of enhancement of this thermal reaction. Irradiations with a low-pressure Hg lamp for up to 8 hr produced no spectral changes other than those expected for the thermal reaction. Thus, although VO(acac)$_2$ would seem to be a logical choice for observation of photooxidation by CTTL, it does not take place under the conditions employed. Interestingly, the pale-yellow alcoholic solution of V(V) produced by the thermal reaction is photosensitive and reacts to reform VO(acac)$_2$ when exposed to 254-nm radiation. A similar reaction between V(V) and acetylacetone in aqueous solutions irradiated with visible light has been used analytically to determine the concentration of acetylacetone.[30]

B. d^3: Cr(1,3-diketonate)$_3$

1. *Absorption Spectra.* There have been several studies dealing with the absorption spectra of Cr(1,3-diketonate)$_3$ chelates.[9,25,31-34] A summary of the absorptions in the 10 to 40 kK region for twelve chelates is given in Table 7-1.[9,35] The assignments are those proposed by Fatta and Lintvedt.[9] The energy ranges are due to the different electronic effects of the substituent groups in the various chelates. The assignments of the charge-transfer bands and the $^4A_{2g} \rightarrow {}^4T_{1g}$ transition are not unambiguous. Indeed, the transitions between 20 and 30 kK have been the subject of some controversy. The assignments given in Table 7-1 are, however, internally consistent with ligand-field predictions. It is interesting that, on

Table 7-1. Summary of Visible and Near-Ultraviolet Spectra of Cr(1,3-diketonate)$_3$ Chelates

Assigned Transition	Energy Range Observed, kK	Intensity, ε_{max}	Nature of Band
$\pi \to \pi^*$	35–39.5	3×10^4	Intraligand
$\pi \to \pi^*$	31–36	3×10^4	Intraligand
$\pi \to \pi^*$	32.3–33.3	6×10^3	Intraligand
$t_{2g} \to \pi^*$ or $\pi \to e_g$	26.5–29.8	5×10^4	CTTM or CTTL
$t_{2g} \to \pi^*$ or $\pi \to e_g$	22–26	3×10^2	CTTM or CTTL
$^4A_{2g} \to {}^4T_{1g}$	20.8–23.6	60	d–d
$^4A_{2g} \to {}^4T_{2g}$	17.5–18.0	75	d–d
$^4A_{2g} \to {}^2E_{2g}$	12.1–13.0	0.5	d–d

the basis of simple one-electron energy-level diagrams, predicted CTTM and CTTL transitions are at nearly the same energy. Thus, although charge-transfer bands are expected in the 25 kK region, it is not possible at this time to know the direction of the transition.

2. *Emission Spectra.* One reason for interest in the Cr(III) chelates is that they are apparently unique among first-row complexes in that they luminesce in the visible region of the spectra. The phosphorescence observed is due to $^2E_g \to {}^4A_{2g}$ emission, which occurs at about 12.5 kK. This emission in Cr(acac)$_3$ has been studied by several groups.[25,35–37] The emission is observed using several different exciting wavelengths indicating rapid intersystem crossings to the ^2E_g level. The emission from ^2E_g has been used to study the lifetime of the $^4T_{2g}$ state by exciting into the $^4T_{2g}$ and observing emission from ^2E_g.[36,38a] In the most recent investigation of Cr(acac)$_3$, Yardley and Beattie[36] observed a phosphorescence decay lifetime of 431 μsec, corresponding to a rate constant of 2.3×10^4 sec^{-1} at 77°K. This is in general agreement with previous studies.[25,38] The fact that they observed no luminescence maximum led them to the conclusion that the ^2E_g state is populated with a rate constant of 10^7 sec^{-1} or greater. Thus the intersystem crossing is considered to be much more rapid than reported by Chen and Porter.[38a] Yardley and Beattie point out, however, that rapid $^4T_{2g} \to {}^2E_g$ intersystem crossing does not preclude $^4T_{2g}$ from being an important photochemical state since back-thermal-intersystem crossing could be equally fascile. In Cr(acac)$_3$, the $^4T_{2g}$–^2E_g energy separation, as calculated by Fleischauer, Adamson, and Sartori,[38b] is 3.3 kK. This separation is probably too great for thermal back population of $^4T_{2g}$ from the ^2E_g level.

The relative lifetimes and populations of these states are of particular importance to Cr(III) photochemical reactions since they have their origins in strong-field states that differ greatly in the spatial distribution of the d electrons. The ^2E_g state arises from the t_{2g}^3 ground state by simply changing the spin of one electron. Thus, the overall d electron distribution in ^2E_g is similar to the ground state. For this reason, one would not expect a great deal of photochemical activity arising from it. The $^4T_{2g}$ state, on the other hand, has its origin in the $t_{2g}^2 e_g^1$ strong-field configuration. The extent to which this state is populated represents increased electron density in strongly antibonding e_g orbitals. The effect should be a considerable labilization of the ligands. Therefore, the $^4T_{2g}$ state is intrinsically a photoactive state. Several authors have suggested recently that it is the photochemically reactive state.[38,39,40] The tremendously short lifetime attributed to $^4T_{2g}$,[36] however, casts some doubt on how chemically important this state can be. There are, of course, several other spectrally unobserved states, for example, spin-forbidden charge-transfer states and spin-forbidden states originating in the $t_{2g}^2 e_g^1$ configuration, that may be reactive.

3. **Photochemistry.** The photochemistry of Cr(1,3-diketonate)$_3$ chelates has not been investigated to a significant degree at this time. Two quantitative studies have been reported: one deals with the photoinduced isomerization of *cis-trans* bis(1,1,1-trifluoro-5,5-dimethylhexane-2,4-dionato) Cr(III),[41] and the other deals with the partial photoresolution of Cr(acac)$_3$.[42] A third preliminary report[15a] discusses the 254-nm irradiation of Cr(acac)$_3$, Cr(tfac)$_3$, and Cr(hfac)$_3$. In all cases, the quantum yields are very low, indicating efficient mechanisms other than reaction for dissipating the absorbed energy.

An interesting difference between these neutral chelate systems and most other coordination complexes of Cr(III) is their solubility in polar or nonpolar organic solvents. The photochemistry of Cr(III) coordination compounds has been the subject of extensive investigation.[43] In almost all of these studies the solvent is water, and water is intimately involved in the photochemical reaction. In the work cited above,[15a,41,42] the solvents are all organic solvents with varying degrees of coordinating ability, for example, alcohols,[15a] acetone,[42] chlorobenzene,[42] and hexane.[41] In the case of hexane, there can be no doubt that the solvent is not actively participating in the photoisomerization. Perhaps this accounts for the low quantum yields in certain reaction modes. At any rate, the 1,3-diketonates present an effective system for delving into solvent effects in nonaqueous media in Cr(III) photochemistry.

Koob et al.[41] showed that the irradiation of optically dense solutions of pure *cis* or *trans* isomers of unsymmetric Cr(1,3-diketonate)$_3$ chelates leads to isomerization. Since the solvent was hexane no ligand substitution step is possible. The isomerization is wavelength independent from

trans   cis

254 to 366 nm with an initial quantum yield of 0.003 ± 0.002. This region undoubtedly includes intraligand and charge-transfer bands. It is significant to note that no isomerization occurred on direct irradiation of the $^4T_{2g}$ state. No additional reactions were observed at any wavelength.

A photostationary state with $K = trans/cis = 1.6 \pm 0.2$ was achieved upon irradiation, which is essentially identical with thermal isomerization. The authors suggest that the similarity between photoinduced and thermal isomerization is due to a common intermediate. The wavelength dependence, however, rules out the $^4T_{2g}$ state as the common intermediate.

In addition to direct isomerization by irradiation, Koob et al. were able to sensitize the isomerization with methyl-*o*-benzyloxyphenylglyoxylate. Irradiation of the sensitizer at 313 and 366 nm resulted in isomerization quantum yields comparable to the direct irradiation. It was accompanied by Cr(acac)$_3$ quenching of the sensitizer triplet state.

Stevenson[42] has reported the partial resolution of Cr(acac)$_3$ by irradiation with circularly polarized light at 546 nm which corresponds to the

$^4T_{2g} \leftarrow {}^4A_{2g}$ transition. The quantum yield for inversion in chlorobenzene at 25°C is 0.0055. The partially resolved solutions were irradiated at other wavelengths in order to measure photoracemization at several wavelengths. The quantum yield for inversion is relatively independent of wavelength from 365 to 651 nm. Stevenson suggests that the $^4T_{2g}$ state may be the excited state responsible for inversion. The reported energy of the vibronic ground state for $^4T_{2g}$ [38b] would appear to support this suggestion since it is only about $900\,cm^{-1}$ higher in energy than the irradiating photons at 651 nm.

It is interesting to compare the results of Koob et al. and Stevenson with respect to the wavelength dependence of the two photoinduced reactions. The quantum yields in both studies were of the same order of magnitude. However, the wavelength dependence of ϕ is much different for cis-trans isomerism than for racemization. The cis $\rightleftharpoons$ trans ϕ is zero at energies lower than the intraligand and charge-transfer bands. The racemization, ϕ, however, does not diminish down to 650 nm. These data indicate a considerable difference between the excited states or intermediates responsible for the reactions. The cis-trans isomerism that takes place at high-energy irradiations is consistent with a dissociative mechanism. The racemization that takes place during low-energy irradiations appears more compatible with intermediates associated with twist mechanisms.

Photoredox reactions of Cr(III) complexes are apparently rare. This is undoubtedly due to efficient relaxations out of charge-transfer states and into d–d excited states. Nonetheless, one would expect to observe small quantum yields of photoredox, especially in noncoordinating organic solvents where ligand exchange is not fascile. In a search for photoredox, Cr(acac)$_3$, Cr(tfac)$_3$, and Cr(hfac)$_3$ were irradiated at 254 nm in alcoholic solvents.[15a] Solutions saturated with air showed no reaction after 6-hr irradiation time. Solutions thoroughly degassed by freeze-pump-thaw cycles, however, exhibit considerable reaction in 2-hr irradiations. If oxygen is an efficient quencher, it explains the observations of Stevenson[42] that no photoredox takes place, since he states that his solutions were not degassed.

One of the irradiation products in degassed ethanolic solutions is the protonated ligand. This perhaps implies reduction of Cr(III); however, qualitative tests to determine the oxidation state of the metal were inconclusive. The approximate quantum yield for the decomposition of Cr(tfac)$_3$ was determined by measuring the decrease in absorbance of the 343-nm charge-transfer band. The value obtained at 254 nm was approximately 7×10^{-5}.

As mentioned, there is insufficient information for any clear trends to have emerged. Nonetheless, the studies to date have demonstrated interesting photochemical behavior in organic solvent systems. Although the quantum yields are very low, it does appear that selective reactions of Cr(1,3-diketonate)$_3$ chelates can be carried out in inert organic solvents by selective irradiation.

C. d^4: Mn(1,3-diketonate)$_3$

1. *Absorption Spectra.* The Mn(1,3-diketonate)$_3$ chelates exist as high-spin complexes and, as such, have an idealized ground state of ^5E_g, whose strong-field origin is in the $t_{2g}^3 e_g^1$ configuration. On the basis of ligand-field theory one spin-allowed $d-d$ transition is expected in the visible region to the $^5T_{2g}$ level that arises from the $t_{2g}^2 e_g^2$ strong-field state. It is, therefore, surprising that Mn(acac)$_3$ exhibits two bands, one at about 550 nm assigned to $^5T_{2g} \leftarrow {}^5E_g$ and another at 1050 nm. The intensity of the 1050-nm band is approximately one-half the $^5T_{2g} \leftarrow {}^5E_g$ transition intensity.[44,45]

The low-energy band has been assigned to splitting of the $^5T_{2g}$[46] and a transition within the ^5E_g state due to a tetragonal distortion.[47] Both possibilities depend on severe distortions that were considered probable through Jahn–Teller splitting. The crystal structure of Mn(acac)$_3$ has recently been reported by Fackler and Avdeef.[47] Their results show the expected Jahn–Teller tetragonal distortion with two Mn—O bond distances about 0.2 Å longer than the other four. These structural results are in good agreement with a previous spectral assignment by Fackler and co-workers,[44] in which the two bands are assigned to $^5A_{1g} \leftarrow {}^5B_{1g}$ and $^5E_g \leftarrow {}^5B_{1g}$ transitions in a strong tetragonal field.

In earlier work,[48] rationalization was given for assigning the low-energy visible band to a charge-transfer transition. As evidence for a low-energy charge-transfer state, the authors cite the known thermal decomposition of the Mn(III) chelates to Mn(II) 1,3-diketonates in addition to "photodecomposition." They conclude that the band is charge transfer in nature, either $\pi^* \leftarrow e_g$, $t_{2g} \leftarrow O_n$, or $t_{2g} \leftarrow \pi$. The energy of such transitions based on simple ligand-field theory is consistent with a band at about 9000 cm^{-1}.

If the assignment of the low-energy band to a charge-transfer transition is correct, the photochemical implication is clear; it would be possible to populate a reactive, low-energy state that will result in photoredox upon irradiation in the ultraviolet or the visible regions. Based on the recent structural and spectral results, however, it is much more reasonable to assign the band to a $^5A_{1g} \leftarrow {}^5B_{1g}$ transition.

2. *Emission Spectra.* Apparently no emission from Mn(1,3-diketonate)$_3$ chelates has been observed. If the band at 1050 nm discussed above is a charge-transfer band, and especially if it is spin-forbidden, these compounds may exhibit emission in the near infrared. It is doubtful that this spectral region has been investigated.

3. *Photochemistry.* The photodecomposition of Mn(1,3-diketonate)$_3$ chelates has been known for some time,[49] but no quantitative reports appear

Table 7-2. Quantum Yield for Reduction of Mn(tfac)$_3$ at 27°C

Initial concn, moles liter^{-1}	Rate of Reaction, molecules sec^{-1} [a]	Intensity, quanta sec^{-1}	Φ, 254 nm
2.88×10^{-3}	8.18×10^{14}	4.37×10^{15}	0.184 ± 0.009

[a] Corrected for thermal reaction.

in the literature. Gafney and Lintvedt[15a,50] have studied the photoreduction of Mn(tfac)$_3$ in alcoholic solvents. On irradiation of Mn(tfac)$_3$ in degassed, anhydrous ethanol with a medium-pressure Hg lamp, two major products were isolated. One was the protonated ligand, H(tfac), and the other was Mn(tfac)$_2$. The spectra of the two products were identical with authentic samples. Thus, the reaction may be written

$$\text{Mn(tfac)}_3 \rightarrow \text{Mn(tfac)}_2 + \text{H(tfac)} \qquad (7\text{-}1)$$

The free-ligand hydrogen atom is presumed to come from the solvent.

In determination of the quantum yield, complications arise from thermal decomposition. For this reason, this rate was measured and the photochemical rate corrected for thermal decomposition. The measured thermal decomposition rate for a 1.78×10^{-3} M solution at 27 ± 1°C is 2.79×10^{-5} moles min^{-1}. The quantum yield at 254 nm was determined by measuring the decrease in absorbance of the 550 nm d–d transition. The results of several determinations are presented in Table 7-2. Unfortunately, the wavelength dependence of Φ is not known.

D. d^5: Fe(1,3-diketonate)$_3$

The only photochemical studies reported to date on a d^5 1,3-diketonate chelate involve Fe(hfac)$_3$.[23] There are no reports for Mn(1,3-diketonate)$_2$, although this system is of interest since Mn(II) complexes often exhibit emission in the visible region. In addition, Mn(II) chelates seem ideal for observing photooxidation.

1. *Absorption Spectra.* The Fe(III) chelates have spectra that are completely dominated by intraligand and charge-transfer bands, since all d–d bands for this high-spin ion are spin forbidden. Spectrally, Fe(1,3-diketonate)$_3$ chelates present an interesting case since one $\pi^* \leftarrow \pi$ and two charge-transfer bands occur as three well-separated, distinct bands. The assignment of these bands in Fe(acac)$_3$ has been the subject of some controversy[32,51,52]; however, on the basis of spectral shifts in a series of Fe(1,3-diketonate)$_3$ chelates, Lintvedt and Kernitsky[53] have assigned the

high-energy charge transfer to $\pi^* \leftarrow t_{2g}$ and the lower-energy band to $e_g \leftarrow \pi$. The absorption maxima and intensities for a series of chelates are shown in Table 7-3.

The band occurring from 30.0 to 36.6 kK is $\pi^* \leftarrow \pi$, while the bands occurring from 24.5 to 28.3 kK and 20.5 to 22.9 kK are assigned $\pi^* \leftarrow t_{2g}$ (CTTL) and $e_g \leftarrow \pi$ (CTTM), respectively. On the basis of the photochemical data discussed below, it appears that the assignment of the charge-transfer bands should be reversed.[23] Either assignment is consistent with a simple one-electron molecular orbital scheme that predicts that CTTM and CTTL bands should occur near the energies of the observed transitions.[53] The Fe(III) chelates are one of the few cases in

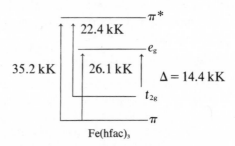

Fe(hfac)₃

which charge-transfer transitions of two types, CTTM and CTTL, may be assigned with some confidence.

Table 7-3. Electronic Absorption Spectra in the 20 to 38 kK Region for Various Tris(1,3-diketonato) Fe(III) Chelates (in Chloroform)

	Wave Number, kK,	Log ε		Wave Number, kK	Log ε
Fe(acac)₃	36.6	4.5	Fe(DBM)₃	30.0	4.1
	28.3	3.7		24.5	3.3
	22.9	3.7		20.5	3.1
Fe(tfac)₃	36.1	4.1	Fe(bztfac)₃	32.7	4.4
	27.2	3.7		25.0	3.4
	22.8	3.6		21.7	3.4
Fe(hfac)₃	35.2	4.4	Fe(3-Pacac)₃	35.2	4.4
	26.1	3.6		26.9	3.6
	22.4	3.6		21.9	3.6
Fe(bzac)₃	33.3	4.2	Fe(3-Bacac)₃	35.0	4.4
	25.9	3.4		26.5	3.6
	22.2	3.1		21.6	3.6

2. Emission Spectra. No emission spectra have been observed in which emission is attributable, in total or in part, to Fe(III).

3. Photochemistry. The first report of a photochemical nature concerning Fe(1,3-diketonate)$_3$ chelates actually discusses a preparation rather than a reaction. In this paper, Goan et al.[54] prepared Fe(hfac)$_3$ from Fe(CO)$_5$ and the ligand in benzene solution by irradiation in the ultraviolet region. In 95% ethanol it appeared that the ligand may be sensitizing the reaction. However, on further investigation, Gafney[15a] determined that the rate of reaction in benzene increases with increasing concentration of Fe(CO)$_5$. The rate for benzene solutions 3.5×10^{-2} M in Fe(CO)$_5$ and 3.3×10^{-3} M in H(hfac) is twice as great as solutions 2.9×10^{-3} M in Fe(CO)$_5$ and 1.8×10^{-2} M in H(hfac). This result suggests the reaction is initiated by photoreaction of Fe(CO)$_5$. Furthermore, the reaction occurs when irradiating at 312 nm, where Fe(CO)$_5$ absorbs strongly.

The only report of Fe(1,3-diketonate)$_3$ photochemical reaction deals with photoreduction of Fe(III) to Fe(II).[23] The reaction is extremely solvent dependent, taking place in alcohols, but not in CHCl$_3$, C$_6$H$_6$, (C$_2$H$_5$)$_2$O, or alcohol/water mixtures. Thus, the solvent plays a significant role in the reaction. In alcoholic solvents the rate of photoreduction increases with increasing stability of the solvent radical, that is, t-butyl alcohol > isopropyl alcohol > ethanol. The quantum yield for Fe(II) production decreases markedly with the addition of water, leading to the conclusion that there is significant ligand replacement by H$_2$O in 95% ethanol.

The fact that different species exist in the different solvents is confirmed by the absorption spectra. In CHCl$_3$, Fe(hfac)$_3$ exhibits three distinct bands at 284, 383, and 447 nm.[53] In ethanol, however, bands occur at 279, 304, and 342 nm. It is interesting that photoredox is not observed in solvents in which the 447-nm band is present. If this band is CTTL, it may provide an efficient mechanism for dissipating excitation energy into the solvent.

The emission spectrum of Fe(hfac)$_3$ in ethanol provides evidence for the presence of species such as [Fe(hfac)$_2$(C$_2$H$_5$OH)$_2$]hfac. The excitation and emission spectra of the free ligand, H(hfac), and Fe(hfac)$_3$ in ethanol are shown in Fig. 7-4. The interpretation of the emission is that it is due to hfac$^-$, which may be associated with Fe(hfac)$_2$(C$_2$H$_5$OH)$_2^+$ in an outer-sphere manner.

The major products of the photoredox reaction were isolated and identified as Fe(hfac)$_2$(C$_2$H$_5$OH)$_2$ and Hhfac by comparison with authentic samples. Thus the data are consistent with the following reaction.

$$[\text{Fe(hfac)}_2(\text{C}_2\text{H}_5\text{OH})_2]\text{hfac} \xrightarrow[\text{C}_2\text{H}_5\text{OH}]{h\nu} \text{Fe(hfac)}_2(\text{C}_2\text{H}_5\text{OH})_2 + \text{Hhfac} \quad (7\text{-}2)$$

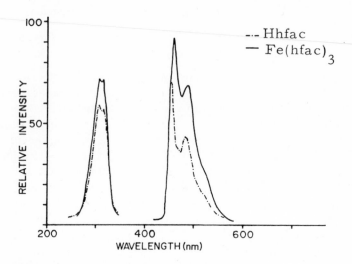

Fig. 7-4. Phosphorescence spectra of hexafluoroacetylacetone and tris(hexafluoroacetylacetonato)iron(III) at 77°K.

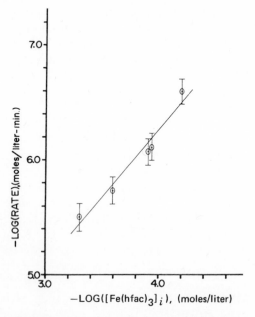

Fig. 7-5. Log of the initial rate versus log of the initial concentration of tris(hexafluoroacetylacetonato)iron(III) for irradiation in ethanol at 345 nm.

Table 7-4.   Quantum Yield for Photoreduction as a Function of Irradiating Wavelength

Irradiating Wavelength, nm	Initial concn of Fe(hfac)$_3 \times 10^4$, M	Av Quantum Yield	Variation, %
253	5.91	0.14	±5
275	5.91	0.12	±6
302	5.91	0.10	±5
312	5.91	0.08	±10
345	5.91	0.13	±7
450	5.91	0.0	

The rate of the reaction is linear with time up to 80% reaction for optically dense solutions in which 99% of the light is absorbed. Results of concentration and intensity variations are consistent with the rate equation,

$$\frac{d(Fe^{2+})}{dt} = \Phi I_a, \qquad I_a = \text{intensity} \qquad (7\text{-}3)$$

The plot of initial rate versus initial concentration is shown in Fig. 7-5.

The dependence of the quantum yield on irradiating wavelength is given in Table 7-4. These data suggest the presence of a reactive state populated by $\pi^* \leftarrow \pi$ irradiation and by charge-transfer irradiation. Since reduction takes place, it is presumably a CTTM state that is efficiently populated by intersystem crossing from π^* levels.

Unlike the Mn(1,3-diketonate)$_3$ system, reduction is not facile in Fe(1,3-diketonate)$_3$ chelates. In fact, apparently only the hfac$^-$ chelate of Fe(II) is stable in air as Fe(hfac)$_2$(solvent)$_2$. The Fe(acac)$_2$ derivatives oxidize readily in air. Thus, similar photoredox studies employing other 1,3-diketonates would be complicated by thermal back-reactions.

E. d^6: Co(1,3-diketonate)$_3$

1. *Absorption Spectra.*   In principle, the absorption spectra of Co(III) octahedral complexes should be readily interpretable since the ground state of this low-spin d^6 ion is the uncomplicated $^1A_{1g}$ state. In this respect, Co(III) systems should be somewhat simpler or at least comparable to Cr(III) and Fe(III) chelates. Barnum[31] reported the absorption maxima for Co(acac)$_3$ in CHCl$_3$ and C$_2$H$_5$OH given in Table 7-5. As was true in Fe(III) and Cr(III) spectra, the assignment of an intense band in the near-ultraviolet-to-visible region creates some difficulties. The assignment of this band at 25.0 kK in Co(acac)$_3$ to the second spin-allowed d–d transition is not entirely reasonable based on simple ligand-field predictions and on its intensity (log $\varepsilon_{max} = 2.5$).

Table 7-5. Spectral Assignments for Co(acac)$_3$

Assignments due to Barnum[39]	Co(acac)$_3$ in CHCl$_3$, kK	log ε_{max}	Co(acac)$_3$ in C$_2$H$_5$OH, kK	log ε_{max}
$\pi \rightarrow \pi^*$	—	—	43.8	4.6
$\pi \rightarrow \pi^*$	39.3	4.54	39.3	4.53
$\pi \rightarrow \pi^*$	33.9	4.0	33.9	4.0
$t_{2g} \rightarrow \pi^*$(CTTL)	30.9	3.9	30.9	3.9
$^1A_{1g} \rightarrow {}^1T_{2g}$	25.0	2.5	25.0	2.5
$^1A_{1g} \rightarrow {}^1T_{1g}$	16.8	2.10	16.8	2.10
Spin-forbidden	12.5	0.5	—	—
d–d bands	9.1	0.28	—	—

For Co(III) chelates a simple ligand-field approach ignoring configuration interaction may be used since such interactions are very small. The dependence of the spin-allowed and spin-forbidden d–d transitions on $10\,Dq$ and interelectronic repulsion parameters is given below.

$$\Delta E(^3T_{1g} \leftarrow {}^1A_{1g}) = 10\,Dq - 3C$$

$$\Delta E(^3T_{2g} \leftarrow {}^1A_{1g}) = 10\,Dq + 8B - 3C$$

$$\Delta E(^1T_{1g} \leftarrow {}^1A_{1g}) = 10\,Dq - C$$

$$\Delta E(^1T_{2g} \leftarrow {}^1A_{1g}) = 10\,Dq + 16B - C$$

The first three absorptions have been observed for Co(acac)$_3$.[31]

	ΔE, kK	log ε_{max}
$^3T_{1g} \leftarrow {}^1A_{1g}$	9.0	0.28
$^3T_{2g} \leftarrow {}^1A_{1g}$	12.5	0.5
$^1T_{1g} \leftarrow {}^1A_{1g}$	16.8	2.1

Using the energy equations and the experimental absorption maxima

$$C = 3.9\,kK$$

$$10\,Dq = 20.7\,kK$$

$$B = 0.44\,kK$$

With these values the calculated $^1T_{2g} \leftarrow {}^1A_{1g}$ transition energy is

$$\Delta E(^1T_{2g} \leftarrow {}^1A_{1g}) = 10\,Dq + 16B - C = 20.7 + 16(0.44) - 3.9 = 23.8\,kK$$

Thus the calculated maxima is at 23.8 or 1.2 kK away from maxima observed at 25.0 kK. The discrepancy may be explained by the presence of a charge-transfer band near 25 kK. Using a simple, one-electron

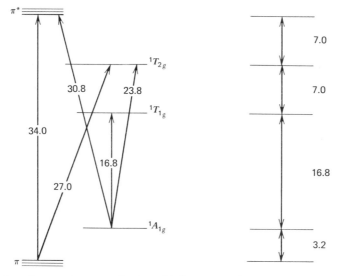

Fig. 7-6. A crude, one-electron molecular orbital representation of the relative energies in Co(acac)₃. Transition energies are in kiloKaysers. Values at right were calculated from the differences in transition energies.

molecular orbital scheme, the prediction of a CTTM band in this region can be rationalized. The calculated energy level scheme is shown in Fig. 7-6.

Interestingly, on this basis the CTTM transition is $^1T_{2g} \leftarrow \pi$. No evidence of the lower-energy $^1T_{1g} \leftarrow \pi$ is found. Although this approach cannot be considered quantitative since more than one π and π^* level may be involved, it does point out that charge-transfer transitions can be expected in the spectral region near 25.0 kK. Indeed, one interpretation of the 25.0 kK band is that it is a combination of $^1T_{2g} \leftarrow \pi$ and $^1T_{2g} \leftarrow {}^1A_{1g}$, in which the predicted energies for the "pure" transitions are 27.0 and 23.8 kK, respectively.

The spectra of other Co(1,3-diketonate)₃ chelates are quite similar to Co(acac)₃. For example, the phenyl-substituted complex Co(DBM)₃ exhibits the absorption maxima shown in Table 7-6.

2. Emission Spectra. No emission has been observed from Co(1,3-diketonate)₃ chelates.

3. Photochemistry. Although many Co(III) coordination complexes have been studied in aqueous media,[55] there is little photochemical data in organic solvents. In a recent paper, Filipescu and Way[56] report the photoredox reactions of Co(acac)$_3$, Co(bzac)$_3$, and Co(DBM)$_3$ in diethyl ether and other organic solvents. The major products were identified as the Co(II) chelate and biacetyl. The following reaction scheme was postulated. The organic products depend on whether or not the solutions were degassed.

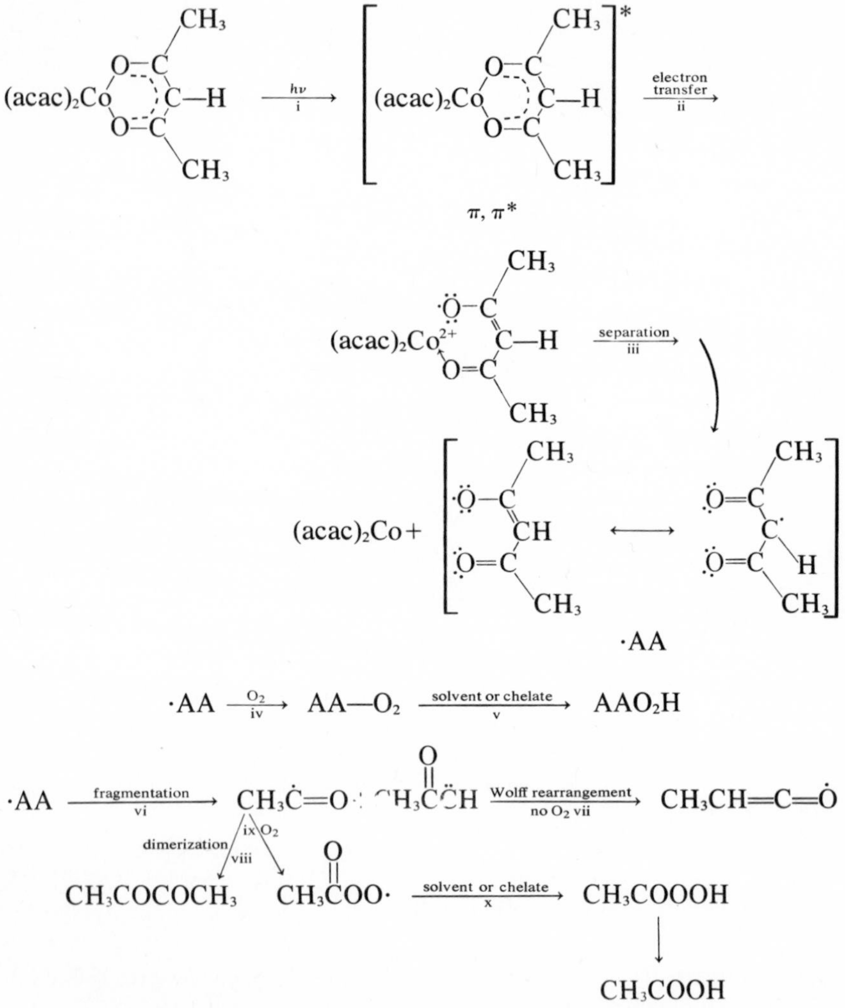

$$(acac)_2Co\,[\text{acac}] \xrightarrow[i]{h\nu} \Big[(acac)_2Co\,[\text{acac}]\Big]^* \xrightarrow[ii]{\text{electron transfer}}$$

$$\pi, \pi^*$$

$$(acac)_2Co^{2+}\,[\text{acac}] \xrightarrow[iii]{\text{separation}} \Big)$$

$$(acac)_2Co + \Big[\text{·}\ddot{O}\text{–C···CH···O=C} \Big] \longleftrightarrow \Big[\ddot{O}=C···C\text{·}···\ddot{O}=C \Big]$$

$$\text{·AA}$$

$$\text{·AA} \xrightarrow[iv]{O_2} \text{AA–O}_2 \xrightarrow[v]{\text{solvent or chelate}} \text{AAO}_2\text{H}$$

$$\text{·AA} \xrightarrow[vi]{\text{fragmentation}} CH_3\dot{C}{=}O \text{·: } \widehat{}H_3C\overset{O}{\overset{\|}{C}}CH \xrightarrow[\text{no O}_2\ vii]{\text{Wolff rearrangement}} CH_3CH{=}C{=}\dot{O}$$

$$\text{dimerization}\diagdown_{viii} \quad \diagup^{ix\,O_2}$$

$$CH_3COCOCH_3 \qquad CH_3\overset{O}{\overset{\|}{C}}OO\text{·} \xrightarrow[x]{\text{solvent or chelate}} CH_3COOOH$$

$$\downarrow$$

$$CH_3COOH$$

The quantum yields for Co(II) show a marked wavelength dependence. As summarized in Table 7-7, the values are about 0.5 for irradiations in the region of the strong intraligand $^*\pi \leftarrow \pi$ transition; they decrease to about 0.1 in the charge-transfer region, and then further to 10^{-3} for

Table 7-6.   Absorption Maxima for Co(DBM)$_3$

Probable Assignment	Co(acac)$_3$, kK[a]	Co(DBM)$_3$, kK[a]
$*\pi \leftarrow \pi$	38.6(4.48)	38.5(4.2)
$*\pi \leftarrow \pi$	33.9(4.0)	33.8(4.2)
$*\pi \leftarrow {}^1A_{1g}$	30.6(3.9)	29.4(4.4)
${}^1T_{2g} \leftarrow \pi$ and ${}^1T_{2g} \leftarrow {}^1A_{1g}$	25.0(2.5)	25.0(3.3)
${}^1T_{1g} \leftarrow {}^1A_{1g}$	16.9(2.1)	16.7(1.7)

[a] $\log \varepsilon_{max}$ given in parentheses.

irradiation of the $d-d$ band at 590 nm. The quantum yields are not dependent on the presence of dissolved O_2 to any appreciable degree, as shown in Table 7-7.

The quantum yield results are strong indications that the photoreduction takes place from relatively high-energy charge-transfer bands, presumably of the CTTM type. If charge-transfer triplet states are present at lower energies, they are apparently not photoreactive. One interpretation of the wavelength dependence is that one of the intense bands occurring at 295 and 325 nm in Co(acac)$_3$ is a CTTM transition. However, the 295-nm band occurs in acac$^-$ itself and, therefore, is almost certainly the $*\pi \leftarrow \pi$ transition. The 325-nm band was originally assigned by Barnum[31] to a $*\pi \leftarrow t_{2g}$ transition, that is, CTTL, but, in view of the photochemical results, it is perhaps more logical to assign it to ${}^1T_{2g} \leftarrow \pi$, that is, CTTM. A redrawn one-electron MO diagram to reflect this assignment change is shown in Fig. 7-7.

Table 7-7.   Quantum Yields for Co(1,3-diketonate)$_3$ Chelates in Ether

Compound	Excitation Wavelength, nm	Co(II) Quantum Yield Nondegassed	Degassed
Co(acac)$_3$	265	0.49	0.52
	325	0.14	0.17
	590	$\sim 2 \times 10^{-3}$	$\sim 4 \times 10^{-3}$
Co(bzac)$_3$	260	0.31	0.55
	355	0.11	0.18
	595	$\sim 1.6 \times 10^{-3}$	$\sim 6 \times 10^{-3}$
Co(DBM)$_3$	285	0.40	0.57
	380	0.11	0.11
	598	$\sim 2 \times 10^{-3}$	$\sim 5 \times 10^{-3}$

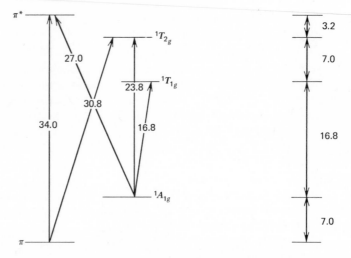

Fig. 7-7. Energy-level diagram in which the 325 nm band is assigned to the $^1T_{2g} \leftarrow \pi$ (CTTM) transition.

If the diagram in Fig. 7-7 is a reasonable representation, the combination band at 400 nm is $*\pi \leftarrow {}^1A_{1g}$ and $^1T_{2g} \leftarrow {}^1A_{1g}$ or, in other words, of the CTTL type. Such an assignment may well explain the sharp decrease in Co(II) formation at irradiations approaching 400 nm. Similar observations have been made[57] concerning the photoreduction of $Co(oxalate)_3^{3-}$, whose wavelength dependence is nearly identical to the $Co(1,3\text{-diketonate})_3$ chelates studied by Filipescu and Way.

The *cis* and *trans* isomers of $Co(1,3\text{-diketonate})_3$ chelates containing unsymmetrically substituted ligands are readily separated by column chromatography,[58] a property due to the relatively large difference in dipole moments. It is this difference that prompted Gafney and Lintvedt[15a] to investigate the photoredox behavior of the *cis*- and *trans*-$Co(tfac)_3$ chelates. As a result of the difference in the position of the polar CF_3 groups, the orientation of polar solvent molecules about the *cis* and *trans* isomers should be considerably different. In addition, if there are significant changes in the polarity or charge of the transition states in the isomers, the photochemical and spectral properties would be affected. Fay and Piper[58] have reported that the spectra in polar (ethanol) and nonpolar (hexane) solvents are identical. It was also determined that the spectra are not affected by ionic-strength charges or by the addition of water.[15a]

The photochemical studies[15a] indicate that optically dense solutions of

Table 7-8. Quantum Yields for the Photoreduction of *cis*- and *trans*-Co(tfac)$_3$ at 254 nm

Solvent	Compound	Initial concn $\times 10^{-3}$, moles/liter	Intensity $\times 10^{15}$, quanta/sec	Rate of Reaction $\times 10^{14}$, moles/sec	Φ
abs ethanol	*cis*-Co(tfac)$_3$	4.30	4.32	5.99	0.122 ± 0.017[a]
abs ethanol	*trans*-Co(tfac)$_3$	4.30	4.83	6.12	0.124 ± 0.003[b]
hexane	*cis*-Co(tfac)$_3$	1.0	3.83	1.69	0.044 ± 0.010[c]
hexane	*trans*-Co(tfac)$_3$	1.0	4.03	1.94	0.038 ± 0.013[c]

[a] Average of ten determinations.
[b] Average of five determinations.
[c] Each value is the result of three independent determinations.

each isomer follow the rate equation

$$\frac{-d[\text{Co(III)}]}{dt} = \Phi I_a \qquad (7\text{-}4)$$

Quantum yields for the disappearance of the Co(III) species during irradiation at 254 nm are shown in Table 7-8 in ethanol and hexane. All solutions were thoroughly degassed by freeze-pump-thaw cycles. The reactions are independent of temperature between 0 and 24°C. In addition, the reactions are independent of ionic strength (as evidenced by the addition of LiCl to ethanol solutions) and of the water concentration. It is also apparent from the data in Table 7-8 that the quantum yields for the *cis* and *trans* isomers are the same within experimental error. The differences in Φ for the polar and nonpolar solvent systems are most logically due to subsequent thermal reactions. For example, in ethanol the isolated Co species is Co(tfac)$_2$(EtOH)$_2$, while in hexane a precipitate of composition Co(tfac)$_2$ forms, which is almost certainly polymeric [Co(tfac)$_2$]$_x$. The important fact, however, is that the quantum yields for Co(III) reduction are the same for both isomers.

The independence of the quantum yields and the charge-transfer transitions of the two isomers to changes in solvent polarity and ionic strength is consistent with a transition state that is not polarized or charged to any significant extent.

F. d^8: Ni(1,3-diketonate)$_2$

Only one quantitative study of Ni(II) photochemistry has appeared, and it reports the photoredox behavior of Ni(II) in bis(1,3-diketonato)nickel(II) chelates.[59]

1. *Absorption Spectra.* Ni(II) complexes are known to exhibit several different geometries, for example, tetrahedral, square planar, octahedral. With 1,3-diketonate ligands, however, the predominate geometry is octahedral, especially in coordinating solvents in which solvent molecules occupy the fifth- and sixth-coordination sites. In such solvents, the $d-d$ absorption spectra is explainable on the basis of pseudooctahedral Ni(II) coordination. In d^8, octahedral symmetry, three spin-allowed and several spin-forbidden $d-d$ transitions are expected.[60] In many complexes, the third spin-allowed transition is observed in the 25- to 30-kK spectral region. In the 1,3-diketonates the strong intraligand $*\pi \leftarrow \pi$ transitions obscure this transition. However, ligand-field predictions based on the observed energies of other $d-d$ transitions place the third spin-allowed transition at energies near or greater than 30 kK.[61] Other spin-forbidden $d-d$ transitions are expected at even higher energies, based also on ligand-field predictions.[62]

A relatively unusual feature of the Ni(II) 1,3-diketonates is the absence of absorptions assignable to charge-transfer transitions at energies below the $*\pi \leftarrow \pi$ transition. This may indicate that charge-transfer absorptions are at energies higher than or comparable to $*\pi \leftarrow \pi$ bands. The ultraviolet spectrum of Ni(acac)$_2$ in ethanol, for example, consists of a complex, strong absorption whose maximum is at 295 to 300 nm or 34 kK ($\varepsilon_{max} = 3 \times 10^4$ M^{-1} cm^{-1}). The maximum corresponds to the familiar $*\pi \leftarrow \pi$ transition in chelated acac$^-$. This band has a low energy shoulder at 312 nm and a high energy shoulder at about 255 nm. No other intense bands are observed at lower energies. There are several possible explanations for the shoulders; other components of the $*\pi \leftarrow \pi$ transition, charge-transfer bands, or the third spin-allowed $d-d$ transition.

2. *Emission Spectra.* Although phosphorescence emission has been observed for NiO,[63] Ni(H$_2$O)$_6^{2+}$, Ni(NH$_3$)$_6^{2+}$, and Ni(en)$_3^{2+}$,[64] there apparently are no reports of emission from Ni(1,3-diketonate)$_2$. For the complexes in which phosphorescence is observed, the emission is due to the second spin-forbidden state, $^1T_{2g} \rightarrow {}^3A_{2g}$. This is a rare example in which phosphorescence does not originate from the lowest excited spin-forbidden state.

3. *Photochemistry.* As mentioned above, only one report of photochemical activity involving Ni(II) complexes has appeared.[59] In this paper, two Ni(1,3-diketonate)$_2$ chelates were irradiated at 254 nm in ethanol and in ethanol/water. The isolated products were metallic nickel and the protonated ligand. The nickel was identified chemically and the ligand spectrally. Isolation of the protonated ligand implies hydrogen abstraction from the solvent or another Ni(II) chelate. The photoredox is extremely

sensitive to oxygen. Solutions that were not thoroughly degassed showed no photochemical activity even after prolonged irradiation.

During the reaction a black intermediate solid forms as a dispersion in the irradiated solution. This product is air sensitive. Opening the cell to the atmosphere results in an instantaneous reaction of the black dispersion and reformation of the green Ni(1,3-diketonate)$_2$ solution. The kinetics of Ni^0 formation follow the rate equation

$$\frac{d(\text{Ni}^0)}{dt} = \Phi I_a^{0.8\pm0.2} \tag{7-5}$$

The deviation from a first-order dependence on light intensity is undoubtedly the result of difficulties in correcting for dispersed light in these heterogeneous systems.

The observations are consistent with a two-step mechanism in which Ni(I) is formed in a primary photochemical process. It is believed that the black dispersion is the Ni(I) chelate. The second step is a thermal reaction producing Ni^0 by oxidation of the ligand or by disproportionation.

$$\text{Ni(acac)}_2(\text{EtOH})_2 \xrightarrow[\text{Ethanol}]{h\nu, 254\,\text{nm}} \text{Ni(acac)} + \cdot\text{acac} \tag{7-6}$$

$$\text{Ni(acac)} \xrightarrow{\hspace{3cm}} \text{Ni}^0 + \cdot\text{acac} \tag{7-7}$$

$$2 \cdot \text{acac} \xrightarrow[\text{abstraction}]{\text{H}} 2\text{Hacac} \tag{7-8}$$

The quantum yield data for Ni^0 formation are given in Table 7-9.

The wavelength dependence of the quantum yield was determined by irradiating ethanolic solutions of Ni(acac)$_2$ at 312 nm, which corresponds to the low-energy shoulder of the $^*\pi \leftarrow \pi$ band. The degassed solution showed no photochemical reaction after 17.5-hr irradiation at 312 nm. As seen above, the photoreduction does take place with appreciable quantum yields upon irradiation at 254 nm, which corresponds to the high-energy shoulder of the 295 to 300 nm band. These observations indicate reaction from a specific high-energy state. In addition, while reduction may be

Table 7-9. Quantum Yield of Photoreduction for Irradiation at 254 nm[a]

Compound	Solvent	Initial concn, M	Intensity, quanta/sec	Rate, R, molecules/sec	Φ_{av}[b]
Ni(acac)$_2$	Ethanol	5.20×10^{-3}	1.98×10^{16}	5.64×10^{13}	1.8×10^{-2}
Ni(acac)$_2$	95% ethanol/ water	5.20×10^{-3}	2.37×10^{16}	2.35×10^{14}	2.3×10^{-2}
Ni(tfac)$_2$	Ethanol	5×10^{-3}	2.35×10^{16}	5.64×10^{13}	4.6×10^{-3}

[a] Results obtained at $23 \pm 1°C$ in ethanol. The ethanol was degassed by freeze-thaw techniques.
[b] The quantum yields shown are calculated from $R/\{(I_i^0)(I_i^f)\}^{1/2}$, where I_i^0 is the incident light intensity and I_i^f is the final incident corrected for the absorbance of the deposited nickel film.

initiated by $*\pi \leftarrow \pi$ absorption, it does not appear that the $*\pi$ vibrational ground state is involved since $\Phi = 0$ at 312 nm. It does appear that the acceptor is either a vibrationally excited $*\pi$ state or a charge-transfer, CTTM, state near 254 nm. If a $*\pi$ state is the acceptor, it is quite feasible that intersystem crossing to the $^3T_{2g}(P)$ state or higher-energy singlet d states results in Ni(II) reduction. As mentioned above, such singlet states are predicted (based on ligand-field theory) to be of approximately the correct energy. Whatever the nature of the initially populated state, it seems clear that metal acceptor state is of quite high energy.

G. d^9: Cu(1,3-diketonate)$_2$

Only a few photochemical studies of Cu(II) complexes in general have been reported.[65] One of these deals with photoredox reactions of 1,3-diketonate chelates.[66]

1. **Absorption Spectra.** Since the only photochemical reactivity of Cu(II) complexes observed to date is photoreduction initiated by high-energy radiation (254 nm), only the ultraviolet absorption spectra is considered. Absorptions in this region for the bis(1,3-diketonato) Cu(II) chelates have been assigned by Cotton and co-workers.[67,68] There are two bands at wavelengths shorter than 350 nm whose positions depend on the groups substituted on the chelate rings. The first band, which occurs between about 280 and 310 nm, is assigned to the $*\pi \leftarrow \pi$ transition within the ligand. The second band at 230 to 255 nm is assigned to a charge-transfer transition of the ligand-to-metal type (CTTM).

2. **Emission Spectra.** No emission from the Cu(1,3-diketonate)$_2$ chelates has been observed.

3. **Photochemistry.** The overall reaction that takes place during 254-nm irradiation of ethanolic solutions of Cu(1,3-diketonato)$_2$ chelates is represented below. The isolated products are metallic copper and the protonated ligand.

$$CuL_2 \xrightarrow[\text{EtOH}]{h\nu} Cu^0 + 2HL \qquad (7\text{-}9)$$

The reaction is air sensitive and proceeds smoothly only in carefully degassed solutions. Several experimental observations (including a temperature dependence of the quantum yield) are explained by an initial photochemical reduction to Cu(I) followed by thermal reduction to Cu^0.

$$CuL_2 + h\nu \xrightarrow[k_{-1}]{k_1} [CuLL\cdot]^* \qquad (7\text{-}10)$$

$$[CuLL\cdot]^* \xrightarrow{k_2} CuL + \cdot L \qquad (7\text{-}11)$$

$$CuL \xrightarrow{k_3} Cu^0 + \cdot L \qquad (7\text{-}12)$$

Table 7-10. Quantum Yields for Reduction of Cu(1,3-diketonate)$_2$ Chelates[a] at 254 nm

Compound	Initial concn, M	Intensity, quanta/sec	Reaction rate, molecules/sec	Φ_{av} [b]
Cu(acac)$_2$	1.0×10^{-3}	3.23×10^{15}	4.52×10^{13}	0.018 ± 0.005
	4.26×10^{-5}	3.09×10^{15}	1.97×10^{13}	0.017 ± 0.004
Cu(tfac)$_2$	1.0×10^{-3}	3.40×10^{15}	6.97×10^{13}	0.027 ± 0.007
	6.0×10^{-5}	3.35×10^{15}	2.90×10^{13}	0.028 ± 0.004
Cu(hfac)$_2$	1.0×10^{-3}	2.99×10^{15}	1.04×10^{14}	0.036 ± 0.006
	5.95×10^{-5}	3.79×10^{15}	3.23×10^{13}	0.034 ± 0.005
Cu(t-buac)$_2$	1.0×10^{-3}	4.30×10^{15}	5.16×10^{13}	0.012 ± 0.004
	6.50×10^{-5}	4.83×10^{15}	2.83×10^{13}	0.012 ± 0.001
Cu(bzac)$_2$	1.0×10^{-3}	4.11×10^{15}	7.36×10^{13}	0.018 ± 0.004
	6.50×10^{-5}	4.58×10^{15}	4.47×10^{13}	0.017 ± 0.003

[a] Results obtained at $27 \pm 1°C$.
[b] Each value is the result of at least 4 and as many as 12 independent determinations.

There is some evidence for the CuL intermediate since a black copper-containing species appears during photolysis. Nast[69] has reported the preparation of Cu(acac)$_1$, which is black and readily decomposes to Cu^0 and Cu(acac)$_2$. The mechanism of the thermal reaction is not known. However, rate studies indicate that there is an induction period consistent with a disproportionation reaction.

$$2CuL \rightarrow Cu^0 + CuL_2 \qquad (7-13)$$

In this case, the induction time would be necessary to build a steady-state concentration of CuL.

The quantum yields for five different Cu(1,3-diketonate)$_2$ chelates are given in Table 7-10. In all cases, irradiation at wavelengths longer than 260 nm resulted in no observable reaction. Therefore, reaction is assumed to be initiated only by irradiation of the band near 250 nm, which is assigned to a CTTM transition. Indeed, the photochemical results lend strong support to this assignment.

The differences in the quantum yields for these chelates are significantly outside the error limits and appear to be dependent on the ligand substituent groups. If the reduction is caused by migration of an electron mainly associated with the ligand to the metal, then there are obvious similarities between the reduction potential of the chelated Cu(II) ions and the CTTM photoreduction. The polarographic reduction potentials of these chelates have been measured in 75% dioxane.[70] The relationship between the polarographic $E_{1/2}$ values and the quantum yields is shown in Fig. 7-8. The complexes are identified by the groups present at the 1 and 3

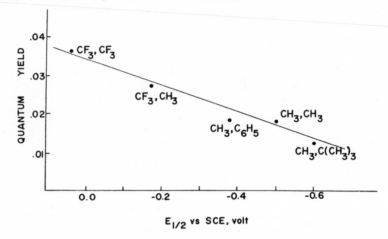

Fig. 7-8. Quantum yields of Cu(II) chelates versus the half-wave potentials of reduction.

positions of the chelate rings. The relationship is quite linear, showing that as the ease of reduction increases the quantum yield increases.

IV. SUMMARY AND CONCLUSIONS

Although relatively few photochemical studies of 1,3-diketonate chelates have been reported, there appear to be some generalized observations, which are summarized below. Since the only comparative studies reported involve photoredox, only those results are summarized.

A. Photochemical Reactions

The rate laws and photoredox reactions for optically dense (greater than 99% of the light is absorbed by the chelates) ethanolic solutions are given in Table 7-11. The photoreduction of the chelates is consistent with the following scheme.

$$ML_n + h\nu \underset{k_{-1}}{\overset{k_1}{\rightleftharpoons}} [ML_{n-1}L\cdot]^* \tag{7-14}$$

$$[ML_{n-1}L\cdot]^* \xrightarrow{k_2} ML_{n-1} + \cdot L \tag{7-15}$$

The ligand radical then undergoes further reaction, and, when $M = Cu^{2+}$ or Ni^{2+}, ML_{n-1} reacts thermally yielding M^0. In optically dense solutions, the rates of chelate disappearance and of product appearance are given by

$$-\frac{d(ML_n)}{dt} = \left(1 - \frac{k_{-1}}{k_2 + k_{-1}}\right) I_a \tag{7-16}$$

Table 7-11. The Photochemical Reactions Occurring in Ethanol and the Rate Equations for Optically Dense Solutions

---

1. $Mn(tfac)_3 \xrightarrow[\lambda=254nm]{} Mn(tfac)_2(EtOH)_2 + Htfac$

2. $Fe(hfac)_3 \xrightarrow[\lambda<350nm]{} Fe(hfac)_2(EtOH)_2 + Hhfac$

$$\frac{d[Fe(hfac)_2(EtOH)_2]}{dt} = \Phi I_a^{1.0\pm0.1}$$

3. $cis\text{-}Co(tfac)_3 \xrightarrow[\lambda<366nm]{} Co(tfac)_2(EtOH)_2 + Htfac$

$$\frac{-d[cis\text{-}Co(tfac)_3]}{dt} = \Phi I_a^{1.0\pm0.1}$$

$trans\text{-}Co(tfac) \xrightarrow[\lambda<366nm]{} Co(tfac)_2(EtOH)_2 + Htfac$

$$\frac{-d[trans\text{-}Co(tfac)_3]}{dt} = \Phi I_a^{0.9\pm0.1}$$

4. $NiL_2 \xrightarrow[\lambda=254nm]{} Ni^0 + 2HL$

$$\frac{d[Ni^0]}{dt} = \Phi I_a^{0.8\pm0.2}$$

$$L = acac^-, tfac^-$$

5. $CuL_2 \xrightarrow[\lambda=254nm]{} Cu^0 + 2HL$

$$\frac{-d[CuL_2]}{dt} = \Phi I_a^{0.9\pm0.1}$$

$$L = acac^-, tfac^-, hfac^-, t\text{-}buac^-, bzac^-$$

---

and

$$\frac{d(ML_{n-1})}{dt} = \left(\frac{k_2}{k_2+k_1}\right)I_a \qquad (7\text{-}17)$$

Since

$$-\frac{d(ML_n)}{dt} = \Phi I_a \qquad (7\text{-}18)$$

Rearranging

$$\frac{k_2}{k_{-1}} = \frac{\Phi}{1-\Phi} \qquad (7\text{-}19)$$

The ratio, k_2/k_{-1}, is a measure of the number of molecules reacting and, therefore, is called the reactivity ratio. Although it is not possible to measure either k_2 or k_{-1} directly in these systems, k_2 can be estimated by

Table 7-12. Quantum Yields and Reactivity Ratios for Irradiations at 254 nm in Degassed Ethanol

Chelate	Quantum Yield for Metal-Ion Reduction	k_2/k_{-1}	Number of d Electrons
VO(acac)$_2$	0	0	1
Cr(tfac)$_3$	?	?	3
Mn(tfac)$_3$	0.18	0.23	4
Fe(hfac)$_3$	0.14	0.16	5
Co(tfac)$_3$	0.12	0.14	6
Ni(tfac)$_2$	0.005	0.005	8
Cu(tfac)$_2$	0.027	0.03	9

the Gaussian analysis of the charge-transfer band by the method of Strickler and Berg.[71] In the Cu(II) and Fe(III) complexes, the bands are sufficiently well resolved to calculate the lifetimes of the charge-transfer states. Using this approximate method, the lifetime of this state for Cu(tfac)$_2$ was found to be 1×10^{-9} sec.[76] Thus $k_2 \geq 1 \times 10^9$ sec^{-1} and $k_{-1} \geq 3 \times 10^{10}$ sec^{-1}. A similar analysis of Fe(hfac)$_3$ spectral data yields an identical lifetime for the change-transfer state and, therefore, an identical k_2. This suggests that the differences in reactivity of these chelates is not due to different rates of reaction, k_2, but rather to different rates of relaxation, k_{-1}, from the excited state. Unfortunately, the spectra of other chelates are not sufficiently well resolved to calculate k_2.

A tabulation of quantum yields and reactivity ratios for different metal chelates with similar ligands is given in Table 7-12. All results are for 254-nm irradiations in degassed, dry ethanol.

B. Role of the Solvent

The best information to date indicates that the photoreduction of 1,3-diketonate chelates is quite solvent dependent. The most comparable results are for the photoreduction of Co(tfac)$_3$ in ethanol and in hexane. In the hydrocarbon solvent, the quantum yield is only about 25% of the value, $\Phi = 0.12$, found in ethanol. Interestingly, Filipescu and Way[56] report a photoreduction Φ of 0.52 for Co(acac)$_3$ in diethyl ether irradiated at 265 nm. These large differences are most reasonably explained by differing solvent interaction.

In another solvent-dependent study, Cu(tfac)$_2$ was photoreduced in the alcohols, C_2H_5OH, n-C_3H_7OH, iso-C_3H_7OH, and n-C_4H_9OH.[15a] The quantum yield of photoreduction increases in the order $C_2H_5OH < n$-$C_3H_7OH <$ iso-$C_3H_7OH < n$-C_4H_9OH. If formation of ligand radicals is followed by hydrogen atom abstraction from the solvent, then the Φ trend is understandable since the ease of solvent radical formation follows the same trend.[72]

It is reasonable to expect important solvent dependence, particularly when dealing with hydrogen-bonding solvents since the electronic and molecular structures of these chelates are well suited to strong hydrogen-bonding interactions of the type shown below.

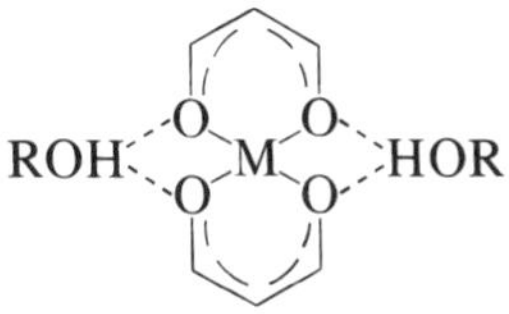

$$\text{ROH} \cdots \overset{\cdots O \quad O \cdots}{\underset{O \quad O}{M}} \cdots \text{HOR}$$

C.   Mechanism

For the photoreduction reactions in alcoholic solvents, the similarity of products, the reaction kinetics, and the effect of solvent and ionic strength changes indicate a common mechanism. All available data is consistent with the following

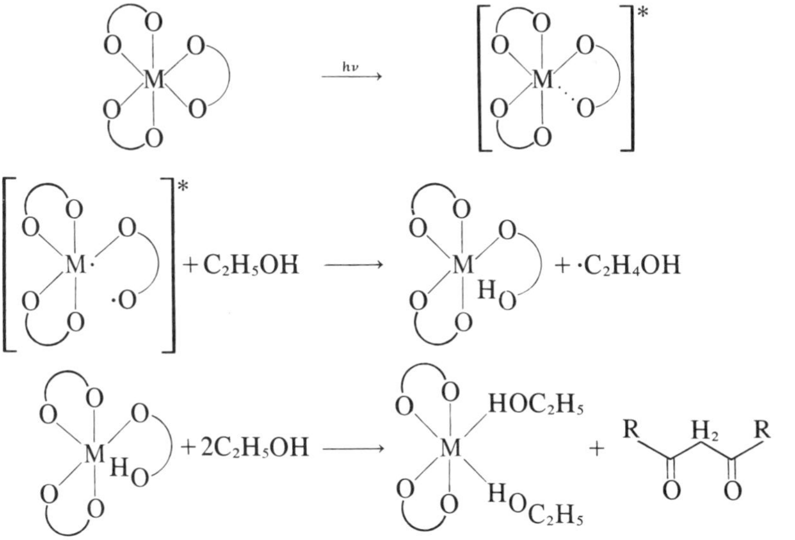

$$\left[\text{M chelate}\right] \xrightarrow{h\nu} \left[\text{M chelate}\right]^{*}$$

$$\left[\text{M}\cdot \text{ chelate} \cdot\text{O}\right]^{*} + C_2H_5OH \longrightarrow \text{M chelate (HO)} + \cdot C_2H_4OH$$

$$\text{M chelate (H}_2\text{O)} + 2C_2H_5OH \longrightarrow \text{M(HOC}_2\text{H}_5\text{)(HOC}_2\text{H}_5\text{)} + \overset{R \quad H_2 \quad R}{\underset{O \quad O}{}}$$

For Cu(II) and Ni(II) chelates a similar mechanism can be written, followed by the thermal reduction to the M^0 state.

Much of the work reported here may be considered to be the initial efforts in a difficult, but exceedingly interesting area of inorganic photochemistry. The versatility of the systems discussed is sufficiently great to warrant enduring the difficulties of a relatively complex ligand system. It is hoped that this discussion will generate additional interest by showing that photochemical investigations of neutral chelates can produce results that complement more traditional aqueous studies of complex ions.

LIST OF ABBREVIATIONS

acac	2,4-pentanedionate anion
tfac	1,1,1-trifluoro-2,4-pentanedionate anion
hfac	1,1,1,5,5,5-hexafluoro-2,4-pentanedionate anion
bzac	1-phenyl-1,3-butandionate anion
DBM	1,3-diphenyl-1,3-propandionate anion
bztfac	1-phenyl-4,4,4-trifluoro-1,3-butanedionate anion
3-Pacac	3-phenyl-2,4-pentanedionate anion
3-Bacac	3-benzyl-2,4-pentanedionate anion
t-buac	2,2-dimethyl-3,5-hexandionate anion
en	ethylenediamine
CTTM	charge-transfer-to-metal
CTTL	charge-transfer-to-ligand

REFERENCES

1. J. P. Fackler, *Progr. Inorg. Chem.* **7**, 361 (1966).
2. D. W. Thompson, *J. Chem. Educ.* **48**, 79 (1971), and references therein.
3. J. A. S. Smith and J. D. Thwaites, *Disc. Faraday Soc.* 143 (1962).
4. D. W. Barnum, *J. Inorg. Nucl. Chem.* **21**, 221 (1961).
5. E. C. Lingafelter and R. L. Braum, *J. Amer. Chem. Soc.* **88**, 2951 (1966).
6. R. H. Holm and F. A. Cotton, *J. Amer. Chem. Soc.* **80**, 5658 (1958).
7. A. M. Fatta and R. L. Lintvedt, *Inorg. Chem.* **10**, 478 (1971).
8. L. G. Van Uitert, W. C. Fernelius, and B. E. Douglas, *J. Amer. Chem. Soc.* **75**, 2736 (1953).
9. R. L. Lintvedt and H. F. Holtzclaw, *Inorg. Chem.* **5**, 239 (1966).
10. J. P. Collman, R. A. Moss, H. Maltz, and C. C. Heindel, *J. Amer. Chem. Soc.* **83**, 531 (1961).
11. C. J. Ballhausen, *Introduction to Ligand Field Theory*, McGraw-Hill, New York, 1962.
12. T. M. Dunn, *Modern Coordination Chemistry* (J. Lewis and R. G. Wilkins, Eds.), Wiley-Interscience, New York, 1960.
13. C. J. Balhausen, *Progr. Inorg. Chem.* **2**, 251 (1960).
14. H. L. Schlafer, *J. Phys. Chem.* **69**, 2201 (1965).

15. (a) H. D. Gafney, "The Photochemical Reactions of 1,3-Diketonate Chelates of the First Transition Series," Ph.D. Thesis, Wayne State Univ., 1970; (b) R. L. Belford, A. E. Martell, and M. Calvin, *J. Inorg. Nucl. Chem.* **2**, 11 (1956); (c) F. A. Cotton, C. B. Harris, and J. J. Wise, *Inorg. Chem.* **6**, 909 (1967); (d) D. W. Barnum, *J. Inorg. Nucl. Chem.* **22**, 183 (1961); (e) L. S. Forster, *J. Amer. Chem. Soc.* **86**, 300 (1964).

16. M. Kasha, *Disc. Faraday Soc.* **9**, 14 (1950).

17. H. H. Jaffe and M. Orchin, *Theory and Applications of Ultraviolet Spectroscopy*, Wiley, New York, 1962, Chapter 7.

18. L. W. Orgel, *Quart. Revs.* **8**, 422 (1956).

19. C. K. Jørgensen, *Orbitals in Atoms and Molecules*, Academic, New York, 1962, Chapter 7.

20. C. K. Jørgensen, *Absorption Spectra and Chemical Bonding in Complexes*, Addison-Wesley, Reading, Mass., 1962.

21. A. J. Fry, R. S. H. Lin, and G. S. Hammond, *J. Amer. Chem. Soc.* **88**, 4781 (1966).

22. A. Vogler and A. W. Adamson, *J. Amer. Chem. Soc.* **90**, 5943 (1968).

23. H. D. Gafney, R. L. Lintvedt, and I. S. Jaworiwsky, *Inorg. Chem.* **9**, 1728 (1970).

24. W. Siebrand, *J. Chem. Phys.* **47**, 2411 (1967).

25. K. DeArmond and L. S. Forster, *Spectrochim. Acta* **19**, 1393, 1403, 1687 (1963).

26. G. A. Crosby and R. E. Whan, *J. Chem. Phys.* **36**, 863 (1962).

27. J. R. Leto, *J. Chem. Phys.* **40**, 2790 (1964).

28. G. A. Crosby, R. E. Whan, and R. M. Alire, *J. Chem. Phys.* **34**, 743 (1961).

29. D. Ogden and J. Selbin, *J. Inorg. Nucl. Chem.* **30**, 1227 (1968).

30. K. S. Panwar and J. N. Gaur, *Talanta* **14**, 127 (1967).

31. D. W. Barnum, *J. Inorg. Nucl. Chem.* **21**, 221 (1961); D. W. Barnum, *ibid.* **22**, 183 (1961).

32. I. Hanajaki, F. Hanajaki and S. Nagakura, *J. Chem. Phys.* **50**, 265, 276 (1969).

33. T. S. Piper and R. L. Carlin, *J. Chem. Phys.* **36**, 3330 (1962).

34. P. R. Singh and R. Sahai, *Aust. J. Chem.* **22**, 1169 (1969).

35. L. S. Forster and P. X. Armendarez, *J. Chem. Phys.* **40**, 273 (1964).

36. J. T. Yardley and J. K. Beattie, *J. Amer. Chem. Soc.* **94**, 8925 (1972).

37. H. L. Schlafer, H. Gausmann, and H. Witzke, *J. Chem. Phys.* **46**, 1423 (1967).

38. (a) S. N. Chen and G. B. Porter, *J. Amer. Chem. Soc.* **92**, 2189 (1970); (b) P. D. Fleischauer, A. W. Adamson, and G. Sartori, *Prog. Inorg. Chem.* **17**, 1 (1972).

39. G. B. Porter, S. N. Chen, H. L. Schlafer, and H. Gausmann, *Theor. Chim. Acta* **20**, 81 (1970).

40. N. A. P. Kane-Maguire and C. H. Langford, *J. Chem. Soc. D*, 895 (1971).

41. R. D. Koob, J. Bensen, S. Anderson, D. Gerber, S. P. Pappas, and M. L. Morris, *J. Chem. Soc., Chem. Commun.*, 966 (1972).

42. K. L. Stevenson, *J. Amer. Chem. Soc.* **94**, 6652 (1972).

43. V. Balzani and V. Carassiti, *Photochemistry of Coordination Compounds*, Academic, New York, 1970, Chapter 7.

44. (a) J. P. Fackler, T. S. Davis, and I. D. Chawla, *Inorg. Chem.* **4**, 130 (1965); (b) T. S. Davis, J. P. Fackler, and M. J. Weeks, *ibid.* **7**, 1994 (1968).

45. C. K. Jørgensen, *Acta Chem. Scand.* **16**, 2406 (1962).

46. R. Dingle, *J. Mol. Spectry.* **9**, 426 (1962).

47. J. P. Fackler and A. Avdeef, Paper No. 164, 165th ACS National Mtg., Dallas, Texas, April 1973.
48. T. S. Piper and R. L. Carlin, *Inorg. Chem.* **2**, 260 (1963).
49. V. Balzani and V. Carassiti, *Photochemistry of Coordination Compounds*, Academic, New York, 1970, Chapter 9 and references therein.
50. H. D. Gafney and R. L. Lintvedt, unpublished data.
51. D. W. Barnum, *J. Inorg. Nucl. Chem.* **21**, 22 (1961); **22**, 183 (1961).
52. Y. Murakami and K. Nakamura, *Bull. Chem. Soc. Jap.* **39**, 901 (1966).
53. R. L. Lintvedt and L. K. Kernitsky, *Inorg. Chem.* **9**, 491 (1970).
54. J. Goan, G. Huether, and H. Podall, *Inorg. Chem.* **2**, 1078 (1963).
55. See, for example, V. Balzani and V. Carassiti, *Photochemistry of Coordination Compounds*, Academic, New York, 1970, Chapter 11.
56. N. Filipescu and H. Way, *Inorg. Chem.* **8**, 1863 (1969).
57. V. Balzani, L. Moggi, F. Scandola, and V. Carassiti, *Inorg. Chim. Acta Rev.* **1**, 7 (1967).
58. R. C. Fay and T. S. Piper, *J. Amer. Chem. Soc.* **85**, 500 (1963).
59. H. D. Gafney and R. L. Lintvedt, *J. Amer. Chem. Soc.* **92**, 6996 (1970).
60. See, for example, A. B. P. Lever, *Inorganic Electronic Spectroscopy*, Elsevier, New York, 1968, Chapter 9.
61. J. R. Kline and R. L. Lintvedt, unpublished results.
62. C. K. Jørgensen, *Absorption Spectra and Chemical Bonding in Complexes*, Pergamon, London, 1962, Chapter 5.
63. F. A. Kroger, H. J. Vink, and J. vanden Bloomgaard, *Physica* **18**, 77 (1952).
64. L. Ancarani-Rossiello, *Ric. Sci. Rend.* **A6**, 437 (1964).
65. V. Balzani and V. Carassiti, *Photochemistry of Coordination Compounds*, Academic, New York, 1970, Chapter 13.
66. H. D. Gafney and R. L. Lintvedt, *J. Amer. Chem. Soc.* **93**, 1623 (1971).
67. J. P. Fackler, F. A. Cotton, and D. W. Barnum, *Inorg. Chem.* **2**, 97 (1963).
68. J. P. Fackler and F. A. Cotton, *Inorg. Chem.* **2**, 102 (1963).
69. R. Nast, R. Mohr, and C. Schultz, *Chem. Ber.* **96**, 2127 (1963).
70. R. L. Lintvedt, H. D. Russell, and H. F. Holtzclaw, Jr., *Inorg. Chem.* **5**, 1603 (1966).
71. S. J. Strickler and R. A. Berg, *J. Chem. Phys.* **37**, 814 (1962).
72. P. Gray and A. Williams, *Chem. Rev.* **59**, 239 (1959).

8

THE PHOTOLYSIS OF SIMPLE INORGANIC IONS IN SOLUTION*

Malcolm Fox

School of Chemistry
Leicester Polytechnic
Leicester, United Kingdom

I. INTRODUCTION

Ultraviolet irradiation of deaerated aqueous sulfate at pH 4 evolves hydrogen, whereas identical treatment of aqueous perchlorate produces oxygen. The former involves a charge-transfer-to-solvent (CTTS) inter-molecular process, while the latter proceeds through an internal transition of the ion. These examples illustrate the boundaries of interest for this chapter. Simple anions are regarded as the halides/pseudohalides and oxyanions of Groups 4 to 7. Metal-cation spectra and photolyses will also be considered.

With the benefit of hindsight, we can now see that the major interpretative difficulties in early work lay with identification of the reacting species. In early work, the initial step in the irradiation of aqueous iodide was proposed[1] as the formation of an iodine radical and a "free-solution electron." Following photoionization the electron was thought to be ejected into the solution, where it was bound by the electric polarization of its surrounding water molecules, its solvation energy, B. At that time (1928), there was no experimental evidence for free-solution electrons, and other species were proposed as initial intermediates. The free-solution electron concept was discarded until the increasing evidence in the 1950s led to the discovery of the solvated electron by fast kinetic spectroscopy in 1961.

---

*In the original manuscript of this chapter the S.I. unit system was used throughout. However, to maintain consistency with the other chapters all entries have been changed to the more customary units.

The nature of the radical produced was correctly deduced from aqueous iodide spectroscopy, where the separation of the two bands observed were recognized as being the $^2P_{1/2}-^2P_{3/2}$ doublet splitting of the gaseous iodine radicals. The free-solution electron model was supplanted by one in which the electron was transferred to an intimately contiguous water molecule following photoionization, the water molecule dissociating immediately, for example, for iodide[2]

$$I^-(H_2O) \xrightarrow{h\nu} \cdot I + \cdot H + OH^- \qquad (8\text{-}1)$$

The above scheme does not allow an explanation of the marked effect of pH on the product quantum yield. Moreover, at low pH a plateau effect is found for the product quantum yield that is well below unity. A new model accounted for the pH effects by proposing a competition between thermal deactivation of the excited state and reaction with a proton to give a hydrogen atom.[3]

$$I^-(H_2O) \underset{\ }{\overset{h\nu}{\rightleftharpoons}} \cdot I(H_2O)^- \xrightarrow{H^+} \cdot I + H_2O + \cdot H \qquad (8\text{-}2)$$

Molecular hydrogen and iodine would be formed by dimerization of the radicals. This mechanism, in turn, did not explain the subsequently observed dependence of the product quantum yield upon both proton and iodide concentration, and an additional mechanism[4] proposed a new intermediate, H_2^+

$$\cdot H + H^+ \rightarrow H_2^+ \qquad (8\text{-}3)$$

$$H_2^+ + I^- \rightarrow \cdot I + H_2 \qquad (8\text{-}4)$$

The various mechanisms proposed above have tried to rationalize the available experimental data in the absence of the solvated electron concept, whose reactivity leads to subtly different results under slightly varied conditions. To understand simple inorganic-ion photolysis, the spectroscopic transition must be related to the nature of the excited state, which in turn must be related through scavenging processes to the final product formation.

Experimental photochemical techniques have been described in general by Pitts and Calvert[5]; considerations more specific to simple-ion photolysis have been reviewed with particular reference being given to actinometry problems at 185 nm.[6] The quantum yield of 5 M aqueous ethanol, the most convenient actinometer at 185 nm, is now established as 0.40. All quantum yields from other systems photolysed at 185 nm are referred here to the above value; apparent discrepancies between published values are then reduced.

II. SPECTROSCOPY OF SIMPLE INORGANIC IONS

A. Anion Charge-Transfer-To-Solvent Spectra

Understanding simple inorganic ion photochemistry may only be achieved through knowledge of the relevant spectroscopy. It is important to recognize that various fundamentally different excited states can be formed on irradiation of the ground state of an ion.

Charge-transfer-to-solvent (henceforth CTTS) spectra of simple anions comprise intense ultraviolet absorption bands (oscillator strengths of the order of 0.1) that are broad and unstructured[7] ($\nu_{1/2}$ approximately 4000 cm^{-1} for aqueous iodide). Anion CTTS spectra are unique in their high sensitivity to environmental perturbations such as temperature, pressure, change of solvent, and addition of electrolyte or nonelectrolyte. At suitably low concentrations, generally below 5×10^{-3} M, the spectra are independent of cation. Information from CTTS studies thus belongs to the small group of single-ion properties represented by ionic conductances derived from equivalent conductivities and transport numbers,[8] electron-spin-resonance spectra of radical ions,[9] magnetic-resonance spectra of ions,[10] and nuclear-magnetic-resonance spectra of solvent molecules in ionic solvation shells.[11] The solvent defines the excited state—there are no stable higher states for isolated iodide—and the ion acts as a microprobe of its environment to give information on solute-solvent interactions. Trends in CTTS spectra of solvated anions correlate well with trends in other physical measurements of the same system.[7] One correlation of particular note is that found between the absorption maxima of the solvated electron and iodide in a large range of solvents.[12,102] Polyatomic anions such as thiosulfate and azide show internal transitions in addition to CTTS processes and have quite complex, overlapping band spectra.[7]

The nature of CTTS transitions is inferred from the spectroscopy[7] and photochemistry[6] of the halide ions in mainly aqueous solution. Thermodynamic cycles have been used to derive the spectroscopic transition energies, and the initial act leads to the formation of a halogen radical and a solvated electron, $X^-_{solv} \xrightarrow{h\nu} (\cdot X + e^-)_{solv}$.

Intermediate states have been written into reaction schemes, generally as X^*_{solv}. Fluorescence has never been observed from solvated halide systems, and the transition must be regarded as predominantly dissociative. The nature of the radical is confirmed by the observed band separations in the ultraviolet-solution spectroscopy of halide ions being close to the doublet splitting of the respective halogen radicals in the gas phase. Thus, for chloride, where the gas phase splitting (Δ) is 880 cm^{-1}, the close overlap of the broad band components gives only one band overall, whereas for bromide $(\Delta = 3660 \text{ cm}^{-1})$ and iodide $(\Delta = 7603 \text{ cm}^{-1})$ the increased doublet splitting gives progressively better resolution of two bands arising from the $^2P_{3/2}$, $^2P_{1/2}$ atomic states.[13]

Recognition of the above features led to the Franck–Scheibe model for the transition energy, $h\nu_{max}$, using the free-solution electron concept[1]

$$h\nu_{max} = I_{X^-} - \Delta H_s^{\ominus-} + \Delta H_s^{\ominus} + \chi - B \qquad (8\text{-}5)$$

where I_{X^-} is the ionization potential of the ion and $\Delta H_s^{\ominus-}$, $\Delta H_s^{\ominus}$ the standard enthalpies of solvation for the ion and radical, respectively. The Franck-Condon energy, χ, accounts for the strain caused by the relative immobility of the solvent molecules during the transition to form a radical. The solvation energy of the free-solution electron, B, arises from the polarization of water molecules around the electron, but could not be estimated at the time.

The Franck–Platzman model[13] uses an excited-state orbital for the electron that was centrosymmetric upon the original anion site to overcome the difficulty of calculating B. Two terms, E_e and S_e, contribute to B. E_e arises from the electric polarization of the solvent molecules around the ion, giving an effective coulomb field with effective charge $Z_{eff} = 1/D_{op} = 1/D_s$, a ground-state effect, D_{op} and D_s being the optical and static dielectric constants of the medium, respectively. S_e is an excited-state effect from the electric polarization of the solvent molecules by the solvated electron. B was found to be effectively independent of anion in aqueous solution at 146 kJ mole^{-1} for a hydrogen-like $2s$ excited-state orbital and 298 K. The χ term of the Franck–Scheibe model was redefined as the energy contribution required to reproduce the same degree of solvent polarization around the atom in the excited state as had existed around the ion in the ground state and depends only slightly on the nature of the halide. Calculation of $\Delta H_s^{\ominus-}$ and $\Delta H_s^{\ominus}$ gives all of the terms in Eq. (8-5), and the calculated values of $h\nu_{max}$ agree well with experiment.

Detailed temperature and pressure studies of iodide spectra in a large range of solvents and binary systems showed that, while the Franck–Platzman theory predicts $h\nu_{max}$ well, it does not predict the second-order spectroscopic shifts following environmental perturbation.[14] When allowance is made for the effect of temperature on D_{op} and D_s, a small and positive shift of $h\nu_{max}$ is predicted. However, a large and negative shift is found in practice.

The Smith-Symons treatment[15] proposed the "confined" model, using an electron in a spherical well of variable radius, for the excited state, giving a much simpler final expression

$$h\nu_{max} = I_{X^-} + \frac{h^2}{8mr_0^2} \qquad (8\text{-}6)$$

The radius parameter obtained from this model does not correlate readily with known ionic properties, because it is the distance between the center of the ion and the inner edge of the potential well. One example of the difficulties experienced with this model is given by the requirement that the wall of the potential well be infinitely steep, while this is probably not the case in practice. The treatment is elegantly simple: while $h\nu_{max}$ cannot be calculated *a priori*, the effects of environmental perturbation are well explained by the variable-radius concept, an *excited* state phenomenon.

Combination of the Franck-Platzman model with the variable-radius concept in the "diffuse" model[16] gives good absolute values for $h\nu_{max}$ and environmental effects for halides in solution. The radius parameter, r_d, is related to the crystallographic radius, r_i, of the ion by a simple factor of 1.25 and is thus a ground-state parameter. Derived values of r_d agree well with the values obtained from analyses of partial molar ionic volumes.[17] The diffuse model was generalized by Treinin[18] for multivalent polyvalent ions, the transition now being represented as $X_{aq}^z \rightarrow (X^{z-1} + e^-)_{aq'}$. Furthermore, the effect of environmental perturbation on $h\nu_{max}$ for multivalent ions was shown to be more difficult to initiate because of the more tightly held solvation layer, but when the barrier is overcome then $d(h\nu_{max})/dt$ should be greater than for monovalent ions of equivalent size.

The models described thus far have been developed primarily for halide ions, but extension of CTTS studies to other ions shows some shortcomings in the treatments described. First, to a first approximation, $h\nu_{max}$ is linearly related to the ionization potential of the ion, I_{X^-}, but as I_{X^-} decreases below 300 kJ mole^{-1}, $h\nu_{max}$ becomes effectively independent of I_{X^-} and for $I_{X^-} < 200$ kJ mole^{-1}, $h\nu_{max}$ increases.[19] It is fortuitous that the predominantly electrostatic approach of the above models appears to work when I_{X^-} is the major contributor to $h\nu_{max}$. The corollary is that I_{X^-} values calculated from $h\nu_{max}$ for the pseudohalides differ markedly from those obtained from other approaches, such as the magnetron technique or quantum mechanics, and they pose ambiguities for the thermodynamic cycles used to calculate $h\nu_{max}$. Moreover, interaction between the ion and the solvent or between radical (ion) and solvent has been considered as a possibility but not pursued quantitatively.[18] Recent detailed work on iodide in a range of powerful red- and blue-shifting solvents (relative to water) shows that the doublet splitting of the bands, $A_1-A_2(\Delta)$ varies linearly with $h\nu_{max'}$; the oscillator strength of band A_1, also varies from 0.36 to 0.24 as ν_{max} is shifted from 48,141 to 38,300 cm^{-1} for iodide in hexafluoro-2-propyl alcohol and hexamethylphosphoramide, respectively, at 273°K.[20] Such interactions strongly imply ion-solvent interactions of a manner as yet unencompassed by the previous theories.

An alternative view of CTTS spectra looked at a much larger range of anions in water and acetonitrile solution for a total of twelve transitions of nine anions in each solvent.[19] The difficulty of describing the precise nature of the excited-state orbital was circumvented by proposing a simple spherical-shell acceptor orbital with an electron affinity, E_{solv}. The electron affinity of the oriented solvent-molecule shell is quite different from that of an isolated solvent molecule. The anion is regarded as an electron donor, and together with the solvated electron-acceptor shell, the solvated anion is treated as a donor-acceptor charge-transfer complex. Application of Mulliken charge-transfer theory, only modified to include a radial dependence term, gives

$$h\nu_{max} = I_{X^-} - E_{H_2O} + \frac{\sigma^2}{I_{X^-} - E_{H_2O}} \frac{1}{r_i^2} \qquad (8\text{-}7)$$

where σ is the overlap term, depending upon the overlap integral and polarization terms. Good agreement between predicted and observed transition energies is found for the set of anions studied. However, detailed descriptions of solute-solvent interactions have not yet been developed for this model.

Theories for CTTS spectra given thus far have been confined to dealing with the relatively low-energy bands, A_1 and A_2, in Fig. 8-1 because of limitations imposed by atmospheric oxygen, increasing solvent absorptions, and decreasing spectrophotometer performance above 50,000 cm^{-1}. Such limitations may be circumvented by various means to obtain spectra to 61,000 cm^{-1} in some solvents.[20]

Far-ultraviolet spectra of halide ions show further absorption bands,[21] the overall spectra being analogous in complexity to crystalline rare-gas[22] and alkali-halide spectra[23] in the far-ultraviolet region. Bands A_1 and A_2 must be discussed in the context of additional band pairs and also transitions of different character[21] (see Fig. 8-1). The relevance of the rare-gas crystal/alkali-halide far-ultraviolet spectra is underlined by noting that the lower energy transitions are proposed as $np^6 \xrightarrow{h\nu} np^5(n+1)s$ and the higher energy bands as arising from $np^6 \xrightarrow{h\nu} np^5(n+1)d$ transitions, particularly as Jorgenson has suggested the former transition for the halide A_1 and A_2 bands.[24]

The B_1 and B_2 band pair appears to have similar, but accentuated CTTS character compared to the A_1 and A_2 pair, for example, the temperature sensitivities, $d\nu_{max}/dT$, are greater and the solvent sensitivities of the B and A bands have a ratio of approximately $3:1$.[21] The spectra of halides are complicated by the doublet splitting of the radical, but the radicals derived from other anions do not show this feature, and the higher-energy CTTS states of hydrosulfide and hydroxide are more readily observed.[21]

Polyatomic anions show internal transitions in addition to CTTS

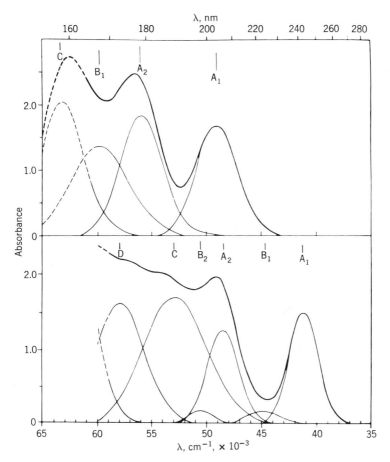

Fig. 8-1. Absorption spectra of 5×10^{-3} M I^- in hexafluoro-2-propyl alcohol (top, tetramethylammonium salt) and in acetanitrile (bottom, potassium salt) at 274°K. Heavy lines indicate the spectra obtained experimentally. The continuous narrow lines represent the Gaussian analog analysis of the experimental curve, and the broken lines the necessary continuation of the Gaussian analysis beyond the limit of measurement.

processes that are considerably overlapped in the absorption envelope.[7] Internal transitions have been studied theoretically for some simple anions, but the possibility of CTTS processes occurring from the same ground state has not been considered. Thus, in the absorption spectrum of thiosulfate, three bands have been observed, to a current limit of approximately 51,000 cm^{-1}, labeled A, B, and C, respectively.[26] Band A is an internal transition of moderate intensity, $\varepsilon_{max} = 2 \times 10^2$ M^{-1} cm^{-1} and $\nu_{max} = 41,000$ cm^{-1} assigned to an $n-\pi^*$ transition on oxygen. Band B is a

CTTS transition with $\varepsilon_{max} = 4 \times 10^3\,M^{-1}\,cm^{-1}$ and $\nu_{max} = 46,500\,cm^{-1}$ in aqueous solution. For band C, ν_{max} is $>50,000\,cm^{-1}$ with $\varepsilon_{max} > 4 \times 10^3\,M^{-1}\,cm^{-1}$, the uncertainty arising from band C being beyond the limit of measurement. Current work indicates that band C possibly arises from a CTTS transition.

The implications for the photochemistry of anions are, first, that the spectroscopic evidence indicates an excited state predominantly defined by the solvent and centered on the parent anion site. The large range of electrolytes and nonelectrolytes used to give large red and blue shifts of ν_{max} show that even at very high concentrations such as 12 M LiCl the nature of the transition is unaltered. The added solutes act indirectly through solvent-solvent interactions to stabilize/destabilize the ground and/or excited states.[7] The added solutes do not interact directly with the excited state.

Second, photolysis of anions absorbing by CTTS processes in different solvents at a given wavelength should take account of the solvent-induced shift of the transition energy. Thus, 229-nm photolysis of aqueous iodide gives irradiation almost exclusively into the center of band A_1,[6] whereas 229-nm photolysis of iodide in acetonitrile irradiates bands A_1, B_1, and the low-energy edge of A_2.

Finally, for the polyatomic anions there is the probability of internal and CTTS transitions with strongly overlapped absorption bands. Both the anion radical and the associated solvated electron are produced together with products resulting from bond fission of the internal-transition excited state. Irradiation sources in anion photolysis are usually element-resonance lamps with high-emission intensities concentrated into relatively narrow bandwidths at 185, 229, and 254 nm.[5,6] The technical situation is now perhaps sufficiently advanced for studies to be made at other wavelengths using high-intensity continuum sources and monochromators. Simple-anion spectra are summarized in Table 8-1.

B. Cation Spectra

The spectra of cations are a much less complete story. First, it must be emphasized that in the context of this chapter we are only concerned with the spectra of hydrated cations, complex formation photochemistry being dealt with in other chapters. Second, it must be emphasized that it is experimentally difficult to obtain only the hydrated ion since ion pair/complex formation for cations occurs much more readily than for anions. The complexes thus formed have quite distinct spectra, for example, for ferric ion complex formation, Fig. 8-2a. Finally, little recent work has been done on hydrated-cation ultraviolet spectra to support the

Table 8-1.   Absorption Spectra of Simple Anions in Aqueous Solution at 298°K[a]

Anion	ν_{max}, kK	ε_{max}, $M^{-1}\,cm^{-1}$	Assignment
I^-	43.9	1.29×10^4	CTTS, $^2P_{3/2}$ iodine, $2s$-type electron
	51.1	1.32×10^4	CTTS, $^2P_{1/2}$ iodine, $2s$-type electron
	54.6	4×10^3	CTTS, $^2P_{3/2}$ iodine, $3s$-type electron
	58.5	2×10^4	CTTS, unknown, d-type electron?
Br^-	50.6	1.30×10^4	CTTS, $^2P_{3/2}$ bromine, $2s$ electron
	53.9	1.35×10^4	CTTS, $^2P_{1/2}$ bromine, $2s$ electron
	57.1	5×10^3	CTTS, $^2P_{3/2}$ bromine, $3s$-electron
Cl^-	57.5	1.69×10^4	CTTS, both $2P$ chlorine terms, $2s$-electron
OH^-	53.5	3.85×10^3	CTTS
SH^-	43.5	7×10^3	CTTS
	47.5	1×10^3	Probably CTTS
	54.1	5×10^3	Probably CTTS
	61.5	8×10^3	Probably CTTS
S^{2-}	27.8	$\sim2\times10^2$	CTTS
Se^{2-}	24.0	Unknown	Unknown
CN^-	>55.0	Unknown	Unknown
CNO^-	>55.0	Unknown	Unknown
CNS^-	45.0	3.5×10^3	CTTS
$CNSe^-$	42.5	3×10^3	CTTS
$CNTe^-$	37.0	$\sim3\times10^3$	CTTS
N_3^-	43.0	4×10^2	Internal
	49.3	4×10^3	CTTS
	53.0	$>1\times10^4$	Internal
HCO_3^-	37.0	<1	Internal, forbidden $n-\pi^*$
	45.0	<1	Internal, forbidden $n-\pi^*$
	>52.0	$>4\times10^2$	Internal
CO_3^{2-}	37.0	<1	Internal, forbidden $n-\pi^*$
	46.1	<1	Internal, forbidden $n-\pi^*$
	50.0	$>4\times10^2$	Internal
CS_3^{2-}	20.0	30	Internal
	29.8	1×10^4	Internal
	44.8	1×10^4	Internal
CSe_3^{2-}	15.8	80	Internal
	21.3	8×10^2	Internal
	26.3	2×10^4	Internal
	41.5	1×10^4	Internal
NO_2^-	28.2	22.5	Internal, $n_0-\pi^*$
	35.0	9.4	Internal, $n_0-\pi^*$
	47.6	5.4×10^3	Internal, $\pi-\pi^*$
	~52.0		Possibly CTTS

Table 8-1. (Cont'd)

Anion	ν_{max}, kK	ε_{max}, $M^{-1}\,cm^{-1}$	Assignment
NO_3^-	33.0	7	Internal, $n-\pi^*$
	49.5	9.9×10^3	Internal, $\pi-\pi^*$
HPO_4^{2-}	37.0	<1	Internal
	48.0	<1	Internal
	56.0	1×10^3	CTTS
$H_2PO_4^-$	59.0	$>1 \times 10^2$	CTTS
HSO_3^-	>55.0	weak	Unknown
SO_3^{2-}	44.0	$>10^3$	Probably internal
HSO_4^-	~60.0	$>10^2$	Probably CTTS
SO_4^{2-}	57.1	3×10^2	CTTS
$HSeO_3^-$	~48.0		Internal
	~54.0		CTTS
SeO_3^{2-}	~48.0		CTTS
$HSeO_4^-$	~48.0		Internal
SeO_4^{2-}	~48.0		Internal
	~55.0		CTTS
$S_2O_3^{2-}$	41.0	2×10^2	Internal
	46.5	4×10^3	CTTS
	~52.0	$>4 \times 10^3$	Internal, but possibly CTTS
ClO^-	34.4	3.6×10^2	Internal
BrO^-	30.0	3.3×10^2	Internal
IO^-	27.4	31	Internal
	39.2	4×10^2	Internal
	>42.0	$>10^3$	Internal
ClO_2^-	~38.0	1.5×10^2	Internal
	~47.0	2×10^2	Internal
	>50.0	$>5 \times 10^2$	Internal
ClO_3^-	>50.0	$>10^3$	Internal
BrO_3^-	>50.0	$>10^3$	Internal
IO_3^-	>50.0	$>10^3$	Internal
ClO_4^-	>56	20	Internal
XeO_4^-	40.0	6×10^3	Unknown
NH_2^-	30.1	~3.5×10^3 (at 224°K)	CTTS
NHEt$^-$ (in liquid ammonia/ethylamine)	31.5	2.7×10^3	CTTS
Ferrocyanide	39.0	—	CTTS

[a] Based on Table IV of Ref. 7. Data for selenium oxyanions taken from A. Treinin and J. Wilf, *J. Phys. Chem.* **74,** 4131 (1970).

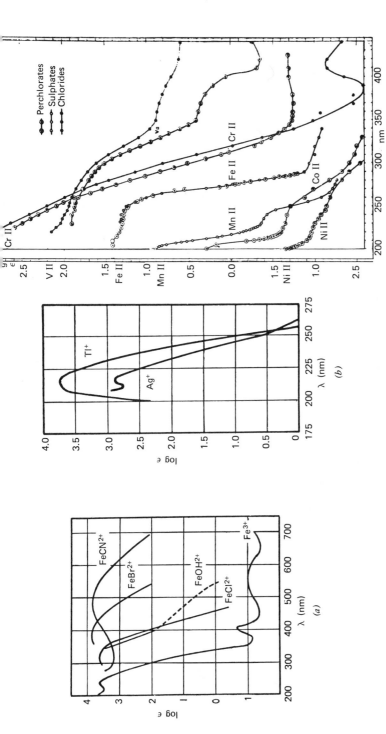

Fig. 8-2. (a) Absorption spectra of Fe^{3+} and of $Fe^{3+}X^-$ complexes in aqueous solution. [From E. Rabinowitch and W. Stockmayer, *J. Amer. Chem. Soc.* **64**, 335 (1942).] (b) Absorption spectra of monovalent cations in aqueous solution. [From E. Robinowitch, *Rev. Mod. Phys.* **14**, 112 (1942).] (c) Absorption spectra of some cations in aqueous solution. [From L. E. Orgel, *Quart. Rev.* **8**, 429 (1954).] Reprinted by kind permission of the copyright holders.

343

assignment of the absorption bands as charge-transfer processes in a modern context.

The spectral absorptions of closed-shell cations are intense and at high energy, Fig. 8-2b, and have not been investigated in depth.[27] Transition-metal cation spectra are shown in Fig. 8-2c.[28] The spectra have been described as arising from the photodetachment of an electron from the hydrated ion

$$M_{aq}^{x+} \rightarrow M_{aq}^{(x+1)+} + e^-$$ (8-8)

by analogy to

$$A_{aq}^{x-} \rightarrow A_{aq}^{(x-1)-} + e^-$$ (8-9)

where M^{x+} and A^{x-} are solvated cations and anions, respectively. The cation photoionization process assignment is validated by linear plots of the onset energy of the cation absorption bands against the free energy of the respective (thermal) redox process for the couple, $M^{x+}/M^{(x+1)+}$, from the relation[29]

$$h\nu = \text{constant} + E_0^h F$$ (8-10)

where ν is the frequency of the absorption band onset, E_0^h is the redox potential for the cation couple, and F is the Faraday. The relationship was also shown for a range of anions using ν_{max} instead of the absorption-band onset energy.[30] For cations the current poor characterization of the spectra enforces the use of absorption-edge values; the main difficulties with accepting the cation spectra as charge-transfer transitions lies with the absence of environmental-perturbation studies of the type applied extensively to the study of anions in solution. Transition-metal ions show high intensity for ultraviolet absorption bands with ε_{max} of the order $10^3 M^{-1} cm^{-1}$ and characteristic values of ν_{max}, Table 8-2. The nature of the transitions is not clear since the results from comparison of spectra in water and D_2O,[31] a standard test for CTTS spectra, are ambiguous. The most probable transitions are $3d^n \rightarrow 3d^{n-1}4s$ for the metal with both ground and excited states perturbed by interaction with the surrounding water molecules. Moreover, solvated electron production has never been observed in the flash photolysis of solvated cations. Much further work is needed to characterize the solvated cation spectra further.

Detailed studies of the absorption spectra of some complexes have demonstrated CTTS transitions underlying other transitions. In aqueous solution ferrocyanide ion has four resolved absorption bands with a weak band at 422 nm ($f = 2 \times 10^{-5}$, $^1A_{1g} \rightarrow ^3T_{1g}$), 323 nm ($f = 8.4 \times 10^{-3}$, $^1A_{1g} \rightarrow ^1T_{1g}$), 270 nm ($f = 4.7 \times 10^{-3}$, $^1A_{1g} \rightarrow ^1T_{2g}$), and 218 nm ($f = 5.35 \times 10^{-1}$, $^1A_{1g} \rightarrow ^1T_{1u}$).[32] The first three bands are assigned to ligand-field $d-d$ transitions, and the last to a charge-transfer band, probably from the

Table 8-2.   Far-Ultraviolet Transitions of Metal Ions[a]

Cation	Band Type	ν_{max}, cm^{-1}	ε_{max}, M^{-1} cm^{-1}
Mn^{2+}	Shoulder	58,300	10^2
	Strong	63,300	5–10×10^3
Fe^{2+}	Shoulder	58,140	10^2
Co^{2+}	Strong	62,500	10^4
	Medium strong	60,750	1–3×10^3
Ni^{2+}	Medium strong	62,300	2.5×10^3
Cu^{2+}	Shoulder	48,500	1.1–1.3×10^3
	Several close and unresolved bands	59,000	10^4

[a] From the thesis of J. T. Shapiro, Bryn Mawr College, Bryn Mawr, Pa., 1965.

metal to a ligand [charge-transfer-to-ligand (CTTL)]. Solvent-perturbation studies to the 310 to 355 and 210 to 240 nm regions show only small changes in the absorption spectrum of ferrocyanide. In the 245 to 300 nm region large solvent effects are observed, characteristic of a CTTS transition. However, the oscillator strength of the CTTS band is low in comparison with the intensities of the halides, and a complex mechanism involving vibrational rearrangement is invoked to explain the discrepancy. Such explanations are unnecessary because the intensities of CTTS transitions decrease with increases in both molecular size and charge.[7] The details of these variations are not well understood at present, but, in this context, the low intensity of the ferrocyanide CTTS absorption band is quite within character for an ion of large size and high charge.

It is clear from the foregoing discussion that the detailed nature of simple-ion spectra, particularly for cations, is not well understood in comparison with the theory of inorganic complex spectroscopy. The lack of detailed knowledge for the excited state in some areas of the following discussion on photochemistry must be recognized as an inhibiting factor.

III.   THE PRIMARY PHOTOLYTIC PROCESS

Thermodynamic cycles used in CTTS spectroscopic models propose transfer of an electron from the anion to the associated solvent shell. Photolysis studies on aqueous iodide show that the electron and iodine atoms are formed in a solvent cage such that a fraction of these species diffuse out into the bulk solvent.[33-36] The hydrated electron was shown to be produced in the pulse radiolysis of water in 1961. An earlier study had shown that X_2^- radical ions were formed from the flash photolysis of aqueous halide ions, the transient X_2^- radical ions having characteristic main absorption bands at 385, 360, and 355 nm for $X = I$, Br, and Cl, respectively.[38] There is a common band at 300 nm for these radicals. Using 0.2 M aqueous methanol to scavenge the halogen radicals, and extending the study to longer wavelengths, gives a transient absorption with λ_{max} at 670 nm.[39] The transient absorption band is very similar to that of the hydrated electron produced by pulse radiolysis of water[37] and, moreover, is attenuated or removed by addition of such specific electron scavengers as oxygen, nitrous oxide, or hydrogen ions. Similar hydrated electron spectra are obtained by the flash photolysis of bromide, chloride, hydroxide, thiocyanate, and ferrocyanide, each of which absorbs at least in part by a CTTS process. Production of solvated electrons from aqueous chloride, λ_{max} 175 nm, shows the flash energy to extend into the far-ultraviolet region. The molar absorption coefficients of the ions photolyzed are very high (see Table 8-1), and ion concentrations are adjusted such that the incident flash energy is almost entirely absorbed by the ions. The solvated electron is produced from the CTTS excited state, written simply as

$$X_{aq}^- \rightarrow \cdot X + e_{aq}^- \qquad (8\text{-}11)$$

Flash photolysis of iodide in methanol produces solvated electrons and I_2^- with characteristic peaks at 290, 385, and 750 nm.[40] For deoxygenated solutions the system is reversible, but when oxygen is present solvated electrons are not observed and $I_2(I_3^-)$ is a product. In the absence of oxygen, the solvated electron and I_2^- decay by first- and second-order kinetics, respectively. The alcohol radicals (or the products from their dissociation) react with I_2^- to regenerate I^-. Pulse radiolysis of methanolic I^- solutions has shown I_2^- to react competitively with $CH_2OH \cdot$ radicals or itself[41]

$$I^- \rightarrow I + e_{solv}^- \xrightarrow{O_2} O_2^- \qquad (8\text{-}12)$$

$$\downarrow I^- \qquad \searrow MeOH$$

$$I_2^- \qquad H_2 + \cdot CH_2OH + OH^-$$

followed by

$$I_2^- + \cdot CH_2OH \rightarrow CH_2O + 2I^- + H^+ \qquad (8\text{-}13)$$

$$I_2^- + I_2^- \rightarrow I^- + I_3^- \qquad (8\text{-}14)$$

$$I_3^- + \cdot CH_2OH \rightarrow CH_2O + I_2 + H^+ + I^- \qquad (8\text{-}15)$$

No distinction may be made in the flash photolyses of the halides and pseudohalides between direct photoionization of the electron into the solvent or a solvent-relaxation mechanism that releases only a certain fraction of the radical and electron into the bulk solvent.[39] If the solvated electron and a radical are produced in close proximity, and such a juxtaposition may be observed, then the spectrum of either species, or both, would be modified by the intense potential field of the other.

Spectra of both solvated electrons and associated radicals have been obtained both together and separately in the photolysis of several anions. Thus, steady-state photolysis of the phosphate anions evolves hydrogen from deaerated solutions when electron and hydrogen atom scavengers are present.[42] The spectra of phosphate anions are summarized in Table 8-1 and are recalled here as extremely weak internal transitions above 200 nm and medium-intensity CTTS bands below 200 nm. At pH 8.9 the predominant species present in phosphate solutions is the mono-hydrogen phosphate anion, HPO_4^{2-}. When oxygen-saturated phosphate solutions at pH 8.9 are irradiated, a transient absorption band is observed with λ_{max} at 480 nm (shoulder at 575 nm) Fig. 8-3a, assigned to the HPO_4^- radical anion

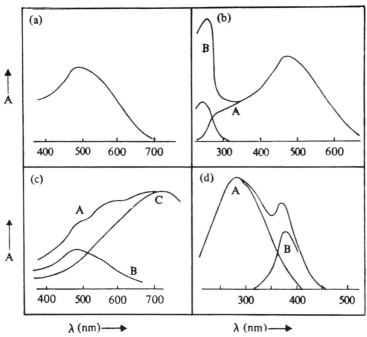

Fig. 8-3. (a) Phosphate radical-ion spectrum from oxygen-saturated aqueous 6.7×10^{-3} M Na_2HPO_4, pH 8.9; (b) Transient absorption spectrum from A, air-free and B, oxygen-saturated $H_2PO_4^-$; (c) transient absorption from A, argon-saturated 6.7×10^{-3} M Na_2HPO_4, pH 8.9, B, HPO_4^- radical spectrum from (a) above, and C, difference spectrum, $A - B$; and (d) transient absorption spectrum from the flash photolysis of thiosulfate ion, A, $S_2O_3^{2-}$, and B, $S_2O_3^-$. Reprinted by permission from *Q. Rev.* **24**, 565 (1970).

$(\varepsilon_{max} = 600 \pm 50 \, M^{-1} \, cm^{-1})$.[43,44] When argon-saturated HPO_4^{2-} solutions are irradiated, a more complex spectrum is observed, Fig. 8-3c. When the normalized spectrum of the hydrated electron is removed from Fig. 8-3c, then the absorption spectrum of HPO_4^- remains. Pulse radiolysis of aqueous HPO_4^{2-} gives an identical spectrum of HPO_4^-, Fig. 8-3b, confirming the assignment of the 480-nm absorption to this radical.[43]

In pH 4.5 solutions $H_2PO_4^-$ is the dominant ionic species present, and photolysis of these solutions gives a transient absorption with λ_{max} at 500 nm and $\varepsilon_{max} = 650 \pm 50 \, M^{-1} \, cm^{-1}$, again corroborated by pulse radiolysis.[43] There is a close similarity between the transient spectra produced in the photolysis of HPO_4^{2-} and $H_2PO_4^-$ in aqueous solution, which extends also to their decay rates, indicating the radicals to be the same, the H_2PO_4 radical dissociating through the equilibria

$$H_2PO_4 \xrightleftharpoons{pK_a = 5.9} H^+ + HPO_4^- \xrightleftharpoons{pK_a = 10.7} 2H^+ + PO_4^{2-} \qquad (8\text{-}16)$$

The pK_as of the radicals are substantially different from those of the associated ions, a feature also observed in peptides and their associated radicals.[45] Irradiation of deaerated aqueous $H_2PO_4^-$ does not give the transient spectrum of the solvated electron because the solvated electron reacts with $H_2PO_4^-$, $k(e^- + H_2PO_4^-) = 7.7 \times 10^6 \, M^{-1} \, sec^{-1}$, giving a lifetime of $<10^{-3}$ sec for hydrated electrons in these solutions.[43] Aqueous pyrophosphates give similar results when irradiated.[43]

The spectra of the hydrated electron and the HPO_4^- radical ion are clearly unaltered whether they are produced by photolysis or pulse radiolysis, indicating that the two species are well separated in solution. The lifetime of the excited state, or the time necessary for the formation of the solvated electron is, therefore, taken to be less than 10^{-10} sec.

Solvated electrons are also produced in the photolysis of aromatic compounds.[46] The quantum yield of electrons is dramatically increased when the molecules are ionized as naphtholate or phenolate. The absorption spectra of phenolate or naphtholate do not show CTTS transitions above 48,000 cm^{-1} (the limit of measurement), and the bands observed are assigned to allowed $\pi-\pi^*$ processes. The ions fluoresce strongly in solution, and, as the pH of solutions is raised, the fluorescence intensity decreases and the solvated electron production increases to a constant value between pH 12 and 14. The solvated electron and fluorescence originate from the same excited state.

Whereas $\pi \rightarrow \pi^*$ transitions of the organic anions lead to solvated electron formation, $n \rightarrow \pi^*$ and $\pi \rightarrow \pi^*$ transitions of inorganic anions lead to dissociation.

The solvated halate anions, XO_{3solv}^-, where X = Cl, Br, or I, do not absorb by CTTS processes; an earlier CTTS assignment was shown to be

Table 8-3.   Dissociation Modes of XO_n^{m-} [a]

I. $XO_n^{m-} \rightarrow XO_n^{(m-1)-} + e_{aq}^-$
   SO_4^{2-}, SO_3^{2-}, $S_2O_3^{2-}$, $H_2PO_4^-$, HPO_4^{2-}, $P_2O_7^{2-}$, CO_3^{2-}, OH^-

II. $XO_n^- \rightarrow XO_{n-1}^- + O(^3P)$
   ClO^-, BrO_3^-, BrO_2^-, BrO^-

III. $XO_n^{m-} \rightarrow XO_{n-1}^{(m-1)-} + O^-$
   ClO_3^-, ClO^-, BrO_3^-, BrO_2^-, BrO^-, IO_3^-, IO^-, $S_2O_3^{2-}$, NO_3^-, NO_2^-, HO_2^-

IV. $XO_n^- \rightarrow XO_{n-2}^- + O_2$
   possibly MnO_4^-, NO_3^-

Isomerization
   NO_3^-, $ONOO^-$

---

[a] From Ref. 48.

---

incorrect for the intense bands with $\lambda_{max} = 200$ nm and $\varepsilon_{max} = 10^3$ M^{-1} cm^{-1}.[47] Photolysis of aqueous bromate using specific electron scavengers shows a very low quantum yield of solvated electrons, $\phi_e = 5 \times 10^{-4}$, attributed to thermal ionization of the excited state. The main photochemical reaction is evolution of molecular oxygen, $\phi_{O_2} = 0.15$, initially interpreted as[47]

$$BrO_3^- \xrightarrow{h\nu} BrO^- + O_2 \qquad (8\text{-}17)$$

indicating that no important CTTS bands underlie the intense intramolecular absorption band. Further investigation has shown the photochemistry of oxyanions for absorption in other than CTTS bands to be more complex; several initial processes lead to dissociation of the ion. Primary processes for the oxyanions are summarized in Table (8-3).[48]

The CTTS photoionization process designated I in Table 8-3 is included not only on formal grounds, but also to emphasize that several processes can occur simultaneously from the same ground state. Thus, thiosulfate is included under processes I and III and is described later. It is not clear at present whether the different dissociation processes characterized as II and III originate from the same excited state.

The process generating O^-, designated III, is the most common dissociation mechanism of XO_3^{-*}. O^- forms $\cdot OH$ radicals by reaction with water unless previously scavenged by reaction with another species. The presence of $\cdot OH$ radicals is readily demonstrated by their characteristic reaction with Br^-, CO_3^{2-}, or CNS^- to give the distinctive spectra of Br_2^-, CO_3^-, or $(CNS)_2^-$. In oxygenated solution at pH 12, O^- reacts with an oxygen molecule to give O_3^-, and the overall reaction has been developed as a convenient method to study the reaction of O^-/OH with oxyanions, the "ozonide method" [49,50]

$$O^- + O_2 \underset{k_{-18}}{\overset{k_{18}}{\rightleftharpoons}} O_3^- \tag{8-18}$$

$$O^- + H_2O \underset{k_{-19}}{\overset{k_{19}}{\rightleftharpoons}} \cdot OH + OH^- \tag{8-19}$$

$$\cdot OH + XO_n^- \overset{k_{20}}{\longrightarrow} \cdot XO_n + OH^- \tag{8-20}$$

$$O^- + XO_n^- + H_2O \longrightarrow \cdot XO_n + 2OH^- \tag{8-21}$$

Kinetic analysis gives

$$\frac{-d(O_3^-)}{dt} = k_0(O_3^-) \tag{8-22}$$

where

$$\frac{1}{k_0} = \frac{1}{k_{-18}} + \frac{k(O_2)}{k_{-18}k^*(XO_n^-)} \tag{8-23}$$

and

$$k^* = k_{18} + \frac{k_{20}K_w}{K_{OH}(OH^-)} \tag{8-24}$$

K_w and K_{OH} are the dissociation constants for water and $\cdot OH$, respectively.

Rate constants, k_{20} and k_{21}, have been obtained for the reaction of a range of anions with O^- or $\cdot OH$; good agreement is found between different reports.[49-55]

It has been pointed out that formation of O^- has not been proven conclusively. An alternative explanation based on the direct formation of hydroxide radicals from the excited state of the *hydrated* ion

$$XO_n^{m-} \cdot H_2O \overset{h\nu}{\longrightarrow} XO_{n-1}^{(m-1)-} + \cdot OH + OH^- \tag{8-25}$$

has been studied thermodynamically,[48,56] together with reaction (8-21) and the other modes of photodissociation of XO_n^- in Table 8-3. Ignoring Franck–Condon energies (essentially comparing equilibrium states), one finds the molar enthalpies of reactions (8-21) and (8-25) to be comparable, but they are less than the energy required for oxygen atom generation in the absence of spin reversal.

Spin reversal from the excited state of the oxyanions was first shown for the oxybromine anions[57] and later for ClO^-;[58] $O(^3P)$ was identified by conversion to O_3 in the presence of oxygen. $O(^1D)$ can be generated when ClO^- is irradiated at 290 nm.

O_3 has a characteristic transient absorption band with λ_{max} at 260 nm.

Many other transients absorb in the 260-nm region, and formation of O_3 must be further demonstrated through its first-order decay, $k_{O_3} = (600 \pm 100)(OH^-)$. Since $O(^3P)$ reacts with other oxyanions,[57] the determination of the O_3 decay constant must be extrapolated to zero oxyanion concentration.

Oxygen molecule generation, previously regarded as the principal mechanism of oxygen evolution from irradiated oxyanions, is supported by little evidence. While energetically favorable,[48] the change in oxidation number of the central atom, X, is large and unlikely to occur in one step. Photolysis of bromate gives oxygen molecules as a final product, but addition of suitable scavengers for $O(^3P)$, such as allyl alcohol, completely suppresses oxygen evolution.[59] There is some preliminary and circumstantial evidence that only NO_3^- and MnO_4^- evolve molecular oxygen when irradiated.[60-62] The "residual yield" from nitrate photolyzed in acid solution has still not been explained satisfactorily. Irradiation within the high-energy (195 nm) band also gives isomerization of nitrate to pernitrite, a metastable isomeric species being involved to account for both the isomerization and residual yields.[62] A mechanism, operating through an intramolecular recombination, proposes a stepwise change in oxidation number. Permanganate photolyzed in rigid alcohol glasses does not give alcohol radicals at low temperatures[63] and points to an absence of radicals produced from permanganate directly in the initial step. The mechanism has not been further tested in solution.

To complete the pattern, we now consider the act(s) involved when both intramolecular and CTTS transitions occur for an ion. The group of anions, loosely called the pseudohalides, such as thiocyanate and azide together with thiosulfate, have solution spectra composed of overlapping bands.[7] Detailed studies have found CTTS bands in the presence of internal transitions through the very much higher sensitivity of the former to environmental perturbations. When irradiated at a given frequency, two transitions may be simultaneously excited through the overlap of broad absorption bands of different character.

The spectrum of aqueous thiosulfate has been described in Section II, and flash photolysis of aqueous thiosulfate gives two transients with different decay characteristics, Fig. 8-3d.[69] Both transients are formed in the presence or absence of oxygen and for solutions up to 1×10^{-2} M thiosulfate concentration.

The shorter-lived transient, A, has λ_{max} at 380 nm and is intensified by 50% when nitrous oxide-saturated solutions are irradiated relative to nitrogen- or oxygen-saturated thiosulfate solutions. The increase arises from the reaction of hydroxyl radicals produced through reaction of nitrous oxide with the solvated electrons from the original photoionization, with additional thiosulfate;

$$S_2O_{3aq}^{2-} \xrightarrow{h\nu} S_2O_3^- + e_{aq}^- \tag{8-26}$$

$$e_{aq}^- + N_2O \longrightarrow N_2 + OH^- + \cdot OH \tag{8-27}$$

$$\cdot OH + S_2O_3^{2-} \longrightarrow S_2O_3^- + OH^- \tag{8-28}$$

The role of hydroxyl radicals was confirmed by virtue of a reduction of the intensity of transient A by 50%, relative to oxygen- or nitrogen-saturated solutions, when 1×10^{-1} M ethyl alcohol was added to nitrous oxide-saturated thiosulfate solutions. The hydroxyl radical is removed by

$$\cdot OH + CH_3CH_2OH \rightarrow H_2O + CH_3\dot{C}HOH \tag{8-29}$$

The relative increase or decrease obtained for the use of different scavengers is not consistent with reaction (8-26) as the *only* initial source of transient A, the $S_2O_3^-$ radical ion. If reaction (8-26) is the sole source of $S_2O_3^-$, then, assuming that the solvated electron does not normally react with thiosulfate to produce transient A, the addition of nitrous oxide should increase the intensity of transient by 100%. As this is not so, then another source must be found for the production of $S_2O_3^-$ radical ion. The lowest level of transient intensity corresponds to reaction (8-26) and is consistent with an initial CTTS transition.

The decay of $S_2O_3^-$ radical ion is bimolecular in neutral or alkaline deoxygenated solution and first order in the presence of oxygen. The bimolecular decay is unaffected by the addition of carbonate ion, allyl alcohol, or ethyl alcohol and points to the formation of $S_4O_6^{2-}$, but this product has not been identified.

The absorption band producing $S_2O_3^-$ is found by using filters with cutoffs at 215 and 237 nm, respectively, which reduce the intensities of $S_2O_3^-$ by 50 and 20%, respectively. $S_2O_3^-$ is therefore confirmed to arise from the CTTS transition at 210 nm. Pulse radiolysis of aqueous thiosulfate gives an almost identical transient, $\lambda_{max} = 375$ nm, the radical being produced from hydroxyl radical attack on thiosulfate, reaction (8-28).

Use of the 215- and 237-nm filters, to study the relatively long-lived transient B, reduces the 280-nm intensity maximum by 80 and 100%, respectively, placing the absorption band leading to radical B formation below 200 nm. Such a conclusion rules out transients produced from the 240-nm band, which does not appear to form radicals.

The absorption band below 200 nm is assigned to an allowed $n-\pi^*$ transition, the nonbonding electron being on oxygen. The dissociation of the excited state of thiosulfate produced by photolysis below 200 nm, therefore, would be strictly analogous to that described previously for the

oxyanions
$$S_2O_{3aq}^{2-} \rightarrow S_2O_2^- + \cdot OH + OH^- \qquad (8\text{-}30)$$
that is,
$$XO_n^{m-} \cdot H_2O \rightarrow XO_{n-1}^{(m-1)-} + \cdot OH + OH^- \qquad (8\text{-}25)$$

The possibility of O^- being a precursor of the hydroxyl radical was not examined. The hydroxyl radical formation provides for the additional source of $S_2O_3^-$ through Eq. (8-30) and thus contributes 50% of transient intensity for photolysis in oxygen- or nitrogen-saturated solutions. The relative contributions of the two processes to $S_2O_3^-$ intensities indicate that the respective quantum yields are approximately equal.

The $S_2O_2^-$ radicals decay by a second-order process, which is unaffected by addition of alcohol, carbonate, or oxygen, through unknown intermediates to give sulfur, hydrogen sulfide, and sulfite ion. Similar products are found in the steady-state photolysis of thiosulfate, where the quantum yield of electrons is found to be low, $\phi_e = 0.04$.[26] The overall quantum yield must necessarily arise from a finite photolysis time and is almost certainly decreased by secondary reactions. Current work casts doubt on the internal transition nature of band C. It is possible that band C is a CTTS transition, with a high-energy electronic state and the same solvated electron state as band B.[103]

Internal transitions, both $\pi-\pi^*$ and $n-\pi^*$, of inorganic anions lead to dissociation, whereas $\pi-\pi^*$ transitions of organic anions can give solvated electrons. Simple anions give solvated electrons only through CTTS transitions. The spectra of the solvated electron and the associated radical ion are independent, whether formed from pulse radiolysis or photolysis. The solvated electron spectrum has been shown to be very sensitive to environmental perturbations such as temperature, solvent, and solvent composition. The intense potential field of the radical ion would distort the potential well of the solvated electron if they were in close proximity during the period available for observation. However, since the solvated electron spectrum does not depend on the method of generation, the radical ion and the electron are conclusively shown to be well separated in the bulk of the solution.

The primary quantum yields for various modes of photodecomposition for hypochlorite have been carefully studied as a function of irradiation frequency,[65] Table 8-4. The secondary reactions are complex and were investigated in a separate pulse-radiolysis study.[66] Production of $O(^3P)$ dominates for low-energy irradiation. The process is spin-conservation forbidden; hence, $O(^3P)$ may result from a triplet state formed by intersystem crossing from an excited singlet state of ClO^-. If the process is forbidden, then an explanation must be sought for the relatively high quantum yield of $O(^3P)$. One explanation has pointed to the kinetic energy

Table 8-4. Product Yields for Photolysis of Hypochlorite and Chlorite Ions

(a) Hypochlorite			
λ, nm	365	313	254
pH	12.1	12.0	11.5
$[ClO^-]M$	$(0.7-1) \times 10^{-2}$	1×10^{-3}	1×10^{-3}
$\phi(-ClO^-)$	0.60	0.39	0.85
	(0.64)	(0.43)	(0.86)
$\phi(ClO_2^-)$	0.16	—	—
$\phi(O_2)$	0.04	0.07	0.20
$\phi(ClO_3^-)$	0.08	0.08	0.15
$\phi(Cl^-)^a$	0.36	0.27	0.70
	(0.40)	(0.28)	(0.68)
(b) Chlorite at pH 10			
$[ClO_2]M, \times 10^3$	(5–10)	(1.4–2.5)	1
$\phi(-ClO_2^-)$	1.20	0.42	0.50
	(1.23)	(0.42)	(0.54)
$\phi(ClO^-)$	0.32	0.12	0.14
	(0.38)	(0.13)	(0.16)
$\phi(O_2)$	0.27	0.03	0.04
$\phi(ClO_2)$	0.23	0.20	0.24
$\phi(ClO_3^-)$	0.35	0.07	0.11
$\phi(Cl^-)^b$	0.38	—	0.07

[a] Obtained from $[\phi(-ClO^-) - \phi(ClO_3^-) + \phi(ClO_2^-)]$. Values given in parentheses are calculated from the reaction schemes.
[b] Obtained from $[\phi(-ClO_2) - \phi(ClO^-) + \phi(ClO_2)]$ (taken from Refs. 65 and 67).

given to Cl^- and $O(^3P)$ because of the significant differences between the electron affinities of Cl^- (366 kJ mole^{-1}) and $O(^3P)$ (141 kJ mole^{-1}).

No CTTS processes, or products resulting from such processes, are observed[65] from hypochlorite photolysis in contrast to a previous report.[67] This result agrees with an assignment of the hypochlorite absorption spectrum to internal transitions.[7] The quantum yields of O^- and $O(^1D)$ formation increase with increase in irradiation quantum energy, Table 8-4. Formation of $O(^1D)$ does not occur for 365-nm photolysis, as the process is only thermodynamically feasible below 320 nm. Photolysis of aqueous chlorite ion[65,68] shows similar features to hypochlorite, Table 8-4. Again, $O(^3P)$ and $O(^1D)$ production depend on the quantum energy of irradiation, that is, upon the excited state produced. The major reactions fit the overall reaction scheme described for the photolysis of oxyanions. An additional feature is the proposal of excited chlorite ion formation, which reacts with a second chlorite to give ClO_2, ClO^-, and O^-. This mechanism is necessary to account for an additional and otherwise unexplained yield of ClO_2. The process requires an efficient electron transfer from excited chlorite to another chorite ion. It appears to be specific to chlorite since it occurs even when other efficient electron acceptors are present in excess, for example, BrO_3^- or IO_3. Support for the electron-transfer mechanism is the very high rate of reaction for chlorite with hydrated electrons $(4.5 \times 10^{10} \text{ M}^{-1} \text{ sec}^{-1})$ when compared with bromate $(4.1 \times 10^9 \text{ M}^{-1} \text{ sec}^{-1})$ and iodate $(8 \times 10^9 \text{ M}^{-1} \text{ sec}^{-1})$.[65,68]

The primary act in the photolysis of solvated, but not complexed, cations has been described as an electron transfer to the cation from a solvent molecule.[70] The situation also exists where the cation is oxidized, for example, $Fe_{aq}^{2+} \xrightarrow{h\nu} Fe_{aq}^{3+}$, and an electron might be expected to be ejected into the solution, where it would be observed. The solvated electron has never been observed in solution following flash photolysis of ferrous ion, and this is theoretically plausible because the ferric ion, produced as the remaining center, has increased positive charge that would make escape more difficult for the electron.[70] Photolysis of deaerated ferrous ion gives hydrogen and ferric ion through formation of an excited ferrous ion that reacts with an electron acceptor such as H^+

$$Fe_{aq}^{2+} \xrightarrow{h\nu} (Fe_{aq}^{2+})^* \tag{8-31}$$

$$(Fe_{aq}^{2+})^* + H^+ \longrightarrow Fe^{3+} + \cdot H \tag{8-32}$$

The solvated electron is not observed because it is not formed in the normally accepted definition of a solvated electron. More recent and extensive investigations of ferrous sulfate photolysis show, at high acidity in the absence of oxygen, that the excited state solvated ferrous ion produced in the photolysis will give $(Fe^{3+}OH^- \cdot H)$ in a solvent cage[71,74]

$$(Fe_{aq}^{2+})^* \to [Fe^{3+}OH^- \cdot H] \text{ solvent cage} \tag{8-33}$$

$$[Fe^{3+}OH^- \cdot H] \to Fe_{aq}^{2+} \tag{8-34}$$

$$[Fe^{3+}OH^- \cdot H] \to FeOH^{2+} + \cdot H \tag{8-35}$$

$$[Fe^{3+}OH^- \cdot H] + H^+ \to [Fe^{3+}OH^- H^+] + \cdot H \tag{8-36}$$

$$Fe^{2+} + \cdot H \leftrightarrows FeH^{2+} \tag{8-37}$$

$$FeH^{2+} + H^+ \to Fe^{3+} + H_2 \tag{8-38}$$

In the presence of excess oxygen, the initial quantum yield of ferric ion is doubled, through

$$\cdot H + O_2 \rightarrow \cdot HO_2 \tag{8-39}$$

$$Fe^{2+} + \cdot HO_2 + H^+ \rightarrow Fe^{3+} + H_2O_2 \tag{8-40}$$

$$Fe^{2+} + H_2O_2 \rightarrow Fe^{3+} + OH^- + \cdot OH \tag{8-41}$$

$$Fe^{2+} + \cdot OH \rightarrow Fe^{3+} + OH^- \tag{8-42}$$

The rate constant ratio for reactions (8-37) and (8-39), k_{39}/k_{37}, is 10^3, and thus excess oxygen effectively cuts out reaction (8-37) in favor of (8-39). Photolysis of ceric ion gives cerous ion and oxygen, explained by an increased electron affinity of excited ceric ion, leading to reduction[75]

$$Ce_{aq}^{4+} \xrightarrow{h\nu} (Ce_{aq}^{4+})^* \tag{8-43}$$

$$(Ce_{aq}^{4+})^* \longrightarrow Ce^{3+} + \cdot OH + H^+ \tag{8-44}$$

the further reactions of $\cdot OH$ leading to oxygen evolution.

Interest in the photochemistry of solvated cations had waned in favor of the photochemistry of anions and inorganic complexes until recently, when the spectroscopy and photochemistry of aqueous Eu(III) solutions were reexamined using modern techniques.

Aqueous Eu(III) solutions, such as the perchlorate, show a broad absorption band in the ultraviolet with a maximum in the region of 190 nm with $\varepsilon_{max} = 215\ M^{-1}\ cm^{-1}$ (Fig. 8-4). The transition is assigned to a charge-transfer process from the effect of various ligands on the absorption band. The absorption spectra are independent of concentration up to $1 \times 10^{-1}\ M$.[76] Fluorescence spectra show two bands, I and II, Fig. 8-4, the first with $\lambda_{max} = 350$ nm for excitation below 260 nm. The excitation

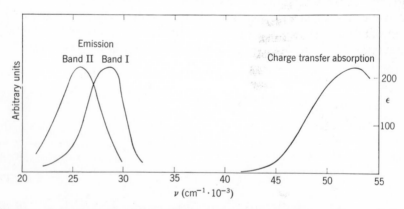

Fig. 8-4. Absorption and emission CT spectra of aqueous europium perchlorate solutions. The spectra normalized to unit height, [From Y. Haas, G. Stein, and M. Tomkiewicz, *J. Phys. Chem.* **74**, 2558 (1970).] Reprinted by permission of the copyright holders.

spectrum of band I follows the absorption spectrum of Eu(III) as far as can be measured to 250 nm. Normalized intensities are dependent on concentration in the range, 2×10^{-1} to 1.0 M, and independent for $<2 \times 10^{-1}$ M. The Stern–Volmer law for concentration quenching is not followed in the concentration-dependent region. The presence of oxygen does not affect the fluorescence of either bands I or II. Solutions of Eu(II) perchlorates do not show a detectable fluorescence for excitation at 260, 280, or 310 nm, the absorption maxima for aqueous Eu(II).

The absorption process is assigned to a charge-transfer process in which an electron from an orbital centered mainly on a water molecule is transferred to an orbital centered mainly on Eu(III), producing an excited $(Eu^{2+} \cdot H_2O)^*$ species.[76] The proposed process assumes a one-electron transfer and ignores any contributions from other water molecules of the solvation shell of Eu^{3+}_{aq}. The excited state may be deactivated by several mechanisms, Fig. 8-5. The first mechanism proceeds through radiative coupling to the ground state. In Fig. 8-5, the absorption band is given as a transition from the ground-state charge-transfer complex, $Eu^{3+} \cdot H_2O$ to the excited state of the same complex, $G \rightarrow F$. Potential energy curves for the two states are drawn in the "critical vibration approximation" and excitation below 270 nm, as shown for 254 and 185 nm, would give dissociation within the lifetime of a few vibrations of the excited state. Subsequent intermolecular crossing will give the excited state of a $(Eu^{2+} \cdot H_2O)$ species, which fluoresces to give band I and the $Eu^{2+}H_2O$ ground state, D. The assignment agrees with the wavelength dependence of band I and also accounts for the deviation from Stern–Volmer concentration-quenching kinetics. Band II arises from radiative coupling of $(Eu^{3+} \cdot H_2O)^*$ to $(Eu^{3+}H_2O)$, where insufficient energy exists for inter-molecular crossing to occur.

The formation of the excited Eu(II) ion is accompanied by the formation of H_2O^+, equivalent to a hydroxyl radical and hydrogen ion

$$H_2O^+ \rightarrow H^+ + \cdot OH \qquad (8\text{-}45)$$

As these species are formed in a solvent cage, recombination of the hydroxyl radical with excited Eu(II) ion is possible

$$(Eu^{2+})^* \cdots OH \rightarrow (Eu^{3+}OH^-) \qquad (8\text{-}46)$$

Alternatively, the Eu(II) ion may be oxidized by other oxidizing agents present in solution, such as H_3O^+

$$(Eu^{2+})^* + H_3O^+ \rightarrow Eu^{3+} + \cdot H + H_2O \qquad (8\text{-}47)$$

Reaction (8-47) provides a check on the proposed mechanism since

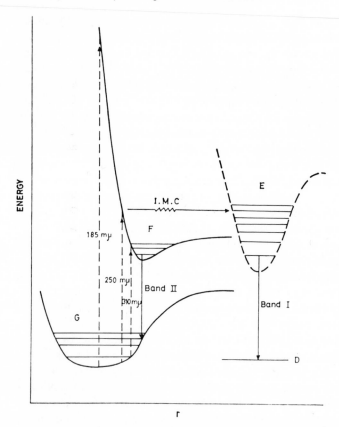

Fig. 8-5. Energy-level diagram for the system $Eu^{3+}-H_2O$. r is the internuclear distance along the critical coordinate responsible for the transitions observed. G is the ground-state level of the charge-transfer complex ($Eu^{3+}-H_2O$). F is an excited state of the same charge-transfer complex. D is the ground-state level of Eu^{2+} hydrate. *I.M.C.* denotes intermolecular crossing of the excited state of the charge-transfer complex into the excited state of Eu^{2+}-hydrate. E is an excited state of Eu^{2+} hydrate. The transitions observed in the spectrum are denoted by arrows. [From Y. Haas, G. Stein, and M. Tomkiewicz, *J. Phys. Chem.* **74**, 2558 (1970).] Reprinted by permission of the copyright holders.

photolysis of aqueous Eu(III) solutions should evolve molecular hydrogen by dimerization of the hydrogen atoms produced. Irradiation of deaerated Eu(III) solutions at 254 nm, where $\varepsilon_{Eu^{3+}} = 0.3\ M^{-1}\ cm^{-1}$ does not evolve hydrogen, probably because the hydrogen atoms formed in the primary act are scavenged by solution impurities to give non-gas-forming products. However, irradiation at 254 nm ($Eu^{3+} = 0.048\ M$, $HClO_4 = 2.38\ M$) with ethanol as a scavenger (7×10^{-2} and $7 \times 10^{-3}\ M$) gives quantum yields of 6 and 3, respectively (based on the contribution of the charge-transfer band at 254 nm.) The presence of ethanol at varying concentrations does not affect either the absorption or fluorescence spectra of Eu(III) significantly. The rather high quantum yields of hydrogen are explained by the formation of Eu^{2+}, which absorbs strongly at 254 nm (ε_{254} for $Eu^{2+} = 2 \times 10^3 \ M^{-1} \ cm^{-1}$).

The observed evolution of hydrogen following photolysis of Eu^{3+} proves the validity of the proposed mechanism.[76] It is now possible to see why solvated electron formation is not observed in cation photolysis since the electron-transfer process is inwards and a very much more intimate process than, in contrast, CTTS processes.

IV. SECONDARY REACTIONS

The cataclysmic processes of flash photolysis and pulse radiolysis used to generate, and subsequently study, reactions of solvated electrons initially engendered the feeling that the species reacted both rapidly and indiscriminately. Developments in detection and time-resolution techniques have shown the solvated electron to be a discriminating reactant, for in the final analysis it is the simplest nucleophile. The reaction rates of solvated electrons with substituted benzenes indicate that electron-withdrawing groups, such as $-NO_2$, markedly increase the reactivity of the aromatic system. A very large number of bimolecular rate constants for the solvated electron reacting with other species have been determined, and a representative sample is shown in Fig. 8.6.[29]

Specific scavengers for solvated electrons include such Brønsted acids as H^+, HF, $H_2PO_4^-$, and NH_4^+. Other specific scavengers include N_2O, acetone, monochloracetic acid, and 1,2-dichloroethane. The first group react with the solvated electron in accord with the Brønsted general acid-catalysis law.[33] An attempted extension of this correlation to include other Brønsted acids within the same pK range found that the solvated electron did not react at all with the compounds studied.[79] Clearly, a proton-transfer mechanism is invalid and is rejected in favor of an intermediate radical or radical ion formed through incorporation of the solvated electron into the scavenger. The intermediate subsequently decomposes to give other radicals that will then be scavenged in turn by other specific scavengers. An excellent example is that of solvated electrons produced in the photolysis of $H_2PO_4^-$, reacting with $H_2PO_4^-$ to give a hydrogen atom,[42,80] the latter then being scavenged by a specific hydrogen atom scavenger such as an aliphatic alcohol.

While some overall patterns are emerging for solvated-electron reaction-rate constants, it would be premature at this time to make firm

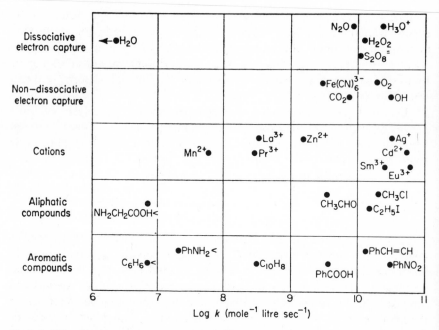

Fig. 8-6. Bimolecular rate constants for e_{aq}. [From D. M. Brown and F. S. Dainton, in *Energetics and Mechanisms in Radiation Biology*, Academic, London, 1969.] Reprinted by permission of the copyright holders.

generalizations about the molecular factors involved.[29] In Fig. 8-6, it may be noticed that many rate constants lie in the range, 10^{10} to 10^{11} M^{-1} sec^{-1}, values that would be construed as being diffusion controlled from use of the appropriate equation,[29] for example, for where the reactant is uncharged, the Smoluchowski equation

$$k = \frac{4\pi\sigma DN}{10^3 \text{ M}^{-1} \text{ sec}^{-1}} \tag{8-48}$$

or for where the reactant, A, is charged, the Debye equation

$$k = \frac{4\pi z_e z_A DN}{10^3 k' Kt(e^{ze}A/kKt\sigma^{-1}) \text{ M}^{-1} \text{ sec}^{-1}} \tag{8-49}$$

where D is the sum of the diffusion constants of the two reactants, e_{aq}^- and A, z_e is the charge on the electron, K is the permittivity of the medium, and k' is Boltzmann's constant.

Assuming that the faster group of rate constants are diffusion controlled gives the effective radius of the solvated electron as 3 Å, thus being a larger species than the crystallographic radii of the halide ions.[29]

This is not really a fair comparison since the solvated electron radius is measured, by definition, for a solvated species, whereas similar measurements for the halides are overshadowed (1) by the confusion as to whether the halides in solution are solvated and (2) as to what is the effective radius of a halide ion in solution.[7] Comparison with H_3O^+ bimolecular reaction-rate constants[81] shows the highest values observed for solvated electron rate constants to be less by an order of magnitude. Proton tunneling in a hydrogen-bonded network has been suggested for the high-proton bimolecular reaction rates, from which it may be inferred that tunneling processes for the solvated electron are not major contributors to its diffusion mechanism. The electron must be regarded, instead, as being bound to a group of water molecules that move with it during the diffusion process. However, the question of diffusion control is not fully explained since the temperature dependence of the diffusion equations have not been tested. Moreover, the bimolecular rate constants for the solvated electron + hydrogen ion reaction in both water and ethanol are the same and an order of magnitude less than would be predicted if diffusion controlled.[79] Inspection of the diffusion equations shows that both include a solvent-permittivity term that is quite different for the respective solvents. The absence of a solvent effect shows that the solvated electron + hydrogen ion reaction is not diffusion controlled, even when the high diffusion coefficient of the hydrogen ion in these solvents is considered. However, the reaction between solvated electrons, $e_{aq}^- + e_{aq}^-$, is diffusion controlled.[29]

The spread of values for the bimolecular rate constants in Fig. 8-6 emphasizes, on the other hand, that the reactions of the solvated electron are not necessarily encounter controlled, implying a sensitivity to the nature of the reacting partner. The sensitivity of the electron, as the simplest nucleophile, has already been demonstrated in the four orders of magnitude difference between the bimolecular rate constants for solvated electrons reacting with benzene and with nitrobenzene. Another striking feature is the lower reactivity of hyper-reduced cations such as Mn^+, readily produced in pulse radiolysis, with solvated electrons when compared to reaction with ions such as Eu^{3+}, which has a well-known reduced form as Eu^{2+}.[29]

When reactions of solvated electrons are encounter-controlled, the activation energy for the process is found to be low. When solvated electrons are produced and immobilized in glasses at low temperatures, warming the glass slightly mobilizes the electrons that then react, possessing low kinetic energies. Encounter-controlled reactions of solvated electrons with solutes in glasses proceed as effectively as in solution at room temperature.[29]

Scavenging systems that are particularly efficient for solvated electrons include hydrogen ion/aliphatic alcohols and $H_2PO_4^-$/aliphatic alcohols, provided that the latter system does not absorb in the same region as the compound studied.[80] The acids convert the solvated electron to a hydrogen atom that is then efficiently scavenged by the alcohol. The concentrations of alcohol required for complete scavenging of hydrogen atoms should vary inversely as the ratio of the rate constants for the hydrogen + alcohol reaction. Thus, in the photolysis of bisulfate ion at 185 nm,[82] the required concentrations of methyl, ethyl, and 2-propyl alcohols are 1×10^{-1}, 3×10^{-2}, and 1×10^{-2} M, the relative ratios of which are the inverse of the reaction rate constant ratios.[82,83] The correlation between the photochemical[82] and radiolysis[84] kinetics indicates that hydrogen atoms are scavenged in the bulk of the solution. In contrast, when alcohols alone are used as scavengers in neutral solution, the quantum yield of hydrogen evolved depends on the alcohol used. In the photolysis of aqueous sulfate at pH 7, methyl alcohol as scavenger gives $\phi_{H_2} = 0.36$, ethyl alcohol gives $\phi_{H_2} = 0.14$, and 2-propyl alcohol gives $\phi_{H_2} = 0$.[82] In the last case, all hydrogen evolved comes from photolysis of the alcohol and water. The reason for the variation is not clear since the quantum yield of electrons obtained, using nitrous oxide as a scavenger, is 0.64,[85] a value also obtained as the pH of alcohol/sulfate solutions are decreased, Fig. 8-7. Production of hydrogen atoms may arise in a solvent cage from which the alcohols are progressively excluded with increase in size of the alkyl group. When methanol is used as scavenger in the photolysis of hydroxide ion at 185 nm in solutions of pH 12, the quantum yield of hydrogen is 0.105, essentially the value obtained for the same photolysis using nitrous oxide as a scavenger.[82,85]

Hydrogen molecules are formed by abstraction from the α-carbon atom of the alcohol, the alcohol radicals either combining to give a glycol or disproportionating to give an aldehyde/ketone and the original alcohol.[42,82,85,86] The trend lies toward disproportionation, for example, methyl alcohol gives 96% glycol and 2% formaldehyde, whereas 2-propyl alcohol gives 50% acetone.[82]

Addition of alcohols to aqueous solutions of anions absorbing by CTTS processes shifts the absorption bands to higher energies. This effect is particularly important in the far-ultraviolet region, where almost all components of the solutions to be photolyzed absorb, and also interact, such that Beer's law is not obeyed.[20,82] In aqueous sulfate photolysis, addition of alcohol as a scavenger alters the absorption coefficients of the alcohol, often in a dramatic manner, and also that for sulfate at 185 nm.[82] The latter effect arises from observing the anion absorption coefficient on the low-energy edge of its absorption band at a fixed wavelength, while

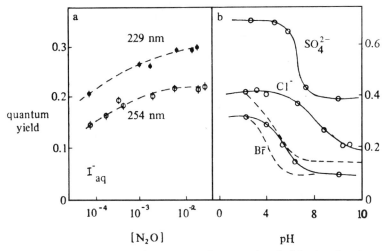

Fig. 8-7. Plot of quantum yield for (*a*) 5×10^{-3} M potassium iodide photolyzed at 229 nm and 10^{-1} M potassium iodide photolyzed at 254 nm in the presence of N_2O, and (*b*) Cl^- (0.04 M), Br^- (0.01 M), and SO_4^{2-} (0.1 M) in the presence of H^+ and 0.4 M methyl alcohol. Reprinted by permission from *Q. Rev.* **24**, 565 (1970).

the band is steadily shifted to higher energies by progressive addition of alcohol. The quantum yield of hydrogen arises from photolysis of all the solution components

$$\phi_{H_2} = \frac{1}{\alpha_{tot}} [\phi_{H_2O} \alpha_{H_2O} + \phi_{MeOH} \alpha_{MeOH} + \phi_{SO_4^{2-}} \alpha_{SO_4^{2-}}] \qquad (8\text{-}50)$$

where the α's represent the absorbances of the solution components. An example of the extent of the variation in the molar absorption coefficients of the respective solution components is given in Table 8-5. The quantum yield for sulfate photolysis, after complete scavenging for this system is

Table 8-5. Variation of Molar Absorption Coefficients with Concentration for Photolysis Solutions[a]

Concn of 2-propyl alcohol, M	0.27	0.53	0.80	1.07	1.60	2.13	2.67	4.0	5.33
$\varepsilon_{2\,PrOH}$, $M^{-1}\,cm^{-1}$	47.2	44.3	44.0	40.8	36.8	33.2	30.2	26.6	22.0
ε_{SO_4}, $M^{-1}\,cm^{-1}$	194	188	183	179	172	166	159	135	130

[a] Mixed solvent, 2-propyl alcohol (PrOH)/water. $[Na_2SO_4] = 0.1$ M.
[b] Measured at 185 nm. 2-PrOH = 2-propyl alcohol.

achieved, is constant when the variations in absorption coefficients are used in Eq. (8-50) above. When constant values for the absorption coefficients are assumed, the quantum yields for sulfate photolysis first increase with increasing alcohol concentration and then decrease.[83] Similar effects are observed in the photolysis of aqueous alcohols at 185 nm, a nonlinear quantum yield being observed after complete scavenging is reached if Beer's Law is assumed.[88] On the other hand, use of the actual absorption coefficients found at each concentration of alcohol gives a constant quantum yield value once complete scavenging is achieved.[82]

"Secondary" reactions of the solvated electron with products must be considered. It is useful that the free radicals produced by hydrogen abstraction from aliphatic alcohols, for example, $\cdot CH_2OH$ from CH_3OH, CH_3CHOH from CH_3CH_2OH, have a low reactivity toward solvated electrons.[79] Such behavior correlates with the electron-donor activity of $\cdot CH_2OH$ toward alkyl bromides, giving debromination.[89] The electron-donor character indicates a preference for carbonium-ion formation, rather than accepting an electron to form a carbanion. Simple aliphatic alcohols are good scavengers for hydrogen atoms and hydroxyl radicals in the presence of solvated electrons, the latter reacting only slightly with alcohols or alcohol radicals.

The hydrogen ion/aliphatic alcohol or dihydrogen phosphate/alcohol scavenger systems do not discriminate between hydrogen molecules formed by abstraction or from the dimerization of hydrogen atoms or solvated electrons. If deuterated aliphatic alcohols, such as CD_3OH or C_2D_5OH, are used in H^+/H_2O solutions, then HD will appear as the hydrogen molecules arising from hydrogen abstraction and H_2 from dimerization of either H atoms or solvated electrons.[29]

A primary quantum yield for anion photolysis is defined as the quantum yield of electrons (radicals) that escape primary recombination in the original solvent cage and diffuse into the bulk of the solution (these terms are discussed extensively in Section V). The primary quantum yield is calculated from the limiting initial slopes of product concentration against time. For solvated electron and radical formation (and also in radical pair formation, where the original pair can be identified), there is a marked dependence on concentration that increases with scavenger concentration up to a limiting value, Fig. 8-7. The limiting quantum yields obtained are independent of irradiation intensity and, together with the observed scavenger concentration effect, favor a cage, inhomogenous, radical scavenging mechanism and reject a homogenous radical-recombination/radical-scavenging competition in solution.

A good specific scavenger for solvated electrons is nitrous oxide, which

has a high rate constant for reaction with solvated electrons, $k(e_{aq}^- +$
$N_2O) = 8.7 \times 10^9$ M^{-1} sec^{-1}, giving an inert product, nitrogen, according
to[29,79]

$$e_{aq}^- + N_2O \rightarrow (N_2O^-) \rightarrow N_2 + O^- \qquad (8\text{-}51)$$

$$\downarrow H_2O$$

$$\cdot OH + OH^-$$

The yield of nitrogen is directly related to the yield of solvated
electrons. The use of N_2O as a scavenger in the photolysis of iodide at
254 nm is illustrated in Fig. 8-7, where the yield of nitrogen increases with
increase in nitrous oxide concentration up to a plateau value when
complete scavenging of available electrons is attained.[90] The N_2O^-
intermediate has a finite lifetime of at least 10^{-7} sec and can act as a strong
oxidizing or mild reducing agent. Intermediate formation indicates that
the hydroxyl radicals are not formed from an immediately dissociative
electron capture by N_2O.

Various scavenger systems used in sulfate, hydroxide, bromide, and
iodide photolysis give the same individual limiting quantum yields,[6]
which, together with spectroscopic studies of an anion/scavenger systems
(I), confirms that the scavenger does not interact directly with the
spectroscopic excited state. Furthermore, if complete dissociation of the
excited state is accepted, then because the limiting quantum yield is
always less than unity, the scavenger does not react with the first
dissociation process of the spectroscopic excited state.

When iodide is photolyzed in solvents other than water, different
quantum yields are obtained, Table 8-6.[35]

There are several underlying points which require expansion: first, the
summary of Section II (concerning spectroscopy) is recalled; that is, the
position of the absorption band with respect to the irradiating wavelength

Table 8-6. Quantum Yields of Iodide Photolysis
(254 nm) in Various Solvents[a]

Solvent	Limiting Quantum Yields
Water	0.29
	0.23
Methyl alcohol	0.60
Ethyl alcohol	0.66
2-Propyl alcohol	0.70
Methyl cyanide	0.75

[a] From Ref. 6. Values for 298°K.

must be considered. Photolysis of iodide at 254 nm in the simple aliphatic alcohols, methyl, ethyl, and 2-propyl, increases the quantum yield in that order; yet the absorption band is steadily shifted away from the irradiation wavelength by these solvents. On the other hand, photolysis of iodide in methyl cyanide at 254 nm, where λ_{max} is 247 nm, gives the highest quantum yield of all, at $\phi = 0.75$, where irradiation occurs close to the center of the absorption band.[35]

When aqueous iodide is photolyzed at 254 nm, the irradiating wavelength is more than two half-band widths to lower energies from the Gaussian absorption-band center at 226 nm. The absorption coefficient of aqueous iodide at 254 nm is 3% of that at 226 nm (at 298°K). The low-energy limit for iodide photolysis has not been explored. When irradiated close to the band center (for aqueous iodide) at 229 nm, an increased quantum yield of 0.31 is obtained (298°K). The half-widths at the iodide bands are large, for example, 4200 and 3000 cm^{-1} in water and methyl cyanide, respectively,[7] and are attributed to a distribution of radical-solvated electron separations within the solvent potential well. The ratio of quantum yields at 254 and 229 nm for aqueous iodide photolysis is constant for N_2O concentrations up to, and including, the limiting quantum yields.[90] The fraction of solvated electrons scavenged by N_2O appears to be independent of the number of solvated electrons and iodine radicals produced at each wavelength. From this it is deduced that the separation of electron and radical produced at each wavelength must be the same. This conclusion is reached despite the quantum difference of 50 kJ mole^{-1} between the two wavelengths and points to radical formation in a solvent cage defined essentially by the ground-state solvent configuration.[90] The Franck-Condon principle evidently extends its influence to include radical formation.

The quantum yield for 185-nm photolysis of aqueous iodide, Fig. 8-8, is 0.25,[36] a wavelength where there is considerable overlap between the $I(^2P_{1/2})$ component of the doublet and a higher-energy component. The lower quantum yield at 185 nm is provisionally attributed to the mixed states of the solvated electrons produced. The iodine radical, produced by 185-nm photolysis, should be in the $^2P_{1/2}$ atomic state, by analogy to those produced in the $^2P_{3/2}$ state produced from lower-energy photolysis since the next energy level of atomic iodine lies at very much higher energies above the ^2P states.[21]

The higher-energy iodine radical should fit the correlation between the energy of the halide radicals produced in photolysis and their ability to abstract hydrogen from a C—H bond. Thus $^2P_{3/2}$ iodine radicals do not dehydrogenate alcohols, as is also found for $^2P_{3/2}$ bromine radicals produced by 229-nm photolysis. Chlorine radicals, probably both the $^2P_{1/2}$

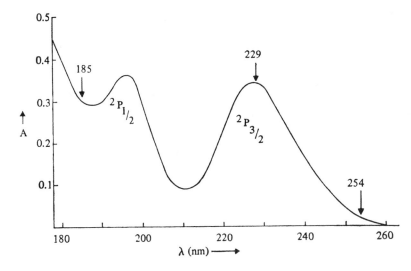

Fig. 8-8. A typical CTTS absorption spectrum, that of iodide in aqueous solution at 274°K, showing irradiation frequencies. (M. F. Fox, unpublished work.)

and $^2P_{3/2}$ (because of the small separation of the chlorine doublet) and bromine $^2P_{1/2}$ radicals, produced by 185-nm photolysis of aqueous bromide, do dehydrogenate alcohols.[36] On this basis, the iodine $^2P_{1/2}$ radical produced by photolysis of aqueous iodide solutions at 185 nm, having the highest energy of all, should fit the correlation and dehydrogenate alcohols but do not, and the discrepancy has not been explained.

The wavelength dependence of quantum yields has only briefly been touched upon in investigations and has raised more questions than provided solutions. Clearly, the wavelength dependence of quantum yields needs further investigation in depth. On the other hand, it must again be emphasized that the wavelength dependence of ferrocyanide-photolysis quantum yields arises from varying contributions of ligand field (d–d) and CTTS/CTTL transitions with wavelength, as previously described.[32]

Increasing the temperature increased the photolysis quantum yields of aqueous anions at both 254 and 229 nm, although the range of temperatures studied was not extensive, 278 to 328°K. There is a tendency for the quantum yields to become independent of temperature as the temperature increases, Fig. 8-9.[90]

A competition between two deactivation modes of the excited state of iodide, reactions (8-52) and (8-53) is proposed to account for the variation

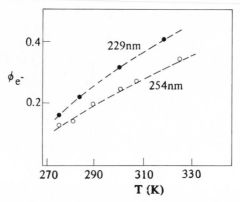

Fig. 8-9. Plot of quantum yield against temperature for 5×10^{-3} M potassium iodide photolyzed at 229 nm and 10^{-1} M potassium iodide photolyzed at 254 nm. [From F. S. Dainton and S. R. Logan, *Proc. Roy. Soc. A* **287**, 281 (1965).] Reprinted by permission from *Q. Rev.* **24**, 565 (1970).

of the quantum yields with temperature

$$I_{aq}^{-}* \xrightarrow{k_{52}} I_{aq}^{-} \tag{8-52}$$

$$I_{aq}^{-}* \xrightarrow{k_{53}} \cdot I + e_{aq}^{-} \tag{8-53}$$

The first deactivation mode is envisaged as a collisional deactivation, (radiative processes may be ruled out as fluorescence has never been observed from a CTTS absorption band). The collisional thermal deactivation process will have a $T^{1/2}$ temperature dependence. For the second process, it is proposed that the fragments of the dissociation diffuse out of the solvent cage at a rate related to the bulk viscosity of the solution, η, where[91]

$$\eta = Ae^{-B/RT} \tag{8-54}$$

and B is the activation energy for viscous flow. Relating reaction (8-53) to Eq. (8-54) gives

$$k_{55} = k_{53} \exp\left(\frac{-B}{RT}\right) \tag{8-55}$$

The limiting quantum yield, ϕ_e, results from a simple competition between reactions (8-52) and (8-53), assuming complete photodissociation

$$\phi_e = \frac{k_{55}}{k_{55} + k_{52}} \tag{8-56}$$

An Arrhenius plot can be plotted for the limiting quantum yields obtained, and activation energies found for the halide ion photolyses are given in Table 8-7, together with B for water.

Table 8-7.   Activation Energies for Halide Photolysis[a]

Anion	Irradiation Wavelength, nm	B, kJ mole^{-1}
I$^-$	254	19.7
	254	20.5
	229	20.9
Br$^-$	229	16.8
	185	14.2
Cl$^-$	185	4.2
Water	(Viscosity)	18.9

[a] From Ref. 6.

Values of B for the photolyses of anions, except that for chloride, agree quite well with the water value and point to the role of molecular relaxations in releasing radicals and solvated electrons to the bulk solvent.[91] It is both surprising and interesting to find the correlation between B for water, which is a bulk solvent property, and the B values for radical release to the solvent, which would be expected to reflect the localized properties of the solvated ion. Furthermore, as the fluidity of the solvent increases with temperature, it would be expected that eventually $k_{55} \gg k_{52}$ and ϕ_e would tend toward a constant value of unity. The predicted behavior is only observed in part since the quantum-yield values do tend toward constant values at the higher temperatures, but in no case do they approach unity. The low value for chloride has not been explained and may be related to the small size of the ion. It is noticeable that the B values do decrease with decrease in the size of the ion, Table 8-7, but the problem has not been explored further.

The limiting quantum yield, ϕ_e, results from a simple competition between reactions (8-52) and (8-53), assuming complete photodissociation,

$$\phi_e = \frac{k_{55}}{k_{55} + k_{52}} \tag{8-56}$$

An Arrhenius plot can be plotted readily for the limiting quantum yields obtained over the temperature-range studies, and activation energies found for the halide ion photolysis are in Table 8-7, together with B for water.

The values of B obtained for the photolysis of anions, except that for chloride, agree quite well with the values for water and point to the role of molecular relaxations in releasing radicals and solvated electrons to the bulk solvent.[91] It is both surprising and interesting to find the correlation between B for water, which is a bulk solvent property, and the B values for radical release to the solvent, which would be expected to reflect the localized properties of the solvated ion. Furthermore, as the fluidity of the solvent increases with temperature, it would be expected that eventually $k_{55} \gg k_{52}$ and ϕ_e would tend toward a constant value of unity. The predicted behavior is only observed in part since the quantum yield values do tend toward constant values at the higher temperatures, but in no case do they approach unity. The low value for chloride has not been explained and may be related to the small size of the ion. It is noticeable that the B values do decrease with decrease in the size of the ion, Table 8-7, but the problem has not been explored further.

V. THE SCAVENGING PROCESS

Before discussing the application of scavenging kinetics to solution photochemistry, it is an opportune moment to step back and consider the fundamental processes that occur. Photolysis of gaseous molecules produces two fragments that separate with little chance of recombination. Separation of the fragments in the gas phase by one mean free path reduces the chance of original pair recombination (geminate recombination) to a negligible amount.

In a liquid the photolyzed molecule, now existing as a pair of radicals, is surrounded by a cage of closely packed molecules. For photolysis of a liquid compound the radicals are surrounded by their closely packed parent molecules. When a solute species is photolyzed in solution, the duration of the solvent cage depends on the interactions between the solute and solvent. Ionic solutes dissolved in polar media have a solvation layer around the ions, the Frank-Wen model,[92] and the solvent molecules contiguous to the ion are relatively immobilized by the intense potential field of the ion relative to those in the bulk solvent, zone A of the model. Zone B is further out from the ion and represents an intermediate region where solvent molecules are neither influenced by the potential field of the ion nor by the bulk solvent. Zone C is the bulk solvent. The relative importance of the zones is a matter of debate for various ions in different solvents.[93] The radicals formed are, therefore, initially confined in a structured solvent cage.

It is not surprising that photolytic production of radicals within a solvent cage often leads to recombination of the original partners within a typical vibration period of 10^{-13} sec. When the radicals diffuse away from each other, the chance is still high that the original partners will recombine. Original-pair combination for iodine radicals produced by photolysis of iodine in hexane is estimated to be as high as 0.5 of radical pairs produced.[94]

Introduction of a scavenger as a species to compete with the radicals recombination process is only effective in certain concentration ranges since there are several processes leading to recombination. The reaction sequence given below includes not only the photodissociation, recombination, and scavenging processes, but also the physical diffusive processes that occur.

Process	Description	
$AB \xrightarrow{h\nu} (A+B)$	Dissociation to give A, B radicals in a solvent cage	(8-57)
$(A+B) \longrightarrow AB$	Primary recombination within the solvent cage	(8-58)
$(A+B) \longrightarrow A+B$	Diffusive separation	(8-59)
$A+B \longrightarrow AB$	Diffusive geminate recombination of original partners	(8-60)
$A+B \longrightarrow AB$	Steady-state recombination of unoriginal partners in the bulk solvent	(8-61)
$\left.\begin{matrix} A+B \\ B+B \end{matrix}\right\} \longrightarrow \left\{\begin{matrix} A_2 \\ B_2 \end{matrix}\right.$	Dimerization	(8-62)
$\left.\begin{matrix} A+S \\ B+S \end{matrix}\right\} \longrightarrow$ intermediates	Scavenging	(8-63)

The quantum yield from the overall process depends on the scavenger concentration, Fig. 8-10. The treatment is that of Noyes,[95,96] and it has been extended to the photolysis of aqueous anions.[33-36]

Radicals formed in the solvent cage may undergo a primary recombination, probability, β', reforming the ground-state solvated anion, which is kinetically indistinguishable from an excited-state thermal deactivation, reaction (8-52). The relative importance of the two processes has not been explored, but singly or together they reduce the observable quantum below unity. For a scavenger to compete effectively with primary recombination, (1) it would have to constitute the pure solvent and (2) have a very high bimolecular rate constant (k_s) for reaction with at least one of the radicals produced. If reaction occurred at every encounter, then the curve at highest concentrations of scavenger in Fig. 8-10 might be realized. It is emphasized by Noyes that this section of the graph is a subjective estimate.

Radicals that do not recombine within the original solvent cage separate with different diffusive properties. Separation is taken to be movement of more than one molecular diameter between the radicals. If radicals escape

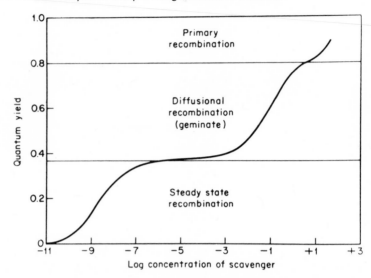

Fig. 8-10. Quantum yield for photolysis in liquid solution in the presence of a scavenger for radicals. [From R. M. Noyes, *J. Amer. Chem. Soc.* **81,** 999 (1950); *ibid.* **77,** 2042 (1955); *ibid.* **78,** 5486 (1956).] Reprinted by permission of the copyright holders.

the solvent cage, they undergo random diffusive displacements of the order of one molecular diameter with a frequency of approximately 10^{-11} sec in liquids of normal viscosity. Original partners from a dissociation may reencounter each other during the random-diffusion process, and geminate recombination takes place. The asymmetrization process that gives rise to the radicals from the excited state of a CTTS transition results in the solvation shell moving with the electron from the initial anion site. A detailed Debye–Hückel treatment for iodide photolysis in aqueous solutions of varying ionic strength showed the electron to move rapidly to a distance of at least 1 nm from the initial anion site.[90] The movement must involve rotation of water molecules and thus provides the link between quantum yields and the activation energy of viscous flow for bulk water. Scavenger must be present in sufficient concentration to encounter, and react with, a radical in the time scale of 10^{-9} to 10^{-11} sec from the formation of the radicals in the solvent cage.

If secondary recombination has not taken place within approximately 10^{-9} sec, the chance of geminate recombination is negligible. The process cannot be treated using conventional rate constants since, if many isolated pairs are produced at zero time and survive a few diffusive displacements, then the rate of pair recombination is proportional to $t^{-3/2}$. The limiting quantum yield at a given temperature, ϕ_e, and the quantum yield, ϕ, at scavenger concentration, [S], are related through

$$\ln(1 - \phi/\phi_e) = \ln \beta' - \frac{2a}{\beta'(\pi k_s [S])^{1/2}}$$

where the parameter, a, is related to the frequency of diffusive motions of the liquid and k_s is the bimolecular rate constant for the radical + scavenger reaction. At higher scavenger concentrations, the scavenging efficiency reaches a limiting value at which the scavenger is catching all radicals that escape primary and secondary recombination.

If the scavenger concentration is very low, it competes with radical recombinations or dimerizations in the bulk solvent. As the scavenger concentration is increased, it progressively competes with these processes until all bulk solvent reactions other than scavenging are eliminated, giving the lower plateau of Fig. 8-10. The competition between scavenging and recombination/dimerization then obeys conventional kinetics.

There is very good agreement between the predicted and observed photochemical kinetics of the aqueous halides,[33-36] nitrate photolyzed at 313 nm, and thiocyanate photolyzed both at 229 and 254 nm in aqueous solution. The treatment is known as the Jortner–Ottolenghi–Stein (JOS) model.[34]

Some doubts have arisen whether such inhomogenous scavenging processes do in fact occur. Flash photolysis of aqueous iodide showed that the primary yields were independent of scavenger concentration, suggesting that a cage effect was not being observed, but that instead electron scavenging was being observed in the bulk solvent.[97]

There are several discrepancies between experiment and prediction. The first, and principal, discrepancy is that the scavenger appears to be extraordinarily efficient since efficient scavenging is reached at a concentration of 5×10^{-3} M for iodide photolysis using either hydrogen ion or nitrous oxide as scavengers, Fig. 8-7. These values are two orders of magnitude less than predicted by theory and must be compared with incomplete scavenging at 0.5 M (10^{-2} mole fraction) in systems other than anion photolysis.[97]

The second problem concerns the method of calculating primary quantum yields. The standard practice has been to plot product quantum yield/time graphs and to take the initial slope of the graph as a measure of the primary quantum yield. However, the final products are often formed in secondary or even tertiary steps.

A useful cross-check exists between rate constants derived from photolysis-competition kinetics and those obtained for the same reaction studied through pulse radiolysis. The advantage of the latter method is that it is a direct method, and significant discrepancies are found between rate constants determined using the two methods. The initial slope of the product concentration/time plot does not necessarily give the true quantum yield and may introduce misleading scavenger concentration dependences unless certain conditions are observed.[97]

In the flash photolysis of aqueous iodide previously cited, the intensity of the primary intermediates increased slightly for the initial flashes, probably due to "impurity cleanup," and then decreased. The decrease in primary yield is assigned to reaction of the radicals with reaction products. Therefore, the additional information required is that of the relative magnitude of $k_s c_s$ (the product of the scavenging rate constant and the scavenger concentration) and $k_b c_p$ (the product of the radical-reaction product rate constant and the product concentration). The former product must be of the order of 10^9 to 10^{10} sec^{-1} and $k_s c_s \gg k_b c_p$ for the experimental yields to be taken as the true primary yields and then used subsequently to test cage-scavenging mechanisms.[97]

A detailed study of the photolysis of ferrocyanide in various aqueous nonelectrolyte solutions[98] compared the quantum yields of electrons produced in the solvent systems with the predicted values of the escape probability.[99] The motion of the thermalized electron after ejection, following photolysis or by diffusion from a "spur" in pulse or steady-state radiolysis, is controlled by coulombic attraction to the positive charge. For nonpolar media the ultimate escape probability of the electron from charge neutralization is given by the reciprocal of the Boltzmann factor with respect to the initial potential energy.

Mozumder has extended the treatment to charged species in polar solvents.[100] The photolysis of ferrocyanide in aqueous nonelectrolyte mixtures shows that the results obtained cannot be interpreted through one single solvent parameter, such as permittivity, diffusion coefficient, or dielectric relaxation time.

Applying Mozumder's treatment of the escape probability of a pair of oppositely charged species in polar media gives good agreement between predicted and experimental results, provided that care is taken in choosing the relaxation times of the solvent. The "a fortiori" step is necessary because of the lack of a suitable theory in which appropriate weights may be given to the relaxation time, diffusion coefficients, or permittivities of the system.[98] The thermalization length for the electron from application of Mozumder's treatment, 3 nm, is independent of the (rather limited) nature of the solvent systems used.[98] The thermalization length also agrees well with the value of not less than 1 nm for movement of the electron after photodissociation of aqueous iodide, using a Debye-Hückel approach.[90] The thermalization length may only depend on the solute and the excitation wavelength, the escape probability appearing to be determined before the medium has fully relaxed.

With these discrepancies in mind, aqueous iodide and bromide have been photolyzed at 254 and 229 nm, respectively, in the presence of high concentrations of nitrous oxide under pressure.[101] The quantum yields were calculated from the slopes of the straight line plots of trihalide ion product against irradiation time. The method is valid since the condition, $k_s c_s \gg k_b c_p$, is obeyed because the rate constant of electrons with trihalides is only four to eight times that for solvated electrons reacting with nitrous oxide, and the concentration of nitrous oxide is at least two to three orders of magnitude greater than trihalide. The iodide quantum yield increases with increase in nitrous oxide concentration until there is little variation in quantum yield over a fourfold increase in nitrous oxide concentration, within experimental error. Quantum yields are independent of halide concentration and irradiation dose. Bromide quantum yields follow the same pattern. The maximum quantum yield observed for iodide, 0.36 ± 0.06, is close to the highest value of the quantum yield observed in the temperature study, indicating an upper limit for the quantum yield of iodide photolysis at 254 nm.

Three mechanisms were considered for the increased iodide quantum yield at high nitrous oxide concentrations.[101] A charge-transfer complex between nitrous oxide and halide is ruled out by the quantum yields being independent of halide ion concentration at constant nitrous oxide concentration. An excited-state halide ion reacting directly with nitrous oxide is also ruled out by the wealth of spectroscopic evidence that indicates that scavengers do not interact with CTTS excited states. Moreover, the lifetime of the excited state of iodide is too short, being below an upper limit of 10^{-11} sec, to react directly with nitrous oxide. Only the possibility of a cage-scavenging mechanism remains

$$I_{aq}^- \longrightarrow \frac{(\cdot I + e^-)_{aq}}{(or\ I_{aq}^{-*})} \xrightarrow{N_2O} \cdot I + N_2O^- \longrightarrow \cdot I + N_2 + O^-$$

In the above scheme, the iodide excited state is given alternatively as $(\cdot I + e^-)_{aq}$ or I_{aq}^{-*}, as it seems that the evidence for the existence of I_{aq}^{-*} is not conclusive.

The concentration range for which bromide quantum yields are found to be independent overlaps slightly with a region, 1.9×10^{-7} to 1.67×10^{-2} M,[85] for which a limiting quantum yield is observed. The limiting quantum yield for bromide is reached at a lower concentration of nitrous oxide than for iodide, an effect assigned to a larger initial separation of radical and electron. Extrapolating the assignment to chloride photolysis predicts a further lower level of scavenger required to obtain limiting quantum yields, and the predicted behavior is observed, Fig. 8-7. The variation in radical-electron separations may arise from a constant electron radius and a decreasing halogen radical radius for the series, iodine → chlorine.

A further correlation may be drawn between the residual yield at pH 7 with no electron scavenger present, but with a hydrogen atom scavenger present and the limiting quantum yield with electron scavenger present, Fig. 8-7, as a measure of the initial-electron separation.

Quantum yields have been redefined as $\Gamma\phi_e$,[101] where Γ is the fraction of radicals produced in the initial photoact. The introduction of Γ is a very promising development because it has become increasingly clear from this chapter that the initial quantum yield may be less than unity. The highest quantum yields recorded are 0.64 and 0.67 for aqueous sulphate and ferrocyanide, respectively. There is a somewhat paradoxical relationship between the oscillator strengths of CTTS bands and the limiting quantum yields derived therefrom on photolysis. Thus the highest oscillator strength for the first transition of a CTTS band of an anion, for example, I^-, gives $\phi_e = 0.36 \pm 0.06$, whereas the much lower oscillator strengths of such anions as sulfate have the higher quantum yields described. Such a trend is borne out by considering the higher-energy CTTS transitions of iodide, for example, at 185 nm the higher-intensity band, C, contributes considerably to the intensity, as well as bands B_1 and A_2. The quantum yield at 185 nm for iodide photolysis is 0.25.

Little is known at present about Γ; initially, its value was unfortunately taken to be unity for aqueous chloride photolysis at 185 nm, based on a quantum of $\phi_e = 0.98$, instead of an actinometry-corrected 0.43. ϕ_e is then the fraction of radicals that escape cage recombination and may be scavenged. ϕ, the product quantum yield, then varies between ϕ_e and ϕ_r, the residual quantum yield. The scavenger concentration required to compete with recombination and give plateau values of quantum yields depends on the initial separation distance between electron and radicals.

VI. FUTURE PROBLEMS

The photochemistry and spectroscopy of inorganic anions are strongly interdependent, and progress in one either rests on or leads to advances in the other. The possibility of ϕ_e being less than unity is of prime importance and must be tested by photolysis of a simple anion, such as a halide, in a solvent that is a very efficient scavenger of radicals and solvated electrons. One such system might be iodide photolysed in acetone at 229 nm, the photolysis wavelength being in the spectroscopic window between the $n-\pi^*$ and $\pi-\pi^*$ transitions of acetone.

However, more information is required on the detailed photochemistry of anions other than the halides. There is a tendency for interest to become fixated upon halides, a problem common to other areas such as electrochemistry. In order to probe the effects of temperature, ion size, very high scavenger concentrations, and so forth, studies must be regarded as incomplete unless extended to both large- and small-radii anions outside of the halide series.

High-energy photolysis is another area of interest, the higher-energy excited states probably giving different products and reaction pathways; a change in products with irradiating wavelength was shown in Table 8-3 for some of the oxyanions of chlorine. An immediate question arises as to the products for, say, 185-nm photolysis of these and other anions. It is emphasized again that the photochemical processes must be related to the initial spectroscopic processes and the excited states produced for the subsequent reactions to be correctly interpreted.

REFERENCES

1. J. Franck and G. Scheibe, *Z. Phys. Chem.* A139, 22 (1928).
2. J. Franck and F. Haber, *Sitzgsber. Preuss. Akad. Wiss.*, 250 (1931).
3. A. Farkas and L. Farkas, *Trans. Faraday Soc.* 34, 1113, 1120 (1938).
4. T. Rigg and J. J. Weiss, *J. Chem. Soc.*, 4198 (1952).
5. J. N. Pitts and J. G. Calvert, *Photochemistry*, Wiley, New York, 1966.
6. M. F. Fox, *Quart. Rev. (London)* 24, 565 (1970).
7. M. J. Blandamer and M. F. Fox, *Chem. Rev.* 70, 59 (1970).
8. F. Franks, Ed: *Water—A Comprehensive Review*, Vol. III, Plenum, New York, 1973.
9. M. C. R. Symons, *J. Phys. Chem.* 71, 172 (1967).
10. J. A. Burgess and M. C. R. Symons, *Quart. Rev. (London)* 22, 276 (1968).
11. R. N. Butler, E. A. Philpott, and M. C. R. Symons, *Chem. Commun.*, 371 (1968).
12. M. Anbar and E. J. Hart, *J. Phys. Chem.* 69, 1244 (1965); D. M. Brown, F. S. Dainton, J. P. Keene, and D. C. Walker, *Proc. Chem. Soc.*, 266 (1964).
13. J. Franck and R. L. Platzman, *Z. Phys.* 138, 411 (1953).
14. M. Smith and M. C. R. Symons, *Discuss. Faraday Soc.* 24, 206 (1957).
15. M. Smith and M. C. R. Symons, *Trans. Faraday Soc.* 55, 1087 (1959); *ibid.* 56, 1393 (1960).
16. G. Stein and A. Treinin, *Trans. Faraday Soc.* 55, 1087 (1959); *ibid.* 56, 1393 (1960).
17. A. Treinin, *J. Phys. Chem.* 68, 893 (1964).
18. S. P. Ionov, *Zh. Neorg. Khim.* 12, 1995 (1967).

19. M. F. Fox and T. F. Hunter, *Nature* (*London*) **223,** 177 (1968).

20. M. F. Fox, *Appl. Spectrosc.* **27,** 155 (1973).

21. M. F. Fox and E. Hayon, *Chem. Phys. Letters* **14,** 442 (1972).

22. G. Baldini, and B. Bosacchi, *Phys. Status Solidi* **38,** 325 (1970).

23. K. Teegarden and G. Baldini, *Phys. Rev.* **155,** 896 (1967).

24. C. K. Jorgenson, *Advan. Chem. Phys.* **5,** 33 (1963).

25. M. F. Fox and E. Hayon, *J. Phys. Chem.* **76,** 2703 (1972).

26. R. Sperling and A. Treinin, *J. Phys. Chem.* **67,** 897 (1963).

27. E. Rabinowitch, *Rev. Mod. Phys.* **14,** 112 (1942).

28. F. S. Dainton, quoted in L. E. Orgel, *Quart. Rev.* (*London*) **8,** 422 (1954).

29. D. M. Brown and F. S. Dainton, in *Energetics and Mechanisms in Radiation Biology,* pp. 35–39, Academic, London, 1971.

30. E. Gusarsky and A. Treinin, *J. Phys. Chem.* **69,** 3176 (1965).

31. J. T. Shapiro, Ph.D. Thesis, Bryn Mawr College, 1965.

32. G. Stein, *Israel J. Chem.* **8,** 691 (1970).

33. J. Jortner, M. Ottolenghi, J. Rabani, and G. Stein, *J. Chem. Phys.* **37,** 2488 (1962).

34. J. Jortner, M. Ottolenghi, and G. Stein, *J. Phys. Chem.* **66,** 2029, 2037, 2042 (1962).

35. J. Jortner, M. Ottolenghi, and G. Stein, *J. Phys. Chem.* **67,** 1271 (1963).

36. J. Jortner, M. Ottolenghi, and G. Stein, *J. Phys. Chem.* **68,** 247 (1964).

37. E. J. Hart and J. W. Boag, *J. Amer. Chem. Soc.* **84,** 4090 (1962).

38. L. I. Grossweiner and M. S. Matheson, *J. Phys. Chem.* **61,** 1089 (1957).

39. M. S. Matheson, W. A. Mulac, and J. Rabani, *J. Phys. Chem.* **67,** 2613 (1963).

40. D. H. Ellison, G. A. Salmon, and F. Wilkinson, *Proc. Roy. Soc. Ser. A* **328,** 23 (1972).

41. F. S. Dainton, I. V. Janovsky, and G. A. Salmon, *Proc. Roy. Soc. Ser. A* **327,** 305 (1972).

42. M. Halmann and I. Platzner, *J. Phys. Chem.* **70,** 2281 (1966).

43. J. R. Huber and E. Hayon, *J. Phys. Chem.* **72,** 3820 (1968).

44. R. Devonshire and J. J. Weiss, *J. Phys. Chem.* **72,** 3815 (1968).

45. E. Hayon and M. Simic, *Intra-Science Chem. Rep.* **5,** 357 (1971).

46. G. Stein, in *The Solvated Electron,* Advances in Chemistry Series, No. 50, American Chemical Society, Washington, D.C., 1965.

47. A. Treinin and M. Yaacobi, *J. Phys. Chem.* **68,** 2487 (1963).

48. A. Treinin, *Israel. J. Chem.* **8,** 103 (1970).

49. O. Amichai, G. Czapski, and A. Treinin, *Israel J. Chem.* **7,** 351 (1969).

50. D. Behar and G. Czapski, *Israel J. Chem.* **6,** 43 (1968).

51. O. Amichai and A. Treinin, *J. Phys. Chem.* **74,** 830 (1970).

52. A. Treinin and E. Hayon, unpublished work.

53. G. V. Buxton and F. S. Dainton, *Proc. Roy. Soc. A* **304,** 427 (1968).

54. G. V. Buxton, *Trans. Faraday Soc.* **65,** 2150 (1969).

55. W. D. Felix, B. L. Gall, and L. M. Dorfman, *J. Phys. Chem.* **71,** 384 (1967).

56. Thermodynamic values taken from (a) *Selected Values of Chemical Thermodynamic Properties,* N.B.S. Circular No. 500; (b) H. L. Friedman, *J. Chem. Phys.* **21,** 319 (1953);

(c) V. I. Vedeneyev, L. V. Gurvich, V. N. Kondrateyev, and Ye. L. Frankevich, *Bond Energies, Ionisation Potentials and Electron Affinities*, Edward Arnold, London, 1966.

57. O. Amichai and A. Treinin, *Chem. Phys. Letters* **3**, 611 (1969).

58. O. Amichai, Z. Karni and A. Treinin, quoted in A. Treinin, *Israel J. Chem.* **8**, 103 (1970).

59. O. Amichai and A. Treinin, quoted in A. Treinin, *Israwl J. Chem.* **8**, 103 (1970).

60. G. Zimmerman, *J. Chem. Phys.* **23**, 825 (1955).

61. M. Daniels, R. V. Meyers, and E. V. Belardo, *J. Phys. Chem.* **72**, 389 (1968).

62. U. Shuali, M. Ottolenghi, J. Rabani, and Z. Yelin, *J. Phys. Chem.* **73**, 3445 (1969).

63. U. Klaning and M. C. R. Symons, *J. Chem. Soc.*, 3269 (1959).

64. J. O. Edwards, *Inorganic Reaction Mechanisms*, Benjamin, New York, 1964.

65. G. V. Buxton and M. S. Subhani, *J.C.S. Faraday Trans. I* **68**, 958 (1972).

66. A. Prokopcikas, J. Janickis, and M. Salkauskas, *Liet. TSR Mosklu Akad. Darb. Ser. B*, **3**, 49 (1964).

67. G. V. Buxton and M. S. Subhani, *J.C.S. Faraday Trans. I* **68**, 970 (1972).

68. G. V. Buxton and M. S. Subhani, *J.C.S. Faraday Trans. I* **68**, 947 (1972).

69. L. Dogliotti and E. Hayon, *J. Phys. Chem.* **72**, 1800 (1968).

70. J. J. Weiss, *Ber. Bunsenges. Phys. Chem.* **73**, 131 (1969).

71. J. Jortner and G. Stein, *J. Phys. Chem.* **66**, 1258, 1264 (1962).

72. E. Hayon and J. J. Weiss, *J. Chem. Soc.*, 3866 (1960).

73. M. Lefort and P. Douzou, *J. Chim. Phys.* **53**, 536 (1956).

74. L. J. Heidt, M. G. Mullin, W. B. Martin, Jr., and A. M. J. Beatty, *J. Phys. Chem.* **66**, 336 (1962).

75. P. Porret and J. J. Weiss, *Nature (London)* **139**, 1019 (1937).

76. Y. Haas, G. Stein, and M. Tomkiewicz, *J. Phys. Chem.* **74**, 2558 (1970).

77. C. K. Jorgenson, *Mol. Phys.* **5**, 271 (1963).

78. J. C. Barnes and P. Day, *J. Chem. Soc.*, 3886 (1963).

79. M. Anbar, *Advan. Chem.* **50**, 55 (1965).

80. J. Rabani, *Advan. Chem.* **50**, 242 (1965).

81. M. Eigen and J. Schoen, *Z. Electrochem.* **59**, 483 (1955).

82. J. Barrett, M. F. Fox, and A. L. Mansell, *J. Chem. Soc. A*, 489 (1966).

83. M. F. Fox, Ph.D. Thesis, University of London, 1966.

84. A. Appleby, G. Scholes, and M. Simie, *J. Amer. Chem. Soc.* **86**, 1643 (1964).

85. F. S. Dainton and P. Fowles, *Proc. Roy. Soc. A* **287**, 295 (1965).

86. J. H. Baxendale and J. A. Wilson, *Trans. Faraday Soc.* **53**, 344 (1957).

87. J. Barrett and J. H. Baxendale, *Trans. Faraday Soc.* **56**, 37 (1960).

88. C. von Sonntag, *Z. Phys. Chem. (Frankfurt am Main)* **69**, 292 (1970).

89. M. Anbar and P. Neta, quoted in M. Anbar, *Advan. Chem.* **50**, 55 (1965).

90. F. S. Dainton and S. R. Logan, *Proc. Roy. Soc. A* **287**, 281 (1965).

91. D. Eisenberg and W. Kauzmann, *The Structure and Properties of Water*, Oxford U.P., London, 1969.

92. H. S. Frank and W.-Y. Wen, *Discuss. Faraday Soc.* **24**, 133 (1957).

93. For a survey of this field, see Refs. 7 and 8 and the bibliographies therein.
94. R. M. Noyes, *J. Amer. Chem. Soc.* **81**, 999 (1950).
95. R. M. Noyes, *J. Amer. Chem. Soc.* **77**, 2042 (1955).
96. R. M. Noyes, *J. Amer. Chem. Soc.* **78**, 5486 (1956).
97. G. Czapski and M. Ottolenghi, *Israel J. Chem.* **6**, 75 (1968).
98. M. Shirom and M. Tomkiewicz, *J. Chem. Phys.* **56**, 2731 (1972).
99. A. Mozumder, *J. Chem. Phys.* **45**, 1659 (1968).
100. A. Mozumder, *J. Chem. Phys.* **50**, 3153 (1969).
101. G. Czapski, J. Ogdan, and M. Ottolenghi, *Chem. Phys. Letters* **2**, 383 (1969).
102. M. F. Fox and E. Hayon, *Chem. Phys. Letters*, in press (1974).
103. M. F. Fox and E. Hayon, unpublished work.

9

PHOTOCHEMISTRY
IN THE SOLID STATE

Paul D. Fleischauer

Chemistry and Physics Laboratory
The Aerospace Corporation
El Segundo, California

I. INTRODUCTION

The study of the photochemistry of solids has progressed at a much slower rate than that of solutions, not so much because the properties of solids are not well understood, but because they have not been of general interest to most inorganic photochemists. The interionic and intermolecular forces of an ordered, periodic crystal lattice can be treated theoretically with considerably more certainty than the constantly changing conditions of fluid solutions. But, the analytical tasks, so necessary for the solution of even simple reaction schemes, are unreasonably complicated because the rigid structures prevent facile separation of reactants from products.

Nonetheless, considerable advances in the areas of energetics and photoelectrochemical and photophysical properties of solids have been made by solid-state physicists and materials scientists, as well as by chemists. In most cases, even when it is appropriate, these advances have not been interpreted in terms of metal-ion complexes in rigid environments. For the coordination chemist this approach is more meaningful, and the results, therefore, are more significant. Because of the proliferation of practical applications and because there are important fundamental concepts with roots in solid systems, a discussion of some of the interesting aspects of solid-state photochemistry is presented in this chapter. The emphasis of the discussion is on concepts of the coordination chemistry of solids and on the similarities and differences between solid-state systems and corresponding solutions of the same materials.

The chapter content is organized in three separate sections. The first part (Section II) is a discussion of special physical properties of solids that make their reactions different than those treated in other chapters, namely reactions in solution. Section III deals with the photochemistry of robust molecular complexes. These are the complexes that maintain their coordination sphere when dissolved in fluid solution. The same classes of reactions that are observed for these complexes in solution are also observed in the solid.

The photoprocesses of ionic complexes tend to be physical rather than chemical interactions. Thus photocurrents and photovoltages are produced upon irradiation in charge-transfer absorptions of ionic complexes. The chemical reactions that are light induced are usually related to crystal defects, such as vacancies or impurity centers, or they are specifically attributable to interfacial (solid-solution or solid-gas) phenomena. Consequently, the final section of the chapter is devoted to a discussion of experimental studies of the photochemistry of solid-solution interfaces.

II. SPECIAL FEATURES OF THE SOLID STATE

Historically, the solid state has been of interest to inorganic coordination chemists mostly for being the birthplace of crystal-field theory[1,2] and its descendent, ligand-field theory.[3-6] Discussions of the type of splitting of energy levels (primarily d electron levels in transition metals) produced by fields of various symmetries have appeared in previous chapters. In solids geometrical structures are rigid and fields are well defined compared to those in solution, making theoretical treatments of localized effects more precise and accurate. There also are effects other than those treated in normal-symmetry point-group considerations, such as next-nearest-neighbor interactions and lattice periodicities that can be treated for crystalline materials but that do not exist or are not amenable to analysis for ions in solution. For these reasons and because of the vast applications of solid-state materials in electronics,[7] photography,[8] and other industries, solid-state photochemistry is a fascinating subject worthy of closer consideration by inorganic photochemists.

Photochemical reactions in crystalline solids are, to a large extent, very similar to those that are observed for solutions, namely, substitutions and electron transfers. However, because of the rigidity of solids and their periodicity, unusual adaptations of these reactions, such as the excitation of an electron at one site and its facile migration to another site several lattice constants removed, can be observed. In this section, some properties of solids that affect their behavior under illumination are discussed. The treatment is illustrative; it is not designed to be exhaustive, especially in complicated areas such as reaction mechanisms or energetics of solids.

A.   Crystal Structures and Energetics

For the purpose of this presentation, solids are divided into two classes: those of molecular complexes and those of ionic complexes.[9,10] The behavior of complexes in rigid glasses is largely ignored since most of the work on these materials involves their luminescence properties that are treated in Chapter 2. The complex ions that are most familiar to us are the molecular type. They are generally stable in solution, at least in the kinetic sense, and they crystallize to form arrays of these polyatomic ions that may have some of the properties of molecular crystals [e.g., low melting (decomposition) points and high compressibility] but that are often combined with comparatively small (simple) counter ions. In certain cases, complexes are neutral, such as $Cr(acac)_3$; they presumably form true molecular crystals.

The absorption and emission spectra of solid molecular complexes are fundamentally the same as for the complexes in fluid solution.[9-16] There are discrete molecular energy states with internal d or ligand-field transitions and charge-transfer transitions. Figure 9-1 shows the room- and low-temperature absorption (diffuse reflectance) and the low-temperature emission of a neutral molecular complex of Cr(III) in solution (glass) and as the neat solid.[11] It is evident from the narrowing of the absorption bands at low temperature that the L_1 and L_2 absorptions in *both* solution and solid have some vibronic components, but there are only small differences in the absorption features between the solution and the solid species.

It is difficult to determine from the absorption spectra of Fig. 9-1 whether there are differences in the vibronic contributions to the solution species compared to the solid. There are, however, some apparent differences in the energies of the lowest thermally equilibrated spin-allowed quartet and spin-forbidden doublet excited states (i.e., the thexi states) for the complex in the different environments, because the emission of the neat solid consists of primarily fluorescence, while that of the complex in glassy solution, has a large phosphorescence peak. Such differences can result from differences in the geometries and energies of the thexi states of Cr(III) that are not well represented by the positions of the absorption maxima.[12]

There are other examples of Cr(III) complexes whose emission spectra show considerable variations that depend on the crystalline environment of the complex.[13,14] There are also some examples of variations in absorption peaks measured in solutions with different solvents.[15] However, these differences in the energy states of octahedral complexes are small compared, for example, to the ones observed for the square planar

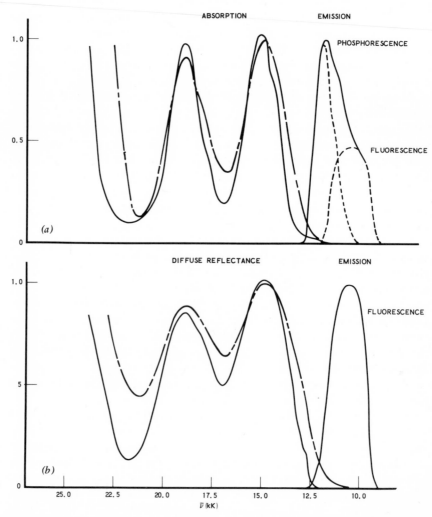

Fig. 9-1. (a) Absorption [(– – – –) 298°K, (————) approximately 80°K] and Emission (77°K) Spectra of Cr(dsc)₃ in CHCl₃/CH₃OH (1:3) glass (dsc = diethyldiselenocarbamate). Approximate resolution of emission (– – –). (b) Diffuse reflectance [(– – – –) 298°K, (————) approximately 100°K] and emission (77°K) spectra of solid (powder) Cr(dsc)₃.

$Pt(CN)_4^{2-}$ complex upon changes of its cation of crystallization, namely, K^+, Rb^+, Cs^+, Ca^{2+}, Ba^{2+}, and so forth.[16,17] In the case of the O_h complexes, the chromophore is a relatively well-shielded structure; there are only minor interactions with the surroundings unless ligands are shared by more than one metal. Conversely, the substantial variations in the maxima of absorption and emission spectra (up to 4000 cm⁻¹) for the square planar complex result from the fact that the responsible electronic transition(s) involves orbitals along the unshielded molecular z axis. The crystal consists of stacks of planar molecules, and the orbitals with z character in one molecule can interact with those of an adjacent one. Presumably, the cations of crystallization cause varying interplanar spacings and, therefore, varying degrees of orbital interaction.

Although these perturbations to the energy states of the molecular complex, $Pt(CN)_4^{2-}$, are large compared to those of the more common octahedral or tetrahedral complexes, they are relatively minor if compared to the circumstances for solid ionic complexes. The classical example for the application of crystal field theory is a metal-ion impurity in a solid matrix, such as an oxide or sulfide. The coordinating atoms are the same as for many molecular complexes, for example, oxygen, sulfur, or halogens, and the crystal (ligand) field transitions are analogous to those of the molecular species. The fundamental difference between the ionic and molecular materials, though, is the coupling of the charge-transfer states in the former to reduce their degeneracy and produce energy bands in the solid in place of discrete energy levels.[18-20] A CT transition in this case can be visualized as the transfer of an electron from a ligand-like orbital (a valence electron) to a metal-like orbital (a conduction electron), that is, a CTTM transition, but the coordinates of the orbitals are such that there is considerable overlap, ligand to ligand and metal to metal, forming extensive resonance structures or energy bands.[18] The net result is that electrons are delocalized throughout the crystal; irradiation of the crystal can result in electrical conduction over several unit cells (photoconductivity) and chemical reaction at a considerable distance from the point of initial excitation.

Interesting examples of this delocalization phenomenon for the coordination chemist are again the square planar Pt(II) complexes. Specifically, the tetracyanide with the crystal slightly oxidized by bromine crystallizes as[21] $K_2[Pt(CN)_4]Br_{0.3} \cdot 3H_2O$. Although the complex is of the molecular type, the crystal is described as a one-dimensional conductor with metallic like strands of interacting platinum ions forming the conduction network. This is a compound with very interesting physical properties; it is a good example of the delocalization phenomenon even though it is not an ionic complex. (See also Section III.D.)

Many "simple" salts and oxides are examples of ionic solids that may contain ionic complexes of transition metals. These materials have broad uses in the electronics industry and in image recording because of their semiconductive, photoconductive, and photovoltaic properties. A metal-ion impurity (an ionic complex) is often responsible for the desirable property. For example, the energy separation between the band consisting of valence electron states and that consisting of conduction electron states for many metal oxides and sulfides is wide enough that the pure

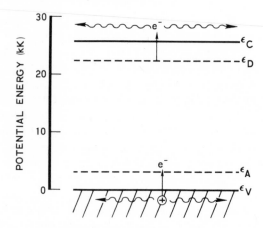

Fig. 9-2. Schematic potential energy diagram showing donor and acceptor impurity levels in insulator or semiconductor host.

materials are insulators at room temperature, that is, there would be no thermal excitation of valence electrons into the conduction band and hence no dark conductivity. In practice, however, the materials are rarely pure; instead, they contain small concentrations of ions with greater (donor ions) or smaller (acceptor ions) numbers of valence electrons than the host.[22] If the impurity is easily oxidized it will be a donor, the energy of its uppermost electron will be just slightly below the conduction band, and this electron will be easily thermally stimulated thus increasing the conductivity of the host. Such a situation is illustrated diagramatically in Fig. 9-2.

The above-described impurity center may be introduced into the host during preparation of the material, after preparation by diffusion into the material, or simply by changing the stoichiometry of the host as in the case of the reduction of TiO_2. If it is stoichiometric, this material is an insulator; however, it is generally found as the slightly reduced TiO_{2-x} containing oxygen vacancies and Ti^{3+} ions.[23] The single d electron of the Ti^{3+} is easily excited into the conduction band, but there is evidence from electron paramagnetic resonance data for some discrete trivalent ions, especially at low temperatures.[24,25] These different types of electronic properties that relate to the photochemistries of these solids are discussed in more detail in Section IV.

B. Optical Excitation of Crystals and Experimental Procedures

A typical ligand-field absorption peak for a molecular complex in solution has a molar extinction coefficient of the order of $100 \, M^{-1} \, cm^{-1}$. This value corresponds to an absorption coefficient, k, for the pure solid (the density is estimated to be $5 \, g \, cm^{-3}$) of approximately $1 \times 10^4 \, cm^{-1}$; it means that 99% of light incident on the solid surface (assuming no reflection loss) is absorbed within a layer of the material that is about 5 to 10 μm thick. The severity of this absorption limitation and the attendant experimental difficulties in analyzing photochemical reactions in such thin layers account for the relative dearth of studies involving pure solid materials. Conversely, the concentration of the reactive portion of solids to the surface region also explains much of the interest in the reactions of adsorbed species.

The various methods employed for the analysis of solid-state photochemical reactions have been discussed in three separate reviews.[26-28] The most popular techniques for molecular complexes have been the measurement of diffuse-reflectance spectra[29] and of infrared spectra of materials dispersed in alkali halide pellets.[30-33] Other techniques such as thermogravimetric analysis, differential thermal analysis, and X-ray powder diffraction techniques have been used successfully in the study of thermal reactions of solid complexes[34,35] and may have application in certain photochemical systems. The measurement of polarized spectra has been useful in characterizing energy levels[36] and photochemical processes[37] in pure materials and in hosts doped with molecular complexes.

Two techniques, luminescence spectroscopy and electron paramagnetic resonance, have been popular in studies of photochemical electron-transfer processes in ionic complexes and simple salts,[38,39] but they have been applied only recently to studies of molecular complexes.[40-43] Certainly, the luminescence properties of complexes have been known for many years,[9,10] but there are few examples of the use of emission spectra to characterize a photochemical reaction. These two techniques, like the reflectance measurements, are well suited to the qualitative analysis of the photochemistry of solids because they are sensitive to changes on surfaces.[44] Unfortunately, it is exceedingly tedious to calibrate these methods for the determination of quantum yields, and there are few of such results in the literature.

Of all the techniques, electron paramagnetic resonance combined with appropriate crystallographic data may offer the greatest potential for quantitative measurements. In theory, the absolute number of spins can be calculated from the resonance data and used to get the percent conversion of the reactant based on the absorbed light intensity. In doing the spin calculation, as in other interpretations, it is necessary to consider possible crystal structure differences between the surface region and the bulk material.[44]

Another method of determining quantum efficiencies for solid-state reactions involves the use of reaction-rate plots from optical transmittance data for thin layers of material.[28,45] The method combines the Beer–Lambert equation for light absorption with the appropriate reaction rate equation. Integration of the resulting equations provides for exact determinations (in the case of transparent products) of the light intensity and the reactant concentration at a given penetration depth and reaction time. If the products absorb incident radiation, similar, approximate solutions are obtained.[45] Using these two parameters, light intensity and reactant (product) concentration, one can calculate quantum efficiencies at specific times and positions for relatively simple reaction systems in solid materials. Unfortunately, for reactions of coordination compounds, especially photosubstitutions, the products often absorb as efficiently as the reactants; therefore, only the approximate treatments that are valid for early stages of reaction are applicable. There have been no attempts to analyze further complications that may result from the photosensitivity of absorbing products.

C. Solid-State Reaction Mechanisms

Details of specific reaction mechanisms are discussed in appropriate later sections, but some general information pertaining to similarities and differences between solid-state and solution reactions is given here. The more or less classical treatment of solid-state reactions divides them into two categories: (1) changes of phase that involve atomic (sometimes intramolecular) rearrangement and (2) composition changes that involve chemical-diffusion processes.[46] The latter category encompasses most of the photochemical processes that have been studied for solid coordination compounds. Reactions of species adsorbed on solid surfaces sometimes involve diffusion steps, but they can also entail phase changes as in the photochemical deposition of metals by electron transfer from solids (see Section IV). In principle, photoinduced molecular symmetry changes, similar to the intramolecular isomerization that occurs for solutions of bis-glycinato Pt(II),[47] could induce crystalline phase changes, especially if the starting material is not the thermodynamically stable configuration. This type of effect should be significant in surface regions where absorbed light intensities are high or in laser-induced "damage" reactions. A variation of it has, in fact, been observed for the X-irradiation of complexes such as $[Co(NH_3)_5(NCS)]Cl_2$[48] and $K_4[Ru(CN)_6]$,[49] wherein linkage isomerizations have produced measurable crystal-structural changes. (See also Section III.C.)

Unfortunately, very few investigations of solid coordination compounds have involved any consideration of crystal structures in the determination of reaction mechanisms. The proximity of potential substituents and their availability through diffusional transfer are surely important parameters in, for example, the photolyses of $[Cr(en)_3]X_3$.[40,50] As opposed to the study of solutions, however, where questions of ion pairing, solvent participation, and geometrical distortions of excited states have played a major role in unraveling reaction mechanisms, there has been little attempt to relate environmental factors to mechanisms of solid reactions. In addition, in most instances the electronic states that are precursors to the photochemical reactions of solid complexes have not been identified.

Both substitution and electron-transfer reactions usually involve some form of diffusion process; the latter reactions frequently result in decomposition or loss of a volatile reagent. But crystal parameters that govern these diffusional processes, such as vacancy and impurity concentrations, the presence of other crystal defects, and diffusion constants, are rarely known for solid molecular complexes; though better understood, they are infrequently applied to photochemical studies of ionic complexes. In the following discussion of the work that has been done on the photochemistry of solids, an attempt is made, therefore, to illustrate the usefulness of crystallographic information for specific examples and to suggest when additional data are necessary in order to fully understand these complex systems.

III. MOLECULAR COMPLEXES IN THE SOLID STATE

The qualitative information that is available, primarily stoichiometry information and, in the few cases for which determinations have been made, the quantum efficiencies for reactions of solid molecular complexes, have been reported in some detail in reviews that have already been mentioned.[26-28] The actual number of observations of the photosensitivity of solids that have been reported is quite surprising,[26] considering the almost total absence of relevant mechanism studies involving analysis of excited-state processes. Certainly, the experimental difficulties in working with solids have limited the interest in studying their reactions, but the discovery and understanding of some unusual properties of certain complexes[17,21] and the realization that many of the concepts of the far better understood solution reactions also probably apply to solids, hopefully, will stimulate an increase in research efforts in this area. The emphasis of this section is speculative in an attempt to illustrate, from the small data base available, situations in the photochemistry of solids that are analogous to those of solutions and then to point out the properties of the solids that can modify their behavior.

A. Photosubstitution Reactions

Substitution and substitution-related reactions in solution involve solvation, ligand exchange (as in anation), isomerization, and racemization (Chapters 3, 4, and 5). Similarly, for solids such compositional changes in the first-coordination sphere can be made to occur, and with the exception of photoracemization all have been reported. For the most part, these substitution processes are expected to result in relatively minor changes in the symmetry of the complex, and, therefore, one might predict only small phase or structure changes. There do not appear to have been any systematic investigations of photoinduced crystallographic structural changes for molecular complexes. However, in a study of the X-ray, infrared, and thermally induced reaction

$$[Co(NH_3)_5SCN]Cl_2 \cdot H_2O \rightarrow [Co(NH_3)_5NCS]Cl_2 + H_2O \qquad (9\text{-}1)$$

a structural change from the orthorhombic S-bonded isomer into the face-centered cubic N-bonded one was documented.[48] Although the SCN^- salt of the isothiocyanato isomer of this compound has been reported to be decomposed by ultraviolet irradiation,[26] the S-bonded isomer has not been investigated. In view of the other solid-state photolinkage isomerizations that have been observed (Section C), it would seem that the thiocyanate isomer would be a likely candidate for photochemical study.

All solid-state reactions involve restrictions (activation energies) due to composition, diffusion rates, and reordering phenomena that are more severe than in solution. For photosubstitution reactions, either the environment of an excited complex must react to replace a ligand within the lifetime of the excited state or an intermediate must be produced in a multistep process. In solution, solvent relaxation can occur extremely rapidly, and intermediates may exist long enough to be the determinants of the stereochemical course of reaction. (See Chapter 4, Section III.A.4.) In solids, the entering group must be in close proximity to the leaving group so that trivial recombination does not overshadow substitution. An ideal system would be one in which a volatile ligand could be photo-detached to form a relatively long-lived intermediate. (In this context, only one study[26] is known where photochemistry has been performed with evacuated samples in an effort to assist outgassing of detached ligands.)

Unfortunately, it may not be sufficient to have the entering group near the point of substitution. Referring to the $[Co(NH_3)_5SCN]Cl_2 \cdot H_2O$ work[48] again, we note that no Cl^- substitution was observed, even though the Cl^- ions are in close contact with the complex. The reason suggested for the lack of substitution was that the Cl^- are probably hydrogen bonded (to the NH_3 hydrogens) and not free to attack the Co(III) intermediate, while the NCS^- is presumably directed away from NH_3 hydrogens and is able to attack a neighboring Co(III). Intramolecular isomerization is not favored because the large motion involved in turning the thiocyanate would probably result in crystal fracture, which was not observed.[48]

Significantly, the photosubstitution of NH_3 by Cl^- in $[Cr(NH_3)_6]Cl_3$ has been observed.[51] But, there have been no studies to evaluate the importance of anion or water of crystallization placement within crystal lattices in photochemical reactions. Furthermore, there have been no efforts to determine whether the rigidity of the solid matrix precludes formation of the intermediate that is apparently involved in the photosubstitution of Cr(III) complexes in solution.[52] One might conclude from the absence of reports of photoreactions that the solids are indeed rigid and even less photosensitive than the cyclam-Cr(III) complex (see Chapter 4, Section III.A).

As shown in Fig. 9-1, there is a vibronic contribution to the quartet absorption bands of some crystalline Cr(III) complexes. There is presumably considerable coupling of the electronic transitions to crystal-lattice vibrations; where detected, fluorescence emissions have Stokes shift comparable to those of complexes in glassy solution.[12] For some molecular complexes fluorescence appears to be favored, relative to phosphorescence, for crystalline samples.[11] These observations suggest that the distortions of the $^4L_1^0$ state of crystalline Cr(III) complexes are similar to those of the complexes in glassy solution. That the neat crystals may be significantly less photosensitive than the fluid solutions suggests that (1) there is a more efficient nonreactive deactivation of $^4L_1^0$ for the crystal, (2) the $^4L_1^0$ state of the crystal is less severely distorted than that in fluid solution, or (3) an intermediate, other than the $^4L_1^0$ excited state, is involved in photosubstitutions of Cr(III) complexes in fluid solution; because of the rigidity, this intermediate cannot be formed with high efficiency in the crystals.

Different salts of $[Cr(en)_3]^{3+}$ are photosensitive and result in different types of products, depending on the identity of the anion.[40,50] In the earlier report,[50] it was stated that the Cl^- and Br^- salts produce cis-$[Cr(en)_2X_2]^+$, while the NCS^- salt gave the *trans* isomer and the I^- salt did not react. More recently, it has been shown that the I^- salt does react, but that at normal temperatures the reverse thermal reaction is so fast that products could not be detected.[40] No quantum yields were reported, but the order of reactivity was given as $NCS^- > Cl^- > Br^-$.

These reactivity differences were determined from the number of flashes (xenon flash tube) required to convert each compound completely to its corresponding product. They include secondary reactions, as well as the primary photochemical processes. The data for the low-temperature irradiation of the I^- salt show that it is probably the secondary processes that have the greatest influence on the stated order of reactivity. However, even at low temperature the chloride is more reactive than the iodide, and the perchlorate salt shows no spectral changes (absorption or emission) upon irradiation in the $^4L_1$ band.[40]

It was suggested[40] that the primary process is the heterolytic fission of a Cr—N bond to produce a unidentate ethylenediamine ligand analogous to the photolysis[53,54] of aqueous $[Cr(en)_3]^{3+}$. Immediately following detachment of the en group the anion enters the coordination sphere of Cr(III), resulting in the following reaction sequence.

$$[Cr(en)_3]X_3 \xrightarrow{h\nu} [Cr(en)_3]^*X_3 \qquad (^4L_1) \qquad\qquad (9\text{-}2)$$

$$^4L_1 \longrightarrow \qquad\qquad X_3 \qquad\qquad (I) \qquad\qquad (9\text{-}3)$$

$$I \longrightarrow [Cr(en)_3]X_3 \qquad\qquad\qquad (9\text{-}4)$$

$$I \longrightarrow \qquad\qquad X_2 \qquad\qquad (II) \qquad\qquad (9\text{-}5)$$

$$II \longrightarrow [Cr(en)_3]X_3 \qquad\qquad\qquad (9\text{-}6)$$

$$II \longrightarrow [Cr(en)_2X_2]X + en \qquad\qquad (9\text{-}7)$$

Both the efficiencies of the formation of II and the final product and the stereochemical course of the reaction depend on the location in the crystal lattice of the anions, X. By analogy with the $[Co(NH_3)_5\text{-}SCN]Cl_2 \cdot H_2O$ reaction,[48] one might expect that the degree of hydrogen bonding between X and the amine hydrogens would also have a significant effect on this reaction. The conclusions of this line of reasoning, however, are ambiguous since one might argue that the greater preference of Cl^- to form hydrogen bonds compared to I^- would facilitate detachment of the en group, but, conversely, this hydrogen bonding would tend to fix the Cl^- in place, preventing it from bonding to Cr(III).

Without comparative structural data for the different salts it is impossible to explain the preference of the NCS^- salt for formation of the *trans* isomer while the others yield *cis* species. The thermal chemistry of the $[Cr(en)_3]X_3$ complexes follows the same stereochemical course for the Cl^- and NCS^-, but not for the Br^- salts.[50] The latter suggests the possibility of some mechanistic differences between the thermal and photochemical reactions.

In comparison again to solution photochemistry, there have been some recent studies of *cis* and *trans* isomers of $[Cr(en)_2XY]^+$ ions, specifically of the mixed complex, $[Cr(en)_2(NCS)Cl]^+$.[55–57] The *cis* isomer[57] gives ethylenediamine aquation predominantly, and from the lowest-excited-quartet state, $\phi = 0.1$ to 0.2, while for the *trans* isomer[55,56] the predominant reaction mode is thiocyanate aquation, $\phi = 0.05$, under irradiation of a higher-quartet band.[56] The reaction products for the *trans* starting material have the *cis* configuration.

There are no comparable data for irradiation of these compounds as solids. However, the nature of the terminal products in the photolysis of the $[Cr(en)_3]X_3$ compounds gives some indication of possible differences of stereochemical specificity for the solids. As mentioned, the experiments were performed by flashing the compounds until no further reaction was observed.[50] The fact is that different isomers were formed for different anions of substitution, so either these isomers were formed directly in the initial substitution act and they are photoinert, or different isomers were formed initially, and they were photoisomerized to the final photoinert products. In any case, *trans*-$[Cr(en)_2(NCS)_2]NCS$ does not react to produce the *cis* isomer as it no doubt would in solution (granting that said *cis* isomer would involve substitution by solvent). It may well be that the bulkiness of the NCS^- ions, both coordinated and as anions of crystallization, prevents relaxation of the Cr(III) excited species to give the *cis* configuration. However, again without detailed crystallographic data such rationalization is purely speculative. On the other hand, since the Cl^- and the NCS^- salts of $[Cr(en)_3]^{3+}$ give opposite isomers, it would be very interesting to see if either the *cis*- or *trans*-$[Cr(en)_2(NCS)Cl]ClO_4$ solids show any photosensitivity.

It is worth noting here that the effect of low concentrations of impurities in solids on the course of photosubstitution reactions is expected to be confined to interactions with the electronic states of the complex ions. Such interactions probably would decrease quantum yields but not change the path of reactions (i.e., whether a *cis* or *trans* isomer was produced), unless the various possible paths originated from different excited states and then only if the states were coupled from molecule to molecule throughout the crystal. The reactions that have been studied all involve substitution by a major component of the crystal, that is, an anion or water of crystallization present in at least a one-to-one ratio with the complex. Impurities, whether substitutional or interstitial, on the other hand, would be expected to be present to the extent of only a few mole percent. The statistics for their direct interaction with a single excited molecule are very unfavorable. If excitation energy can migrate through a crystal and if the migration mechanism preserves the energy in states of differing reactivity, then preferential quenching of a specific state could,

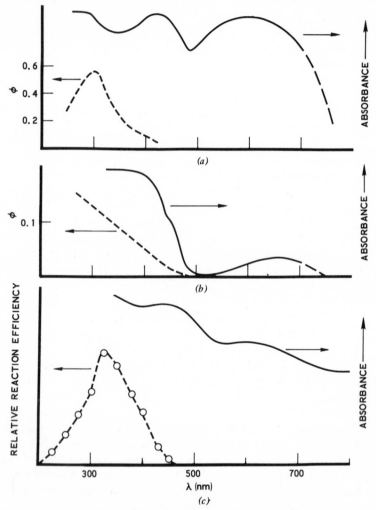

Fig. 9-3. (a) Reflectance spectrum and quantum yields for Co(II) formation for solid $K_3Co(C_2O_4)_3 \cdot 3H_2O$.[58] (b) Absorption spectrum and quantum yields (light intensity $= 1 \times 10^{15}$ photons sec^{-1}), for Fe(II) formation for solid $K_3Fe(C_2O_4)_3 \cdot 3H_2O$.[58] (c) Absorption spectrum and relative reaction efficiencies for $K[Mn(mal)_2(H_2O)_2]$ (mal $= H_2C_3O_4^{2-}$).[60]

for example, change the stereochemical course of a reaction. Such a set of circumstances is considered extremely improbable for molecular complexes in which intermolecular interactions are usually quite small (but see Section III.D for some examples of compounds where the situation might be different than normal).

B. Charge-Transfer Reactions

There is, perhaps, a greater understanding of a few solid-state charge-transfer reactions than there is for substitutions, probably because the mechanisms of the former are more like the corresponding processes in solution. Photochemical charge-transfer reactions result in compositional changes, as is the case for substitutions. In addition, there are frequently structural and phase changes involved when a metal ion is reduced or oxidized. The preponderance of data on these systems has involved reactions of oxalate complexes, and one of the reaction products is obtained from the conversion of solid reactant into gaseous product, CO_2. It is surprising that, with the exception of some considerations of differences between dilute and pure crystals[37] and between pure crystals and solutions,[37,58] once again there has been very little attempt to correlate solid-state reactions to material properties, such as crystal structure and structural and phase changes.

One's intuition suggests that in order to promote photochemical redox reactions, excitation of charge-transfer transitions is required. It is gratifying that even for solid materials this naive prediction seems to be valid. Studies of the wavelength dependence of quantum yields for solids are complicated by the fact that the reaction zone becomes thinner and nearer to the surface as the absorption coefficient increases,[59] but in at least three different studies it has been shown that metal d–d bands are photoinactive with respect to reduction of the metal center, while CT bands are active.[37,58,60] These results are summarized in Fig. 9-3.

The photochemical behavior of solid oxalate complexes[37,58,61–63] is much like the behavior of the complexes in solution.[64,65] ultraviolet irradiation promotes the transfer of an electron from an oxalate ligand to the metal center in $K_3[M(C_2O_4)_3]\cdot3H_2O$ (where M = Mn, Fe, Co) complexes. The stoichiometry of the net reaction for M = Mn and Co is[61,62]

$$2K_3[M(C_2O_4)_3]\cdot3H_2O \xrightarrow{h\nu} 2K_2[M(C_2O_4)_2] + K_2C_2O_4 + 2CO_2 + 6H_2O \quad (9\text{-}8)$$

while for M = Fe, it is[63]

$$2K_3[M(C_2O_4)_3]\cdot3H_2O \xrightarrow{h\nu} 2MC_2O_4 + 3K_2C_2O_4 + 2CO_2 + 6H_2O \quad (9\text{-}9)$$

the difference being attributed to the instability of $K_2[Fe(C_2O_4)_2]$.

The mechanism of the reaction, at least for the case of M = Co, is believed to involve the formation of the $C_2O_4^-$ radical,[37] probably through a primary step that is the homolytic detachment of one end of an oxalate ligand, as in the analogous solution reaction.[65] The quantum yield, using 355 to 385 nm irradiation, for Co(II) formation for $[Co(C_2O_4)_3]^{3-}$ doped into the host crystal, $NaMg[Al(C_2O_4)_3]\cdot 9H_2O$, $\phi = 0.005$, is substantially lower than that of the pure Co(III) solid, $\phi = 0.16$, or of the complex in solution, $\phi = 0.44$. If the mechanism is

$$[Co(C_2O_4)_3]^{3-} \xrightarrow{h\nu} {}^*[Co(C_2O_4)_3]^{3-} \tag{9-10}$$

$$^*[Co(C_2O_4)_3]^{3-} \longrightarrow \left[\begin{array}{c} O \\ \| \\ O-C \overset{\text{II}}{-}O-C \overset{\| \;\; \|}{-}C-O \\ O-O \end{array} \right]^{3-} \text{(I)} \tag{9-11}$$

$$I \longrightarrow [Co(C_2O_4)_2]^{2-} + C_2O_4^- \tag{9-12}$$

$$I \longrightarrow [Co(C_2O_4)_3]^{3-} \tag{9-13}$$

$$[Co(C_2O_4)_2]^{2-} + C_2O_4^- \longrightarrow [Co(C_2O_4)_3]^{3-} \tag{9-14}$$

$$[Co(C_2O_4)_3]^{3-} + C_2O_4^- \longrightarrow [Co(C_2O_4)_2]^{2-} + 2CO_2 + C_2O_4^{2-} \tag{9-15}$$

the reduction in quantum yield for the dilute solid can be attributed to an increased rate of either (9-13) or (9-14) relative to that of (9-15). It was argued[37] that, indeed, the rate of (9-15) in the dilute crystal should be slowed because of the lower concentration of starting material.

In the study of the dilute crystal an intermediate was observed that absorbed near 320 nm; it reached an apparent steady-state concentration as irradiation was continued. This intermediate was believed to be the $C_2O_4^-$ radical. However, in view of the similarities between these observations and those of a recent flash-photolysis study of the same $[Co(C_2O_4)_3]^{3-}$ ion in aqueous solution,[66] it seems more likely that the intermediate may be species I of reaction (9-11).

The quantum efficiencies of solid-state photochemical redox reactions depend on the facility with which intermediates, such as species I, can relax within the crystal lattice to form the thermodynamically stable products; in this case, the tetrahedral, high-spin Co(II) complex. Crystal properties, such as cation and water placement, and defects, such as vacancies or impurity centers, can alter relaxation rates and rates of trapping of free radical products. To increase the yield of Co(II) formation in the dilute crystal, one might try co-doping to form acceptor states (i.e., easily reducible ions) in the aluminum oxalate host to quench the $C_2O_4^-$ radical and prevent reaction (9-14). Under these conditions, of course, the net stoichiometry of the reaction (9-8) would change to

$$K_3[Co(C_2O_4)_3] \xrightarrow{h\nu} K_2[Co(C_2O_4)_2] + [K^+, C_2O_4^-] \tag{9-16}$$

where $[K^+, C_2O_4^-]$ represents a charge compensated quenched oxalate radical. The important point is that the maximum theoretical yield is reduced from 2 to 1, but the observed yield may be increased from 0.005 to (?).

The photochemical reduction[60] of $M[Mn(H_2C_3O_4)_2(H_2O)_2]$ (where $M = NH_4$, Na, K), believed to be the *trans* isomer,[66] proceeds first by elimination of coordinated H_2O and then by photolysis of $M[Mn(H_2C_3O_4)_2]$ to give the corresponding simple malonate salts. This is an interesting reaction because it appears to be an example of photochemical reduction in coordination number without a change in the charge of the central metal. The parent complex in this manganese-malonate series, $K_3[Mn(H_2C_3O_4)_3]\cdot3H_2O$, which is also photoreduced by ultraviolet radiation,[67] has its lowest d–d absorption band shifted into the infrared, $\lambda = 1.23\ \mu m$, compared to 625 nm for the $K[Mn(H_2C_3O_4)_2(H_2O)_2]$ complex. The wavelength dependence shown in Fig. 9-3c is for the complete decomposition reaction.[60] It was not stated whether longer-wavelength irradiation might have initiated the deaquation portion of the reaction, but it would be interesting to examine this point and to examine the tris-malonato complex for sensitivity in the 1.23-μm band.

C. Linkage Isomerization

Recent work, discussed in Chapter 3, has shown that in the photolysis of $[Co(NH_3)_5NO_2]^{2+}$ (where there is a competition between photoredox, giving Co(II), and radical recombination, giving $[Co(NH_3)_5ONO]^{2+}$), an increase in the viscosity of the solvent increased the proportion of Co—ONO relative to Co(II) in the photolysis products.[68] These results were explained on the basis of the formation of two radicals in the primary photochemical step; in the more viscous solvent the radicals could not diffuse apart readily, so they recombined giving the nitrito isomer. In the solid the separation of the radicals, if formed, is apparently totally restricted because the photoinduced linkage isomerization is completely reversible upon warming the products, that is, there is no Co(II) formation.

$$[Co(NH_3)_5NO_2]X_2 \xrightarrow{h\nu} [Co(NH_3)_5ONO]X_2 \qquad (9\text{-}17)$$

where $X = Cl$, Br, I, NO_3.[69]

There are other examples of photochemical linkage isomerization for solid nitro complexes of various metals, namely, Co(III),[70,71] Ir(III),[72] and Pt(IV)[73] and the X-ray-induced isomerizations of $[Co(NH_3)_5\text{-}SCN]Cl_2\cdot H_2O$ and $K_4[Ru(CN)_6]$ have been mentioned.[48,49] One study[70] is particularly interesting because it again shows the dramatic effects of crystal properties on photosensitivity. The compounds, *trans-*[Co(en)$_2$(NO$_2$)(NCS)]X, where X = ClO$_4$, NO$_3$, and NCS, were irradiated with sunlight or artificial laboratory light. The ClO$_4^-$ and NO$_3^-$ salts showed reversible linkage isomerization of the CoNO$_2$ bond, while the NCS$^-$ salt was photoinert. Interestingly, the NCS$^-$ in the coordination sphere does not prevent isomerization, but as the anion of crystallization it apparently introduces a steric restriction that confines the movement of the NO$_2$ group.

Photolinkage isomerization of NO$_2$ is also observed for [Co(acac)$_2$-(A)NO$_2$] (where A = pyridine, piperidine, and NH$_3$, and acac is acetylacetonate).[71] Quantum yields were not reported, nor was it stated whether the starting material was the *cis* or *trans* isomer. However, there did not appear to be gross differences in the reaction for the different amines. Again, the reaction was totally reversible; it was stated that the absence of redox products even for ultraviolet irradiation was not surprising in view of the large cage effects expected for the solid.

The implication of an environmental effect on the quantum yield for linkage isomerization of CoNO$_2$ complexes suggests the desirability of a more detailed study of known *cis* and *trans* isomers of, for example, the [Co(acac)(A)NO$_2$] series. In particular, one would like to know about the steric effects of ligands coordinated *cis* to the NO$_2$ and about the electronic effects for *trans* substituents. For *d–d* excitation, *trans* substituents that direct the antibonding electron toward the nitro group should increase the reaction efficiency, while one might expect electron-withdrawing substituents to enhance the quantum yield under CT excitation.

D. Cooperative Electronic Phenomena

In this section, I should like to make brief mention of a subject that may have considerable future photochemical significance. Recently, there has been a noticeable increase in interest in the spectral properties of solid d^8 metal complexes that crystallize to form layers of square planar molecules (ions).[74] The nonbonding d electrons on each molecule (ion) can interact with those on an adjacent one with the end result being substantial electron delocalization and, in some cases, one-dimensional electronic conductivity. Many solid Pt(II) complexes, Magnus' green salt (MGS, [Pt(NH$_3$)$_4$][PtCl$_4$]) being perhaps the most studied,[17,74] have intense electronic transitions in the visible spectrum (i.e., they are richly colored) that disappear if the material is dissolved in solution or, as in the case of MGS, that are not present for the individual salts of the anion, K$_2$[PtCl$_4$], or cation, [Pt(NH$_3$)$_4$]Cl$_2$. In some cases, namely, Cossa's salt and K$_2$[Pt(CN)$_4$]Br$_{0.3}$·3H$_2$O, intense spectral absorptions are produced by

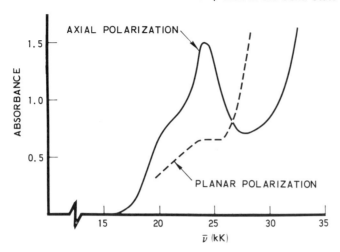

Fig. 9-4. Polarized crystal absorption spectrum for red Cossa's salt,

$$K[Pt(NH_3)Cl_3]_{1-x}[Pt(NH_3)Cl_5]_x \cdot H_2O.[36]$$

partial oxidation of the Pt(II) ion.[21,36] The intense visible absorptions are strongly polarized along the z molecular axis, as shown in Fig. 9-4.

Cossa's salt crystallizes in two forms, a yellow form believed to be pure $K[Pt(NH_3)Cl_3] \cdot H_2O$ and a red form believed to contain oxidized Pt(II).[36] There are other examples of Pt(II) complexes that give both yellow and red salts even though chemical analyses indicate no differences in composition, $Pt(bipy)Cl_2$ being one for which emission spectra of both forms have been reported.[75]

There are no known reports of photochemistry of these variably colored compounds, although the photosensitivity of $[Pt(bipy)_2](ClO_4)_2$ was inferred from changes in luminescence spectra with time[75] and $[Pt(NH_3)Cl_3]^-$ is photosensitive in solution.[76] Since the variation in the color of these solids appears to be due to varying amounts of Pt(IV) impurity in the Pt(II) structure, it seems reasonable to expect that properly designed photochemical experiments might be useful in (1) the synthesis of the colored forms or (2) modification of their electronic properties. For example, ultraviolet irradiation of $trans$-$[Pt(A)_4X_2]X_2$ (X = Cl, Br) under vacuum might result in the reduction of some of the Pt(IV) ions and the release of X_2 gas. Under such conditions, a mixed-valence compound would be obtained that should have electronic properties similar to those of the red Cossa's salt. The degree of mixing of Pt(II) and Pt(IV) in the structure should in theory be adjustable, depending on the length of time of irradiation. One potential difficulty with such a procedure is that the CT bands that give the desired product are very intense, and, therefore, the reaction layer would be very thin and confined to the surface of the material.

The conductive properties associated with the cooperative effects in solid square planar complexes have been studied with the anticipation of new applications for electronic devices. Surely, these unusual materials will prove to have interesting photochemistries as well.

IV. PHOTOCHEMICAL REACTIONS ON SOLID SURFACES

Problems associated with the study of the photochemistry of solids that are caused by high concentrations of excited species near the solid surface have been discussed in earlier sections of this chapter. In this section, the emphasis is changed from an interest in reactions of bulk material, perturbed by surface inconsistencies, to a discussion of various aspects, both physical and chemical, of some specific interfacial reactions. In principal, the subject of interfacial photochemistry could include all heterogeneous catalytic processes that are enhanced by irradiation. However, this discussion is selective in presenting highlights of experimental work that illustrate the basic concepts; it deals with reaction systems in which a metal ion is a principle participant, either in the solid or as the absorbate.

Photochemical reactions on surfaces have tremendous practical importance. For example, they are critical in many industrial catalytic processes, in material degradation problems (such as paint deterioration, yellowing), and in photographic imaging phenomena. Considerable research effort in both industrial and academic laboratories is being directed toward understanding the mechanisms of many surface reactions, so that improvements in the applicable systems can be effected.

The following discussion is divided into two sections classified according to the nature of the electronic transition involved in the absorption process. First, reactions are considered that result after the solid absorbs a photon in a valence to conduction-band (i.e., charge-transfer) transition. This category includes practically all of the catalytic and degradation processes that have been studied. The second section treats photosensitized reactions that occur after an adsorbed molecule absorbs light and transfers energy or an electron to the solid or to another adsorbate on the solid surface. This category includes most of the known visible sensitive imaging materials.

A. Light Absorption in the Solid Substrate

1. *Reaction Systems.* The type of reaction that is of interest is one in which a solid absorbs a photon, creating a mobile electron or hole (the positively charged residue from whence the electron was removed) that

can be captured by an adsorbate. Following this secondary charge-transfer process, the reduced (oxidized) adsorbate can react further to produce the final reaction products, as in the case of the photooxidation of alcohols and acids,[77-80] or it may be stable as in the photoreduction of certain metal ions.[81-84]

A number of different pairs of solids and reactants have been studied. In most cases, the solids are oxides or sulfides of metals such as Ti, Co, Zn, Cd, or Pb, and the reactants are ionic or molecular oxidizing or reducing agents. The metals may have octahedral or tetrahedral coordination; the important condition is that the materials are either simple salts or ionic complexes in which the ligands are coordinated to more than one metal to form delocalized energy bands rather than discrete levels. (It is also possible that one-dimensional conductors such as the materials discussed in Section III.D. would make good substrates for surface charge-transfer reactions.) In addition to the above-mentioned organic-molecule and metal-ion adsorbates, N_2O gas has been employed as an electron-quenching reactant,[85,86] and I^- as a reducing agent (hole quencher).[87] $[Fe(CN)_6]^{3,4-}$ ions are either hole or electron acceptors on ZnO surfaces.[39a] The net photochemical reactions are formally similar to electrode reactions; however, there are no closed external circuits to allow for charge flow between half cells, and, as a consequence, the solid is often partially decomposed during the overall process.[83,87-89] In certain instances, a second electrode has been added to the system and a cell completed in order to examine a particular effect, such as that due to a potential field across the interface. But the reactions of interest occur spontaneously after light absorption; they do not require an additional driving force.

The molecular (ionic) compositions of solid surfaces in contact with electrolyte solutions are not well understood. The surfaces may be viewed, in an ideal sense, as cleavage planes of the bulk material. Metal ions at the ideal surface have unfilled coordination spheres and resort to some form of structural reordering or adsorptive coordination to attain a stable configuration. These surface transformations influence both further adsorption and electronic transport, the two parameters that govern the course of the photochemical reactivity. Hence it is easy to see that the conditions of surface preparation determine not only reaction stoichiometries but also quantum efficiencies.

Solution parameters, such as ionic composition and nonparticipating quencher concentrations, (i.e., O_2 in the case of metal-ion reduction) also affect the qualitative and quantitative results of interfacial photochemical reactions. A good example of the way that changes in solution composition alter potential reactions is the photochemical reduction of Ag(I) on ZnS and CdS surfaces.[90] $AgNO_3$ solution in contact with these sulfides forms a black precipitate that is presumably Ag_2S. However, the complexes, $[Ag(S_2O_3)_2]^{3-}$ or $[Ag(CN)_2]^-$, do not spontaneously precipitate; instead, under irradiation in the appropriate wavelength region (350–400 nm for ZnS and 400–500 nm for CdS), they form another black product that is believed to be fine metallic silver. In contrast to the sulfides, oxides like ZnO or TiO_2 may be contacted with free Ag^+ solutions and then irradiated to produce silver-metal deposits directly with quantum efficiencies up to 0.6 for ZnO and 0.1 for TiO_2.[84] The Ag^+ apparently bonds to the oxide surfaces, at least for TiO_2,[91] but the stabilities of the substrates prevent formation of black Ag_2O prior to irradiation.

2. A Typical Reaction Mechanism.

The mechanisms of surface photochemical electron-transfer reactions have been investigated in some detail for certain metal ion oxidants[39a,84,92] and for the N_2O reaction on ZnO.[85,86] A summary of the results for the reduction of Ag(I) on TiO_{2-x} is presented to illustrate some of the complexities of these ground- and excited-state reactions on solid surfaces.

The photochemical electron transfer from excited TiO_{2-x} to adsorbed Ag(I) proceeds according to a type of Stern–Volmer mechanism, in which the measured quantum yield is independent of exciting light intensity $[1.67 \times 10^{-10}$ to 3.34×10^{-8} Einsteins cm^{-2} (surface) $sec^{-1}]$ but is sensitive to various surface and solution parameters, in particular to the stoichiometry of the solid, the solution Ag(I) concentration, the pH, and the dissolved O_2 concentration (see Fig. 9-5). A simplified reaction scheme can be written to describe the different observations

$$TiO_2 \xrightarrow{h\nu} TiO_2^* \qquad (9\text{-}18)$$

$$TiO_2^* \longrightarrow TiO_2 \qquad (9\text{-}19)$$

$$TiO_2^* \longrightarrow e_s^- + TiO_2^+ \qquad (9\text{-}20)$$

$$Ag(I) \xrightleftharpoons{b} Ag(I)_{ads} \qquad (9\text{-}21)$$

$$e_s^- + Ag(I)_{ads} \longrightarrow Ag_s \qquad (9\text{-}22)$$

$$e_s^- \longrightarrow \text{nonreactive state} \qquad (9\text{-}23)$$

The subscript, s, designates a surface species, while ads means that the ion is adsorbed.

The adsorption of Ag(I) obeys a Langmuir isotherm with adsorption constant, b[91]; the observed quantum yield for Ag formation can thus be written as

$$\Phi = \phi_s \phi_m C_{Ag(I)}[(1 - \phi_m)b^{-1} + C_{Ag(I)}]^{-1} \qquad (9\text{-}24)$$

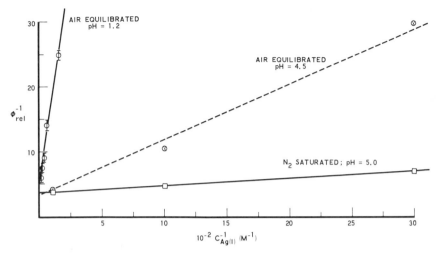

Fig. 9-5. Stern–Volmer plots for photoreduction of Ag(I) on TiO_{2-x} surfaces.

where Φ is the measured yield at Ag(I) concentration $C_{Ag(I)}$, ϕ_s is the fraction of the excited electrons that reach the surface of TiO_{2-x} [reaction (9-20)], and ϕ_m is the fraction of surface electrons that are captured by a monolayer of adsorbed Ag(I) [reaction (9-22)]. From independent evaluation of the adsorption constant, b, and evaluation of the intercepts, $1/\phi_s\phi_m$, and the slope-to-intercept ratios, $(1-\phi_m)b^{-1}$, of plots such as those of Fig. 9-5, one can determine the pertinent quantum-yield information. These data can then be compared for reactions run under different conditions to evaluate the influence of the various environmental effects on the efficiencies of the electron-transfer processes.

It is evident from Fig. 9-5 that both O_2 and H^+ in solution compete with Ag(I) for adsorption sites on TiO_2 and, therefore, reduce the Φ at low $C_{Ag(I)}$. However, if $C_{Ag(I)}$ is increased until each site is occupied by an Ag(I), the limiting yields (i.e., the product, $\phi_s\phi_m$) reach approximately the same value. The small variations observed for different samples are probably due to fluctuations in ϕ_s rather than in ϕ_m because of minor differences in the stoichiometry of the TiO_{2-x} surfaces.[92]

It is hoped that further kinetics studies, along the lines of those described here, will aid in clarifying the mechanisms of these important surface reactions. Also needed is an understanding of the electronic energy levels that are involved in the reactions. It is fairly clear that conduction electrons in the solid are necessary, but electron-trap levels, corresponding to e_s^- in reaction (9-20), appear also to be involved. For the adsorbate, an electron is put into an s orbital to make Ag metal, into a d

orbital to convert $Fe(CN)_6^{3-}$ to $Fe(CN)_6^{4-}$, and presumably into a $p\pi^*$ orbital to reduce leucophthalocyanine dye on TiO_2 surfaces.[93] The interaction of these acceptor orbitals of grossly differing symmetry properties with the electron-trap levels of the solid must influence the electron-transfer efficiencies and account for the quantitative differences in quantum yields that have been reported for similar substrates; namely, intensity-dependent quantum yields ranging from 0.009 to 0.72 were reported for the reduction of Cu leucophthalocyanine on TiO_2 powder; these results differ markedly from ours[84] for Ag formation on a structural variation of the same substrate (i.e. $\Phi \leq 0.1$ with no intensity dependence).

B. Light Absorption in the Adsorbate, Photosensitized Reactions

The history of photosensitization of electron-transfer reactions in solids dates back over one hundred years to the early observations of H. W. Vogel concerning the sensitization of image formation in AgBr crystals.[94,95] Although the primary application of this type of sensitization by a donor molecule adsorbed on the surface of an acceptor has been in silver halide photography, many other solids have been studied together with a variety of adsorbates. Recent books[96] and even more recent review articles[95,97–99] have summarized the status of sensitization of various inorganic substrates, so that here, as in previous sections, the emphasis is on the unusual features of interfacial reactions and on comparisons of the problems and results to photosensitization processes in solution.

The photosensitization of electronic phenomena in solids is a novel process, which, at first, may be viewed with a certain degree of skepticism by the solution photochemist because the sensitizer (donor) absorbs light of much lower energy than the solid (acceptor) and still sensitizes reactions with quantum efficiencies approaching unity in many instances. This apparent energetic anomaly has fascinated workers in the field for many years.[95,100] Today, it is widely recognized that it is the heterogeneity of the donor (D)–acceptor (A) system, that results in a favorable energy-level distribution for efficient energy or electron transfer from D to A.

Although there has been some discussion in the literature whether reductions of some Co(III) complexes in fluid solution are sensitized by $[Ru(bipy)_3]^{2+}$ by electron transfer or electronic energy transfer (Chapter 3, Section III.A.2), it is evident now[101] that the original proposal[102] of the former was justified. This same question has troubled photochemists working on surface reactions; the two different energy-level schemes and transfer processes are shown in Fig. 9-6. The left side of the figure depicts the transfer of an electron from a π^* orbital corresponding to the lowest

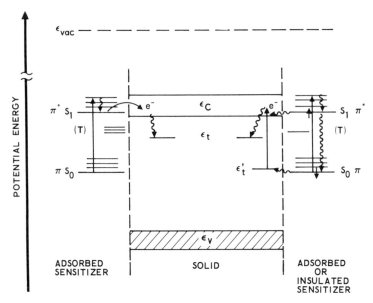

Fig. 9-6. Schematic energy-level diagram for solid acceptor and adsorbed sensitizer; left side shows electron transfer from donor π^* level; right side shows resonance energy transfer to excite a deeply trapped electron, ε_t'.

singlet state of the adsorbed sensitizer to the conduction-band (ε_c) and then to an electron-trap (ε_t) level of the solid. The energetic requirement for this process is that the energy of the π^* orbital be greater, relative to the energy of the electron in a vacuum continuum, than that of the conduction band.

The right side of the figure shows the resonance energy-transfer mechanism. Because of some recent Hückel Molecular Orbital (HMO) calculations of the relative energies of the donor levels in a series of sensitizers and the correlation of the results to sensitization of AgBr,[103] the sensitizer on the right side of the figure is also drawn with the donor level above ε_c, even though there is no electron movement across the interface. The energy-transfer mechanism requires the presence in the solid of a deeply trapped electron before excitation. This electron is not excited in direct irradiation because it has a very low concentration in the crystal and a low absorption coefficient. It can be excited via energy transfer, however, and can then react as a normal conduction electron.

Perhaps the most elegant and definitive experimental work in the field of photosensitization of solid-state reactions has been the study of the energy-transfer mechanism through the technique of manipulative monolayer assemblies.[104] In the initial work, the trough-and-barrier technique[105] was used to deposit monolayers of arachidic acid, a dielectric material, between layers of sensitizers and the surface of a AgBr thin film that had been prepared by sublimation onto an appropriate substrate (see Fig. 9-7). In spite of the fact that the dielectric layers prevented conduction of photoelectrons from the sensitizer to the acceptor film, excitation of the sensitizer still resulted in efficient reduction of Ag^+ to form metallic silver in the AgBr. In other monolayer-deposition experiments,[106] it has been demonstrated that holes in films, such as those of the arachidic acid, were responsible for the observed effects. However, for the situation on AgBr, film-resistance measurements (for a Cd arachidate film between aluminum and mercury or between aluminum and lead electrodes) and fluorescence measurements of superimposed sensitizer dyes have shown that penetration of the sensitizer through the fatty acid layer(s) is extremely unlikely.[107] (Note in particular the discussion of this point in the "Discussion Report" of Ref. 107, pp. 237–238.)

If the distance between the sensitizer and acceptor is increased by the deposition of more than one dielectric layer, the sensitization efficiency decreases and the fluorescence yield of the sensitizer increases in a parallel fashion. The dependence of the sensitization yield on the donor-acceptor separation (d) was found[108] to obey a d^{-4} relationship, in agreement with the predictions of the Förster theory of resonance energy transfer.[109]

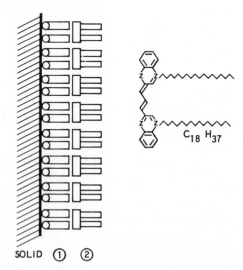

SOLID ① ②

Fig. 9-7. Schematic diagram showing a layer of arachidic acid ① adsorbed on a solid surface and a layer of a typical dye ② adsorbed on ①. Typical dye structure (sensitizer) also shown.

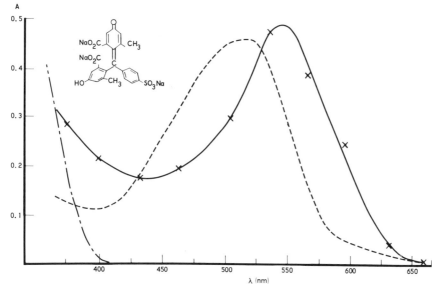

Fig. 9-8. Absorption spectrum [(-----) 2×10^{-4} M] and photochemical action spectrum [(———) arbitrary units] for sensitized reduction of Ag(I) on TiO_{2-x} thin-film surface. Also shown are the action spectrum for unsensitized surface and the structure of the sensitizer.

These results demonstrate that the energy-transfer mechanism *can* be an important one in the sensitization of photographic image formation in AgBr. As of this writing, the consensus seems to be that energy transfer is important when dielectric layers insulate sensitizer from acceptor, but when the two are in direct contact, the electric fields are so strong that upon excitation some degree of electron transfer must take place.[110,111]

The reduction of Ag(I) on TiO_{2-x} surfaces can be photosensitized by certain visible absorbing molecules (ions).[121,113] Figure 9-8 shows the solution-absorption and the photochemical-action spectra for a typical green-sensitive sensitizer; also shown is the spectral dependence of the reaction yield for direct excitation. This system consists of a TiO_{2-x} film in contact with a solution of the sensitizer and Ag(I) ions. The direct electron transfer from donor to Ag(I) in the absence of the TiO_{2-x} has been shown to be negligible.

The path of electron/energy transfer in this system is even more complicated than for the two-component dye/AgBr case. It is believed that the process proceeds according to the scheme presented in reactions (9-19) to (9-23), with the only difference being that the TiO_{2-x} is excited by the sensitizer. It is not known whether the sensitization is via electron or energy transfer. Attempts to use $[Ru(bipy)_3]^{2+}$ as a sensitizer led to very low efficiencies, and when some of this ion was added to a solution of the sensitizer shown in Fig. 9-8, a reduction in the sensitization efficiency was observed.

The types of reactions described in this section combine uncertainties of the structures of solid surfaces with those of the mechanisms of sensitization processes. In spite of these formidable challenges, or perhaps because of them, reactions at interfaces offer a wide variety of interesting problems for future study. The long history of the known photosensitization of these reactions and the potential for the application of well-understood systems contribute to the fascination of such studies.

REFERENCES

1. H. Bethe, *Ann. Phys. (Leipzig)* **3**, 133 (1929).
2. J. H. Van Vleck, *The Theory of Electric and Magnetic Susceptibilities*, Chap. 2, Oxford U.P., Oxford, England, 1932.
3. L. E. Orgel, *An Introduction to Transition-Metal Chemistry Ligand-Field Theory*, Butler and Tanner, London, 1960.
4. C. J. Ballhausen, *Introduction to Ligand Field Theory*, McGraw-Hill, New York, 1962.
5. B. N. Figgis, *Introduction to Ligand Fields*, Wiley-Interscience, New York, 1966.
6. J. S. Griffith, *The Theory of Transition Metal Ions*, Cambridge U.P., New York, 1961.
7. J. P. Suchet, *Crystal Chemistry and Semiconduction in Transition Metal Binary Compounds*, Academic, New York, 1971.
8. C. E. K. Mees and T. H. Janus, *Theory of the Photographic Process*, 3rd ed., Macmillan, New York, 1966.
9. L. S. Forster, in *Transition Metal Chemistry* (R. L. Carlin, Ed.), Vol. V, p. 1, Marcel Dekker, New York, 1969.
10. P. D. Fleischauer and P. Fleischauer, *Chem. Rev.* **70**, 199 (1970).
11. P. D. Fleischauer, E. Cervone, and F. Castelli, unpublished results.
12. P. D. Fleischauer, A. W. Adamson, and G. Sartori, *Progr. Inorg. Chem.* **17**, 1 (1972).
13. J. L. Laver and P. W. Smith, *Aust. J. Chem.* **24**, 1807 (1971).
14. G. Vierke and K. H. Hansen, *Z. Phys. Chem. (Frankfurt)* **59**, 109, (1968).
15. A. W. Adamson, *J. Inorg. Nucl. Chem.* **28**, 1955 (1966).
16. C. Moncuit and H. Poubt, *J. Phys. Radium* **23**, 353 (1962); A. M. Tkachuk, *Izv. Akad. Nauk SSSR, Ser. Fiz.* **27**, 670 (1963).
17. P. Day, *Inorg. Chim. Acta Rev.* **3**, 81 (1969).
18. H. H. Tippins, *Phys. Rev. B* **1**, 126 (1970).
19. C. Kittel, *Introduction to Solid State Physics*, 4th ed., Wiley, New York, 1971.
20. L. V. Azaroff, *Introduction to Solids*, McGraw-Hill, New York, 1960.

21. A. Menth and M. J. Rice, *Solid State Commun.* **11**, 1025 (1972); H. R. Zeller, *Advan. Solid State Phys.* **13**, 31 (1973).

22. L. V. Azaroff, *Introduction to Solids*, Chap. 12, McGraw-Hill, New York, 1960.

23. F. A. Grant, *Rev. Modern Phys.* **31**, 646 (1959).

24. R. D. Iyangar and M. Codell, *Advan. Colloid Interface Sci.* **3**, 365 (1972).

25. M. Che, C. Naccache, B. Imelik, and M. Prettre, *C.R. Acad. Sci. Paris,* (*Ser. C*) **264**, 190 (1967); see also M. Che, C. Naccache, and B. Imelik, *J. Catal.* **24**, 328 (1972).

26. V. Balzani and V. Carassiti, *Photochemistry of Coordination Compounds*, Chap. 18, Academic, New York, 1970.

27. W. W. Wendlandt, in *Analytical Photochemistry and Photochemical Analysis* (J. M. Fitzgerald, Ed.), Chap. 8, Marcel Dekker, New York, 1971.

28. E. L. Simmons and W. W. Wendlandt, *Coord. Chem. Rev.* **7**, 11 (1971).

29. W. W. Wendlandt and E. L. Simmons, *J. Inorg. Nucl. Chem.* **28**, 2420 (1966).

30. J. K. S. Wan, R. N. McCormick, E. J. Baum, and J. N. Pitts, Jr., *J. Amer. Chem. Soc.* **87**, 4409 (1965).

31. L. H. Hall, J. J. Spijkerman, and J. L. Lambert, *J. Amer. Chem. Soc.* **90**, 2044 (1968).

32. G. Lohmiller and W. W. Wendlandt, *Anal. Chim. Acta* **51**, 117 (1970).

33. D. A. Johnson and J. E. Martin, *Inorg. Chem.* **8**, 2509 (1969).

34. E. P. Hertzenberg and J. C. Bailar, Jr., *Inorg. Chem.* **10**, 2371, 2377 (1971).

35. N. Tanaka and K. Nagase, *Bull. Chem. Soc. Jap.* **42**, 2854 (1969); R. Tsuchija, M. Suzuki, and E. Kyuno, *ibid.* **45**, 3105 (1972).

36. P. E. Fanwick and D. S. Martin, Jr., *Inorg. Chem.* **12**, 24 (1973).

37. A. C. Sarma, A. Fenertz, and S. T. Spees, *J. Phys. Chem.* **74**, 4598 (1970).

38. F. Williams, Ed., *Luminescence of Crystals, Molecules, and Solutions*, Plenum, New York, 1973.

39. (a) K. M. Sancier, *Surface Sci.* **21**, 1 (1970); (b) T. Purcell and R. A. Weeks, *J. Chem. Phys.* **54**, 2800 (1971); (c) R. D. Iyengar and M. Codell, *Advan. Colloid Interface Sci.* **3**, 365 (1972).

40. G. Sartori, F. Castelli, and E. Cervone, *Gazz, Chim. Ital.* **101**, 32 (1971).

41. S. T. Spees, Jr., and P. Z. Petrak, *J. Inorg. Nucl. Chem.* **32**, 1229 (1970).

42. M. Fujimoto and Y. Tomkiewiez, *J. Phys. Chem.* **56**, 749 (1972).

43. N. V. Vugman, R. P. A. Muniz, and J. Danan, *J. Chem. Phys.* **57**, 1297 (1972).

44. J. Higinbotham and D. Haneman, *Surface Sci.* **32**, 466 (1972).

45. E. L. Simmons, *J. Phys. Chem.* **75**, 588 (1971); E. L. Simmons and W. W. Wendlandt, *Anal. Chim. Acta* **53**, 81 (1971).

46. L. Nagel and M. O'Keeffe, *Inorg. Chem. Ser.* **10**, 1 (1972).

47. F. Bolletta, M. Gleria, and V. Balzani, *Mol. Photochem.* **4**, 205 (1972), and references quoted therein.

48. M. R. Snow and R. F. Boomsma, *Acta Crystallogr. B* **28**, 1908 (1972).

49. R. S. Eachus and F. G. Herring, *Can. J. Chem.* **50**, 162 (1972).

50. C. H. Stembridge and W. W. Wendlandt, *J. Inorg. Nucl. Chem.* **27**, 129 (1965).

51. D. Berman, G. Bokerman, and R. W. Parry, *Inorg. Synth.* **10**, 41 (1967).

52. C. Kutal and A. W. Adamson, *J. Amer. Chem. Soc.* **93**, 5581 (1971).

53. W. Geis and H. L. Schlafer, *Z. Phys. Chem.* (*Frankfurt*) **65**, 107 (1969).

54. R. Ballardini, G. Varani, H. F. Wasgestian, L. Moggi, and V. Balzani, *J. Phys. Chem.* **77**, 2947 (1973).

55. M. T. Gandolfi, M. F. Manfrin, L. Moggi, and V. Balzani, *J. Amer. Chem. Soc.* **94**, 7152 (1972).

56. M. T. Gandolfi, M. F. Manfrin, A. Juris, L. Moggi, and V. Balzani, *Inorg. Chem.* **13**, 1342 (1974).

57. M. F. Manfrin, M. T. Gandolfi, T. Moggi, and V. Bolzani, *Gazz. Chim. Ital.* **103**, 1189 (1973).

58. H. E. Spencer, *J. Phys. Chem.* **73**, 2316 (1969).

59. H. E. Spencer and M. W. Schmidt, *J. Phys. Chem.* **74**, 3472 (1970).

60. G. D'Ascenzo and W. W. Wendlandt, *Gazz. Chim. Ital.* **102**, 134 (1972).

61. W. W. Wendlandt and E. L. Simmons, *J. Inorg. Nucl. Chem.* **27**, 2317 (1965).

62. E. L. Simmons and W. W. Wendlandt, *J. Inorg. Nucl. Chem.* **27**, 2325 (1965).

63. W. W. Wendlandt and E. L. Simmons, *J. Inorg. Nucl. Chem.* **28**, 2420 (1966).

64. J. I. H. Patterson and S. P. Perone, *J. Phys. Chem.* **77**, 2437 (1973); G. D. Cooper and B. A. De Graff, *J. Phys. Chem.* **76**, 2618 (1972).

65. N. S. Rowan, M. Z. Hoffman, and R. M. Milburn, Abstracts 176th ACS National Meeting, INOR 8, Los Angeles, Calif., April, 1974.

66. J. I. Bullock, M. M. Patel, and J. E. Salmon, *J. Inorg. Nucl. Chem.* **31**, 415 (1969).

67. G. D'Ascenzo and W. W. Wendlandt. *J. Inorg. Nucl. Chem.* **32**, 3109 (1970).

68. F. Scandola, C. Bartocci, and M. A. Scandola, *J. Amer. Chem. Soc.* **95**, 7898 (1973).

69. W. W. Wendlandt and J. H. Woodlock, *J. Inorg. Nucl. Chem.* **27**, 259 (1965).

70. B. Adell, *Z. Anorg. Allg. Chem.* **386**, 122 (1971).

71. D. A. Johnson and J. E. Martin, *Inorg. Chem.* **8**, 2509 (1969).

72. F. Basolo and G. S. Hammaker, *Inorg. Chem.* **1**, 1 (1962).

73. N. Sabbatini, L. Moggi, and G. Varani, *Inorg. Chim. Acta* **5**, 469 (1971).

74. Symposium on Extended Interactions Between Metal Ions in Transition Metal Complexes, 167th ACS National Meeting, Los Angeles, Calif., April, 1974.

75. D. L. Webb and L. Ancarani Rossiello, *Inorg. Chem.* **10**, 2213 (1971).

76. P. Natarajan, Ph.D. Dissertation, University of Southern Calfornia, Los Angeles, 1971.

77. H. D. Muller and F. Steinback, *Nature (London)* **225**, 728 (1970).

78. G. Irick, Jr., *J. Appl. Polym. Sci.* **16**, 2387 (1972).

79. K. Micka and H. Gerischer, *Electroanal. Chem. Interfac. Electrochem.* **38**, 397 (1972).

80. K. M. Sancier and S. R. Morrison, *Surface Sci.* **36**, 622 (1972).

81. G. A. Korsunovskii, *Russ. J. Phys. Chem.* **39**, 1139 (1965).

82. W. C. Clark and A. G. Vondjidis, *Can. J. Phys.* **46**, 1775 (1968).

83. D. N. Goryachev, L. G. Paritskii, and S. M. Ryvlein, *Sov. Phys. Semicond.* **4**, 1354 (1971).

84. P. D. Fleischauer, H. K. A. Kan, and J. R. Shepherd, *J. Amer. Chem. Soc.* **94**, 283 (1972).

85. J. Cunningham, J. J. Kelly, and A. L. Penny, *J. Phys. Chem.* **75**, 617 (1971).

86. K. Tanaka and G. Blyholder, *J. Phys. Chem.* **75**, 1037 (1971).

87. A. Fujishima, E. Sugiyama, and K. Honda, *Bull. Chem. Soc. Jap.* **44**, 304 (1971).

88. R. B. Tal, *Phys. Status Solidi A* **5**, K19 (1971).

89. P. D. Fleischauer and A. B. Chase, *J. Phys. Chem. Solids* **35**, 1211 (1974).

90. P. D. Fleischauer, unpublished results.

91. P. D. Fleischauer and J. R. Shepherd, *J. Phys. Chem.* **78**, 2580 (1974).

92. P. D. Fleischauer and J. R. Shepherd, Abstracts of the 164th ACS National Meeting, New York, September, 1972.

93. E. J. DeLorenzo, L. K. Case, E. M. Stickles, and W. A. Stamoulis, *Photogr. Sci. Eng.* **13**, 95 (1969).

94. H. W. Vogel, *Ber. Deut. Chem. Ges.* **6**, 1302 (1873); *The Philadelphia Photographer* **11**, 25 (1874).

95. W. West, *Photogr. Sci. Eng.* **18**, 35 (1974).

96. H. Meier, *Spectral Sensitization*, Focal Press, New York, 1968; W. F. Berg, U. Mazzucato, H. Meier, and G. Semerano (Eds.), *Dye Sensitization*, Focal Press, New York, 1970.

97. P. B. Gilman, Jr., *Photochem. Photobiol.* **16**, 211 (1972).

98. H. Meier, *Photochem. Photobiol.* **16**, 219 (1972).

99. H. Gerischer, *Photochem. Photobiol.* **16**, 243 (1972).

100. W. West and P. B. Gilman, Jr., *Photogr. Sci. Eng.* **13**, 221 (1969).

101. C. R. Bock, T. J. Meyer, and D. G. Whitten, *J. Amer. Chem. Soc.* **96**, 4710 (1974).

102. H. D. Gafney and A. W. Adamson, *J. Amer. Chem. Soc.* **94**, 8238 (1972).

103. D. M. Sturmer, W. S. Gaugh, and B. J. Bruschi, *Photogr. Sci. Eng.* **18**, 49, 56 (1974).

104. L. v. Szentpoly, D. Mobius, and H. Kuhn, *J. Chem. Phys.* **52**, 4618 (1970); H. Kuhn and D. Mobius, *Angew. Chem. Int.* (Edit.) **10**, 620 (1971), see also D. F. O'Brien, *Photogr. Sci. Eng.* **18**, 16 (1974).

105. G. L. Gaines, Jr., *Insoluble Monolayers at Liquid-Gas Interfaces*, Wiley-Interscience, New York, 1966.

106. A. W. Adamson, *Physical Chemistry of Surfaces*, 2nd. ed., p. 179, Wiley-Interscience, New York, 1967.

107. H. Kuhn, in *Dye Sensitization* (W. F. Berg, U. Muzzucato, H. Meier, and G. Semerano, Eds.), p. 199, Focal Press, New York, 1970.

108. D. Mobius, *Photogr. Sci. Eng.* **18**, 413 (1974).

109. Th. Förster, *Discuss. Faraday Soc.* **27**, 7 (1959).

110. T. Tani, *Photogr. Sci. Eng.* **17**, 11 (1973), and references cited therein.

111. B. Levy and M. Lindsey, *Photogr. Sci. Eng.* **16**, 389 (1972); *ibid.* **17**, 135 (1973).

112. R. H. Sprague and J. H. Keller, *Photogr. Sci. Eng.* **14**, 401 (1970).

113. P. D. Fleischauer, unpublished results.

10

PHOTOCHROMISM AND CHEMILUMINESCENCE

Arthur W. Adamson

Department of Chemistry
University of Southern California
Los Angeles, California

I. INTRODUCTION

The term, *photochromism* (or, sometimes, *phototropism*), has the general definition of being a reversible, light-induced change. This usually includes a visible color change, but need not. It is useful to speak of a *singly photochromic* system as one in which only the forward reaction is photocatalyzed, the return occurring as a purely thermal process. A *doubly photochromic* system is one in which the reaction is photocatalyzed in both directions. In addition, we speak of a *simple* reaction as one in which reactants convert to products in a single reaction step. That is, there are no intermediates and sequential reactions. Thus

$$A \text{ (+ other reactants)} \underset{\phi_B, k_B}{\overset{\phi_A, k_A}{\rightleftarrows}} B \text{ (+ other products)} \qquad (10\text{-}1)$$

where ϕ denotes quantum yield, and k, thermal reaction rate constant. In a singly photochromic system, ϕ_B is either negligible or the absorption by B is negligible.

Chemiluminescence is the emission of electromagnetic radiation as an accompanyment to a chemical reaction. The usual case of interest for us is that for which such emission is in the visible or near-visible wavelength region. Again, a simple chemiluminescent process is defined here as one in which a single step reaction converts initial substance, A, to an emitting excited state of product, B*. No intermediates are involved, other than B*, nor any sequential reactions.

Consistent with the general emphasis in this book, we confine ourselves to solution systems and, furthermore, to ones in which the action centers

413

on a coordination compound. Given these restrictions, the literature is not very extensive on either photochromism or chemiluminescence. Much of it up to 1968–1970 is reviewed in two publications.[1,2] Some more recent papers are noted in this chapter, but the chief purpose is to point out certain aspects of both subjects that have received rather little emphasis. One important such aspect is the strong interaction with conventional transition-state theory.

II. KINETICS OF PHOTOCHROMISM

The kinetic analysis of reaction scheme (10-1) is easily developed if we assume that the thermal interconversion of A and B obeys first-order rate laws. This behavior could either be a consequence of the reaction being unimolecular, such as intramolecular isomerization, or of the constraint that all other species involved in the rate law are held at constant concentration. Strictly speaking, the system should also be dilute in A and in B so that activity coefficient changes can be neglected. As a result of similar constraints, the quantum yields are taken to be independent of the degree of reaction.

We may now write

$$\frac{dC_A}{dt} = -\frac{I_A \phi_A}{C_A V} C_A - k_A C_A + \frac{I_B \phi_B}{C_B V} C_B + k_B C_B \tag{10-2}$$

where I_A and I_B are the einsteins absorbed by species A and B, respectively, per unit time, and V is the volume of the solution. Equation (10-2) may be abbreviated

$$\frac{dC_A}{dt} = -(k_A^p + k_A)C_A + (k_B^p + k_B)C_B \tag{10-3}$$

where $k_A^p = I_A \phi_A / C_A V$ and may be called the photochemical first-order rate constant; k_B^p has the corresponding definition and meaning.

If the irradiation geometry is that of a collimated beam of intensity, I_0, in einsteins per unit time, falling on a cell of length, l, the Beer–Lambert law gives

$$I_A = I_0(1 - e^{-2.303D}) \frac{\varepsilon_A C_A l}{D}$$

$$I_B = I_0(1 - e^{-2.303D}) \frac{\varepsilon_B C_B l}{D} \tag{10-4}$$

where ε denotes extinction coefficient, and D is the total optical density or absorbance of the solution, $D = l(\varepsilon_A C_A + \varepsilon_B C_B)$, if no absorbing species other than A and B are present.

At equilibrium or at the photostationary state we set $dC_A/dt = 0$, and two extreme conditions are easily described. If the k^Ps are negligible, Eq. (10-3) becomes a statement of the ordinary equilibrium constant, $K = k_A/k_B$. If the thermal rate constants, k, are negligible, combination of Eqs. (10-3) and (10-4) gives

$$K^P = \frac{\varepsilon_A \phi_A}{\varepsilon_B \phi_B} = \left(\frac{C_B}{C_A}\right)_\infty \tag{10-5}$$

This last result obtains for any irradiation geometry such that I_A and I_B, while functions of the general optical properties of the solution, are specifically proportional to $\varepsilon_A C_A$ and $\varepsilon_B C_B$, respectively. The more general expression for the steady state equilibrium ratio, K^s, is

$$K^s = \frac{k_A^P + k_A}{k_B^P + k_B} = \frac{k_A^s}{k_B^s} = \left(\frac{C_B}{C_A}\right)_\infty \tag{10-6}$$

The integration of combined Eqs. (10-3) and (10-4) leads to complicated algebra; the result is much simplified, however, if it is assumed that the system is dilute in A and in B so that the exponentials in Eq. (10-4) may be expanded and only the first term retained. I_A and I_B then become $2.303 I_0 \varepsilon_A C_A l$ and $2.303 I_0 \varepsilon_B C_B l$, respectively, and the k^Ps become $2.303 I_0 \varepsilon_A \phi_A l / V$ and $2.303 I_0 \varepsilon_B \phi_B l / V$, again respectively, and are thus independent of the degree of reaction. Equation (10-3) now integrates to give

$$\frac{C_A}{C_A^0} = \frac{1}{1 + K^s} + \frac{K^s}{1 + K^s} \exp\left(-k^s t\right) \tag{10-7}$$

where K^s is given by Eq. (10-6) and $k^s = (k_A^P + k_A) + (k_B^P + k_B) = k_A^s + k_B^s$. The system thus behaves as a reversible first-order one with effective rate constants, k_A^s and k_B^s. Depending on light intensity, for example, the steady-state condition will range from $K^s = K$ at low intensity to $K^s = K^P$ at high intensity. Clearly, K^P also depends on the extinction coefficient and quantum yield ratios, $\varepsilon_A/\varepsilon_B$ and ϕ_A/ϕ_B, and is thus wavelength dependent.

The quantity, K^s, describes the steady-state ratio of C_A to C_B, and it is useful to consider the difference in free energy between this composition and that at thermal equilibrium. This is given by ΔG^{s} [2,3]

$$\Delta G^s = -\left| RT \ln \frac{K^s}{K} \right| \tag{10-8}$$

ΔG^s is always negative when taken in the direction of reaction in which relaxation occurs. We may also write

$$\Delta G^s = \Delta H^s - T \Delta S^s \tag{10-9}$$

Thus the relaxation is accompanied by enthalpy change, ΔH^s, and entropy change, ΔS^s.

ΔG^s is the obvious quantity of interest in practical applications since it expresses the degree of shift from equilibrium that is achieved or achievable. In many cases, such as in imaging and memory-storage devices, ΔG^s is indeed the central quantity. On the other hand, if it is desired to retrieve useful *work* as a photostationary state is allowed to relax back to thermal equilibrium, ΔG^s may not be a reliable indicator of the potential importance of a system.

Consider, for example, the photoracemization of $Cr(C_2O_4)_3^{3-}$.[4] One expects, and it has indeed been shown,[5] that irradiation of a racemic mixture with circularly polarized light produces net optical activity as a consequence of a small preferential racemization of one enantiomorph relative to the other. This preference stems from a small difference in absorption coefficients of the two optically active forms for light of a given circular polarization, and we expect K^p to be given by $\varepsilon_A/\varepsilon_B$. The quantum yields for the two optical isomers should be the same. The point of this example is that, while K^p and hence K^s differ from K, the relaxation of the photostationary state to equilibrium is not accompanied by any appreciable molar heat of reaction. That is, the contribution of ΔH^s to ΔG^s should be negligible; ΔG^s derives primarily from an entropy of mixing of the optical isomers. Clearly, in this somewhat extreme case, it would be very difficult to extract useful work as the photostationary state is allowed to relax. Membranes or electrodes selective to the chirality of the complex would be needed.

Photoreversible substitution reactions represent a more common and less extreme situation. Here again, however, ΔG^s derives mainly from the $T \Delta S^s$ term since enthalpies of ligand substitution reactions in solution are generally small. As a practical matter, it would again be difficult to convert even a very favorable ΔG^s to useful work.

It would appear that obtaining useful work through relaxation of a photostationary state is not easy. The more promising candidate systems are probably those involving a photoredox reaction. Here, at least, there is a more evident possibility of allowing the relaxation to occur as an electrochemical cell reaction.

III. TYPES OF PHOTOCHROMIC SYSTEMS

A number of systems reported in the various preceding chapters are, in principle, photochromic, although they may not have been studied with any thought to this aspect of behavior. Thus a common photoreaction in aqueous solution is that of aquation; if the reverse reaction can occur thermally, as is often the case, the system is at least singly photochromic.

Photoisomerization generally signals a photochromic system. Photochromism may also be implicit in some cases of photoredox reactions. In what follows, some mechanistic categories are outlined with, where possible, illustrations in terms of actual systems.

A. Photosubstitution and Photoisomerization

Reaction (10-1) serves as the simplest mechanism for isomerization photochromism. A family of examples already well discussed in the literature (see Refs. 1 and 2) is that of the dithizonate complexes. In this case, A and B of reaction (10-1) are tautomers, and the isomerization appears to be intramolecular. If A is the more stable isomer of the dithizonate complex, a typical observation is that k_B is large relative to k_B^p so that the system is singly photochromic.

A more recent example of isomerization photochromism is[6]

$$cis\text{-}[Co(en)_2(H_2O)Cl]^{2+} \xrightarrow{h\nu} trans\text{-}[Co(en)_2(H_2O)Cl]^{2+} \quad (10\text{-}10)$$

for which k_B/k_A is about 3 and $\varepsilon_B\phi_B$ is negligible compared to $\varepsilon_A\phi_A$ at the wavelengths studied. Aquation of the chloride ligand is slow and can be repressed by the presence of added free Cl^- ion. No photoaquation of chloride was found, nor any measurable yield of photoredox decomposition (for wavelengths around 500 nm). An analogous system in its behavior is that of $cis\text{-}\beta'\text{-}[Co(trien)(H_2O)Cl]^{2+}$.[7]

While photoisomerization is known for a number of Cr(III) ammines, it is usually accompanied by ligand substitution; such systems are discussed below (see p. 418). However, an apparently simple photoisomerization has been reported for various Pt(II) complexes.[1,2] The case of $[Co(NH_3)_5(NO_2)]^{2+}$ is interesting in that photochemical isomerization of the NO_2^- group occurs with visible light, although, unfortunately, accompanied by much redox decomposition (see Ref. 2). The photoracemization of $Cr(C_2O_4)_3^{3-}$ has already been mentioned.

The simplest mechanism for substitutional photochromism is that of photolabilization of a ligand to give a complex of reduced coordination number. Thus

$$ML_6 \underset{k}{\overset{h\nu}{\rightleftharpoons}} ML_5 + L \quad (10\text{-}11)$$

This mechanism appears to hold for the group VI hexacarbonyl complexes (see Chapter 6); in the case of chromium, the intermediate, $Cr(CO)_5$, is relatively stable and has been studied by visible and infrared spectroscopy. Here, ϕ_B is negligible, as is k_A. Similar examples occur with other carbonyl and arene complexes. Another illustration involves

$Co(CN)_5X^{n-}$ complexes, $X = CN^-$, I^-, or H_2O; here, the reactive intermediate, $[Co(CN)_5]^{2-}$, appears to be produced (see Chapter 4). Photosubstitution of one ligand for another then occurs by the mechanism

$$Co(CN)_5X^{n-} \xrightarrow{h\nu} Co(CN)_5^{2-} + X \qquad (10\text{-}12)$$

$$Co(CN)_5^{2-} + Y \longrightarrow Co(CN)_5Y^{n-} \qquad (10\text{-}13)$$

Thus irradiation of aqueous $Co(CN)_5I^{3-}$ leads to a steady-state mixture of $Co(CN)_5I^{3-}$ and $Co(CN)_5(H_2O)^{2-}$. This type of system may be doubly photochromic, depending on the wavelength of light used. $Mo(CN)_7^{4-}$ is photochromic in aqueous media, the reaction apparently involving loss of cyanide and subsequent dimerization.[8]

Ligand substitution via either the sequence of Eqs. (10-12) and (10-13) or by direct bimolecular substitution may simply be written as

$$ML_6 + L' = ML_5L' + L \qquad (10\text{-}14)$$

if the actual mechanism is uncertain. Corresponding reactions may, of course, be written for other than hexacoordinated compounds. A very common situation is that in which L' is solvent and the solvent is water. Examples include the photolysis of $Cr(CN)_6^{3-}$ and of $Cr(NH_3)_6^{3+}$. The latter photoaquation is irreversible in acidic media, of course. The photoaquation of $PtCl_4^{2-}$ provides an example with coordination number 4 and that of $Mo(CN)_8^{4-}$, of coordination number 8. This general class of ML_n photosolvation reactions is complicated by further photolysis of the $ML_{n-1}S$ (S denoting solvent) product. While photochromic behavior is often present, a mixture of several species is typically present in the photostationary state.

A potentially more complex situation is that of an acidoammine type of complex since more than one photosubstitution mode may be important

$$ML_5X^{n+} + S \xrightarrow{h\nu} \begin{cases} ML_4SX^{n+} + L \\ \\ ML_5S^{n'+} + X \end{cases} \qquad (10\text{-}15)$$

Thus both types of aquation are often observed with Cr(III) and Co(III) ammines. Moreover, compounds such as $Cr(NH_3)_5X^{2+}$ give primarily $cis\text{-}[Cr(NH_3)_4(H_2O)X]^{2+}$, while ones such as $Co(NH_3)_5X^{2+}$ probably give primarily $trans\text{-}[Co(NH_3)_4(H_2O)X]^{2+}$.[9] In the case of Cr(III) ammines, secondary photolysis is apt to be important, giving more highly aquated products. It is not yet established what steady-state conditions may finally obtain in such cases.

In summary, systems showing photosubstitution generally have the potentiality of also showing photochromism. The most common difficulties are those of irreversible loss of ligand (such as of ammonia in acidic media), instability of the complex toward thermal redox decomposition if too much successive solvent coordination occurs through secondary photosolvation reactions (as with Co(III) complexes), and photoredox decomposition or other side reactions of an irreversible nature. Many complicated difficulties of this last nature can occur with carbonyl and related complexes.

From the applied point of view, photochromic systems involving isomerization and substitution reactions have potential usefulness in light control or modification (sunglasses, windows, flash-resistant clothing) and information display or storage. This class of reactions does not generally have a very large ΔH^s, however, and it is apt to be quite difficult to obtain energy from the relaxation of the photostationary state.

B. Photoredox Reactions

Photochromic behavior based on redox reactions has so far been uncommon in coordination chemistry. The products often undergo further, irreversible redox reactions, or the primary photoredox step may be annuled by back-reaction of the products, either as a cage reaction of a geminate pair or as merely a very rapid, ordinary thermal reaction. In such cases, photochromism may in principle be present, but would only be observable on the short time scale of a flash-photolysis experiment. A fast back-reaction also tends to limit useful applications, especially those providing energy production. In general, however, photochromic redox systems are attractive as potential means of converting light into electrical energy.

A general reaction scheme that might permit useful applications would consist of the photoredox reaction, a *transfer* reaction, and a *closing* reaction. The purpose of the transfer reaction is to convert the more reactive product of the photoreaction to some different and more slowly reactive species, thus preventing the direct reverse of the photolysis step, as well as possible irreversible loss of primary product through side reactions. The closure reaction returns the system to the starting condition; it is this reaction that one would try to make of such a nature as to allow the generation of energy.

There are at least three important types of photoredox processes, each possessing different photochromic potentialities. First, the photoreaction may be one of redox decomposition in which the central metal ion is reduced and a ligand is oxidized; such a process is typically a consequence of irradiating a CTTM band (see Chapter 3). An illustrative sequence is the following

$$M^{III}L_5X \xrightarrow{h\nu} M^{II}L_5 + X \qquad (10\text{-}16)$$

$$X + R \longrightarrow X^- + O \qquad (10\text{-}17)$$

$$M^{II}L_5 \ (\text{or } M^{II}L_5S) + O \longrightarrow M^{III}L_5S + R \qquad (10\text{-}18)$$

$$M^{III}L_5S + X^- \longrightarrow M^{III}L_5X \qquad (10\text{-}19)$$

where R and O denote the reduced and oxidized form of some redox agent. The transfer reaction (10-17) should be such as to prevent the thermal reverse of Eq. (10-16) from occurring and also any loss of X by irreversible side reactions. In effect, reaction (10-17) should replace a rapidly reacting oxidant for $M^{II}L_5$ (or $M^{II}L_5S$ if solvent coordination is rapid) by a slowly reacting oxidant, O. This type of requirement clearly is very restrictive. A further and very common difficulty with systems so far studied is that the product, $M^{II}L_5$ or $M^{II}L_5S$, is apt to be substitution labile, so that the principal product is the M^{2+} ion—an ion whose reoxidation to the original complex may not be easy. This is the situation with Co(III) ammines, for example.

Although no functioning photochromic systems of the above class seem to have been devised, there should be possibilities involving ligands that do complex both valence states of the metal ion. This is often true of oxalate and cyanide as ligands; many chelating ligands have this property.

A second type of situation is that in which the primary photochemical process is one of photoelectron production. A possible sequence is now

$$R \rightarrow O + e^-(S) \qquad (10\text{-}20)$$

$$e^-(S) + O' \rightarrow R' \qquad (10\text{-}21)$$

$$O + R' \rightarrow R + O' \qquad (10\text{-}22)$$

Here, for example, R might be $Fe(CN)_6^{4-}$ and O might be $Fe(CN)_6^{3-}$. Again, a transfer reaction, Eq. (10-21), is needed to intercept the rapid reversal of reaction (10-20) or some irreversible reaction of the solvated electron. Reaction (10-22) is the closing reaction, hopefully energetic, yet slow enough to allow the possibility of energy extraction by electrochemical means. The species, O', might be a halogen, such as I_2, in which case, R' would be I^-; the closure reaction of I^- with $Fe(CN)_6^{3-}$ does occur slowly, so this system should be photochromic. This has not actually been demonstrated, however; in fact, no demonstrations of systems conforming to scheme (10-20) to (10-22) seem to have been reported.

A third and newly appreciated type of photoredox process is that in which irradiation produces an excited state capable of acting as a reducing agent. The corresponding reaction scheme is

$$R \xrightarrow{h\nu} R* \qquad (10\text{-}23)$$

$$R* + O' \longrightarrow O + R' \qquad (10\text{-}24)$$

$$O + R' \longrightarrow R + O' \qquad (10\text{-}25)$$

The outstanding current example of species R is $Ru(bipy)_3^{2+}$. Irradiation produces a triplet charge-transfer (CT) excited state that luminesces strongly in room-temperature solution.[10] Of importance in the present context is the finding that the ^3CT state [R* in Eq. (10-23)] can act as a reducing agent to give $Ru(bipy)_3^{3+}$ [O in Eq. (10-24)].[11] The electron accepter, O', may be a Co(III) complex such as $Co(NH_3)_5X^{2+}$ (X = F, Cl, Br)[11] or $Co(C_2O_4)_3^{3-}$.[12] On reduction, free cobaltous ion is produced, the ligands being replaced by solvent water in dilute solution.† Several other species have since been found to function as O' in Eq. (10-24); these include Fe(III) and $Ru(NH_3)_6^{3+}$ among others.[17] There has also been confirmation of the mechanism in the case of the $Co(NH_3)_5X^{2+}$ series.[16]

Note the following interesting situation in the reaction sequence, (10-23) to (10-24). Excited-state species, R*, acts as a *reducing* agent to become an *oxidizing* agent, O. In the case of the ruthenium trisbipyridyl system, the energy difference between R and R* is about 50 kcal mole^{-1},

---

† It should be noted that this mechanism has been disputed, at least in the case of $Co(NH_3)_5Br^{2+}$ [13] (see also Chapter 3). In some early sensitization studies, it was proposed that sensitized redox decomposition of Co(III) complexes occur by means of excitation energy transfer of a ^3CT state of the Co(III) complex,[14] and it has been affirmed that the same mechanism applies when, on irradiation of a solution containing $Ru(bipy)_3^{2+}$ and $Co(NH_3)_5Br^{2+}$, production of $Ru(bipy)_3^{3+}$ and Co(II) occurs. The proposal was that excitation energy transfer from $^3Ru(bipy)_3^{2+}$ to $Co(NH_3)_5Br^{2+}$ produces a ^3CT state of the latter, which then undergoes redox decomposition. The formation of $Ru(bipy)_3^{3+}$ was attributed to oxidation of $Ru(bipy)_3^{2+}$ by radical(s) from the decomposition of the Co(III) complex.

A general objection to the excitation energy-transfer proposal is that the energy available from $^3Ru(bipy)_3^{3+}$ is only about 50 kcal mole^{-1}, and it seems rather unlikely that $Co(NH_3)_5Br^{2+}$ would have so low lying a ^3CT state. More important, however, the same types of experiments that have fairly well established the excited-state reduction mechanism [Eqs. (10-23) and (10-24)] in the case of $Co(C_2O_4)_3^{3-}$ as species, O',[12] have also been carried out with $Co(NH_3)_5Br^{2+}$ as O'. For example, the radicals produced on the direct photoredox decomposition of the bromopentaammine complex have been shown to be reducing toward $Ru(bipy)_3^{3+}$ rather than oxidizing toward $Ru(bipy)_3^{2+}$.[15] Recent results further establish both the general ability of $^3Ru(bipy)_3^{2+}$ to act as a reducing agent,[16] and that this is indeed its action in the case of $Co(NH_3)_5Br^{2+}$ and related complexes.[17]

This writer has no doubts that recognition of excited-state redox reactions will proliferate now that the first examples have been found. In fact, sensitized redox decompositions of Co(III) complexes by various organic sensitizers may turn out generally to involve the excited state redox mechanism rather than excitation energy transfer.

the R*–O couple has a standard potential of about 1 V and the O–R couple, a standard potential of about 1.2 V.

Note also that the geminate pair $(O + R')$ of reaction (10-24) must escape the solvent cage without undergoing reaction (10-25). Since the net of reactions (10-24) and (10-25) is a catalyzed deactivation of R* if such escape does not occur, all that would occur is a quenching of R*. In the case of $^3Ru(bipy)_3^{2+}$ as R*, it would be observed that the luminescence from this species would be quenched with no net reaction taking place. The closure reaction (10-25) has been observed under flash photolysis conditions with O′ as Fe(III), $Ru(NH_3)_6^{3+}$, and certain organic species.[17] In the case of the Co(III) complexes so far studied, the Co(II) product of reaction (10-24) promptly sheds its ligand to give the aquo ion, which is stable toward reaction (10-25).

In summary, at least three types of potentially photochromic redox schemes can be written that involve coordination compounds. Actual examples are rare, however, and so far all require flash-photolytic techniques. Also the actual extraction of electrical energy from the photochemically produced redox couple may present problems. One possibility, suggested by Marcus,[18] involves the use of semiconductor electrodes. In fact, with such electrodes it may be possible to bypass the need for a transfer reaction and obtain electrical work directly from the oxidation or reduction of the excited state of a complex.

IV. CHEMILUMINESCENCE IN COORDINATION CHEMISTRY

Very few examples are known of chemiluminescent reaction in which a coordination compound is the excited-state emitter. Even biological chemiluminescence seems not to use coordination compounds.[19] One very interesting case, however, involves the reverse of Eq. (10-24). The best-known specific example is

$$Ru(bipy)_3^{3+} + R' \rightarrow {}^3Ru(bipy)_3^{2+} + O' \qquad (10\text{-}26)$$

$$^3Ru(bipy)_3^{2+} \rightarrow Ru(bipy)_3^{2+} + h\nu \qquad (10\text{-}27)$$

The reducing agent, R′, may be hydrazine in acid solution, or OH^- ion in alkaline solution.[20] More recently, $e^-(aq)$ and H(aq) have been found to function as well, the efficiency of the former in producing R* being quite high.[21] A very effective laboratory- and lecture-demonstration reducing agent is alkaline sodium borohydride. A quite brilliant emission occurs on mixing $Ru(bipy)_3^{3+}$ in 0.1 N H_2SO_4 (prepared by oxidation of $Ru(bipy)_3^{2+}$ with PbO_2, which is then filtered off) with $NaBH_4$ in 0.1 N NaOH. The mechanism is not known, but could be one of reduction by H atoms generated from the BH_4^- ion.

The same luminescence may be produced by the reaction

$$Ru(bipy)_3^+ + Ru(bipy)_3^{3+} \rightarrow {}^3Ru(bipy)_3^{2+} + Ru(bipy)_3^{2+} \qquad (10\text{-}28)$$

where the reactants are electrogenerated in solvents such as acetonitrile.[22]

A sufficient reason for the present paucity of chemiluminescent systems involving coordination compounds is that very few are known to emit efficiently in room-temperature fluid media. It is very likely that additional families of emitters will be found as the field of inorganic photochemistry continues to grow. To the extent that this occurs, the discovery of new chemiluminescent reactions is also to be expected.

V. INTERRELATIONS BETWEEN PHOTOCHEMISTRY, PHOTOCHROMISM, CHEMILUMINESCENCE, AND ORDINARY CHEMICAL REACTIVITY

Photochemistry has usually been approached by the physical chemist as a kind of spectroscopic phenomenon; chemical change is viewed as a process competing with emission and radiationless deactivation in removing excited-state species. It is conventional to speak of a quantum yield for photochemical product formation, just as for emission and other photophysical processes. Figure 4-3 presents this spectroscopic point of view for the case of a d^3 hexacoordinated transition-metal complex [such as one of Cr(III)]. Absorption in the wavelength region of the first ligand-field or L_1 band maximum gives the ${}^4T_{2g}$ state of ligand-field theory for an O_h complex. Observations such as illustrated in Fig. 2-6 made it clear, however, that the ligand-field excited state is actually not a pure electronic state at all, but rather is one having much superimposed vibrational excitation. This last is currently, at least, believed to be lost easily and rapidly in fluid solution so that in the general time scale of a few dozen molecular vibrations, say 10^{-11} to 10^{-12} sec, the excited molecule has cooled down vibrationally and may then last long enough to sample the Boltzmann distribution of vibrational-rotational energies corresponding to ambient temperature. The complex is now said to be "thermally equilibrated." The complementary aspect is that in the usual photochemical or luminescence experiment a sufficient number of molecules is involved so that an instantaneous collection of excited states would have the approximate Boltzmann distribution of vibrational-rotational energies.

One indication that this cooling down to the thermally equilibrated excited state is rapid in the case of coordination compounds is the observation, noted in Chapter 2, that ligand-field absorption bands are not only generally broad, but are also often quite featureless. There is often little or no indication of any vibrational fine structure, even when the absorption spectra are observed at very low temperatures. This type of behavior is consistent with the idea that the vibrational excited states arrived at following light absorption are not well defined, so rapid is the cooling-down process.

The thermally equilibrated excited state is a *thermodynamic state*. It is not the particular energy state of a particular molecule, but rather, the term refers essentially to an ensemble of molecules in the same electronic quantum state, but otherwise showing the distribution of vibrational-rotation states consistent with the temperature of the medium. By contrast, a *spectroscopic state* refers to a particular energy state of an individual molecule and is usually described in terms of specific electronic, vibrational, and rotational quantum numbers. Thus the term "state" has two rather different current usages, thermodynamic and quantum mechanical.

A.   Thexi States and Franck–Condon States

We polarize the situation here by making a sharp distinction between the spectroscopic state obtained upon absorption of a light quantum and the thermodynamic state of an ensemble of molecules that are in a given electronic quantum state, but which are otherwise in equilibrium with their surroundings. The first may be called a *Franck–Condon state;* that is, absorption of a light quantum produces an electronically excited molecule in an excited vibrational-rotational level. There is distribution of vibrationally excited states if many molecules are excited, but this distribution is determined by the quantum-mechanical workings of the Franck–Condon principle; it is in no way a thermodynamic distribution. Ligand-field theory treats the Franck–Condon state corresponding to absorption of light at the ligand-field band maximum as a simple electronic state. This is a heritage of crystal-field theory in which the transition metal is considered to be imbedded in an electrostatic field of given and fixed geometry.

In the present simplification, we consider that the slower processes, including the chemical ones, occur from the thermally equilibrated excited state. This thermodynamic state is the important one, rather than the Franck–Condon state, and deserves a special name. We have proposed the name "thexi" (*th*ermally *e*quilibrated *ex*cited) state (but we would welcome other suggestions; some esoteric Greek formulations were considered and discarded!). A *thexi* state has an energy, an entropy, a free energy, a structure, and a chemistry. It has, in brief, the attributes of a chemical species. In particular, its energy may be quite different from that corresponding to the absorption band maximum giving the parent Franck–Condon state.

Referring again to Fig. 2-6, the first ligand-field thexi state of $Cr(urea)_6^{3+}$ is evidently lower in energy by several thousand wave numbers than the band maximum Franck–Condon state (we estimate the thexi state energy as given approximately by the crossing of the absorption and emission curves).[23] This thexi-state energy is not predicted and is perhaps not predictable by contemporary ligand-field theory. Furthermore, the geometry of the thexi state need not be and probably is not octahedral. As a minimum, one expects a general bond lengthening of a breathing nature, but, more likely, the distortion from the ground state will not be symmetric. The Franck–Condon state, after all, has acquired electron density in the antibonding e-type orbitals. Actually, the thexi state of a hexacoordinated complex could be of C_{5v} geometry[24] or some other symmetry type quite different from that of the ground state.

In the case of d^8 complexes, such as those of Pt(II) or of Ni(II), a square planar ground-state geometry may, upon excitation, lead to a tetrahedral thexi-state geometry.[25,26] Thus, here again, the point group need not be a subgroup of that of the ground-state molecule. A contemporary problem is that only in a few cases has it been possible even to infer the geometry of a thexi state.

A further characteristic of a thexi state as a thermodynamic state is that its properties should be independent of how the state is produced. This can be shown to be the case in at least one instance, that of $^3[Ru(bipyr)_3^{2+}]^*$. The emission spectrum of this complex, as excited by visible or near-ultraviolet radiation, has been reported.[10] The same excited state has been produced chemically and has the same emission spectrum.[20,22] If the emission were coming from a "hot" or vibrationally excited state, it is unlikely that the same hot state would be produced by the two methods.

In summary, it appears to be both useful and probably correct to suppose that, in the case of coordination compounds in solution, chemical reaction, emission, and radiationless return to the ground state are processes taking place from a thexi state. The state is a thermodynamic one; it may be regarded as an unstable isomer of the ground state.

B. The Transition State as a Thexi State

The transition state of absolute rate theory is a thermodynamic state. That is, while the transition state has extra energy over the reactant or encounter complex of reactants, this energy must be electronic in nature since the transition state is considered to be in thermal equilibrium with its surroundings. It is this assumption that allows its concentration to be formulated in terms of the equilibrium constant, $K^{\ddagger}$, and allows the statistical thermodynamic treatments. Moreover, of course, the translation along the open or reaction coordination is calculated assuming ambient temperature, thus yielding the kT/h frequency factor.

Except for the special and essentially unprovable assumption about the open or reaction coordinate, the transition state seems indistinguishable from a thexi state or, indeed, from simply an isomeric state of the reactant or of an encounter complex of reactants. Thexi and transition states have a specific attribute in common. If they are to be essentially in thermal equilibrium with their surrounding, their lifetime should be significantly greater than the frequency of vibrational exchange of energy with the solvent, that is, greater than perhaps 10^{-12} sec. If this condition does not hold, as in the photochemical situation of predissociation, the thermal equilibration concept is not applicable.

The same condition should apply to the transition state, which implies that its rate of reaction cannot be as fast as the kT/h frequency factor indicates. A commonly discussed way out of this problem has been to suppose that the transition state be regarded as one *almost* ready to react, but requiring a small additional amount of energy to do so. Its lifetime is now great enough to permit thermal equilibration and the use of thermodynamic formulation. The final energy acquisition gives a reacting species of too short a lifetime to be regarded as having well-defined thermodynamic properties, but it is supposed that no great error is introduced if ordinary transition-state formulations are used.

An advantage of this modified formulation of the transition state is that the transition states for the forward and reverse reactions are now different. The paradox is avoided of otherwise having a single transition state affirmed to be in equilibrium both with reactants and with products even though the system is undergoing net reaction.

This modified transition-state concept is indistinguishable from that of a thexi state. An interesting question is whether many chemical reactions may actually proceed through transition states that alternatively may be produced spectroscopically. The spectroscopist and photochemist potentially have much to contribute to the theory of reaction kinetics.

C.   Radiationless Deactivation and Activation

An interesting point of view emerges regarding the nature of activation processes in solution kinetics if one considers the photophysical process of radiationless deactivation. In the case of coordination compounds in solution, usually aqueous solution, there are many instances of compounds whose photochemical reaction, while substitutional, differs from that of the ground state (see Chapter 4), and the photoreaction is regarded, therefore, as reflecting the chemical nature of a certain thexi state. Quantum yields are nearly always less than 0.5, and in the case of Co(III) ammines, they can be quite small. Typically, there is negligible emission in room-temperature solution. It follows that most of the absorbed energy goes into ordinary heating of the medium, and it seems likely that much of this energy dissipation occurs by "radiationless deactivation" of the lowest thexi state. Various mechanisms for this process are discussed in Chapter 2. The general scheme is

$$\text{thexi state} \rightarrow \text{vibrationally hot ground state} \quad (10\text{-}29)$$

$$\text{vibrationally hot ground state} \rightarrow \text{ground state} \quad (10\text{-}30)$$

Process (10-29) appears to be hindered, either quantum mechanically or because of a necessary thermal activation, while process (10-30) should be very rapid—as rapid as is thermal equilibration, generally.

This last conclusion, that process (10-30) is very rapid, is supported by the following observation. It is very striking that the energy dissipated in a radiationless deactivation may amount to 50 to 80 kcal mole^{-1} and yet is able to pass through vibrationally excited states of the ground state without inducing thermal reaction. Most of the complexes involved undergo thermal aquation reactions with activation energies of around 20 to 25 kcal mole^{-1}. Evidently the 50 to 80 kcal mole^{-1} being dissipated is lost too quickly for it to find its way into the labile bond or, alternatively, to form the transition state for the thermal reaction. That is, process (10-30) appears to occur too rapidly for any statistical sampling of equi-energy distributions to be possible.

The reverse of processes (10-29) and (10-30) should also exist, corresponding to thermal population of a thexi state. It seems reasonable to conclude that the reverse of (10-30) is again very rapid. One may think of phonon waves in the medium making a chance convergence on the region containing the coordination compound, suddenly to place in it a large amount of vibrational energy. This energy as quickly dissipates back into the medium, but occasionally the reverse of (10-29) occurs, and a thexi state is produced.

The transition state of reaction kinetics may be formed similarly. We start with the reactant in its solvent cage if the reaction is unimolecular or is one with solvent, or otherwise with the encounter complex of two or more solute species. A chance, rapid aquisition of energy occurs through a confluence of vibrational energy from the medium. If the reactant(s) have a favorable configuration, a prompt, hot ground-state reaction may occur; otherwise, the excess energy quickly dissipates back to the medium. An alternative possibility is that the energetic reactant(s) pass to a new electronic configuration by essentially the reverse of (10-29), and the thermodynamic state we call the transition state is formed. This state may then undergo reaction to products with only minor further activation. These views of reaction kinetics in solution are discussed somewhat further in Ref. 27.

It certainly seems possible that the thexi state of photochemistry and the transition state of thermal reactions are of the same nature and may form and may disappear by essentially the same mechanism.

Figure 10-1 illustrates perhaps the simplest possible scheme embodying the above ideas, and one which shows the close possible relationships between photochemistry, photochromism, chemiluminescence, and thermal reaction. We suppose a simple reaction such as that of Eq. (10-1) and plot energy versus change in nuclear configuration. A small motion along the horizontal coordinate would be a distortion or vibration; a large motion takes us along a reaction coordinate, from isomer A to isomer B. Heavy horizontal lines mark the energies of thermodynamic states; we assume the thexi state, A^*, for photochemical reaction to be the same as the transition state for the corresponding thermal reaction of A to B and make the corresponding assumption about B^*. A degree of generality is included by showing transition states, A'^* and B'^*, for the forward and reverse thermal reactions involving an alternative product, B'.

Several processes may now be traced. We suppose A to be a coordination compound, and absorption at the maximum of the L_1 band is shown by the left vertical solid arrow. The termination is at a Franck–Condon state, A_{FC}^*. Since the ground- and excited-state nuclear positions have not changed during the transition, the nascent Franck–Condon state is the temporary potential well of the solvent cage, as indicated by a parabolic curve. The excited-state compound distorts toward the thexi state

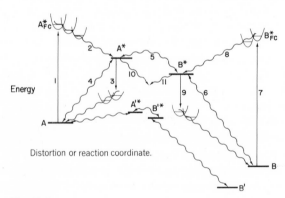

Distortion or reaction coordinate.

Fig. 10-1.

geometry, the solvent cage relaxing as this distortion occurs. At each stage, however, the succession of parabolic curves suggests that solvent-cage restraint provides a restoring force against any extreme libration toward the final geometry. The ladder of vibrational states may thus be regarded as a near continuum of compound-solvent cage configurations. We thus account for the lack of sharp vibronic structure in the ligand-field bands of Cr(III), Co(III), and many other transition-metal complexes. The wavy arrow denotes this above process of thermal equilibration and is used throughout the diagram to denote a radiationless process.

Thexi (and transition) state, A^*, may emit, as indicated by arrow 3, the termination being a Franck–Condon state of the ground state. Thermal equilibration yields the initial state, A. State A^* may also return to A by the radiationless process 4, which, like process 2, is very fast except for a possible initial hindrance as discussed in connection with reaction (10-29). Photochemistry is shown by path 5 as passing through thexi or transition state, B^*, and thence by radiationless step 6 to product B. The "short-cuts" suggested by arrows 10 and 11 encounter a problem with the principle of microscopic reversibility.

Turning to the right-hand side of the figure, excitation of B to B^* is by path 7-8, through a Franck–Condon state, B^*_{FC}. B^* may then disappear by radiationless deactivation, 6, by emission, 9, or by chemical reaction, 5.

The steps covered to this point describe a simple doubly photochromic system with K^P given by Eq. (10-5) and $\phi_A = k_5/(k_3 + k_4 + k_5)$ and $\phi_B = k_{-5}/(k_{-5} + k_6 + k_9)$.

We also can describe a type of photoluminescence not reported so far in coordination chemistry, that is, the photochemical production of a product in an emitting excited state. The process would consist of steps 1, 2, 5, and 9 in Fig. 10-1. It would be interesting to make a planned search for this type of phenomenon in coordination chemistry.

Figure 10-1 also describes thermal reactions. The thermal conversion of A to B is by paths 4, 5, and 6. The activation energy for the forward reaction would be the energy for step 4 plus the small additional energy to pass the barrier of step 5. Since the figure shows B as the stable isomer, a simple singly photochromic system would consist of steps 7, 8, 5, and 4 for the photochemical conversion of B to A, and steps 4, 5, and 6 for the thermal back reaction.

Chemiluminescence is also shown in the figure, a possible path being steps 4, 5, and 9. It would be more representative of actual systems to depress the energy of B and B^* considerably over that shown in the figure so that the activation energy to form A^* could be small, yet the energy available in step 9 large enough to correspond to visible-light emission.

Finally, the figure includes the possibility that the observed thermal reaction of A is to product B′ rather than to B. Thexi or transition states, A'^* and B'^*, are positioned lower in energy to indicate a kinetic preference for the A–B′ reaction over the A–B one, and are positioned along the reaction coordinate so as to suggest that passage from the A^* or B^* states to A'^* and B'^*, respectively, is not important.

Much of the above material is speculative, especially in detail, the main thrust being to emphasize the possible similarities between process and states in photochemistry to these in thermal reactions. It is hoped that the concepts presented are stimulating as well, of course, that they prove to have some experimental validity.

REFERENCES

1. V. Balzani and V. Carassiti, *Photochemistry of Coordination Compounds*, Academic, New York, 1971.
2. A. W. Adamson, W. L. Waltz, E. Zinato, D. W. Watts, P. D. Fleischauer, and R. D. Lindholm, *Chem. Rev.* 68, 541 (1968).
3. A. W. Adamson, A. Chen, and W. L. Waltz, Proceedings, Tenth International Conference on Coordination Chemistry, Tokyo and Nikko, Japan, September 12, 1967.
4. S. T. Spees and A. W. Adamson, *Inorg. Chem.* 1, 531 (1962).
5. K. L. Stevenson and J. F. Verdieck, *J. Amer. Chem. Soc.* 90, 2974 (1968); note also K. L. Stevenson, *ibid.* 94, 6652 (1972), and references therein.
6. P. S. Sheridan and A. W. Adamson, *J. Amer. Chem. Soc.* 96, 3032 (1974).
7. P. S. Sheridan and A. W. Adamson, *Inorg. Chem.* 13, 2482 (1974).
8. H. Gray, private communication.
9. R. Pribush, R. Wright, and A. W. Adamson, unpublished work.
10. J. N. Demas and G. A. Crosby, *J. Amer. Chem. Soc.* 93, 2841 (1971).
11. H. G. Gafney and A. W. Adamson, *J. Amer. Chem. Soc.* 94, 8238 (1972).
12. J. N. Demas and A. W. Adamson, *J. Amer. Chem. Soc.* 95, 5159 (1973).
13. P. Natarajan and J. F. Endicott, *J. Phys. Chem.* 77, 1823 (1973).
14. A. Vogler and A. W. Adamson, *J. Amer. Chem. Soc.* 90, 5943 (1968).
15. H. G. Gafney and A. W. Adamson, unpublished work.
16. G. Navon and N. Sutin, *Inorg. Chem.* 13, 2159 (1974).
17. C. R. Bock, T. J. Meyer, and D. G. Whitten, *J. Amer. Chem. Soc.* 96, 4710 (1974).
18. R. A. Marcus, *J. Chem. Phys.* 43, 2654 (1965); see also Refs. 20.
19. See M. M. Rauhut, *Account. Chem. Res.* 2, 80 (1969); E. H. White and D. F. Roswell, *ibid.* 3, 54 (1970).
20. D. M. Hercules and F. E. Lytle, *J. Amer. Chem. Soc.* 88, 4745 (1966); F. E. Lytle and D. M. Hercules, *Photochem. Photobiol.* 13, 123 (1971); D. M. Hercules, *Account. Chem. Res.* 2, 301 (1969).

21. J. E. Martin, E. J. Hart, A. W. Adamson, H. Gafney, and J. Halpern, *J. Amer. Chem. Soc.* **94,** 9238 (1972).

22. See N. E. Tokel-Takvorgan, R. E. Hemingway, and A. J. Bard, *J. Amer. Chem. Soc.* **95,** 6582 (1973).

23. P. D. Fleischauer, A. W. Adamson, G. Sartori, *Inorganic Research Mechanisms* (J. O. Edwards, Ed.), Wiley, New York, 1972.

24. J. N. Demas and A. W. Adamson, *J. Phys. Chem.* **75,** 2463 (1971).

25. D. S. Martin, Jr., M. A. Tucker, and A. J. Kassman, *Inorg. Chem.* **4,** 1682 (1965); *ibid.* **5,** 1298 (1966).

26. C. J. Ballhausen, N. Bjerrum, R. Dingle, K. Eriks, and C. R. Hare, *Inorg. Chem.* **4,** 514 (1965).

27. A. W. Adamson, *A Textbook of Physical Chemistry*, Academic, New York, 1973.

INDEX

INDEX